Piezoelectric Electromechanical Transducers for Underwater Sound

Boris S. Aronov

ISBN 9781644698235 (hardback)

ISBN 9781644698273 (Open Access)

Published by Academic Studies Press

1577 Beacon Street

Brookline, MA 02446, USA

press@academicstudiespress.com

www.academicstudiespress.com

In memory of

my mentor, Lev Yakovlevich Gutin,

and my friends and colleagues, prominent Russian electroacousticians,

Lev Davidovich Lubavin and Vladimir Igorevich Pozern

PREFACE

This book is initiated by the engineering experience of the author. Throughout his career the author has encountered many of the problems known to others involved in the design of electroacoustic transducers. The fact of the matter is that the complexity of designing electroacoustic transducers is inherent in the multidisciplinary nature of the subject. Therefore, the developers and designers of the transducers must possess the knowledge of several different theoretical disciplines (such as the vibration of mechanical systems, electromechanical conversion by deformed piezoelectric bodies, and acoustic radiation) and be able to actively use this knowledge to derive equations that describe the performance of the transducers. Furthermore, creating practical transducer designs that meet certain requirements and can operate under realistic environmental conditions requires the knowledge of properties of materials used and a certain level of engineering intuition that cannot be developed without a clear understanding of the underlying physics. Hardly anyone may possess all these capabilities without having received a specially targeted education, which, to the best of the author's knowledge, is not commonly available in the academic world. Usually, the necessary skills may be acquired through self-education, which was the case for the author. The main difficulties that arise in this endeavor are not in the lack of available information. On the contrary, the theoretical disciplines listed above are very well developed and are well-represented in the literature. Nevertheless, all these disciplines employ different methods for solving their problems and the results obtained are usually presented in forms not suitable for direct use in concert for synthesizing equations that govern transducer performance. Thus, the results must be tailored accordingly.

Experiencing the above difficulties over several decades, the author gradually developed a special approach to treating transducers problems that allows one to overcome many of the obstacles. The essence of this approach is in the consistent application of the physics-based energy method for solving all the problems that arise in the course of treating electromechanical and electroacoustic transducers. The first attempt to describe this concept was undertaken in *Electromechanical transducers from piezoelectric ceramic* published in 1990 in Russia. This version has now been updated and expanded to the extent that it can be considered a completely different book. Only the underlying energy approach to solving the problems has remained unaltered. This book is written for students, applied scientists and engineers in a way that should prove fruitful both for those who have only begun to chart their careers in electroacoustics as well as for those at a more advanced level. The content of the book is split into four parts.

In Part I, titled "Introduction of energy method of treating the transducers," the main concepts of the method are considered (Chapter 1); applications of the method to calculating properties of transducers with single degree of freedom are illustrated (Chapter 2); and the study of problems for designing the transducers as a part of the transmit/receive channel is made (Chapter 3). The main concept is that of energy and following its transformation. Different types of energies involved in the electro-mechano-acoustic conversion in the course of transducer operation are presented in the generalized coordinates. All the governing equations are derived from the energy principles, that is, from the Law of Conservation of Energy for transducers with a single mechanical degree of freedom, and from the Principle of Least Action for transducers with multiple degrees of freedom. Equations describing the electromechanical part of the problem are reinterpreted as Kirchhoff's equations for the corresponding equivalent electromechanical circuits. In Chapter 2, the general approach is applied towards calculating the properties of transducers of widely used types (spheres, cylinders, bars undergoing extensional vibration and for circular plates and rectangular beams vibrating in flexure) that may be considered as systems with single mechanical degree of freedom. In Chapter 3, the operating properties of transducers as a part of a transmit/receive channel are considered and some recommendations regarding a rational transducer designing are presented. Given that the single degree of freedom approximation covers many practical transducer designs, Part I can be regarded as a self-sufficient study of underwater electroacoustic transduction on a basic level and can be read independently from the rest of the book.

The general treatment of electroacoustic transduction requires an advanced knowledge of the vibration of mechanical systems, electromechanical conversion in deformed piezoceramic bodies and acoustic radiation. Information about these topics, which is necessary for the consideration of virtually all practical transducer types is presented in Chapters 4-6 of Part II under the title: "Subsystems of the Electroacoustic Transducers." All the constitutive equations are derived in these chapters from the Principle of Least Action as Euler's Equations in generalized coordinates. The obtained results are presented in the form of impedances, (including the radiation impedances), electromechanical transformation coefficients and acting forces (including those of acoustic origin) that can be directly substituted into the equivalent electromechanical circuits (multi contour in general) of the transducers. The diffraction coefficients and directional factors for differently configured transducer surfaces are also presented.

In Chapter 4, special attention is paid to the consideration of coupled vibrations in the generally two-dimensional mechanical systems. The results allow determining the range of aspect ratio, at which the system can be approximately considered as one-dimensional, where the problem can be simplified.

In Chapter 5, especial importance is ascribed to the theorem that sets the conditions at which the electromechanical conversion under the longitudinal and transverse piezoelectric effects can be treated qualitatively in the same way. This allows for the unifying calculation technique for the transducers that employ these types of ceramics polarization. Another important subject is the general analysis of optimizing the effective coupling coefficients in nonuniformly deformed piezoceramic bodies.

Chapter 6 touches upon several noteworthy issues. Besides solving the general radiation problems, it provides a detailed consideration of the effects of baffling parts of the surfaces of cylindrical and spherical transducers, which ensures their unidirectionality. The technique for the experimental investigation of the acoustic interaction between transducers (or between the mechanically isolated parts of the same transducer) is also analyzed. Since the baffles have an effect on the acoustic near field, the interactions can rarely be treated analytically for practical transducer configurations, hence more reliable characterization of the interaction can be obtained through an experimental investigation.

The results obtained in the Part II are used in Part III of the book titled "Calculating transducers of different types" for synthesizing equations that describe the detailed operation of transducers of various configurations: cylindrical (Chapter 7), spherical (Chapter 8), plates and beams vibrating in flexure (Chapter 9) and bar transducers (Chapter 10).

Chapter 7 presents a study of cylindrical transducers that employ multimode extensional and flexural vibration of complete and incomplete cylinders (slotted cylinder projectors are also considered) for various practical applications. Different modes of cylinder polarization are considered, including the tangential polarization (with striped electrodes). An extensive study is provided of the effects of coupled vibrations on the electromechanical and acoustic performance of transducers that employ cylindrical piezoelements having finite thickness to diameter aspect ratios. Chapter 8 covers transducers which employ general multimode extensional vibrations of complete and incomplete piezoceramic spherical shells, (hemispherical in particular). The baffling of the parts of the surface that allow using multiple modes of vibration for unidirectional transducer operations is also considered.

In Chapter 9, a general analysis is provided of transducers which feature flexural vibrations of circular and rectangular piezoceramic plates (beams), including non-uniform over thickness and radius (length) transducer designs. Optimizing the effective coupling coefficients of the transducers is considered making use of the nonuniformity of the distribution of deformations in the volume of the plates. Corrections for transducer parameters due to a finite thickness to radius (length) ratio of the plates are considered. It is then concluded that the accuracy with which the wave numbers can be predicted substantially depends on the aspect ratio (especially for the higher modes of vibration) and presenting their values without the notion of the aspect ratios is not appropriate.

In Chapter 10, the length expander bar transducers are considered Transitions of configurations of bars to thickness vibrating plates at different polarizations and related dependencies of their effective coupling coefficients on the aspect ratios are considered using the technique of coupled vibrations. Relatively small attention is paid to the widely used Tonpilz transducer designs because they have already been described in detail in the available literature.

Part IV (Chapters 11 through 15) is titled: "Some aspects of the transducers designing."

In Chapter 11, a review of the existing data and some new results is presented regarding effects of operating environmental conditions, such as the hydrostatic pressure, temperature, and drive level on the parameters of piezoceramics. It is emphasized that, under these conditions, the parameters of ceramics may deviate significantly from those that are given in specifications for normal conditions. Moreover, they may differ for samples of ceramics supplied by different (and even by the same) manufacturers. This must be kept in mind when calculating the operating parameters of transducers under real conditions and in estimating a reasonable accuracy of calculation of the parameters. The variations in the parameters of transducers intended for operating at great depths can be avoided by using designs which incorporate hydrostatic pressure compensation. Issues related to the practical implementation of the pressure compensation are examined in Chapter 12 (more general information), in Chapter 13 (regarding the liquid filled cylindrical projectors) and in Chapter 14 (regarding the hydrophones).

Chapter 13 presents some considerations regarding the practical challenges of the projectors design. Using the concept of the Reserves-of-Strength for improving parameters of the transducers of different types by optimizing their matching with the acoustic field is considered. The possibilities of increasing the dynamic and static mechanical strength of the projectors by prestressing and combining piezoceramic with passive materials in their mechanical systems are analyzed.

Chapter 14 is dedicated to the design of hydrophones and related issues. The hydrophones employing different transducer types are classified by the pressure and pressure-gradient hydrophones of the diffraction and motion types. Their properties as a source of energy of signal and internal noise for a receive channel are considered. Special attention is paid to the response of hydrophones and accelerometers to unwanted actions and to measures aimed at increasing their noise immunity.

Chapter 15 is crucial for the structure of the book because it introduces the practice of combining Finite Element Analysis (FEA) with analytical energy methods. This is illustrated with examples of flextensional and oval transducers. Combining powerful computer-based FEA techniques that are used to obtain results for vibration mode shapes with the energy method that yields great physical insight opens up a new area of research collaboration for many transducer problems. FEA allows the determination of the vibration mode shapes for mechanical systems that cannot be approximated analytically due to the complexities of the mechanical system and its boundary conditions.

The book also contains appendices with information on the properties of the piezoelectric ceramics and passive materials that may be used in transducer designs, and on the properties of the special functions that are referred to throughout the book.

In summary, the book presents methods for calculating the properties of most common electroacoustic transducer problems with particular focus on underwater applications. Moreover, by combining the FEA technique to determine the prerequisite vibration mode shapes with the energy method, virtually any transducer type may be analyzed. Still, however, when it comes to choosing and designing a particular transducer for a particular application under demanding operating and environmental specifications – this remains somewhat of an art. Thus, recommendations of transduction choices for representative problems remain a guide and not a prescription for success.

It is inevitable that the book may contain typographical or content errors and thus the author would welcome the readers' comments and notifications of such.

Boris S. Aronov

TABLE OF CONTENTS

Part III

Calculating Transducers of Different Types

CHAPTER 7

CYLINDRICAL TRANSDUCERS

7.1 Introduction

After vibration of the potential mechanical systems of the transducers, electromechanical energy conversion in the piezoelectric vibrating bodies and conditions of their loading are considered in Chapters 4-6, respectively, the equations that govern operation of the particular electromechanical transducer types can be completed. This will be done in the following Chapters. We will start the treatment with cylindrical transducers. This example is the most typical for illustrating the general procedure of application of the energy method. Mechanical systems of cylindrical transducers are usually composed of a number of coaxial rings mechanically isolated from each other. In calculating vibrations of the comprising rings that constitute the transducers, their joint operation affects only the conditions of their loading. Therefore, the general analysis of the rings vibration that is performed in Ch. 4 remains applicable. The main body of the treatment will be performed under the assumption that the rings have small height and small thickness compared with their diameter in order to avoid unnecessary complications in illustrating the essential concepts, moreover that the most of practical transducer designs fall into this category. The cases that the height of a ring is comparable with its diameter (i. e., the ring becomes a tube of a finite height), or the ring has significant thickness will be considered separately with example of a transducer uniformly vibrating in radial direction.

As it was discussed in Section 4.4.4, the circular ring may vibrate in the extensional and flexural modes that have significantly different ranges of the resonance frequencies spectrum and require different conditions for their electromechanical excitation In order to determine fundamental design differences of the ring transducers that realize extensional or/and flexural vibrations, we will assume at first that the transverse piezoelectric effect is employed for electromechanical conversion in piezoceramic rings. For illustrating the design differences let us place the origin of the coordinate system on the mean circumference of a ring, as it is shown in Figure 7.1(a), and represent the electric field in the ring in the form

$$E_3(z,\varphi)=\frac{2V}{t}\Omega_1(z)\Omega_2(\varphi), \tag{7.1}$$

where $\Omega_1(z)$ and $\Omega_2(\varphi)$ are the functions that depend on configuration and on the manner of connection of electrodes located in the volume of the ring Figure 7.1 (b)-(d)).

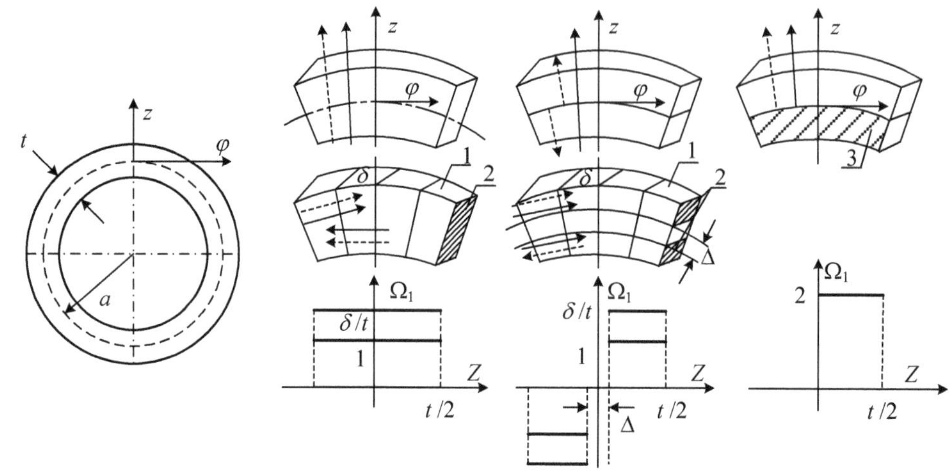

(a) (b) (c) (d)

Figure 7.1: General view of a ring (a) and the electrical field distributions over the ring thickness for excitation of: (b) extensional deformations, (c) flexural deformations, (d) both extensional and flexural deformation. 1 - prisms, from which ring is assembled in the case that the longitudinal piezoeffect is employed, 2 - electrodes, 3 - layer of passive material. Solid arrow shows direction of electric field, dashed arrow – of the field of polarization.

After substituting function $E_3(z,\varphi)$ and the general expression (4.220) for strain in the ring (replacing in it *(r-a)* by *z*) into formula (5.67) and integrating the energy density over the volume of the ring we obtain,

$$\begin{aligned}W_{em}=\frac{d_{31}Vh}{2s_{11}^E t}\Bigg[&\int_0^{2\pi}\int_{-t/2}^{t/2}\Omega_1(z)\Omega_2(\varphi)\left(\frac{\partial\xi_\varphi}{\partial\varphi}+\xi_r\right)dzd\varphi\\ &-\int_0^{2\pi}\int_{-t/2}^{t/2}\Omega_1(z)\Omega_2(\varphi)\frac{z}{a}\left(\frac{\partial^2\xi_r}{\partial\varphi^2}+\xi_r\right)dzd\varphi\Bigg]\end{aligned}. \tag{7.2}$$

Apparently, the first integral is responsible for electromechanical conversion related to the extensional vibration and the second - for those caused by the flexure. The results of calculating W_{em} depend significantly on the properties of functions $\Omega_1(z)$ and $\Omega_2(\varphi)$. The following variants differ in principle regardless of how the function $\Omega_2(\varphi)$ looks: when $\Omega_1(z)$ is an even function of z, $\Omega_1(z)=\Omega_1(-z)$, as shown in Figure 7.1 (b), and when it is an odd function, $\Omega_1(z)=-\Omega_1(-z)$, as shown in Figure 7.1(c). At even $\Omega_1(z)$ the second integral in formula (7.2) vanishes. Electromechanical conversion can occur only under extensional deformations. The basic design of a ring element is presented in Figure 7.1 (b). A typical example of such a transducer is the pulsating piezoceramic ring, considered in Section 2.3. At odd $\Omega_1(z)$ the first integral in expression (7.2) becomes zero. Electromechanical conversion under the flexural deformations may take place. The variants of basic designs of the ring elements, which allow for such conversion, are shown in Figure 7.1 (c). Finally, the function $\Omega_1(z)$ may have such a form that both integrals therein exist, and the electromechanical conversion of both modes of deformation is possible. The example of one of the possible designs of a ring element for this case is given in Figure 7.1 (d). In the second row of the Figure 7.1 the variants of ring design are shown that allow employing the longitudinal piezoeffect for excitation of vibration. In these cases, the rings are composed of prisms that have electrodes on their side surfaces. Getting different patterns of electric field through the thickness of a prism is self-explanatory from the figure.

In accordance with design versions presented in Figure 7.1 (b)-(c) we will consider separately transducers operating in the extensional and flexural vibration modes. It is noteworthy that by switching electrodes in the volume of a ring the transducers can be, in principle, converted from one type to another. Since in the design shown in Figure 7.1 (d) the result of excitation of both types of vibrations can be obtained as their superposition, transducers of this kind will not be considered here as a separate issue.

7.2 Transducers Operating in the Extensional Modes

Considering transducers of this type we will use the results presented in Sections 4.4.4 and 4.5.2, where vibration of the passive rings was discussed.

7.2.1 Generating Different Modes of Vibration

Substituting expressions (4.258) and (4.259) for displacements ξ_r and ξ_φ under the sign of the first integral in relation (7.2) and assuming that in this case $\Omega_1(z)=1$ due to a small thickness of a ring, we obtain

$$W_{em} = \frac{d_{31}Vh}{s_{11}^E}\int_0^{2\pi}\Omega_2(\varphi)\left[\xi_o + \sum_{i=1}^{\infty}\xi_i\left(1+i^2\right)\cos(i\varphi)\right]d\varphi. \tag{7.3}$$

Now the electromechanical transformation coefficients can be found by using formula (5.212) in the form

$$n_i = \left(d_{31}h / s_{11}^E\right)\Omega_i, \tag{7.4}$$

where

$$\Omega_i = \left(1+i^2\right)\int_0^{2\pi}\Omega_2(\varphi)\cos(i\varphi)d\varphi. \tag{7.5}$$

In the variant of the longitudinal piezoelectric effect in a segmented ring

$$n_i = \left(d_{33}th / s_{33}^E\delta\right)\Omega_i. \tag{7.6}$$

Here δ is the average width of the prisms comprising the ring (see Figure 7.1). Thus, the conditions of electromechanical conversion for different modes of vibration depend on how the function $\Omega_2(\varphi)$ looks. By giving this function different forms, it is possible to govern the frequency characteristics and displacements distribution over the ring surface, as it was shown for the bar transducer in Section 2.4. Consider values of coefficients Ω_i for some of configurations of electrodes. At first, we will assume that the electrodes fully cover the entire side surfaces of the ring. In the case that the electrodes are unipolar, as shown in Figure 7.2(a),

$$\Omega_2(\varphi) = 1,\ \Omega_0 = 2\pi,\ \Omega_i = 0 \text{ at } i \ge 1. \tag{7.7}$$

Only the zero mode of vibration that corresponds to the pulsating vibrations of the ring is electromechanically active.

In the case that the electrodes are divided into $2N$ parts, which are alternately connected in phase opposition, as it is shown for $N = 1$ and $N = 2$ in Figure 7.2 (b)-(c),

$$\Omega_i = (-1)^{l+1}\cdot 4\left(1+i^2\right)/(2l-1) \text{ at } i = (2l-1)N \text{ and } l = 1,2,\ldots \tag{7.8}$$

All the other transformation coefficients are zero. The first electromechanically active vibration mode is $\theta_N(\varphi) = \cos N\varphi$. Besides, the active vibration modes are those with numbers $3N$, $5N$, etc. Thus, for $N = 1$

$$\Omega_1 = 8 \text{ and } \Omega_3 = -40/3\,. \tag{7.9}$$

If the electrodes cover only a part of the ring surface at angles $|\varphi| < \pi/m$, i.e., $\Omega_2(\varphi) = 1$ at $|\varphi| < \pi/m$ and $\Omega_2(\varphi) = 0$ at $|\varphi| > \pi/m$, then

$$\Omega_0 = 2\pi/m,\ \Omega_i = \left[2\left(1+i^2\right)/i\right]\sin(i\pi/m)\,. \tag{7.10}$$

Note, that it is possible to govern the resonance frequency spectrum of the transducer by varying the value of m and hence changing relation between transformation coefficients for different modes of vibration. Thus, for example, if half of the electrodes are used ($m = 2$), the zero and all the odd vibration modes are active. Upon changing the position of the electrodes by 180°, the functions Ω_i and transformation coefficients n_i reverse the sign at odd values of i.

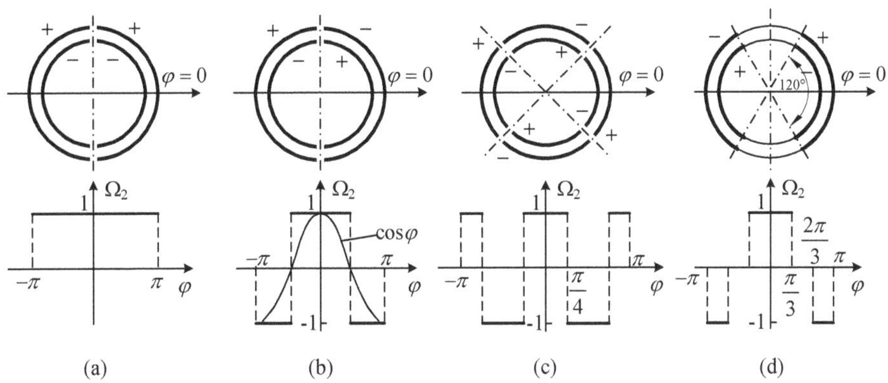

Figure 7.2: Distributions of electric field over circumference of a ring for generating different resonance modes of vibration: (a) zero mode, (b) first mode (full size electrodes), (c) second mode, (d) first mode optimized (with suppressed third mode).

Theoretically one can realize all the above-listed variants of electrodes connection in a single ring and thus to change frequency responses of a transducer by dividing the electrodes into corresponding parts and by switching these parts accordingly. As an example, in Figure 7.2 the qualitative plots are shown for the frequency characteristics of the amplitude of vibration of the reference point on the ring surface, $\xi(0) = \xi_0 + \xi_1$, obtained for the case that electrodes are divided into halves and both halves are connected or in phase (curve ξ_0), or in phase opposition (curve ξ_1), or only one half of the electrodes is used at $|\varphi| < \pi/2$ (curve $\xi_0 + \xi_1$) and at $|\varphi| > \pi/2$ (curve $\xi_0 - \xi_1$). (Note that the frequency responses are shown in vicinity of the resonance frequencies f_0, f_1 and below. In the broader frequency range, the higher modes of vibration become involved).

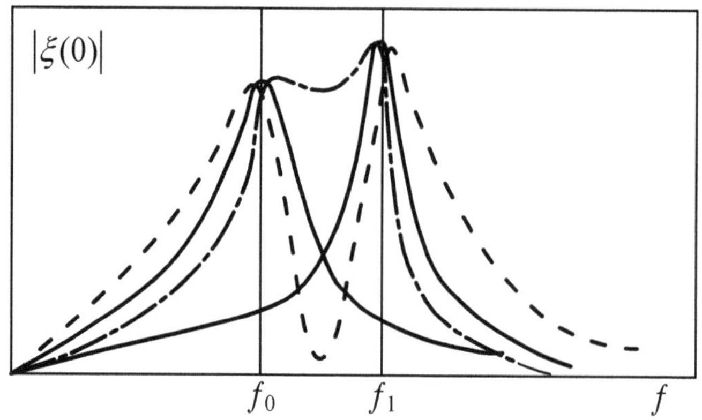

Figure 7.3: Frequency responses of displacement $|\xi(0)|$ of the reference point at different electrodes configuration. Solid lines - halves of the electrodes are connected or in phase (curve ξ_0 with resonance frequency f_0), or in phase opposition (curve ξ_1 with resonance frequency f_1); dashed line - one half of the electrodes is used at $|\varphi| < \pi / 2$ (curve $\xi_0 + \xi_1$); dash-dotted line – the electrodes are used at $|\varphi| > \pi / 2$ (curve $\xi_0 - \xi_1$).

To generate a single mode of vibration, for example, $\theta_1(\varphi) = \cos\varphi$, the electrodes should be designed in such a way as to meet the condition $\Omega_2(\varphi) = \cos\varphi$. However, in practice for operation in a sufficiently broad frequency range it is enough for this purpose to "suppress" the closest of the active vibration modes, $\theta_3(\varphi) = \cos 3\varphi$ in our example. In accordance with formula (7.5) $\Omega_3 = 0$ at $m = 3$, i.e., in this case two electrode segments of size $\Delta\varphi = 2\pi/3$ must be connected in phase opposition (Figure 7.2 (d)), which produces the simplest stepwise approximation to the ideal distribution $\theta_1(\varphi) = \cos\varphi$.

As shown in Section 5.6, applying electrodes to the entire side surface of the ring is not optimal for maximizing the value of its effective coupling coefficient in the case of a nonuniform strain distribution. By the same reason that is discussed in Section 5.6.2 regarding the bars, k_{eff1} for the vibrational mode $\theta_i(\varphi) = \cos i\varphi$ can be increased, if to reduce the size of electrodes appropriately. This is illustrated by plots in Figure 7.4, where dependence of ratio $\alpha_c(\Delta\varphi) / \alpha_c(\pi)$ from angle $\Delta\varphi$ of the electrodes coverage is shown. It is seen that the maximum value of coefficient k_{eff1} for the vibration mode $\theta_1(\varphi) = \cos\varphi$ is reached at $\Delta\varphi = 0.74\pi$. Reduction in electrode size to $\Delta\varphi = \pi/2$ does not lead to a decrease in k_{eff1}, and at $\Delta\varphi = 2\pi/3$, which corresponds to suppression of the vibration mode $\theta_3(\varphi)$ (at this point $k_{eff3} = 0$), it is somewhat higher.

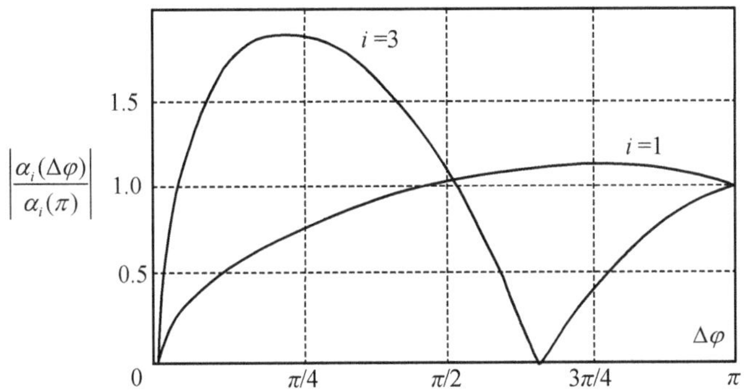

Figure 7.4: On the dependence of k_{eff} from the electrode dimension.

For calculating modal equivalent masses M_{ii} and rigidities K_{ii}^E of the piezoceramic rings relations (4.262) and (4.263) are valid that were obtained for the rings made of a passive material. One has only to replace Young's modulus Y in expressions for K_{ii} with $1/s_{11}^E$ for the transverse piezoeffect, and with $1/s_{33}^E$ for the longitudinal piezoeffect. As for the additional rigidity ΔK_i, which arises in course of electromechanical conversion in the case that strain is nonuniform along the lines of electric field, the following should be noted.

In the variant of the transverse piezoelectric effect in a thin ring $\Delta K = 0$, since the strains do not vary over the ring thickness. This quantity should be considered, would the thickness of a ring be significant (on this issue see Section 7.2.2.1.1). In the variant of the longitudinal piezoelectric effect the segmented ring design is used, and for the modes with nonuniform deformation at number of segments more than 6 on the half wavelength of deformation $\Delta K \approx 0$ by the same reason, as for a segmented bar (see Figure 5.9). Thus, the expressions for the equivalent parameters of a thin ring are as follows.

The mechanical parameters are:

$$M_{eqv0} = M,\ M_{eqvii} = M\left(1+i^2\right)/2\,; \tag{7.11}$$

$$K_{eqv0}^E = 2\pi th / as_{ll}^E,\quad K_{ii}^E = \left(\pi th / as_{ll}^E\right)(1+i^2)^2. \tag{7.12}$$

Here $l = 1$ and $l = 3$ for the transverse and longitudinal piezoeffect, respectively. The modal resonance frequency is

$$\omega_i = \omega_0\sqrt{1+i^2}\,, \tag{7.13}$$

where $\omega_0 = \left(1/a\sqrt{s_{II}\rho}\right)$ is the natural frequency of the zero mode.

The electromechanical transformation coefficients for the transverse and longitudinal piezoeffects in variants of transducers with full size electrodes are according to expressions (7.4), (7.6) and (7.8):

$$n_0 = \left(2\pi d_{31} h / s_{11}^E\right),\; n_0 = \left(2\pi d_{33} th / s_{33}^E \delta\right); \tag{7.14}$$

$$n_i = 4\frac{d_{31}h}{s_{11}^E}\left(1+i^2\right),\; n_i = 4\frac{d_{33}th}{s_{33}^E\delta}\left(1+i^2\right). \tag{7.15}$$

The last relations are for the first active modes (at $l = 1$ in expression (7.8)). Together with expressions for capacitances,

$$C_{el}^{S_1} = C_{el}^T\left(1-k_{31}^2\right) \text{ and } C_{el}^{S_3} = C_{el}^T\left(1-k_{33}^2\right), \tag{7.16}$$

where C_{el}^T is the capacitance of a clamped ring determined at the particular electrodes configuration, the full set of the equivalent parameters that are needed for calculating electromechanical cylindrical transducers is completed.

In further treatment we will title different cylindrical transducers types according to their main active vibrational modes, which are determined by a certain configuration of the electrodes. Additional characterization of the transducers will be made according to their applications.

7.2.2 Zero Mode Transducers

7.2.2.1 Omnidirectional Transducers

Transducers of this kind that employ uniform vibration of uniformly loaded rings are considered in Section 2.32 under the assumption that the rings have small enough height and thickness compared with their diameter. But it remains not clear to what extent "enough is enough" in application to the real rings that have a finite height and thickness, and what corrections should be made to results obtained under this assumption, if the rings are not sufficiently short and thin. This information can be obtained based on the results regarding radial vibration of thick isotropic rings of small height and coupled vibration of the finite height cylinders of small thickness that are presented in Secs. 4.4.2.3 and 4.6.2.4, respectively.

One of the ways of employing longitudinal piezoeffect in the radially vibrating relatively thin rings is in using so called "tangential polarization" that is achieved without segmenting the rings, as shown in Figure 7.1. In this case the electrodes that are used for polarization and operation have form of the stripes (see Figure 7.10). Approximate method of evaluating electromechanical parameters of tangentially polarized piezoelements is considered in Section 7.2.2.1.2

7.2.2.1.1 Equivalent Parameters of the Thick Piezoceramic Rings

In Section 4.4.2.3 the mode shapes of vibration of short isotropic rings were derived as dependencies of ratio of their inner to outer radiuses, *b/a.* The condition of isotropy holds piezoceramic rings axially poled. In the cases that the rings are radially or circumferentially poled properties of piezoceramic material in the plane of deformation are anisotropic, equations of motion differ from the Bessel equation (4.162), and strictly speaking do not allow simple analytical solution. But a very reliable assumption can be made that the mode shapes of deformation remain the same, as for axial poled isotropic ring. Under this assumption all the equivalent parameters for differently poled rings can be determined from expressions for the respective energies, if to use the expressions for stress, charge density, dielectric constant and for electric field specified for a particular mode of polarization. Under the condition that the height of a ring is much smaller than wavelength of deformation, the stress in the axial direction is negligible. Thus, we assume that in geometrical coordinates shown in Figure 7.5 $T_z = 0$.

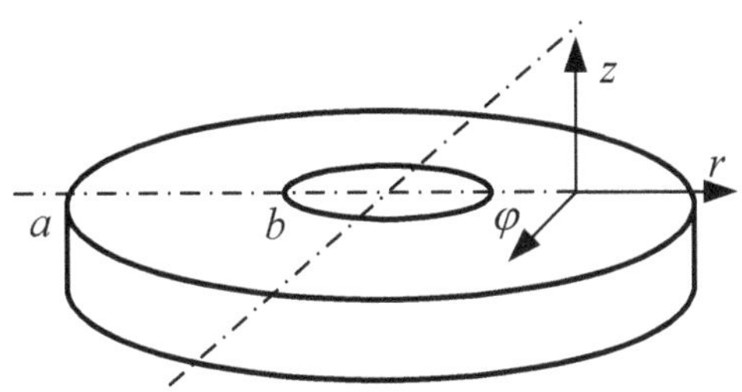

Figure 7.5: Geometry of a thick ring and coordinate system

Correspondence between geometrical and crystallographic coordinates for different modes of polarization will be as follows: for axial polarization $r \to 1, \varphi \to 2, z \to 3$; for circumferential polarization $r \to 1, \varphi \to 3, z \to 2$; for radial polarization $r \to 3, \varphi \to 1, z \to 2$. Subscripts for the strain in the plane of the ring,

$$S_r = \frac{d\xi_r}{dr} = \xi_r(a)\frac{d\theta(r)}{dr}, \quad S_\varphi = \frac{\xi_r}{r} = \xi_r(a)\frac{\theta(r)}{r} \tag{7.17}$$

(and for the stress accordingly) must be changed to numbers that correspond to coordinates r and φ in crystallographic notations for a particular mode of polarization.

In case of the radial and circumferential polarization the piezoelectric equations with stress as independent variables have form (given that $T_2 = 0$)

$$S_1 = s_{11}^E T_1 + s_{13}^E T_3 + d_{31} E_3, \tag{7.18}$$

$$S_3 = s_{13}^E T_1 + s_{33}^E T_3 + d_{33} E_3, \tag{7.19}$$

$$D_3 = d_{31} T_1 + d_{33} T_3 + \varepsilon_{33}^T E_3. \tag{7.20}$$

The stress T_1^E and T_3^E at $E_3 = 0$ being expressed from Eqs. (7.18) and (7.19) in terms of the working strain are

$$T_1^E = \frac{Y_1^E}{1-\sigma_3^E \sigma_{13}^E}\left(S_1 + \sigma_3^E S_3\right), \tag{7.21}$$

$$T_3^E = \frac{Y_3^E}{1-\sigma_3^E \sigma_{13}^E}\left(S_3 + \sigma_{13}^E S_1\right), \tag{7.22}$$

where the notations are introduced

$$Y_1^E = 1/s_{11}^E, \; Y_3^E = 1/s_{33}^E, \; \sigma_3 = -s_{13}^E/s_{33}^E, \; \sigma_{13} = -s_{13}^E/s_{11}^E. \tag{7.23}$$

The charge density obtained from Eq. (7.20) after substituting expressions for the stress will be

$$D_3 = D_3^E\left(S_1, S_3\right) + \varepsilon_{33}^{S_{1,3}} E_3, \tag{7.24}$$

where

$$D_3^E\left(S_1, S_3\right) = d_{31}\frac{Y_1^E}{1-\sigma_3^E \sigma_{13}^E}\left(S_1 + \sigma_3^E S_3\right) + d_{33}\frac{Y_3^E}{1-\sigma_3^E \sigma_{13}^E}\left(S_3 + \sigma_{13}^E S_1\right), \tag{7.25}$$

and $\varepsilon_{33}^{S_{1,3}}$ is the dielectric constant at $S_1 = S_3 = 0, T_2 = 0$,

$$\varepsilon_{33}^{S_{1,3}} = \varepsilon_{33}^T\left(1 - \frac{k_{31}^2 + k_{33}^2 - 2\left|k_{31}\right| k_{33}\sqrt{\sigma_{13}^E \sigma_3^E}}{1-\sigma_3^E \sigma_{13}^E}\right). \tag{7.26}$$

Difference between the variants of the radial and circumferential polarizations is that in all the above relations instead of $S_1 = S_\varphi$, $S_3 = S_r$ for the radial polarization it must be adopted $S_1 = S_r$, $S_3 = S_\varphi$ for the circumferential.

In variant of the axial polarization $T_3 = 0$, $S_1 = S_\varphi$, $S_2 = S_r$ and piezoelectric equations are

$$S_1 = s_{11}^E T_1 + s_{12}^E T_2 + d_{31} E_3 , \tag{7.27}$$

$$S_2 = s_{12}^E T_1 + s_{11}^E T_2 + d_{31} E_3 , \tag{7.28}$$

$$D_3 = d_{31}\left(T_1 + T_2\right) + \varepsilon_{33}^T E_3 . \tag{7.29}$$

The stress T_1^E and T_2^E at $E_3 = 0$ are

$$T_1^E = \frac{Y_1^E}{1-(\sigma_1^E)^2}\left(S_1 + \sigma_1^E S_2\right), \tag{7.30}$$

$$T_2^E = \frac{Y_1^E}{1-(\sigma_1^E)^2}\left(\sigma_1^E S_1 + S_2\right). \tag{7.31}$$

Here $\sigma_1^E = -s_{12}^E / s_{11}^E$. The charge density can be presented as

$$D_3 = D_3^E\left(S_1, S_2\right) + D_3^{S_{1,2}} , \tag{7.32}$$

where

$$D_3^E\left(S_1, S_2\right) = \frac{d_{31} Y_1^E}{1-\sigma_1^E}\left(S_1 + S_2\right), \tag{7.33}$$

$$D_3^{S_{1,2}} = \varepsilon_{33}^{S_{1,2}} E_3 , \text{ and } \varepsilon_{33}^{S_{1,2}} = \varepsilon_{33}^T\left(1-k_p^2\right). \tag{7.34}$$

For determining the equivalent parameters of the rings as functions of their thickness, consider expressions for the energies, in which these parameters are involved. The expressions for energies and for corresponding equivalent parameters in the form suitable for a single degree of freedom system are as follows.

The kinetic energy is

$$W_{kin} = \frac{1}{2}\int_{\tilde{V}} \rho \dot{\xi}_r^2 d\tilde{V} = \frac{1}{2}\dot{\xi}^2\left(a\right) 2\pi h \rho \int_b^a \theta^2\left(r\right) r dr = \frac{1}{2}\dot{\xi}^2\left(a\right) M_{eqv} , \tag{7.35}$$

where from M_{eqv} can be represented as

$$M_{eqv} = 2\pi h \rho \int_b^a \theta^2\left(r\right) r dr = M_{eqv.tr} F_M\left(b/a\right). \tag{7.36}$$

Here $M_{eqv.tr} = \pi(a^2 - b^2)h\rho = 2\pi hta_{av}\rho$ is the equivalent mass determined in the approximation of a thin ring (at $b \to a$), and $F_M(b/a)$ is the correction factor that accounts for nonuniform distribution of displacements in the radial direction. The equivalent mass obviously does not depend on a mode of polarization. Plot of the function $F_M(b/a)$ is shown in Figure 7.6.

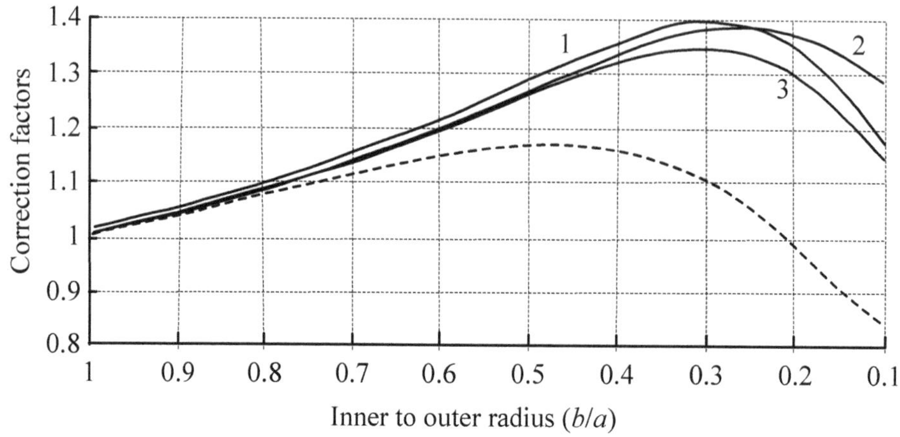

Figure 7.6. Correction factors: $F_M(b/a)$ for the equivalent mass (dashed line); $F_K(b/a)$ for the rigidities at the radial, circumferential and axial polarizations (solid lines numbered 1, 2, 3, respectively.)

The equivalent rigidity is determined from the expression for the potential energy that in the case of piezoelectric mechanical system we denote W_m^E, as follows,

$$W_m^E = \frac{1}{2} 2\pi h \int_b^a [T_r^E S_r + T_\varphi^E S_\varphi] r dr = \frac{\xi^2(a)}{2} K_{eqv}^E . \tag{7.37}$$

We denote in this expression

$$K_{eqv}^E = K_{eqv.tr}^E F_K(b/a), \tag{7.38}$$

where

$$K_{eqv.tr}^E = \left. \frac{4\pi Y_i^E h(a-b)}{(a+b)} \right|_{b \to a} = \frac{2\pi Y_i^E ht}{a_{av}} \tag{7.39}$$

is the rigidity determined by formulas for a thin ring with thickness t and radius a_{av}. In variants of the radial and axial polarization $i = 1$, in the variant of the circumferential polarization $i = 3$. $F_K(b/a)$ is the correction factor accounting for the finite thickness of the axially polarized ring. When calculating rigidity K^E_{eqv} from expression (7.37) for different modes of polarization, the subscripts r and φ must be replaced by the corresponding numbers. Thus, for the variants of the radial and circumferential polarization after substituting expressions (7.21) and (7.22) we obtain

$$W^E_m = \frac{1}{2}\int_{\tilde{V}}\left(S_1T^E_1 + S_3T^E_3\right)d\tilde{V} = \\ = \frac{1}{2}\frac{2\pi h}{1-\sigma^E_3\sigma^E_{13}}\int_b^a\left[Y^E_1S_1^2 + \left(Y^E_1\sigma^E_3 + Y^E_3\sigma^E_{13}\right)S_1S_3 + Y^E_3S_3^2\right]rdr \quad . \tag{7.40}$$

Here in variant of the radial polarization $S_2 = S_\varphi$, $S_3 = S_r$, and in variant of circumferential polarization $S_1 = S_r$, $S_3 = S_\varphi$.

In the variant of axial polarization $S_1 = S_r$, $S_2 = S_\varphi$ and

$$W^E_m = \frac{1}{2}\int_{\tilde{V}}\left(S_1T^E_1 + S_2T^E_2\right)d\tilde{V} = \frac{1}{2}\frac{2\pi hY^E_1}{1-(\sigma^E_1)^2}\int_b^a\left[S_1^2 + 2\sigma^E_1S_1S_2 + S_2^2\right]rdr \, . \tag{7.41}$$

Functions $F_K(b/a)$ for different modes of polarization are plotted in Figure 7.6. After the correction factor is introduced, the mechanical energy can be presented as

$$W^E_m(b/a) = W^E_{m.tr}F_K(b/a), \tag{7.42}$$

where $W^E_{m.tr}$ corresponds to energy calculated for a thin ring.

It is noteworthy that dimensions of the rings at $(b/a) < 0.7$, in which case $(t/a_{av}) > 0.4$, are not practical for cylindrical transducer designs with the radial and circumferential polarization. Not only it is hard to imagine how such piezoelements could be effectively polarized at related nonuniformity of the electric field. Operational characteristics of transducers composed of such thick rings would not have positive features whatsoever. The axial polarized radial vibrating piezoelements can be used as electromechanical transducers at any ratio b/a, for example, as annular disks in piezoelectric transformer designs. However, the data in Figure 7.6 are retained for all the range of ratios *b*/*a* in order to show tendency in changing the equivalent parameters. For the practical range of rings thickness the data will be presented in a scale more appropriate for using in calculations as correction factors. In case of the radial polarization deformation changes along the lines of electric field. At this condition the additional energy term ΔW , strictly speaking, must be taken into consideration, when calculating mechanical energy and hence the equivalent rigidity. But in the practical range of ratios *b*/*the* mode shape of deformation changes not significantly (see Figure 4.12), and this term can be neglected, as it is shown in Section 5.5.2.

The electromechanical transformation coefficient n is determined from the expression for the electromechanical energy,

$$W_{em} = \frac{1}{2}\int_{v} D_3^E\left(S_r, S_\phi\right) E_3 d\tilde{V} = \frac{1}{2} 2\pi h \int_b^a D_3^E\left(S_r, S_\varphi\right) E_3 r d = \frac{1}{2}\xi(a)\cdot Vn, \tag{7.43}$$

where $D_3^E\left(S_r, S_\varphi\right)$ is given by formula (7.25) in variants of the radial and circumferential polarization (with $S_\varphi = S_1$, $S_r = S_3$ for the radial, and $S_r = S_3$, $S_\varphi = S_3$ for the circumferential polarizations), and by formula (7.33) with $S_r = S_1$, $S_\varphi = S_2$ in variant of the axial polarization. Directions of the electric fields in a thick ring (annular disk) at different variants of polarization are illustrated with Figure 7.7.

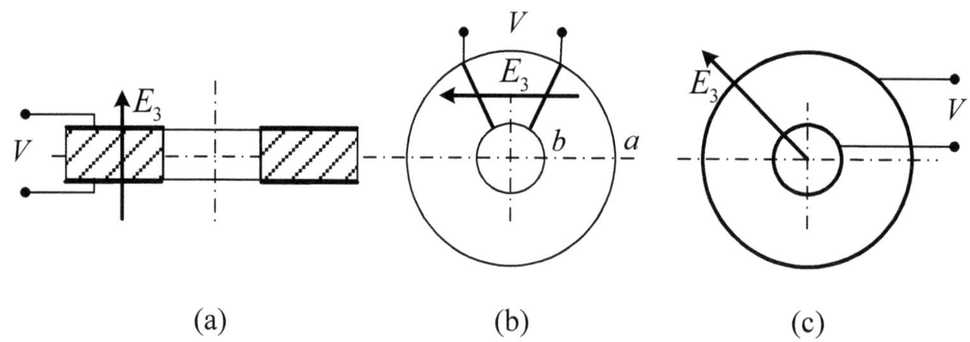

Figure 7.7: Directions of the electric fields at different variants of polarization: (a) axial, (b) circumferential, (c) radial.

Expressions for the strength of electric field at different variants of polarization with voltage V applied to electrodes are:

$$E_3 = V / h \tag{7.44}$$

at axial polarization;

$$E_3 = \frac{VN}{2\pi r}, \tag{7.45}$$

at circumferential polarization (N is the number of segments, of which the piezoelement is composed);

$$E_3 = \frac{V}{r}\ln(b/a), \tag{7.46}$$

at radial polarization. Thus, in the variant of the axial polarization after combining expressions (7.33), (7.43) and (7.44) will be obtain

$$W_{em} = \frac{1}{2}\xi(a)V\frac{2\pi Y_1^E d_{31}}{1-\sigma_1^E}\int_b^a\left[\frac{\theta(r)}{r}+\frac{d\theta}{dr}\right]rdr = \frac{1}{2}\xi(a)Vn_{ax}. \tag{7.47}$$

We denote

$$n_{ax} = n_{tr.ax}F_{n.ax}(b/a), \tag{7.48}$$

where $n_{tr.ax} = 2\pi(a-b)Y_1^E d_{31}$ is determined by formula for a thin axial poled ring, and

$$F_{n.ax}(b/a) = \frac{1}{(1-\sigma_1^E)(a-b)}\int_b^a\left[\frac{\theta(r)}{r}+\frac{d\theta}{dr}\right]rdr \tag{7.49}$$

is the correction factor accounting for the finite thickness of the ring. It can be obtained in analogous way that for the variant of the radial polarization

$$n_{rad} = n_{tr.rad}F_{n.rad}(b/a), \tag{7.50}$$

where $n_{tr.rad} = 2\pi hY_1^E d_{31}$ is determined by formula for a thin radial poled ring and

$$F_{n.rad}(b/a) = \frac{1}{1-\sigma_3^E\sigma_{13}^E}\int_b^a\left[\left(1+\frac{d_{33}}{d_{31}}\sigma_3^E\right)\frac{\theta}{r}+\frac{Y_3^E}{Y_1^E}\left(\frac{d_{33}}{d_{31}}+\sigma_{13}^E\right)\frac{d\theta}{dr}\right]\left[\frac{1}{r\ln(a/b)}\right]rdr \tag{7.51}$$

is the correction factor. And for the variant of the circumferential polarization

$$n_{cf} = n_{tr.cf}F_{n.cf}(b/a), \tag{7.52}$$

where $n_{tr.cf} = 2htY_3d_{33}N/(a+b)$ and

$$F_{n.cf}(b/a) = \frac{(a+b)}{2(a-b)(1-\sigma_3^E\sigma_{13}^E)}\int_b^a\left[\left(1+\frac{d_{31}}{d_{33}}\sigma_{13}^E\right)\frac{\theta}{r}+\frac{Y_1^E}{Y_3^E}\left(\frac{d_{31}}{d_{33}}+\sigma_3^E\right)\frac{d\theta}{dr}\right]dr. \tag{7.53}$$

Plots of the correction factors are presented in Figure 7.8. (See the note made in relation to Figure 7.6 regarding practical values of ratio *b/a*). After the correction factor is introduced, the electromechanical energy can be presented as

$$W_{em}(b/a) = W_{emtr} F_n(b/a), \tag{7.54}$$

where $W_{em\,tr}$ corresponds to energy calculated for a thin ring.

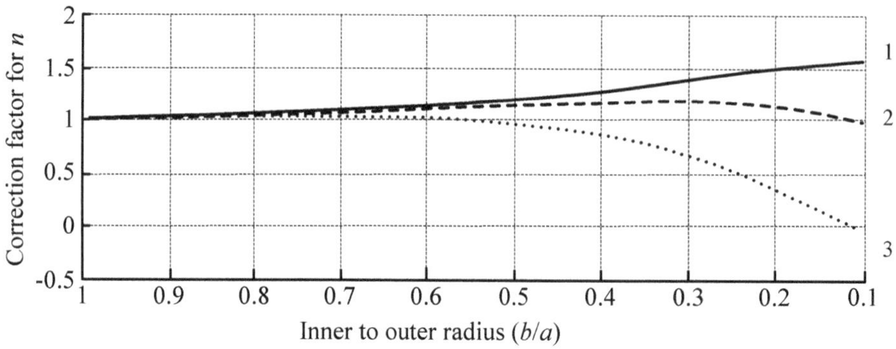

Figure 7.8: Correction factors $F_n(b/a)$ for different variants of polarization: 1 - axial, 2 - circumferential, 3 - radial.

Electrical capacitances of the thick rings clamped in direction of the working strain must be determined from expressions for the electric energy

$$\varepsilon_{33}^{S_{1,3}} = \varepsilon_{33}^{S} + e_{31}^{2} / c_{11}^{E}. \tag{7.55}$$

After substituting corresponding expressions for the dielectric constants (from (7.26), (7.34)) and for the electric field strength (from (7.44) - (7.46)), we will obtain the following expressions for the capacitances related to different modes of polarization:

$$C_{el}^{S_{1,2}} = \varepsilon_{33}^{S_{1,2}} \pi\left(a^2 - b^2\right) / h \tag{7.56}$$

for thee axial polarization;

$$C_{el}^{S_{1,3}} = \varepsilon_{33}^{S_{1,3}} 2\pi h \ln(b/a) \tag{7.57}$$

for the radial polarization;

$$C_{el}^{S_{1,3}} = \varepsilon_{33}^{S_{1,3}} h \frac{N^2}{2\pi} \ln(a/b) \tag{7.58}$$

for the circumferential polarization. Expressions for the dielectric constants in the formulas are:

Expressions for $\varepsilon_{33}^{S_{1,2}} = \varepsilon_{33}^{T}(1 - k_p^2)$ and $\varepsilon_{33}^{S_{1,3}} = \varepsilon_{33}^{S} + e_{31}^{2} / c_{11}^{E}$ (see in Table 5.3 in Section 5.3).

After dependencies of the energies W_m^E, W_{em} and W_{el} from relative thickness of the rings are determined, the effective coupling coefficients of the differently polarized piezoelements can be obtained using the expression (5.126),

$$k_{eff}^2 = \frac{W_{em}^2}{W_{el}^T W_m^E}. \tag{7.59}$$

This formula can be rewritten as

$$k_{eff}^2(b/a) = \frac{n^2(b/a)}{C_{el}^T(b/a)K_{eqv}^E(b/a)} \tag{7.60}$$

considering expressions (7.37), (7.47) and (7.55) for the energies involved and the subsequent expressions for the equivalent parameters. The effective coupling coefficients k_{efftr} of the thin rings for corresponding modes of polarization are k_{31} for axial and radial polarizations and k_{33} for the circumferential polarization. The relative change of effective coupling coefficients vs. ratio *b/a* will be denoted as

$$F_{k_{eff}}(b/a)_i = \frac{k_{eff}(b/a)_i}{k_{efftr\,i}}, \tag{7.61}$$

where number *i* corresponds to a mode of polarization. Plots of these functions are shown in Figure 7.9.

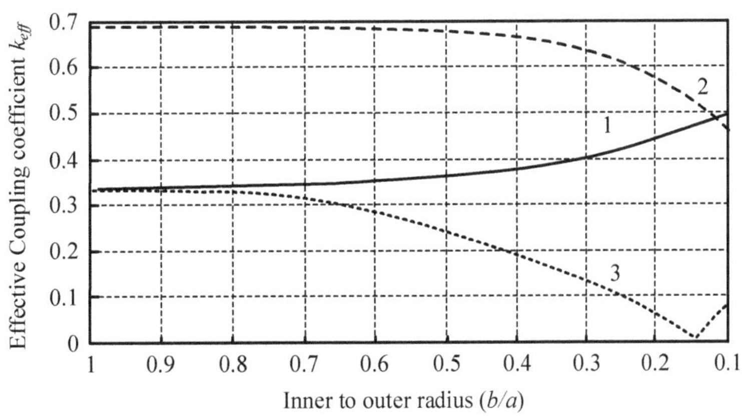

Figure 7.9: Plots of relative change of effective coupling coefficients vs. ratio *b/a* for the different modes of polarization: 1 - axial, 2 - circumferential, 3 - radial.

A qualitative explanation for the trend of the plots is as follows. The effective coupling coefficient of a ring under radial polarization (curve 3) decreases with increase of the thickness due to increase of stress T_3 and thus developing distractive contribution of the longitudinal piezoeffect, which is proportional to d_{33}, to transverse piezoeffect in the thin ring (remember that d_{31} and d_{33} have opposite signs). Similar effect takes place in the variant of circumferential polarization (curve 2). The difference is that now the main is the longitudinal piezoeffect in circumferential direction and due to increasing the radial stress T_1 a distractive contribution of transverse piezoeffect is developing. Thus, the effective coupling coefficient drops with increase of the thickness. In the variant of the axial polarization both stress T_1 and T_2 in the plane of the ring contribute positively that results in increasing the effective coupling. Results of calculating plots in Figure 7.6 through Figure 7.9 were presented in Ref. 1.

Results of the analysis show that values of the correction factors for the equivalent parameters are not significant in the practical range of thicknesses for the rings with radial and circumferential polarization. They can be taken for calculations from the plots presented in the Figure 7.6 and Figure 7.8. The thick ring with axial polarization gradually becomes an annular disk vibrating radially. Similar results for this variant of the ring can be obtained from Ref. 2, though the equivalent parameters are not presented in this work in explicit form.

7.2.2.1.2 Tangentially Polarized ("Stripe-Electroded") Design of a Piezoelement

The most common cylindrical transducers that employ longitudinal piezoeffect utilize the circumferential polarization and are made of segmented rings cemented out of the prisms with electrodes on their sides, as shown in Figure 7.1 (b). Fabrication of such piezoelements is time consuming and expensive. Besides the segmented rings have significantly smaller dynamic mechanical strength compared with solid ones especially in case that the rings are thin. Therefore, they require reinforcement for a heavy duty operation. For some applications an alternative to the segmented is design of piezoelements that employs so called "tangential polarization", in which case the electrodes in the form of stripes are applied to the exterior surfaces of thin walled cylinders or short rings. Such electrodes configuration is also used for polarizing piezoelements in the shape of thin plates, stripes, and rods. Piezoelements design that employs stripe electrodes for performing tangential polarization can be called "stripe-electroded". Thus, the terms "tangentially polarized" and "stripe-electroded" may be used interchangeably for such piezoelements.

Voltage applied to the striped electrodes induce in the body of a piezoelement electric field that has configuration qualitatively shown in Figure 7.10. The electric field in this piezoelement design is inherently nonuniform, therefore the electromechanical parameters of the piezoelement do not reach that of the segmented piezoelement of the same geometry.

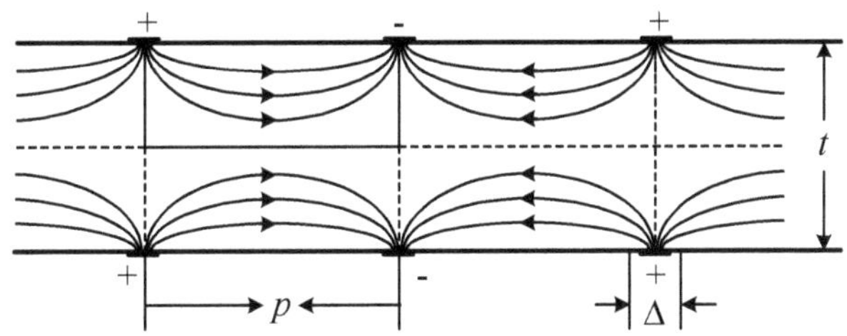

Figure 7.10: Qualitative view of configuration of electric field in a piezoelement with striped electrodes: p characterizes periodicity of the electrodes, Δ is the width of the electrodes.

For example, the effective coupling coefficient is usually about 20% lower than that achievable in the segmented design, in which case the electric field is uniformly aligned with the extensional deformation. Another peculiarity regarding the stripe-electroded cylindrical transducer designs is that diameter of the solid cylinders is typically limited by about 15 cm. from manufacturing considerations. This makes the operational frequency range of the tangentially poled projectors higher than approximately 10 kHz. It can be said that the stripe-electroded design of piezoelements may be a practical alternative to the segmented design, if the performance requirements cannot be met by using solid piezoelements employing transverse piezoeffect (radially poled cylinders in particular).The tangentially poled piezoelements combine simplicity of manufacturing solid transversely poled piezoelements and advantage of employing the longitudinal piezoelectric effect, though for expense of some reduction of electromechanical parameters compared with the segmented piezoelements of the same configuration. Analysis of electromechanical parameters of the tangentially polarized piezoelements from geometry of the elements and configuration of the striped electrodes was reported in papers Refs. 3 and 4. In the first paper an approximate analytical method for evaluating parameters was considered. In the second paper the electromechanical parameters were calculated more accurately by using FEA. Numerical FEA analysis was also used for modeling the piezoelements with "interdigitated electrodes" (analog of striped electrodes in our notation) for application as electromechanical actuators[5,6]. We will present an approximate analytical analysis that provides a simple means for calculating the piezoelectric properties of common elements, following the main results of Ref. 3.

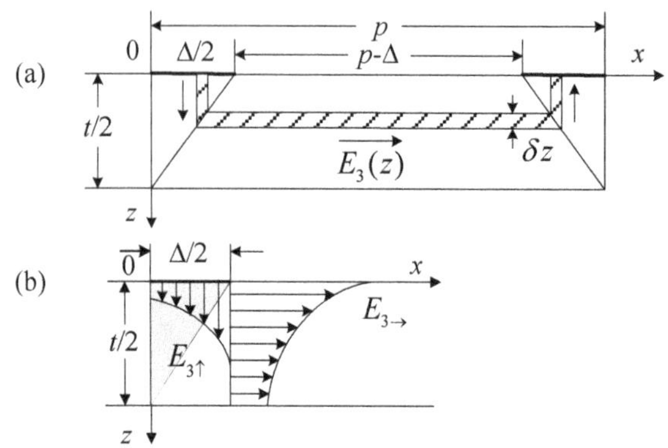

Figure 7.11: (a) Approximation of the real electric field lines by rectangular contours ("piecewise" representation), (b) the corresponding nonuniform distributions of the vertical and horizontal components of the electric field. $E_{3\rightarrow}$ and $E_{3\uparrow}$ are the electric field in the horizontal and vertical directions, respectively.

The main peculiarities of the tangentially poled piezoelements are due to nonuniformity of the electric field in their volume, as it is qualitatively illustrated in Figure 7.10. If the actual electric field distribution in the piezoceramic was known to a high accuracy, analytical or numerical calculations of the electric field related electromechanical parameters could be possible. As this is not readily the case, a simplified model of electric field may be used in order to reveal the effects on the electromechanical parameters, which arise from nonuniformity of the electric field. The following simplifying assumptions will be made in this approach. The representative part of a piezoelement in the shape of a cylinder will be replaced by a volume of piezoelectric ceramic material having a rectangular shape with attached electrodes, as shown in Figure 7.10. This is justified, because the thickness of the stripe-electroded cylinders usually is small relative to their radius. By obvious symmetry considerations it is sufficient to consider the electric field in a representative part of the volume between two electrodes that is limited by the bold solid line in Figure 7.10. Further we assume that the real curved electric field lines will be approximated by rectangular contours tangential to the real lines on their horizontal and vertical parts, as it is shown schematically in Figure 7.11(a).

Nonuniform distributions of the horizontal and vertical components of the electric field are qualitatively shown in Figure 7.11(b). Thus, the volume of a piezoelement can be imagined as subdivided in elemental volumes limited by the electric field lines that are *dz* apart, inside of which the electric field can be considered independent of *z* coordinate. At voltage *V* between the electrodes the electric field in the elemental volume with coordinate *z* of its horizontal part will be found as $E_3(z) = V/l(z)$), where $l(z)$ is the length of corresponding electric field line. Using geometry considerations following from Figure 7.11(a), we obtain that $l(z) = (p-\Delta)(1+2\beta z/t)$, where

$$\beta = \frac{t+\Delta}{p-\Delta}. \tag{7.62}$$

Correspondingly,

$$E_3(z) = \frac{V}{p-\Delta}\cdot\frac{1}{1+2\beta z/t} = E_{max}\frac{1}{1+2\beta z/t}. \tag{7.63}$$

In Eq. (7.63) $E_{max} = E_3(0) = V/(p-\Delta)$ is the electric field between edges of the electrodes. The electric field changes along *z* axis and reaches its minimum value E_{min} at $z = t/2$

$$E_{min} = E_{max}\frac{1}{1+\beta}. \tag{7.64}$$

Thus, a concentration of the electric field takes place around the edges of the electrodes, and the electric field drops to its minimum value, which is determined by the coefficient β. In order to get a quantitative estimate of the field nonuniformity for a practical piezoelement geometry, consider two examples that are representative for the tangentially poled cylinders that have outer diameter D_o = 38.1 mm, thickness t = 3.2 mm, electrode width Δ = 2.0 mm. The samples differ by number of stripes: N = 12 (sample 1) and N = 16 (sample2). The samples of this geometry were used for the experimental investigation in Ref. 3 that was aimed on verifying results of calculation of parameters performed by the method under consideration. It follows from the geometry of the cylinders and from the electrode configurations that: in the first case β = 0.73 (p = 9.1 mm, t = 3.2 mm, Δ = 2.0 mm) and E_{min} = 0.58 E_{max}; in the second case β = 1.07 (p = 6.9 mm, t = 3.2 mm, Δ = 2.0 mm) and E_{min} = 0.48 E_{max}. The same estimates are valid for piezoelements in the shape of rectangular plate having the same parameters p, t and Δ.

Nonuniformity of the electric field can lead to incomplete polarization of a part of its volume during the manufacturing process of polling the piezoelement. We denote the magnitude of electric field that is necessary to fully polarize the piezoelement, i.e., to achieve the maximum piezoelectric modulus of its material, as E_P. If the electric field does not reach this value in some part of the piezoelement, the resulting piezoelectric modulus may depend on the actual field applied. This is shown qualitatively by the idealized plot in Figure 7.12. By no means can this plot be recommended for a quantitative estimation. It is used only to illustrate the reason for possible incomplete polarization. The actual dependence is not known. In fact, the elastic and dielectric constants also depend on quality of poling the piezoceramic. In order to fully polarize a piezoelement, the minimum electric field must be greater than that assumed to be necessary, i.e., the field should be $E_{min} \geq E_P$, or

$$E_{max} = \frac{V}{p-\Delta} \geq E_P(1+\beta). \tag{7.65}$$

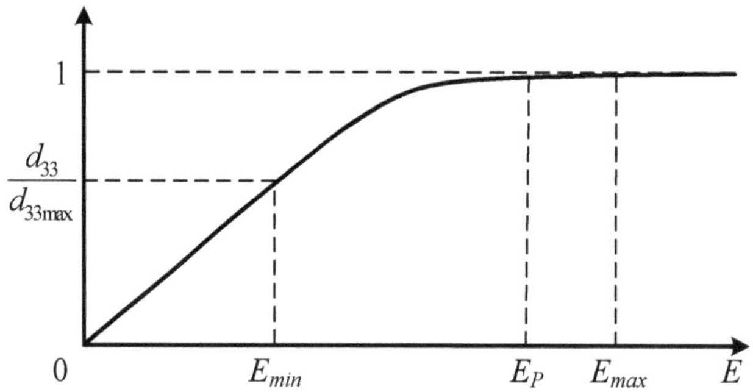

Figure 7.12: Qualitative representation of dependence of the piezoelectric modulus on the electric field of polarization.

The maximum magnitude of the voltage applied to a piezoelement is limited: by the electric field in terms of intrinsic electric and related mechanical breakdown of the ceramic material(denote the magnitude of this field E_l); and by the voltage between the stripe-electrodes in terms of preventing electrical breakdown over the piezoelement surface (denote it V_l). Obviously, the following conditions must be fulfilled

$$E_{max} \leq E_l, \quad E_{max}(p-\Delta) \leq V_l. \tag{7.66}$$

The conditions (7.65) and (7.66) may be unavoidably contradictory. The first condition (7.66) is common for piezoelements of different kinds; the second condition may be more restrictive for the stripe-electroded piezoelements, due to the concentration of electric field around the edges of the electrodes and thus concentration of the mechanical stress in this region. For small values of the coefficient β, the most probable limitation may be the electrical breakdown voltage V_p. In this case the electric field in the piezoelement can be fairly uniform, but the whole piezoelement may not be fully polarized. For the larger values of β, nonuniformity of the electric field may become the main reason for "under-polarizing" a part of the piezoelement volume. In any case, the applied poling voltage must be determined by relation (7.65). Otherwise, the piezoelement will not be fully polarized.

Even if all the volume of the piezoelement is considered as being fully polarized, it can be expected that the electromechanical parameters of the tangentially poled piezoelements should have reduced values (on average) in comparison with those for the analogous piezoelements operating under uniform electric field. The reduction of the electromechanical parameters of a piezoelement with stripe-electrodes may be estimated as follows.

In order to evaluate parameters of a piezoelement under nonuniform electric field in comparison with those that would exist for the piezoelement under uniform electric field, the notion of "effective" parameters for the nonuniform case can be introduced. We define the effective parameters as the analytically or experimentally determined parameters that can be used in the regular formulas for piezoelements with uniform field in order to account for the nonuniformity of the field. Thus, the effective elastic constant, s^E_{33eff}, dielectric constant, ε^T_{33eff}, and piezoelectric modulus, d_{33eff}, of a tangentially poled cylinder having N electrode stripes can be determined from results of measuring the resonance frequency, f_r, the capacitance at low frequency, C_{Lf}, and the effective coupling coefficient, k_{eff}, of the stripe-electroded cylinder. The effective parameters may be determined from results of the measurements using the following formulas, which are valid for segmented cylinders of the same geometry comprised of N staves:

$$s^E_{33eff} = \frac{1}{(2\pi a f_r)^2 \rho}, \tag{7.67}$$

$$\varepsilon^T_{33eff} = C_{Lf}\frac{2\pi a}{N^2 th}, \tag{7.68}$$

$$d_{33eff} = k_{33eff} \cdot \sqrt{s^E_{33eff} \cdot \varepsilon^T_{33eff}}\ . \tag{7.69}$$

Analytically the effective parameters can be determined from considering the corresponding energies, namely the electrical energy, $W_{el}^{S_3}$, potential energy, W_m^E, and electromechanical energy, W_{em}, of the striped-electroded piezoelement. In addition, the formula $k_{eff}^2 = W_{em}^2 / W_{el}^T \cdot W_m^E$ can be used for calculating the effective coupling coefficient.

The effective dielectric constant can be determined from expression for the electric energy of a clamped piezoelement under the boundary conditions typical for a short and thin-walled cylinder (at $T_1 = T_2 = 0$). The total electric energy can be found by integrating the electric energy stored in elemental volumes of the piezoelement along the electric field lines shown in Figure 7.11(a). The energy of an elemental volume per unit height of a cylinder, $\varepsilon_{33}^{S_1}$ $\tilde{V}(dz) = l(z) \cdot dz$, may be represented as follows

$$w_{el} = \frac{1}{2} E_3^2(z) \varepsilon_{33}^{S_1} 2 l_z(z) dx + \frac{1}{2} E_3^2(z) \varepsilon_{33}^{S_3} l_x(z) dz \,. \tag{7.70}$$

Here $l_z(z)$ and $l_x(z)$ are the lengths of the vertical and horizontal parts of the elemental volume, respectively. From the geometry considerations following from Figure 7.11(a),

$$dx = (\Delta / t) dz, \quad l_z(z) = z, \quad l_x(z) = p - \Delta + 2\Delta z / t, \quad l_z(z) + l_x(z) = l(z) \,. \tag{7.71}$$

In the first term of relation (7.70) is taken with subscript S_1, because on the vertical lines the poling direction is perpendicular to the direction of deformation. Therefore, all the conclusions, which may follow from the piezoelectric equations, should correspond to the case of the transverse piezoeffect, whereas on the horizontal lines they correspond to the longitudinal piezoeffect. Thus, on the vertical lines the elastic constant s_{11}^E and piezoelectric modulus d_{31} should be used instead of s_{33}^E and d_{33} on the horizontal lines. Considering (7.71) expression (7.70) can be transformed to

$$w_{el} = \frac{1}{2} E_3^2(z) \varepsilon_{33}^{S_3} l(z) dz + \frac{1}{2} E_3^2(z) \varepsilon_{33}^{S_3} \cdot 2 \left[\frac{\varepsilon_{33}^{S_1}}{\varepsilon_3^{S_3}} \cdot \frac{\Delta}{t} - 1 \right] z dz \,. \tag{7.72}$$

Now the electric energy per unit height of a piezoelement will be found as

$$W_{el}^S = \frac{1}{2} \frac{V^2}{p - \Delta} \varepsilon_{33}^{S_3} \left\{ \int_0^{t/2} \frac{dz}{1 + 2\beta z / t} + \frac{2}{p - \Delta} \left[\frac{\varepsilon_{33}^{S_1}}{\varepsilon_{33}^{S_3}} \cdot \frac{\Delta}{t} - 1 \right] \int_0^{t/2} \frac{z dz}{[1 + 2\beta z / t]^2} \right\} . \tag{7.73}$$

The ratio of the dielectric constants in the brackets may be presented as

$\left(\varepsilon_{33}^{S_1} / \varepsilon_{33}^{S_3}\right) = \left(1-k_{31}^2\right)/\left(1-k_{33}^2\right)$, and for PZT-4 (Navy Type I) it corresponds to a coefficient of 1.75. As a result of integrating in Eq. (7.73), we arrive at the following expression

$$W_{el}^{S_3} = \frac{1}{2}V^2 \frac{\varepsilon_{33}^{S_3} t}{2\beta(p-\Delta)}\left\{\ln(1+\beta)+\frac{t}{\beta(p-\Delta)}\left[\frac{\varepsilon_{33}^{S_1}}{\varepsilon_{33}^{S_3}}\cdot\frac{\Delta}{t}-1\right]\left[\ln(1+\beta)-\frac{\beta}{1+\beta}\right]\right\}. \quad (7.74)$$

In order to determine the effective dielectric constant, expression (7.74) must be equated to the expression

$$W_{el}^{S_3} = V^2 C_{el}^{S_3} / 2 \quad (7.75)$$

for the electrical energy per unit height of an equivalent piezoelement under uniform electric field generated by applying the same voltage to the electrodes imaginary attached to the side surfaces of the piezoceramic element at $x = 0$ and at $x = p$, where

$$C_{el}^{S_3} = \frac{\varepsilon_{33eff}^{S_3} t}{2p} \quad (7.76)$$

is the capacitance per unit height of the real piezoelement expressed through the effective dielectric constant. Thus, for the effective dielectric constant we obtain

$$\varepsilon_{33eff}^{S_3} \approx \varepsilon_{33}^{S_3} \frac{1+\Delta/(p-\Delta)}{\beta}\left\{\ln(1+\beta)+\frac{t}{\beta(p-\Delta)}\left[\frac{\varepsilon_{33}^{S_1}}{\varepsilon_{33}^{S_3}}\cdot\frac{\Delta}{t}-1\right]\left[\ln(1+\beta)-\frac{\beta}{1+\beta}\right]\right\}. \quad (7.77)$$

The effective elastic constant will be found through calculation of the potential energy. As it was noted, for the parts of the piezoelement, in which the direction of poling is vertical, the elastic constant s_{11}^E must be used instead of s_{33}^E. Given that we consider the one-dimensional uniform deformation of a piezoelement, its potential energy can be represented as

$$W_m^E = \frac{1}{2}S_3^2\left(\frac{\tilde{V}_{\rightarrow}}{s_{33}^E}+\frac{\tilde{V}_{\uparrow}}{s_{11}^E}\right) = \frac{1}{2}\frac{S_3^2\tilde{V}}{s_{33}^E}\left(1+\frac{\tilde{V}_{\uparrow}}{\tilde{V}}\frac{s_{33}^E-s_{11}^E}{s_{11}^E}\right). \quad (7.78)$$

Here S_3 is the deformation in the horizontal direction; $\tilde{V}$ is the total volume of the piezoelement; $\tilde{V}_{\uparrow}$ is the part of the volume, in which lines of the electric field go in the vertical direction (shaded area in Figure 7.11(a)); $\tilde{V}_{\rightarrow}$ is the part of the volume, in which direction of deformation and electric field lines go in horizontal. From the geometry of the piezoelement $\tilde{V} = pt/2$ and $\tilde{V}_{\uparrow} = t\Delta/4$. In order to determine the effective elastic constant, expression (7.78) must be equated to

$$W_m^E = S_3^2 \tilde{V} / 2 s_{33eff}^E \tag{7.79}$$

This will result in the following formula for the effective elastic constant

$$s_{33eff}^E = s_{33}^E \left(1 + \frac{s_{33}^E - s_{11}^E}{s_{11}^E} \frac{\Delta}{2p}\right)^{-1}. \tag{7.80}$$

Thus, from formula (7.80) follows that for piezoelements considered as samples 1 and 2 $s_{33eff}^E = 0.97 s_{33}^E$, if they are made of PZT-4. The change of value of elastic constant can be more substantial in case that the piezoelement is not fully polarized.

For determining electromechanical energy of a piezoelement that undergoes one-dimensional uniform deformation the density of this energy is defined by formula (5.67 c)

$$w_{em} = \frac{1}{2} \frac{d_{3i}}{s_{ii}^E} S_3 E_3 d\tilde{V}. \tag{7.81}$$

When integrating this expression over the volume of the piezoelement, for the parts of the volume, in which the electric field is perpendicular to direction of deformation, subscript in formula (7.81) has value $i = 1$ and in the parts of the volume, for which they are parallel, $i = 3$. Thus, we arrive at the following expression for the electromechanical energy

$$W_{em} = \frac{1}{2} S_3 \left\{ \frac{d_{33}}{s_{33}^E} \int_0^{t/2} E_3(z) l_x(z) dz + 2 \frac{d_{31}}{s_{11}^E} \int_0^{\Delta/2} E_3(z) l_z(z) dx \right\}. \tag{7.82}$$

After substituting under integrals the expressions (7.63) for $E_3(z)$ and (7.71) for l_z, l_x, and *dx*, integrating and performing some manipulations the electromechanical energy can be represented in the following form

$$W_{em} = \frac{1}{2} S_3 V \frac{d_{33} t}{2 s_{33}^E \beta} (A + B), \tag{7.83}$$

where

$$A = \frac{\Delta}{p - \Delta} + \left(1 - \frac{\Delta}{(p - \Delta)\beta}\right) \ln(1 + \beta), \quad B = \frac{\Delta}{p - \Delta} \left(\frac{2 d_{31} s_{33}^E}{d_{33} s_{11}^E}\right) \left[1 - \frac{1}{\beta} \ln(1 + \beta)\right]. \tag{7.84}$$

The factor in parathesis of coefficient *B* is $(2 d_{31} s_{33}^E) / (d_{33} s_{11}^E) = -1.07$ for PZT-4.

The effective coupling coefficient of a piezoelement can be determined from the expression (5.127)

$$\frac{k_{eff}^2}{1-k_{eff}^2} = \frac{W_{em}^2}{W_{el}^{S_3} \cdot W_m^E}. \tag{7.85}$$

After substituting expressions (7.73), (7.78) and (7.83) for the energies involved, from this formula will be obtained

$$k_{33eff}^2 = k_{33}^2 \frac{(A+B)^2}{\frac{\beta p}{p-\Delta}\left(1+\frac{s_{33}^E - s_{11}^E}{s_{11}^E}\frac{\Delta}{2p}\right)\ln(1+\beta)}, \tag{7.86}$$

where A and B are given by formulas (7.84). Calculations performed for piezoelements of examples 1 and 2 made of PZT-4 result in the effective coupling coefficients $k_{33eff} = 0.56$ and $k_{33eff} = 0.52$, respectively, instead of $k_{33} = 0.70$ for the case of uniform electric field in the segmented piezoelement. Analysis of formula (7.86) shows that the effective coupling coefficient may be approximated as

$$k_{33eff} \approx k_{33}\sqrt{\frac{\ln(1+\beta)}{\beta p/(p-\Delta)}} \tag{7.87}$$

for values of $\beta \le 1$. This approximate formula gives an underestimated result with less than 5% error. It follows from formula (7.87) that greater values of k_{33eff} can be achieved for smaller values of β and the ratio $\Delta/(p-\Delta)$ (note that $\ln(1+\beta)/\beta \to 1$ at $\beta \to 0$). As $\beta = t/(p-\Delta) + \Delta/(p-\Delta)$, this means that the smaller are the thickness to separation between electrodes ratio (main contributor) and the electrode width to separation between electrodes ratio, the greater is the effective coupling coefficient k_{33eff}. But this conclusion is made under the condition that all the volume of the piezoelement is fully polarized. Although, the larger the separation between electrodes, $p-\Delta$, the harder is to meet this condition due to technological limits of applying high voltages for achieving a required electric field. Thus, there is a tradeoff between the quality of polarization and uniformity of the electric field in a stripe-electroded piezoelement.

After the values of dielectric and elastic constants, and the effective coupling coefficient are determined from the relations (7.77), (7.79) and (7.86), the effective piezoelectric modulus can be calculated using formula (7.69). Note that in this formula the quantity $\varepsilon_{33eff}^T = \varepsilon_{33eff}^{S_3}/\left(1-k_{33eff}^2\right)$ must be substituted, and finally the piezoelectric modulus may be expressed as

$$d_{33eff} = \frac{k_{33eff}}{\sqrt{1-k_{33eff}^2}}\sqrt{s_{33eff}^E \varepsilon_{33eff}^{S_3}}\,. \tag{7.88}$$

For the piezoelement examples 1 and 2 made of PZT-4, we find $d_{33eff} = 225\cdot10^{-12}$ C/N and $207\cdot10^{-12}$ C/N, respectively, compared to $d_{33} = 289\cdot10^{-12}$ C/N for the uniformly polarized segmented piezoelement.

Results of calculating the dependencies of the electromechanical parameters of tangentially poled piezoelement for examples 1 and 2 are presented in Figure 7.13.

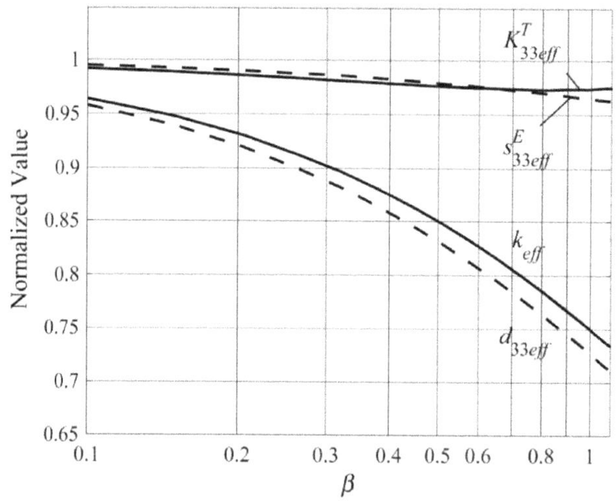

Figure 7.13: Normalized values of electromechanical parameters vs. parameter β for the stripe-electroded rings (examples 1 and 2): effective coupling coefficient, $k_{33eff}(\beta)/k_{33eff}(0)$; piezoelectric constant, $d_{33eff}(\beta)/d_{33eff}(0)$; elastic constant, $s^E_{33eff}(\beta)/s^E_{33eff}(0)$; and dielectric constant $K^T_{33eff}(\beta)/K^T_{33eff}(0)$ [i. e. $\varepsilon^T_{33eff}(\beta)/\varepsilon^T_{33eff}(0)$]. $\varepsilon^T_{33eff}(\beta)/\varepsilon^T_{33eff}(0)$ Note that at β = 1 the poling voltage needs to be twice as large as the voltage required for the case of uniform polarization.

It is noteworthy that the model of the electric field that was used for the calculations was developed for the rectangular piezoelements and thus may be less accurate for the piezoelements in the shape of the cylinders. The electric fields in the outer and inner halves of a hollow cylinder are different because of its curvature. Obviously, the results must be more accurate the smaller the thickness to mean radius ratio, t/a, of a cylinder. Although the tangential polarization is usually used for relatively thin cylinders (with t/a ratio on the order of 0.1 – 0.2), the accuracy of calculating the cylinder parameters performed for only one half of a cylinder is reduced because of curvature considerations. If the geometry of the outer half is taken for calculation, then the requirements for poling electric field are harder and nonuniformity of the electric field is underestimated for the inner half. If the geometry of the inner half is taken, then requirements for the poling electric field are less demanding and nonuniformity of the electric field is overestimated for the outer half of the cylinder. The accuracy related to the curvature effects in the cylinder is improved, if the mean diameter of a cylinder is used when calculating the stripe-electrode periodicity, p. Thus, all calculations in this section consider the mean diameter of the cylinder when calculating p.

The model used may lead to somewhat underestimated values of calculated parameters under the assumption that the piezoelement is fully polarized, because the assumed lines of electric field seem to be longer than the real ones. But, if the piezoelement is not fully polarized, which may be very probable by aforementioned reasons, then the results of calculation can be overestimated. In any case, the results of this section show the tendencies of parameters dependence from geometry of the tangentially poled piezoelement that cannot be ignored in the design and application of the transducers. More accurate estimations of parameters of the tangentially poled piezoelements are presented in Ref. 4.

7.2.2.1.3 Effects of Coupled Vibration on Parameters of the Thin-Walled Cylindrical Transducers

Coupled vibration of the thin-walled cylinders (tubes) made of a passive isotropic material were analyzed in Section 4.6.2.4. Consider now the vibration problem in the case that the tubes are made from piezoelectric ceramics and constitute an electromechanical system. In this section a piezoceramic tube will be considered as an electromechanical transducer without any external load applied (vibrating in air). As the first step the energies involved in the electromechanical conversion should be calculated using information available regarding modes of coupled vibration of analogous passive mechanical systems in accordance with general procedure of application of the energy method. Results of this section were presented in Ref. 7.

The piezoelectric equations of state that must be used for calculating the energies have different form depending on how the piezoceramic tubes are oriented relative to the crystallographic coordinate system, in other words depending on a particular mode of their polarization. Variants of tubes with different electrodes configuration that correspond to different modes of polarization to be considered, and orientations of the crystallographic coordinates are shown in Figure 7.14.

The most often used are the thickness poled tubes especially among those that have relatively large $h/2a$ aspect ratio. Therefore, a detailed analysis will be made for this case. The changes that must be made, when the poling axis is directed circumferentially or axially, will be considered afterwards.

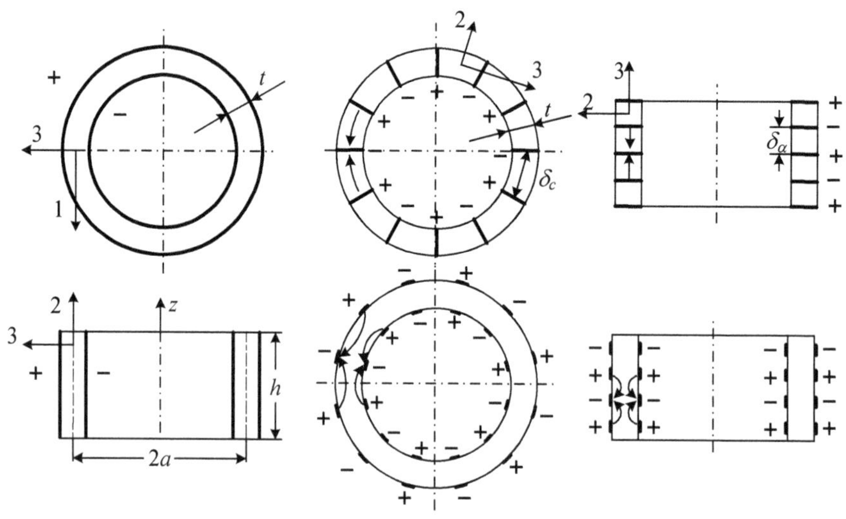

Figure 7.14. Piezoceramic tubes with different electrodes configurations that correspond to their (a) radial, (b) circumferential and (c) axial polarizations, and related orientation of the crystallographic axes.

In the case of the radial (thickness) polarization we have $S_x = S_1$, $S_\varphi = S_2$, $T_r = T_3$ (the crystallographic coordinate system is shown in Figure 7.14 (a)). Remembering that for the thin-walled tube $T_r = T_3 = 0$, the piezoelectric equations simplify to the form of Eqs. (7.27)-(7.29). Therefore the expressions (7.30), (7.31) for stress, expression (7.33) for the charge density at $E_3 = 0$, and dielectric constant of clamped piezoelement $\varepsilon_{33}^{S_{1,2}} = \varepsilon_{33}^{T}\left(1-k_p^2\right)$ are valid for this case. The summary of these expressions is presented here for convenience of further analysis (numbering the formulas is retained), as follows:

$$T_1^E = \frac{Y_1^E}{1-(\sigma_1^E)^2}\left(S_1 + \sigma_1^E S_2\right), \tag{7.30}$$

$$T_2^E = \frac{Y_1^E}{1-(\sigma_1^E)^2}\left(\sigma_1^E S_1 + S_2\right), \tag{7.31}$$

$$D_3 = D_3^E\left(S_1, S_2\right) + D_3^{S_{1,2}}, \tag{7.32}$$

where

$$D_3^E\left(S_1, S_2\right) = \frac{d_{31}Y_1^E}{1-\sigma_1^E}\left(S_1 + S_2\right), \tag{7.33}$$

$$D_3^{S_{1,2}} = \varepsilon_{33}^{S_{1,2}} E_3 \text{, and } \varepsilon_{33}^{S_{1,2}} = \varepsilon_{33}^T\left(1-k_p^2\right). \tag{7.34}$$

Eqs. (7.30) and (7.31) are identical with Eqs. (4.637) and (4.638), if to replace Y_1^E and σ_1^E for the previous Y and σ. Therefore, all the expressions for the equivalent parameters, impedances and resonance frequencies introduced in Section 4.6.2.4 are valid for piezoceramic tubes upon substituting Y_1^E and σ_1^E for Y aid σ. In order to distinguish the mechanical quantities in the case that a piezoelectric material is involved, the notations W_m^E instead of W_{pot} and K_{il}^E instead of K_{il} will be used. The expressions for the rigidities related to the first approximation (at i = 0, 1, 2) are presented in Table 7.1 (expressions for the equivalent masses remain the same independently of the piezoelectric nature of the material). New for the piezoceramic tubes are the electrical, W_{el}^S, and electromechanical, W_{em}, energies. After multiplying both parts of Eq. (7.33) and (7.34) by $E_3 = V/t$, and integrating over the tube volume the following expressions for the energies will be obtained.

For the electrical energy of a clamped tube,

$$W_{el}^{S_{1,2}} = \frac{1}{2}\int_{\tilde V} D_3^{S_{1,2}} E_3 d\tilde V = \frac{1}{2}\int_{\tilde V} \varepsilon_{33}^{S_{1,2}} E_3^2 d\tilde V = \frac{1}{2} C_{el}^{S_{1,2}} V^2 . \tag{7.89}$$

Here the electrical capacitance of a clamped tube is denoted as

$$C_{el}^{S_{1,2}} = 2\pi a h \varepsilon_{33}^{S_{1,2}} / t . \tag{7.90}$$

For the electromechanical energy,

$$W_{em} = \frac{1}{2}\int_{\tilde{V}} D_3^E (S_1, S_2) E_3 d\tilde{V} = \frac{1}{2}\frac{d_{31}}{s_{11}^E + s_{12}^E}\int_{\tilde{V}} (S_1 + S_2) E_3 d\tilde{V} = \frac{V}{2}\sum_{i=0}^{2n-1} n_i \xi_i . \tag{7.91}$$

For calculating the integral and thus determining the electromechanical transformation coefficients, expressions for the deformations $S_1 = S_x$ and $S_2 = S_\varphi$ that follow from Eqs. (4.635), (4.636) to the first approximation (i.e., at $i = 0, 1, 2$) must be used. (It must be remembered that notations ξ_1 (former ξ_{r1}) and ξ_2 (former ξ_{x1}) for generalized coordinates were introduced in Section 4.6.2.4). After performing integration, the electromechanical transformation coefficients n_i are obtained as

$$n_0 = \frac{2\pi h d_{31}}{s_{11}^E}, \quad n_1 = \frac{4 h d_{31}}{s_{11}^E\left(1-\sigma_1^E\right)}, \quad n_2 = \frac{4\pi a d_{31}}{s_{11}^E\left(1-\sigma_1^E\right)} . \tag{7.92}$$

A set of equations of motion of a piezoceramic tube may be represented in the general form of Eqs. (1.101) that incudes mutual impedances due to interaction between the modes of vibration, and Eq. (1.103), namely,

$$\left(j\omega M_{eqvi} + K_{mi}^E / j\omega + r_{mLi} + Z_{aci}\right) U_i = V n_i, \quad (i = 0, 1, 2) . \tag{7.93}$$

$$I = \left(j\omega C_{el}^{S_{1,2}} + \frac{1}{R_{eL}} \right) V + \sum_{i=1}^{2} U_i n_i . \tag{7.94}$$

The equivalent masses and rigidities in Eq. (7.93) in this case are represented by expressions (4.649) and (4.650). Resistances of the electrical and mechanical losses R_{eL}, r_{mLi} in these equations are commonly expressed as

$$R_{eL} = 1/\omega C_{el}^{S_{1,2}} \tan\delta_e \text{, and } r_{mLi} = K_{mi}^E / Q_m \omega . \tag{7.95}$$

The radiation impedances, Z_{aci}, must be set to zero, as in this section transducer is considered as electromechanical without any load. All the other constants are already obtained for the particular case of the thickness poled tube.

The set of Eqs. (7.93) and (7.94) solves the problem of calculating a piezoceramic tube as an electromechanical transducer. Eqs. (7.93) are the equations of forced vibrations of a tube. They become Eqs. (4.648) of free vibrations under the conditions that applied voltage is zero and the resistances of the mechanical losses are neglected. Therefore, the resonance frequencies of a piezoceramic tube may be found from the Eqs. (4.651), if to substitute all the impedances

Z_{il} by Z_{il}^E. The results of calculating the resonance frequencies presented in Figure 4.51 and Figure 4.52 of Section 4.6.2.4 were obtained under exactly this condition. Thus, they are valid for the thickness poled tubes that have parameters of PZT-4 ceramics.

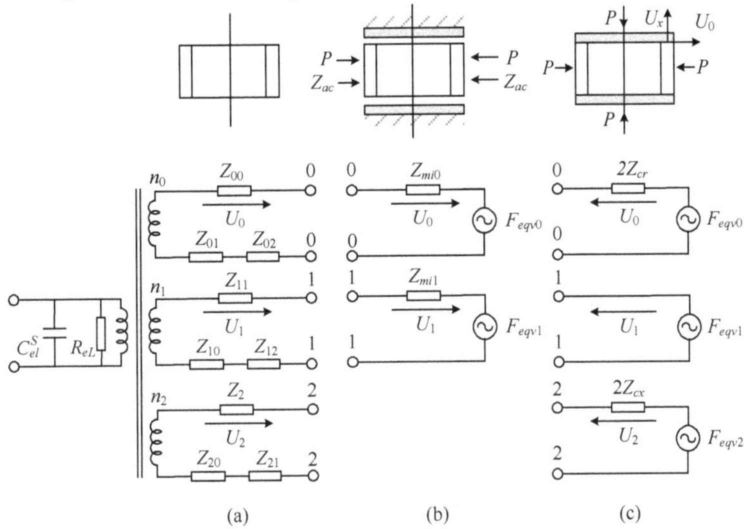

Figure 7.15: The equivalent electromechanical circuit of a transducer made from a piezoceramic tube undergoing the two-dimensional vibration for different transducer modifications: (a) with free ends and without an acoustic load applied to the side surface, (b) with the ends shielded and acoustic load applied to the side surface, and (c) with the caps attached to the ends and sound pressure acting all over the surface (the case typical for hydrophones). In cases (b) and (c) the one-port circuits must be connected to the corresponding terminals of the circuit (a)

Returning to the set of Eqs. (7.93) and (7.94), we note that these equations may be considered as the Kirchhoff's equations for the circuit that is shown in Figure 7.15 (a) for an unloaded transducer.

The circuit is equivalent to these set of equations in terms of calculating the electromechanical parameters of a transducer and distribution of velocity over its surface. The mechanical branches of the equivalent circuit correspond to the zero, first and second modes of vibration. They are coupled. The coupling is introduced through the mutual impedances z_{01}, z_{02}, z_{12} (see Eq. (4.652)) as follows:

$$Z_{01} = z_{01}\left(U_1 / U_0\right),\ Z_{10} = z_{01}(U_0 / U_1),\ Z_{02} = z_{02}(U_2 / U_0), \tag{7.96}$$

$$Z_{20} = z_{02}(U_0 / U_2),\ Z_{12} = z_{12}\left(U_2 / U_1\right),\ Z_{21} = z_{12}(U_1 / U_2). \tag{7.97}$$

After the generalized velocities U_i are calculated, and the admittance of a transducer between input terminals of the equivalent circuit in Figure 7.15 (a) is determined, the calculation of a cylindrical thickness poled transducer as an unloaded electromechanical device may be considered completed. The calculated and measured values of modulus of admittance $|Y|$ are shown in Figure 7.16 for the tube at aspect ratio $h/2a = 1.1$.

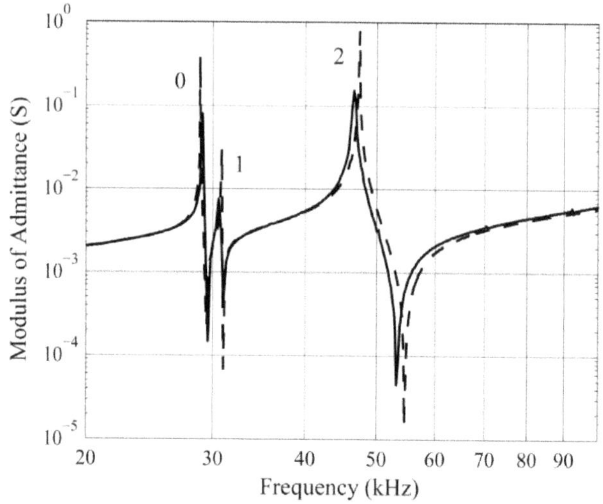

Figure 7.16: Admittances for $h/2a = 1.1$ measured (solid line) and calculated (Refs. 7, 8).

The distribution of radial velocity over surface, $U_r(x)$, and the magnitude of vibration of the ends of a tube, $U_x(\pm h/2)$, are of a great interest for a better understanding of the mechanism of coupled vibration in the tubes and for calculating or predicting the acoustic field related parameters of a transducer. It follows from Eqs. (4.633) and (4.634) rewritten in the complex form that

$$U_r(x) = U_0 + U_1 \cos(\pi x / h), \tag{7.98}$$

$$U_x(h/2) = -(\sigma_1^E h / 2a) U_0 + U_2 . \tag{7.99}$$

The mode shape of the surface vibration defined as the distribution of radial velocity normalized to its value at $x = 0$ is

$$\theta_r(x) = \frac{1}{U_0 + U_1}[U_0 + U_1 \cos(\pi x / h)] . \tag{7.100}$$

The calculated mode shapes corresponding to the resonance frequencies of a tube at $h/2a$ = 1.1 and results of their experimental verification are shown in Figure 7.17. Evidently the mode shape related to branch 1 is typical of the flexural vibration of a bar with free ends at its lowest resonance frequency.

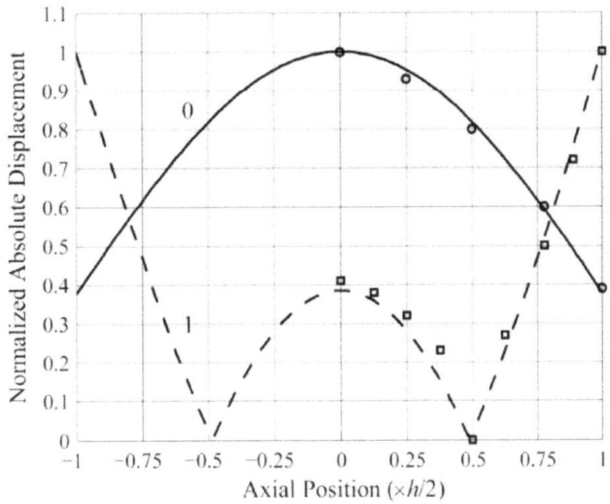

Figure 7.17: Calculated resonance mode shapes of a tube (at $h/2a = 1.1$. For the tube with dimensions $2a$ = 35 mm, t = 3.2 mm at $f_0 = 28.8$ kHz (branch 0, solid line) and at $f_1 = 30.8$ kHz (branch 1, dashed line). Experimental data from Ref. 8 are shown by circles and squares.

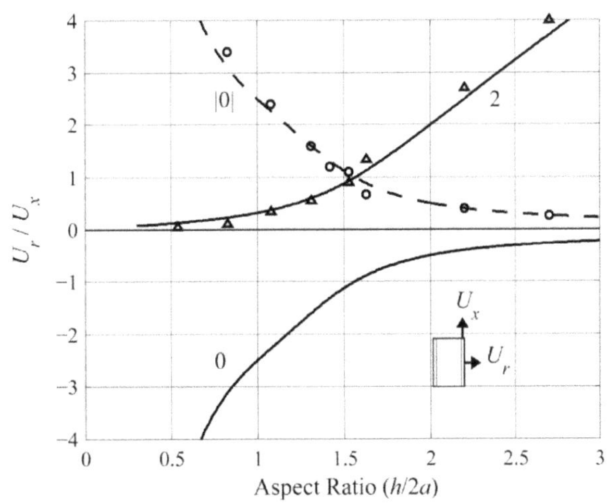

Figure 7.18: The ratio of the magnitudes of vibration in the radial and axial directions $\left[U_r(x=0)/U_x(\pm h/2)\right]$ vs. $h/2a$ along branches 0 and 2. The modulus of $\left[U_r/U_x\right]_0$ is shown by dashed line. Results of measurements[7] are shown by the circles and triangles.

The calculated and experimentally measured dependences of ratio of magnitudes of velocity in the radial direction at x = 0 to velocity of the ends in the axial direction, $U_r(x=0)/U_x(x=\pm h/2)$, along branches 0 and 2 as a function of aspect ratio are shown in Figure 7.18. The dependencies clearly show how the radial component of vibration, being predominant at lower frequency branch 0 below the point of the strongest coupling, becomes predominant at the upper branch 2 above this point. A very important feature resulting from calculations and confirmed experimentally is that U_r and U_x are in anti-phase, when related to branch 0, and in phase, when related to branch 2 (the displacements leading to expansion are conventionally considered as positive). This fact explains peculiarities in the behavior of the effective coupling coefficients, pertaining to the frequency branches.

The effective coupling coefficients that correspond to different frequency branches of the coupled vibrations can be calculated by formula (7.85), in which for this particular case the term $W_{el}^{S_{1,2}}$ must be used, namely,

$$\frac{k_{eff}^2}{1-k_{eff}^2} = \frac{W_{em}^2}{W_{el}^{S_{1,2}} W_m^E}. \tag{7.101}$$

All the energies involved in the formula are determined above. Results of calculating the effective coupling coefficients for the thickness poled tubes of different aspect ratio are shown in Figure 7.19 together with the experimental data from Refs. 7, 8. For the extensional branches 0 and 2 the results do not depend on the wall thickness (so far as the thin-wall assumption remains applicable). For the flexure related branch 1 the results are numerically valid for the tube thickness, for which the calculations are made, at aspect ratios below the strongest coupling point. Above this point they become thickness independent, because the nature of vibration gradually changes to the extensional, as it was pointed out. Given the facts that the thin–walled piezoceramic tube is electromechanically isotropic and that deformations in the circumferential and axial directions are in anti-phase along branch 0 and in phase along branch 2, it could be expected that electromechanical effects being subtracting at branch 0 and adding up at branch 2 will result in $k_{eff\,0}$ dropping to zero and $k_{eff\,2}$ raising to maximum at the point of strongest coupling, where the magnitudes of deformation in the circumferential and axial directions become equal. It is noteworthy that although at this point (at $h/2a = 1.57$) $k_{eff\,2}$ has maximum, the electromechanical energy at this point is equally distributed between vibration in the radial and axial directions, whereas the operational direction is radial. At larger aspect ratios k_{eff2} decreases slightly, but most of the electromechanical energy goes for radial vibration, which makes this range of aspect ratios advantageous for use of piezoceramic tubes, as single cylindrical projectors.

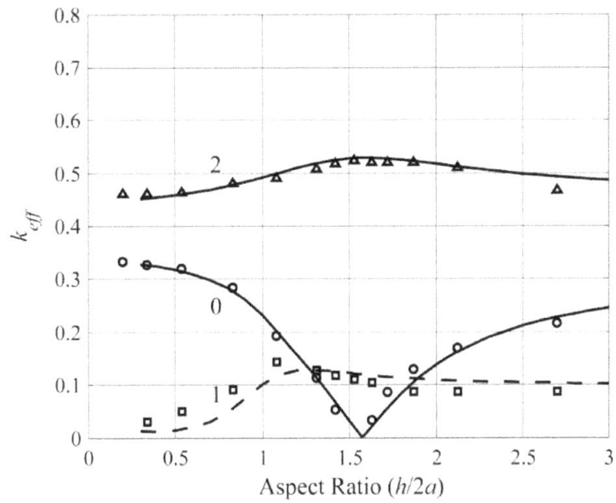

Figure 7.19: Effective coupling coefficients as a function of aspect ratio along the frequency branches 0, 1 and 2 (the plots are labeled respectively). The numerical values for k_{eff1} (dashed line) depend on the wall thickness and are valid for $t = 3.2$mm. Experimental data from Ref. 8 are shown as markers.

Consider now the changes that must be made in calculating transducer parameters in the case that the piezoceramic tubes are poled in circumferential or in axial directions. Orientations of the axes of the crystallographic coordinate system in these cases shown in Figure 7.14 correlate with the geometrical axes as follows: for the circumferential polarization $x \to 1, r \to 2, \varphi \to 3$; for the axial polarization $x \to 3, r \to 2, \varphi \to 1$. In the variants of the circumferential (tangential) polarization we will assume that electrodes are embedded into the body of a tube, as it is shown in Figure 7.14(b), and the electric field is uniform and may be calculated as $E_3 = V / \delta_c$, where δ_c is the separation between electrodes (the segments are supposed to be electrically connected in parallel). However, in reality the stripped electrodes are usually used for tangential polarizing the thin-walled tubes, as shown in Figure 7.10.

In this case the electric field, strictly speaking, cannot be considered as uniform and it is hard to make quantitatively accurate calculations, but qualitatively the results should be similar to those for the segmented design. In the variant of the axial polarization, which can be of interest in application to transducers operating in the axial direction, design of a piezoelement can be imagined, as illustrated in Figure 7.14(c), namely, segmented tubular stack cemented from a number of end-electroded rings, or a tube tangentially poled in axial direction. We will consider the segmented tube design, in which case the electric field may be assumed to be uniform with magnitude $E_3 = V / \delta_a$, where δ_a is the separation between electrodes. The mechanical boundary condition $T_2 = 0$ for circumferential and axial polarization is the same as for the radial and circumferential polarization of the thick rings (Section 7.2.2.1.1). Therefore, the piezoelectric equations (7.18)-(7.20), expressions for strain (7.21) and (7.22), expressions for the charge density (7.25) and dielectric constant (7.26) are valid for these variants. The difference is in correlating expressions for strain S_1, S_3 with expressions for S_x and S_φ by formulas (4.626) and (4.627) according to orientation of the crystallographic axes. Namely, in the variant of the circumferential polarization $S_1 = S_x$, $S_3 = S_\varphi$, and in the variant of the axial polarization $S_3 = S_x$ and $S_1 = S_\varphi$.

Calculating the equivalent parameters using the same general expressions for the energies as in the variant of the radial polarization is straightforward and will be omitted. The resulting expressions for the equivalent rigidities and for the electromechanical transformation coefficients are presented for all the modes of polarization in Tables 7.1 and 7.2.

Note: In the variant of axial polarization, a peculiarity exists in calculating the equivalent rigidities. As is shown in Section 5.5.2, when strains change in the direction of electric field (and in the case of the axial polarization both S_x and S_φ change along the electric field), an additional mechanical energy related term ΔW must be considered, resulting in an increase of the rigidity of mechanical system. If $\delta_a = h$, this effect can be accounted for by introducing the rigidity $\Delta K = 0.2 K_{22}^E k_{S_{1,3}}^2 / \left(1 - k_{S_{1,3}}^2\right)$ in addition to K_{22}^E. For PZT-4 $k_{S_{1,3}}^2 \approx 0.4$, and $\Delta K \approx 0.13 K_{22}^E$. However, when $h / \delta_a > 4$ (and this is usually the case for a range of aspect ratios, for which the effects of the coupled vibrations are significant) the additional rigidities drop and become negligible. Therefore, the corresponding corrections are not included for simplicity.

Table 7.1. Expressions for the equivalent rigidities at different modes of polarization

K_{il}^E Polarization	K_{00}^E, $\times\frac{2\pi th}{a}$	K_{11}^E, $\times\frac{\pi th}{a}$	K_{22}^E, $\times\frac{\pi^3 at}{h}$	K_{01}^E, $\times\frac{4th}{a}$	K_{12}^E, $\times\pi^2 t$
Thickness (Radial)	Y_1^E	$\frac{Y_1^E\left[1+\frac{\pi^4}{48}\left(\frac{t}{h}\right)^2\left(\frac{2a}{h}\right)^2\right]}{1-(\sigma_1^E)^2}$	$\frac{Y_1^E}{1-(\sigma_3^E)^2}$	Y_1^E	$\frac{\sigma_1^E Y_1^E}{1-(\sigma_1^E)^2}$
Circumferential	Y_3^E	$\frac{Y_3^E\left[1+\frac{\pi^4}{48}\left(\frac{t}{h}\right)^2\left(\frac{2a}{h}\right)^2\frac{Y_1^E}{Y_3^E}\right]}{1-\sigma_3^E\sigma_{13}^E}$	$\frac{Y_1^E}{1-\sigma_3^E\sigma_{13}^E}$	Y_3^E	$\frac{\sigma_3^E Y_1^E}{1-\sigma_3^E\sigma_{13}^E}$
Axial	Y_1^E	$\frac{Y_1^E\left[1+\frac{\pi^4}{48}\left(\frac{t}{h}\right)^2\left(\frac{2a}{h}\right)^2\frac{Y_3^E}{Y_1^E}\right]}{1-\sigma_3^E\sigma_{13}^E}$	$\frac{Y_3^E}{1-\sigma_3^E\sigma_{13}^E}$	Y_1^E	$\frac{\sigma_3^E Y_1^E}{1-\sigma_3^E\sigma_{13}^E}$

Table 7.2. Electromechanical transformation coefficients for tubes with different polarizations.

n Polarization	n_0 $\times 2\pi h$	n_1, $\times 4h$	n_2, $\times 4\pi a$
Thickness (Radial)	$\frac{d_{31}}{s_{11}^E}$	$\frac{d_{31}}{s_{11}^E+s_{12}^E}$	$\frac{d_{31}}{s_{11}^E+s_{12}^E}$
Circumferential, $\times t/\delta_c$	$\frac{d_{33}}{s_{33}^E}$	$\frac{d_{33}}{s_{33}^E}\left[1-\frac{d_{31}s_{13}^E}{d_{33}s_{11}^E}\right]$	$\frac{d_{31}}{s_{11}^E}\left[1-\frac{d_{33}s_{13}^E}{d_{31}s_{33}^E}\right]$
Axial, $\times t/\delta_a$	$\frac{d_{31}}{s_{11}^E}$	$\frac{d_{31}}{s_{11}^E}\left[1+\frac{d_{33}s_{13}^E}{d_{31}s_{33}^E}\right]$	$\frac{d_{33}}{s_{33}^E}\left[1-\frac{d_{31}s_{13}^E}{d_{33}s_{11}^E}\right]$

Here $\varepsilon_{33}^{S_{1,3}}$ is given by formula. Capacitances of the tubes (at $S_1 = S_3 = 0$, $T_2 = 0$ and under the condition that all the segments are connected in parallel) are: in the variant of circumferential polarization

$$C_e^{S_{1,3}} = 2\pi aht\varepsilon_{33}^{S_{1,3}} / \delta_c^2 , \tag{7.102}$$

in the variant of the axial polarization

$$C_e^{S_{1,3}} = 2\pi aht\varepsilon_{33}^{S_{1,3}} / \delta_a^2 . \tag{7.103}$$

Here $\varepsilon_{33}^{S_{1,3}}$ is given by formula (7.26)

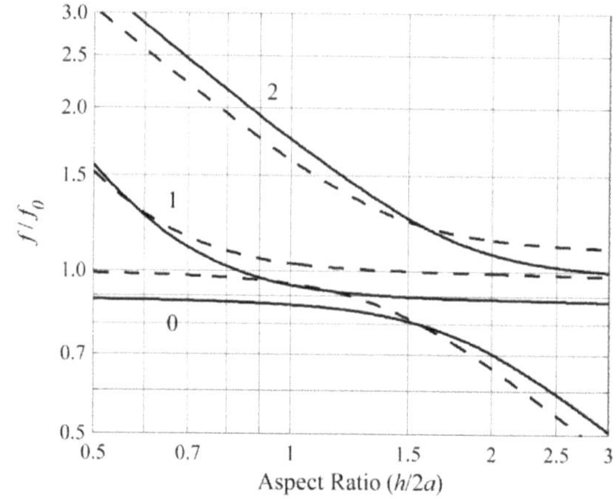

Figure 7.20: The resonance frequencies of the axially (dashed lines) and circumferentially (solid lines) poled tubes normalized by $f_0 = 30$ kHz. Labeling 0, 1 and 2 of the curves corresponds with numbering of the frequency branches.

Table 7.3. Values of calculated resonance frequencies normalized to $f = 30$ kHz.

Polarization	Radial		Circumferential		Axial	
$h/2a$	branch 0	branch 2	branch 0	branch 2	branch 0	branch 2
0.50	1.00	3.35	0.89	3.43	0.99	3.06
0.75	0.99	2.25	0.88	2.30	0.98	2.08
1.00	0.97	1.72	0.87	1.76	0.95	1.61
1.25	0.94	1.42	0.84	1.44	0.89	1.36
1.50	0.89	1.25	0.81	1.25	0.82	1.24
1.75	0.82	1.17	0.76	1.14	0.74	1.18
2.00	0.74	1.13	0.70	1.08	0.66	1.14
2.25	0.67	1.11	0.65	1.04	0.60	1.13
2.50	0.61	1.09	0.59	1.02	0.54	1.12
2.75	0.56	1.08	0.55	1.01	0.50	1.11
3.00	0.51	1.08	0.51	1.00	0.46	1.10

Once the equivalent electromechanical parameters of the circumferentially and axially poled tubes are determined, all the necessary characteristics can be calculated in the same way as it was done in the case of the radial polarization. In particular, the results of calculating the resonance frequencies are shown in Figure 7.20. The plots for the resonance frequencies are normalized to the same value as in the case of the radial polarization, i.e., to the resonance frequency of a short radial poled tube, $f_0 = 30\,\text{kHz}$. The numerical values of the normalized resonance frequencies vs. aspect ratio are presented in Table 7.3.

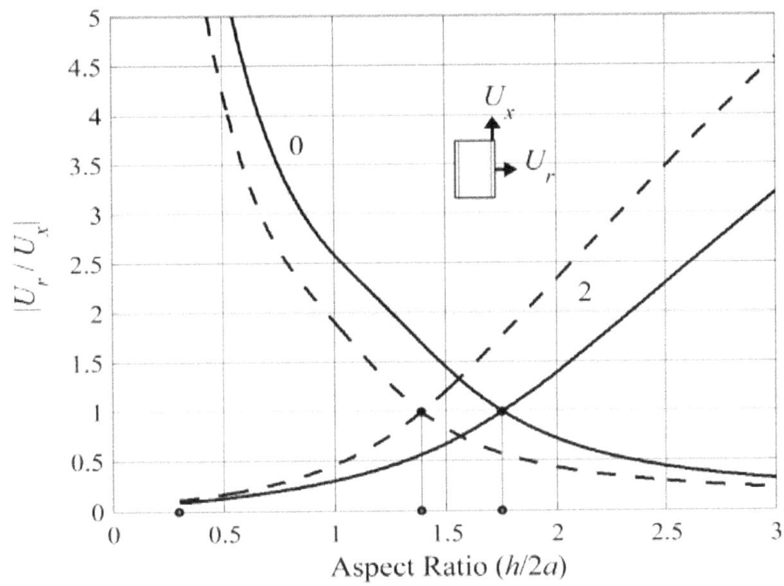

Figure 7.21: Magnitude of radial-to-axial velocity ratios, $[U_r / U_x]_i$, for the axially (dashed curves) and circumferentially (solid curves) poled tubes.

Dependencies of the radial to axial velocities ratios and effective coupling coefficients for the tubes, having the same dimensions, as in the case of the radial polarization, are depicted in Figure 7.21 and Figure 7.22, as function of the aspect ratio *h/2a*.

The results for the effective coupling coefficients show clear distinctions due to difference in the modes of polarization and consequent elastic and piezoelectric anisotropy of the tubes. For example, it can be seen that in terms of a qualitative behavior the plots for k_{eff} related to branches 0 and 2 changed places compared with the case of the radial polarization. This effect could be expected for the following reason. In the cases of the axial and circumferential polarization the piezoelectric moduli effective in the axial and circumferential directions (d_{31} and d_{33}) have different signs, whereas in the case of the radial polarization they (both being d_{31}) are of the same sign. Therefore, the electromechanical effects are subtracting along branch 2 and adding up along the branch 0 on the contrary to the case of the radial polarization. The results of calculating the effective coupling coefficients comply with their expected values for the extreme aspect ratios at $h/2a \to 0$ and $h/2a \to \infty$, in which cases the one-dimensional approximations for the corresponding piezoelements are valid.

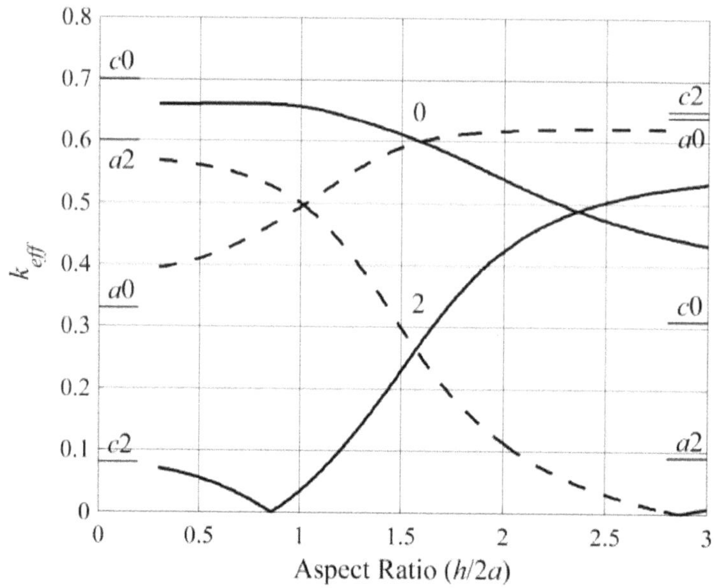

Figure 7.22: Effective coupling coefficients for the axially (dashed curves) and circumferentially (solid curves) poled tubes. Horizontal markers denote the asymptotic limits of the curves as $h/2a \to 0$ and $h/2a \to \infty$.

Expressions for the coupling coefficients of piezoceramic material for the mechanical boundary conditions that correspond to the extreme cases of $h/2a \to 0$ (a low pulsating tube and an infinitely long strip vibrating in the direction of its width with $\cos(\pi x/h)$ distribution) and $h/2a \to \infty$ (a long pulsating tube and a long tube of small diameter axially vibrating with $\cos(\pi x/h)$ distribution) may be found in Ref. 2. They are summarized in Table 7.4. These values for PZT-4 ceramics are shown in Figure 7.22 by the extreme lines, to which the calculated plots must be approaching asymptotically. Note hat in the case that cosine distribution of deformation exists the coupling coefficients of material are multiplied by the corresponding factors in order to get values for the effective coupling coefficients.

Table 7.4. Values of the effective coupling coefficients for extreme mechanical systems.

Polarization	branch 0		branch 2	
	$(h/2a)\to 0$ $T_1 = T_2 = 0,$ $S_3 \neq 0$	$(h/2a)\to \infty$ $T_2 = T_3 = 0,$ $S_1 \neq 0$	$(h/2a)\to 0$ $T_2 = 0, S_3 = 0,$ $S_1 \neq 0$	$(h/2a)\to \infty$ $T_2 = 0, S_3 = 0,$ $S_1 \neq 0$
Circumferential				
k_m 1)	k_{33}	k_{31}	$\lvert k''_{31}\rvert$ 2)	$\lvert k'_{33}\rvert$
k_{eff}, PZT-4	0.70	0.30	0.08	0.65
Axial				
k_m 1)	k_{31}	k_{33}	$\lvert k'_{33}\rvert$ 2)	$\lvert k''_{31}\rvert$ 2)
k_{eff}, PZT-4	0.33	0.64	0.60	0.09
Radial				
k_m 1)	k_{31}	k'_{31}	k'_{31}	k_{31}
k_{eff}, PZT-4	0.33	0.30	0.45	0.50

1) k_m is the coupling coefficient for piezoelectric ceramic material under different boundary conditions.

2) Following Ref. 9, $k'_{31} = \dfrac{|k_{31}|}{\sqrt{1-k_{31}^2}}\sqrt{\dfrac{1+\sigma_1^E}{1-\sigma_1^E}}$, $k''_{31} = \dfrac{-|k_{31}| + k_{33}\sqrt{\sigma_3^E\sigma_{13}^E}}{\sqrt{1-\sigma_3^E\sigma_{13}^E}\sqrt{1-k_{33}^2}}$,

$k'_{33} = \dfrac{k_{33} - |k_{31}|\sqrt{\sigma_3^E\sigma_{13}^E}}{\sqrt{(1-\sigma_3^E\sigma_{13}^E)(1-k_{31}^2)}}$. Note: in Ch.10 it is shown that factors $1/\sqrt{1-k_{31}^2}$ must be waved.

7.2.2.1.4 Effects of the Coupled Vibration on the Electroacoustic Characteristics of the Transducers

All the calculations made so far were concerned with electromechanical parameters of unloaded (vibrating in air) transducers, whereas it is impossible to complete treatment of a transducer for practical applications without considering their acoustic and/or mechanical loading and other external actions typical for the particular applications. The energy method allows acoustic loads and the external actions to be formally included into equations of motion (7.93) A way to combine the results of solving the radiation problems, which we will assume to be known from Chapter 6, with electromechanical part of an unloaded transducer treatment that is presented by the equivalent circuit in Figure 7.15 (a) is illustrated with examples of cylindrical air-backed underwater transducer designs, shown schematically as icons in Figure 7.15 (b)-(c). The variant of transducer design in Figure 7.15 (b) with caps mechanically isolated from the ends of a piezoceramic tube is typical for the projectors and broadband receivers. The ends of the tube are free to vibrate and are isolated from the acoustic field by the caps that are assumed to be absolutely rigid. The design variant in Figure 7.15 (c) with the caps attached to the ends of a piezoceramic tube and exposed to acoustic field are used predominately for the low frequency cylindrical hydrophones. Similar design can be used for radiation along the axis, but in this case, it falls into category of Tonpilz transducers, and usually the thick-walled axially polarized cylinders are used for this purpose.

The radiation problem to be solved in the variant of Figure 7.15 (b) is that for a cylinder of finite height. Analysis of this problem is presented in Section 6.3.1.3. Note that in the case that $h/\lambda > (0.6-0.7)$ a simpler model of the cylinder vibrating between two infinite rigid cylindrical baffles can be used that is considered in Section 6.3.1.2. As the radial velocity distribution over the height exists according to formula (7.98), we will assume that the radiation problem is solved for uniformly vibrating cylinder and for the cylinder vibrating with velocity distribution $U_1 \cos(\pi x/h)$, and therefore the following functions can be considered known: the far field sound pressures $P_0(r,\omega) = A(r)U_0 k_{dif\,0}$ and $P_1(r,\omega) = A(r)U_1 k_{dif\,1}$, where $A(r)$ is a distance depending coefficient and $k_{dif\,i}$ are the diffraction coefficients; the sound pressures on the cylinder surface $P_{\Sigma 0}(ka,x) = B_0(ka,x)U_0$ and $P_{\Sigma 1}(ka,x) = B_1(ka,x)U_1$, where functions $B_i(ka,x)$ are introduced to make it obvious for the further analysis that the sound pressure on the surface is proportional to the velocity. The acoustic power radiated by a transducer can be represented now as

$$\bar{\dot{W}}_{ac} = \int_{\Sigma} (P_{S0} + P_{S1}) U_r^* d\Sigma = \\ = \int_{\Sigma} [B_0(ka,x)U_0 + B_1(ka,x)U_1]\left[U_0^* + U_1^* \cos(\pi x/h)\right] d\Sigma. \tag{7.104}$$

As the functions are assumed to be known, after integrating over the transducer surface in Eq. (7.104) and some manipulations the acoustic power can be expressed as $B_i(ka,x)$ $\bar{\dot{W}}_{ac} = Z_{ac0}|U_0|^2 + Z_{ac1}|U_1|^2$, where

$$Z_{ac0} = Z_{acoo} + z_{ac01} U_1/U_0, \; Z_{ac1} = Z_{ac11} + z_{ac01} U_0/U_1. \tag{7.105}$$

The radiation impedances Z_{ac0} and Z_{ac1} represent the total acoustic loads related to the generalized velocities U_0 and U_1; Z_{ac00} and Z_{ac11} are the modal impedances for the uniform and cosine by height modes of vibration; z_{01} is the mutual impedance between the modes. The impedances Z_{ac0} and Z_{ac1} must be included in Eq. (7.93) and introduced into the equivalent circuit.

For calculating transducer in the receive mode the same equivalent circuit of Figure 7.15 is valid, if the equivalent forces,

$$F_{eqv0} = P_o S_\Sigma k_{dif\,0}\,,\quad F_{eqv1} = P_o S_\Sigma k_{dif\,1}\,, \tag{7.106}$$

that are due to action of acoustic field, are introduced into the contours related to the velocities U_0 and U_1. In the expressions (7.106) P_o is the sound pressure in the free acoustic field, S_Σ is the radiating surface area of the transducer, $k_{dif\,0}$ and $k_{dif\,1}$ are the same diffraction coefficients that are introduced for the radiating mode of operation. The acoustic loads and equivalent forces determined by expressions (7.105) and (7.106) must be introduced into the equivalent circuit between the terminals 0,0 and 1,1 in Figure 7.15 (a) as the one-port networks shown in Figure 7.15 (b) (we will refer to thus obtained circuit as to Figure 7.15 (a)-(b)). After this is done, the equivalent circuit for electroacoustic transducer operating in the transmit and receive modes can be considered completed, and the generalized velocities U_i can be determined. The far field sound pressure, $P(r,\omega)$, generated by a transducer will be found as

$$P(r,\omega) = P_0(r,\omega) + P_1(r,\omega) = A(r)\left(U_0 k_{dif\,0} + U_1 k_{dif\,1}\right), \tag{7.107}$$

so far as the velocities U_0 and U_1 are calculated from Eqs.(7.93) or, alternatively, from equivalent circuit in Figure 7.15 (a)-(b).

The problem of calculating parameters of a transducer with symmetrical caps attached to the ends of a piezoceramic tube and exposed to acoustic field, as shown in Figure 7.15 (c), has both the mechanical and acoustic aspects. The mechanical part of the problem is in formulating the boundary conditions for the joints between the caps and the ends of the tube and in calculating the coupled vibration of the caps and of the tube under these conditions. The acoustic part is in determining the radiation impedances and equivalent forces applied to the surface of the caps and to the side surface of a transducer. Results of solving the mechanical part of the problem can be represented by the input impedances on the contour of the caps in the radial and axial directions, Z_{cr} and Z_{cx}. The impedances are determined in Section 4.7.2.1 and 4.7.2.2.

If to consider the operation of a transducer in the receive mode at frequencies well below its resonance, which is typical for the low frequency hydrophones, then the acoustic part of the problem simplifies to determining the diffraction coefficients only. In addition to the equivalent forces that are defined by expressions (7.106), the equivalent forces $F_{eqv.c}$ applied to the caps must be taken into account, and $F_{eqvc} = P_o S_c k_{dif.c}$, were S_c is the surface area of a cap and k_{difc} is the diffraction coefficient related to the cap radiation. Information regarding the diffraction coefficients for tubes of the finite height with capped ends can be found in Ref. 10. After the input impedances of the caps and diffraction coefficients are determined, the equivalent circuit of a transducer can be completed by connecting the one-port networks, shown in Figure 7.5(c), to the corresponding terminals of the mechanical branches in Figure 7.15 (a). Note that impedances Z_{cr} and Z_{cx} are doubled because of symmetry (associated mechanical energies are doubled), and Z_{cr} is ascribed to the generalized coordinate U_0 only, because according to Eq. (7.98) $U_r(\pm h/2) = U_0$. Also of note is, that velocity U_2 found from the equivalent circuit is the velocity on the contour of a cap. In the case that the cap cannot be considered as absolutely rigid, distribution of velocity on its surface must be determined. These calculations, as well as those for determining the input impedances of the caps, are pure mechanical problems, which can be treated separately. However, methodically the equivalent circuit of Figure 7.15 (a) with the one-port networks of Figure 7.15 (c) included provides the means for calculating parameters of capped cylindrical transducers.

Effects of the coupled vibration on acoustic performance of the cylindrical transducers were estimated by calculation and investigated experimentally in Refs. 11 and 12. Examples of calculated and measured frequency responses of the transducers with different aspect ratios from Ref. 11 are shown in

Figure 7.23 and Figure 7.24.

As can be seen, the results of calculating are in good agreement with the experiment. The conclusions that can be made based on the results of the calculating and experimenting allow several general suggestions regarding the preferable aspect ratios of cylindrical piezoelements used for underwater transducers. It is desirable to have the height to diameter aspect ratio of the radially poled piezoelements $h/2a < 0.5$. A higher effective coupling coefficient can be obtained by using piezoelements of smaller aspect ratio. Piezoelements with larger aspect ratios,

$h/2a > 1.5$, can be used effectively in a frequency band around resonance.

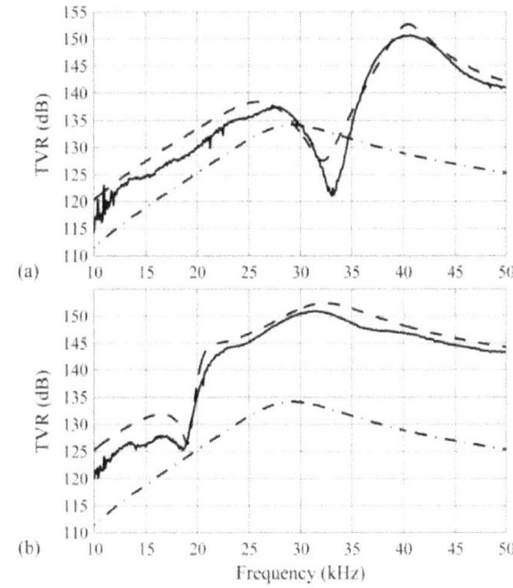

Figure 7.23: Transmit frequency responses (TVR) calculated (dashed curve) and measured (solid curve) for a transducer comprised of a thickness poled tube with (a) h/2a = 1.32 and (b) h/2a = 2.2. Mechanical losses for the fabricated transducers were included in the calculations. For comparison the frequency response without influence of coupled vibration is presented that are calculated for aspect ratio h/2a = 0.5 (dash-dotted line).

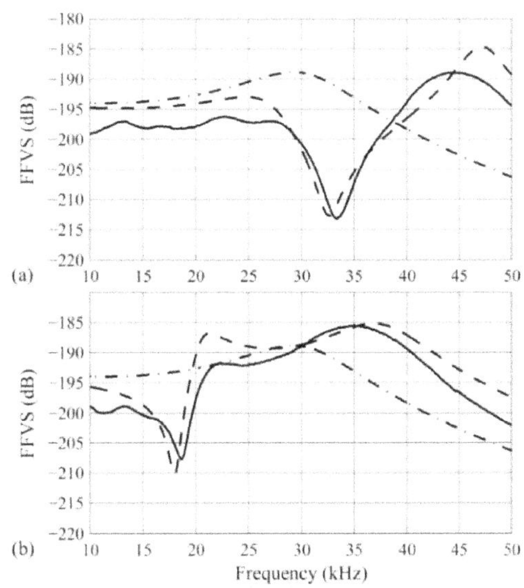

Figure 7.24: Receive frequency responses (FFVS) calculated (dashed curve) and measured (solid curve) for a transducer comprised of a thickness poled tube with (a) $h/2a$ = 1.32 and (b) $h/2a$ = 2.2. For comparison the frequency response without influence of coupled vibration is presented that are calculated for aspect ratio $h/2a$ = 0.5 (dash-dotted line).

In this range they even have an advantage in terms of electroacoustic parameters over transducers of the same height comprised of rings having small aspect ratios. Although, the piezoelements with large aspect ratios cannot be recommended for use as the broadband receivers that operate predominantly below the resonance frequency because of the deep notches that occur in the frequency response due to effects of coupled vibrations. The best results can be achieved for this purpose by employing mechanically separated piezoelements with small aspect ratios. In general, the frequency responses of transducers comprised of individual cylinders having aspect ratios larger than about 0.5 may differ dramatically due to effect of the coupled vibration, especially for the range of aspect ratios $0.5 < (h/2a) < 1.5$.

Circumferentially (tangentially) polarized cylinders operating in the lowest frequency branch can be used effectively up to much larger height to diameter aspect ratios than the radially polarized. There is no apparent need or electromechanical advantage in axially segmenting the tangentially polarized cylinders for improving their performance, as compared with taking such measure for avoiding the effect of coupled vibrations in the case of the radial polarization. It is noteworthy that cylinders of equal dimensions, but different polarizations may have significantly different frequency responses, as is shown with plots in Figure 7.25.

More detailed analysis of effects of coupled vibration on performance of the cylindrical transducers including dependence of their vertical directionality on the aspect ratio can be found in Refs. 11 and 12. Further we will assume that the rings comprising cylindrical transducers that vibrate in zero mode have aspect ratios $h/2a < 0.5$, if it is not noted otherwise.

As to effects of the coupled vibration on parameters of the rings vibrating in higher modes, the exact analysis could be performed in the same way. Though a qualitative conclusion regarding the aspect ratios of the rings, at which they can be considered short enough for neglecting these effects can be made based on the following considerations.

As it was shown, the aspect ratio, at which the strongest coupling between the radial and axial modes of a cylinder vibration takes place, corresponds to condition of equality the resonance frequencies of a thin ring and the thin bar that have dimensions of the cylinder, i.e., $f_{ring0} / f_{bar} = 1$. For radially poled cylinder that vibrates in zero mode the respective aspect ratio is $(h/2a) = 1.5$. For the cylinders vibrating in higher modes $f_{ringi} = f_{ring0}\sqrt{1+i^2}$, therefore the aspect ratio of the strongest coupling reduces in factor of $\sqrt{1+i^2}$, and so should be reduced the value of aspect ratio, at which the effects of the coupling can be neglected.

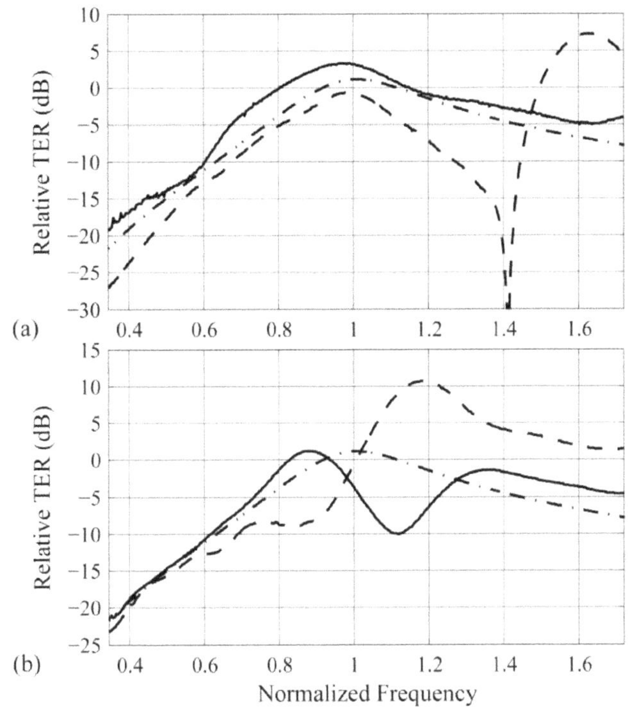

Figure 7.25: Comparison of TER of tangentially (solid line) and radially (dashed line) polarized cylindrical transducers, normalized to the TER of an axially segmented radially polarized cylindrical transducer of the same overall height: (a) aspect ratio $h/2a = 1.1$ and (b) aspect ratio $h/2a = 1.62$. The frequency is normalized to resonance frequency of an axially segmented radially polarized cylindrical transducer in water. For comparison the frequency response without influence of coupled vibration is presented that is calculated for aspect ratio $h/2a = 0.5$ (dash-dotted line).

7.2.2.1.5 Nonuniform (Active-Passive) Ring Transducers

Uniform segmented ring transducers are usually used in order to realize longitudinal piezoelectric effect and to build rings of bigger diameter than it is achievable for manufacturing the solid rings. For some applications it may be required to change the resonance wave size of a transducer. Thus, for example, when employing a transducer in array it is desirable to reduce its wave size from considerations of directivity pattern steering. In situation that rings vibrating in the zero and first modes are used in a single transducer design it is desirable to have the same diameter for both of them, which can be achieved by reducing the zero mode ring wave diameter down to those of the first mode. In the first example the effective sound speed in the ring should be reduced, and in the second example – increased. And in all the cases the mode shape of transducer vibration and its main operating parameters should not change significantly.

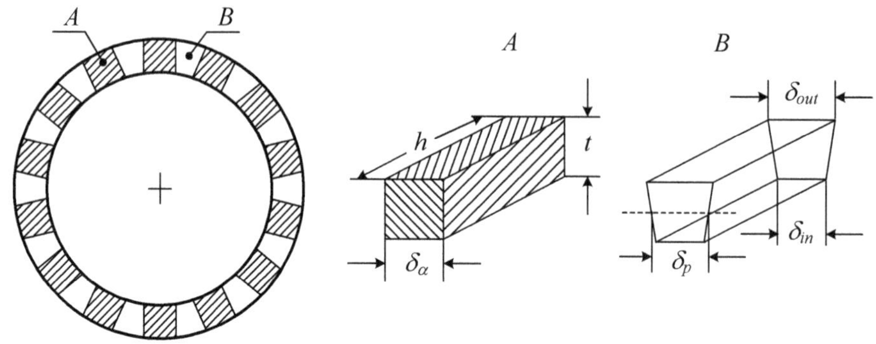

Figure 7.26: Geometry of an active-passive ring.

Changing to some extent the effective sound speed in a ring under this requirement can be produced by combining in its design the active piezoelements and inserts made of a passive material having different properties. If the active and passive parts are situated alternatively and have sufficiently small size in direction of circumference (in other words, number of parts is sufficiently large) the ring can be considered as quasi uniform with averaged elastic properties and density. But what number of parts is sufficient remains to be determined.

Consider an active-passive design, in which case the ring is a combination of piezoelements and of passive inserts in the shape of tapered prisms, as it is shown in Figure 7.26.

We will assume that the thickness of the ring is uniform. The mean widths of the active and passive elements are related as

$$\delta_p + \delta_a = 2\pi a / N \,, \tag{7.108}$$

where N is the number of piezoelements (pairs of active and passive elements) in the ring, the subscripts a and p will stay for all the quantities related to the active and passive parts, respectively. The relative amount of a passive material in the ring,

$$\beta = \delta_p / \delta_a, \tag{7.109}$$

will be referred to as the space factor, i.e., as the ratio of volumes of passive to active material.

Calculation of the active-passive ring transducer parameters will be made under the assumption that the ring vibrates uniformly in the zero mode, and thus can be considered approximately as one degree of freedom system. This assumption is not obvious, and how many of the active-passive segments in the ring may be accepted as sufficient enough for meeting this assumption must be determined. This will be discussed later after the relations for the electromechanical parameters of the rings vs. materials properties and relative amount of passive material will be obtained. Consider deformations in the ring that experiences radial displacement ξ_o under action of applied voltage. The total circumferential elongation of the ring Δl is

$$\Delta l = 2\pi\left(a+\xi_o\right)-2\pi a = 2\pi\xi_o . \tag{7.110}$$

The circumferential stress T^E produced by electrical field in the piezoelements does not change in the volume of the inhomogeneous ring due to the condition of force equilibrium. Therefore, the strains in the active and passive prisms are

$$S_a = s_{33}^E T^E = T^E / Y_3^E ,\ S_p = T^E / Y_p . \tag{7.111}$$

Thus, the total circumferential elongations of prisms are

$$\Delta\delta_a = \delta_a T^E / Y_3^E ,\ \Delta\delta_p = \delta_p T^E / Y_p . \tag{7.112}$$

Elongation of the ring's mean circumference is

$$\Delta l = 2\pi\xi_o = T^E \sum^N \left(\delta_a / Y_3^E + \delta_p / Y_p\right) = \left(l_a / Y_3^E + l_p / Y_p\right) T^E , \tag{7.113}$$

where $l_a = N\delta_a$ is the total length of the active material and $l_p = N\delta_p$ is the total length of the passive material. From Eq. (7.113)

$$T^E = \frac{2\pi\xi_0}{l_a / Y_3^E + l_p / Y_p} . \tag{7.114}$$

The relative elongation of the circumference of the ring is

$$\frac{\Delta l}{2\pi a} = \frac{\Delta l}{l_a + l_p} = \frac{T^E}{Y_{eq}} , \tag{7.115}$$

where Y_{eq} will be considered as the equivalent Young's modulus of the composite ring. Using Eq. (7.114), (7.115), and definition (7.109) we obtain

$$Y_{eq} = \frac{(1+\beta)}{(1+\beta Y_3^E / Y_p)} Y_3^E . \tag{7.116}$$

Now the potential energy of the active-passive ring at $E_3 = 0$ will be found as

$$W_m^E = \frac{1}{2}\int_{\tilde{V}} ST^E d\tilde{V} = \frac{1}{2}[\int_{\tilde{V}_a} S_a T^E d\tilde{V} + \int_{\tilde{V}_p} S_p T^E d\tilde{V}] = \frac{\xi_o^2}{2C_{map}^E} . \tag{7.117}$$

Upon substituting S_a, S_p and T^E from Eq. (7.111) and (7.114) into Eq. (7.117) and integrating over the volumes of the active and passive parts of the composite ring, we obtain

$$W_m^E = \frac{\xi_0^2}{2} \cdot \frac{(2\pi)^2 h \cdot t}{l_a / Y_3^E + l_p / Y_p} . \tag{7.118}$$

Thus, we arrive at the expression for the equivalent compliance of the active-passive ring

$$C_{m\,ap}^E = \frac{a}{2\pi h t Y_{eq}} = \frac{a}{2\pi h t Y_3^E} \frac{1+\beta Y_3^E / Y_p}{1+\beta} = C_{m3}^E \frac{1+\beta Y_3^E / Y_p}{1+\beta} , \tag{7.119}$$

where C_{m3}^E is the equivalent compliance of the segmented ring of the same size made of active prisms only.

When calculating the kinetic energy of the vibrating ring

$$W_{kin} = \frac{1}{2}\int_{\tilde{V}} \rho \dot{\xi}_o^2 d\tilde{V} = \frac{\dot{\xi}_o^2}{2} M_{eqv} , \tag{7.120}$$

it must be considered that the active and passive prisms may have different densities, namely, ρ_a and ρ_p. The integration over the ring volume leads to

$$M_{eqv\,ap} = 2\pi a h t \rho_{ap} = M_{eqv} \frac{1+(\rho_p / \rho_a)\beta}{1+\beta} , \tag{7.121}$$

where the equivalent density of the active-passive ring is

$$\rho_{ap} = \rho_a \frac{1+(\rho_p / \rho_a)\beta}{1+\beta} . \tag{7.122}$$

Thus, the equivalent sound speed in the active-passive ring is

$$c_{ap} = c_3^E \frac{1+\beta}{\sqrt{(1+\beta Y_3^E / Y_p)[1+(\rho_p / \rho_a)\beta]}} \tag{7.123}$$

(note that c_3^E is the sound speed in the segmented active ring), and the resonance frequency of an active-passive ring can be found as

$$f_{rap} = \frac{1}{2\pi\sqrt{M_{eqvap} C_{m\,ap}^E}} = \frac{c_{ap}}{2\pi a}, \tag{7.124}$$

or

$$f_{rap} = f_{ra} \frac{1+\beta}{\sqrt{\left(1+\beta Y_3^E / Y_p\right)[1+\left(\rho_p / \rho_a\right)\beta]}}. \tag{7.125}$$

The electromechanical energy and the electromechanical transformation coefficient n_{ap} will be determined from the expression

$$\dot{W}_{emap} = \frac{1}{2}\int_{\tilde{V}_a} \frac{d_{33}}{s_{33}^E} S_a E_3 d\tilde{V} = \frac{1}{2} V \xi_o n_{ap}, \tag{7.126}$$

where integrating must be performed over the volume of the active prisms. After substituting expressions for the electric field, $E_3 = V / \delta_a$, for the strain S_a from Eq. (7.111) and for the stress T^E from Eq. (7.114) we arrive at

$$n_{ap} = \frac{2\pi ht}{\delta_a s_{33}^E} \frac{1}{1+\beta / s_{33}^E Y_p} = n_a \frac{1}{1+\beta / s_{33}^E Y_p}, \tag{7.127}$$

where n_a is the transformation coefficient of the fully active ring. The electrical capacitance of the active prisms connected in parallel is

$$C_{eap}^{S_3} = \frac{\varepsilon_{33}^T\left(1-k_{33}^2\right) thN_a}{\delta_a} = C_e^{S_3} / \left(1+\beta\right). \tag{7.128}$$

The resistances of the mechanical and electrical losses in the segmented active-passive rings may be determined as follows. The energies of mechanical losses (W_{mL}) in the active and passive parts are proportional to their total potential energies (W_m) and anti-proportional to their quality factors, $W_{mL} = W_m Q$. Using results of calculating the potential energies of active and passive parts from Eq. (12), the resistance of mechanical losses will be represented as

$$r_{mL\,ap} = r_{mL\,a} \frac{1}{1+\beta}\left[1+\beta Q_a / Q_p\right]. \tag{7.129}$$

The resistance of the electrical losses R_{eLap} may be found as

$$R_{eL\,ap} = \frac{1}{\omega C_{e\,ap}^{S_3} \tan\delta} = \frac{1+\beta}{\omega C_e^{S_3} \tan\delta} = R_{eL\,a}\left(1+\beta\right), \tag{7.130}$$

where Eq. (7.128) for $C_{e\,ap}^{S_3}$ is taken into account. Thus, the expressions for all the equivalent parameters of the active-passive ring are determined.

An expected positive effect of combining active and passive materials in the mechanical system of a transducer is that the equivalent sound speed and, respectively, the wave size $2\pi a/\lambda$ of the transducer at resonance frequency can be changed to a degree. This will depend on the parameters of passive and active materials and on the space factor β. However, the tradeoff is that the electromechanical properties of the active-passive transducer decline to some extent. In particular, the effective coupling coefficient as one of the important figure of merits of the transducer may be affected. Another important characteristic of an active-passive transducer is the electromechanical force $F_{em} = n_{ap}V$, or the electromechanical transformation coefficient n_{ap}, as function of the same factors.

The effective coupling coefficient can be determined by formula

$$k_{eff}^2 = 1/\left(1 + C_e^S / n^2 C_m^E\right). \tag{7.131}$$

The combination of the equivalent circuit parameters in formula (7.131) is denoted as

$$\alpha_c = n^2 \frac{C_m^E}{C_e^S} = k_{eff}^2 / \left(1 - k_{eff}^2\right). \tag{7.132}$$

For the completely active ring $k_{eff}^2 = k_{33}^2$ and the coefficient α_c becomes

$$\alpha_m = k_{33}^2 / \left(1 - k_{33}^2\right). \tag{7.133}$$

For the composite active-passive ring $\alpha_c = \alpha_c(\beta)$, and after substituting parameters n_{ap}, $C_{m\,ap}^E$ and $C_{e\,ap}^{S_3}$ into expression (7.132), we arrive at

$$\alpha_c(\beta)/\alpha_m = 1/\left(1 + \beta / s_{33}^E Y_p\right) \tag{7.134}$$

and

$$k_{eff}^2(\beta)/k_{33}^2 = x\left(1+\alpha_m\right)/\left(1 + x\alpha_m\right), \tag{7.135}$$

where $x = \alpha_c(\beta)/\alpha_m$.

Correlation between the equivalent sound speed c_{ap} in Eq.(7.125) and the effective coupling coefficient vs. the relative amount of the passive and active materials, β, depends on the particular combination of these materials properties. Consider as the examples of the active materials PZT-4 ($Y_3^E = 64 \cdot 10^9$ Pa, $\rho_a = 7.5 \cdot 10^3$ kg/m^3) and single crystal (SCr) material PMN-PT ($Y_3^E \approx 10 \cdot 10^9$ Pa, $\rho_p = 8.0 \cdot 10^3$ kg/m^3).

PMN-PT is chosen together with PZT ceramics for illustrating effects of combining active and passive materials due to its exceptionally high coupling coefficient and small Young's modulus (relatively small sound speed). It is often desirable to increase the effective sound speed in a ring made of SCr in order to increase the wave diameter of the ring and thus to achieve its better acoustic loading. The opposite tendency is usually typical for rings made of PZT ceramics. Besides, it is instructive to estimate how a positive effect of changing the effective sound speed may correlate with decline of the effective coupling coefficient with example of combining piezoelectric SCr material with passive material, because an exceptionally large coupling coefficient is exactly what makes SCr materials attractive in comparison with PZT ceramic materials.

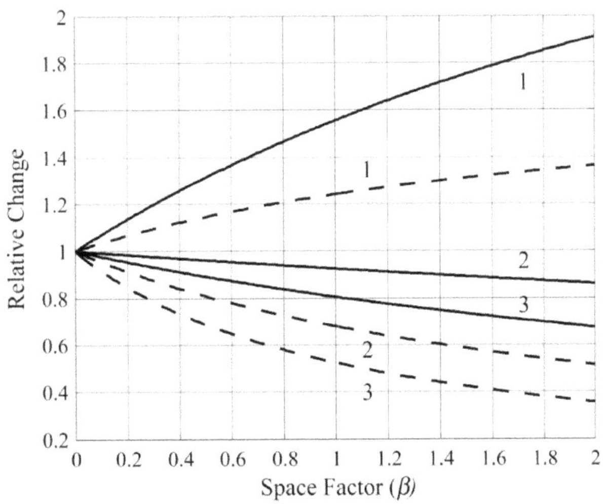

Figure 7.27: The relative changes of parameters of the active-passive rings vs. the space factor β : (1) effective sound speed, $c_{ap}(\beta)/c_3$; (2) effective coupling coefficient squared, $(k_{eff}^2)_{ap}/k_{33}^2$; (3) the effective electromechanical force related parameter $(d_{33}/s_{33}^E)_{ap}/(d_{33}/s_{33}^E)$ of an active-passive segmented ring transducers. Solid lines – combination of PMN-PT and aluminum, dashed lines – PZT-4 and aluminum.

In order to increase the diameter $2a$ of a ring made of the single crystal material such as PMN-PT, the passive material should be lighter and "stiffer." Aluminum with $\rho_p = 2.7\times10^3$ kg/m^3 and $Y_p = 72\times10^9$ Pa makes an almost ideal match in this sense, and it will be considered as the passive counterpart for PMN-PT material. Results of calculation of relative change of the sound speed and effective coupling coefficient of an active-passive ring transducer vs. β by formulas (7.123) and (7.135) are shown in Figure 7.27. Analyzing these results, one can conclude that replacement of half of the volume of single crystal active material by aluminum results in approximately 70% increase of the equivalent sound speed almost without loss of the effective coupling coefficient. Effect of combination of PZT-4 ceramics with aluminum on the same parameters of the ring is shown in the Figure 7.27 for comparison.

In addition to achieving the goal of regulating the wave size of a transducer to a desirable extent, the material cost of the transducer can be reduced significantly. For the rings made of PZT ceramics more often is desirable to reduce their wave size to some extent, i.e., to reduce the effective sound speed in the active-passive structure. This requires using passive materials with large density or/and smaller Young's modulus than those of piezoceramics. Given that the densities and Young's moduli of PZT ceramics are on the order of $\rho \approx 7.5\cdot10^3$ kg/m^3 and $Y \approx 70\cdot10^9$ Pa, it is practically impossible to find passive materials suitable for application in transducers among metals. An alternative is in designing rings nonuniform by thickness (with passive pats thicker than active), or in using modified metal parts with reduced effective Young's modulus (for example, perforated). A number of patents on the related issues exist, but their analysis is out of scope of this treatment.

As the basic assumption behind the above analysis was that the distribution of displacements of the ring surface is uniform, it must be determined how many active-passive segments in a ring is sufficient for meeting this assumption. In other words, what is the minimal number of the active passive segments that ensures close to pulsating resonance mode of vibration and achieving projected value of the resonance frequency. With this goal the Finite Element Analysis of vibration of the active-passive ring was produced for different number of the segments and different combinations of active and passive materials. Results of the analysis are presented in Figure 7.28 and Figure 7.29.

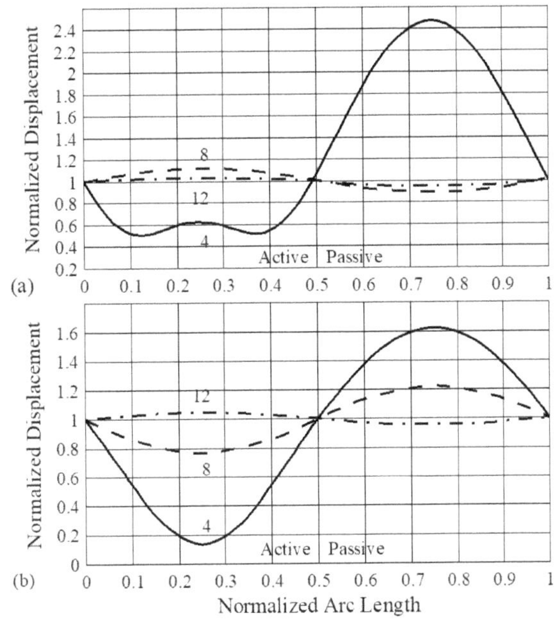

Figure 7.28. Distribution of displacements over a segment of active-passive ring: (a) combination of PMN-PT and aluminum, (b) combination of PZT-8 and aluminum. The curves are numbered according to number of active-passive segments in the rings. The displacement values for each curve are normalized to the displacement at the joint between active and passive segments, which occurs around a normalized arc length of 0.5.

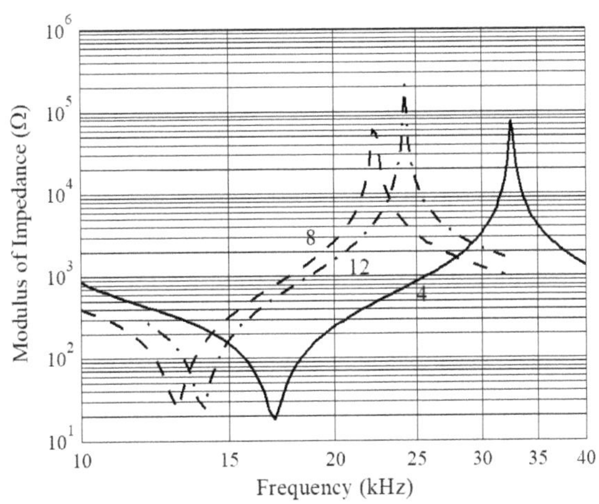

Figure 7.29. The input impedance of the active-passive ring (PMN-PT and aluminum) for four, eight, and twelve active-passive segments in the ring. The projected resonance frequency for the considered ring geometry is 14 kHz.

It can be concluded that in order to obtain practically uniform distribution of the ring, number of active-passive segments must be more than eight (with length of the segment less than $\lambda_{ap}/8$, where λ_{ap} is the wavelength of sound in the active-passive structure).

Similar results regarding calculating parameters of the active-passive rings were reported in Ref. 13 though without mentioning about the underlying assumption that the radial displacement should be uniform, which presupposes that number of active-passive segments in a ring must be sufficiently large for ensuring that the results of calculating are valid.

7.2.2.2 Unidirectional (Baffled) Transducer

Unidirectionality in the horizontal plane of the transducer, in which the zero order mode of vibration is electromechanically active, is achieved by the way of its baffling, as shown in Figure 7.30. The acoustic field related characteristics of the baffled transducers are considered in detail in Chapter 6.

The acoustic load becomes not symmetric due to the baffling, the modes of ring vibration on the radiating part of its surface are not orthogonal anymore, and the general equations of vibration (Eqs. (1.101)) become acoustically coupled. The radiation impedances Z_{aci} must be regarded as

$$Z_{aci} = Z_{acii} + \sum_{l \neq i} z_{acil} \cdot (U_1 / U_i) = Z_{acii} + \sum_{l \neq i} Z_{inil}, \tag{7.136}$$

where the self-radiation impedances Z_{acii} and mutual impedances z_{acil} for the baffled transducers are determined in general form in Section 6.3.1.1.2. And the introduced impedances

$$Z_{inil} = z_{acil} (U_l / U_i) \tag{7.137}$$

depend on solution of the vibration problem for the baffled transducer. In particular, for the

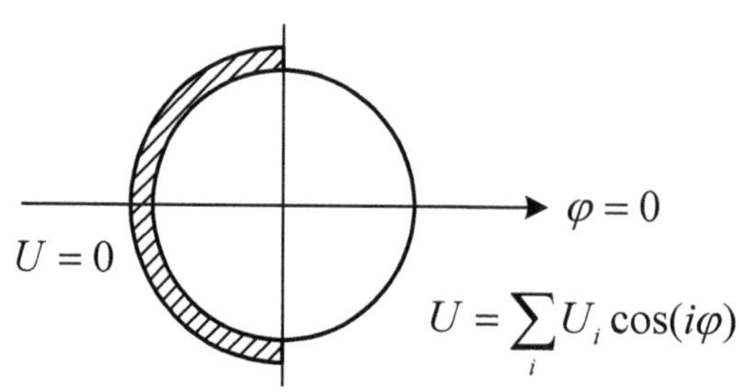

Figure 7.30: Geometry of the baffled transducer with 180° coverage.

zero and first modes of vibration the nondimensional coefficients for the self-impedances are presented in Figure 6.10, and for the mutual impedances between the zero and other modes in Figure 6.11. The last Figure is reproduced here as Figure 7.31.

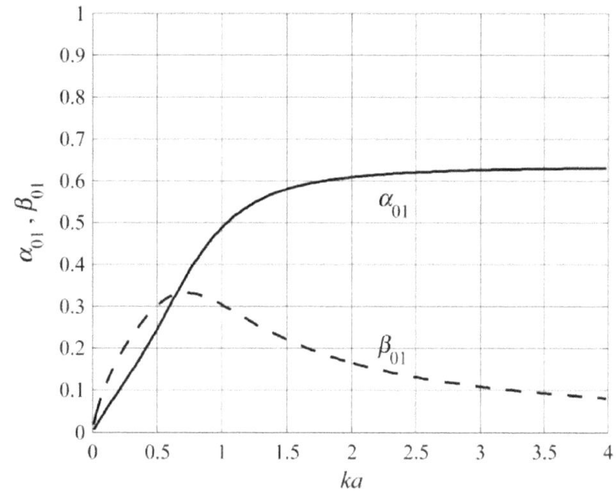

Figure 7.31; Nondimensional coefficients α_{0i} and β_{0i} for the mutual radiation impedances between zero and other modes of the surface vibration of a cylindrical transducer with 180° rigid baffle.

As follows from the plots in Figure 7.31, a significant coupling exists between the zero and first modes of vibration. The mutual impedances with higher modes can be neglected, especially at frequencies around and below the resonance frequency corresponding to the zero mode of vibration, where $(U_l / U_0) << 1$. Thus, the baffled transducer with electromechanically active zero mode, strictly speaking, must be considered as having two mechanical degrees of freedom that correspond to the generalized velocities U_0 and U_1. Its operation in the transmit node can be described in the electromechanical sense by equivalent circuit shown in Figure 7.32. In the equivalent circuit the impedance Z_{in10} introduced from the zero mode contour into contours that correspond to the first mode of the surface vibration is represented in the form of force $F_{il} = z_{acil} U_l$, which is equivalent in terms of the energy balance. Indeed, the energy flow from the contour with velocity U_l into contour with velocity U_i is

$$\bar{\dot{W}}_{ac} = Z_{inil} \left|U_i\right|^2 = F_{il} U_i^* . \tag{7.138}$$

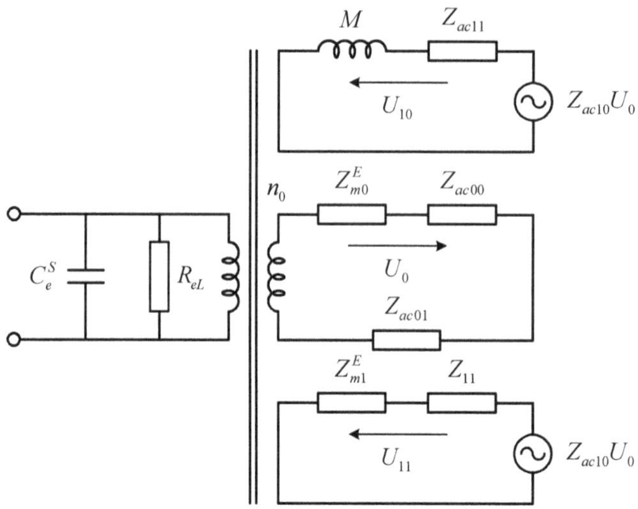

Figure 7.32: The equivalent electromechanical circuit of the baffled cylindrical transducer with the zero mode of vibration active.

Another peculiarity of the equivalent circuit is that the contour that correspond to the first mode of vibration is split into two parts with velocities U_{10} and U_{11}. The reason behind this is that under action of an external force a cylinder can move as a whole (this "transitional" mode of vibration formally corresponds to the resonance frequency $\omega = 0$), in which case the radial component of the surface displacement changes as $\cos\varphi$, i.e., in the same way as in the first mode of the cylinder vibration that corresponds to the resonance frequency ω_1.

If to denote the magnitude of vibration as a whole (as the mass M of the cylinder) U_{10}, and in the first mode as U_{11}, then the magnitude of vibration with cosine distribution of velocity over radiating part of the transducer surface, U_1, will be found as

$$U_1 = U_{10} + U_{11}, \tag{7.139}$$

and expression for the impedance introduced into the zero-mode contour will be

$$Z_{in01} = z_{ac01}\left(U_{10} + U_{11}\right)/U_0. \tag{7.140}$$

After the velocities U_0 and U_1 are determined from the equivalent circuit in the straight-forward way, the sound pressure on the axis and directional factor of a transducer can be found as

$$P(r,0) = P_0(r,0) + P_1(r,0), \tag{7.141}$$

where P_0 and P_1 are the modal sound pressures determined by formula (6.147) with values of the diffraction coefficients $k_{difi}(\pi/2)$ by formula (6.146) (their moduli are presented in Figure 6.8) and

$$H(\varphi,\pi/2)=\frac{U_0H_0(\varphi,\pi/2)+U_1H_1(\varphi,\pi/2)}{U_0+U_1}, \qquad (7.142)$$

where H_0 and H_1 are the modal directional factors determined by formula (6.148) (their moduli are presented in Figure 6.9).

It is noteworthy that calculation the effects of the modes interaction in practical baffled transducers should not be significant in the frequency range around and below the resonance frequency of the zero mode, ω_0, because the introduced impedances are small. Effect of the first mode can be noticeable in the range around the resonance frequency of the first mode, ω_1, due to the fact that the ratio U_1/U_0 becomes large. Thus, to the first approximation the directional factors in the horizontal plane of the baffled transducer are as shown in Figure 6.9 and reproduced in Figure 7.33.

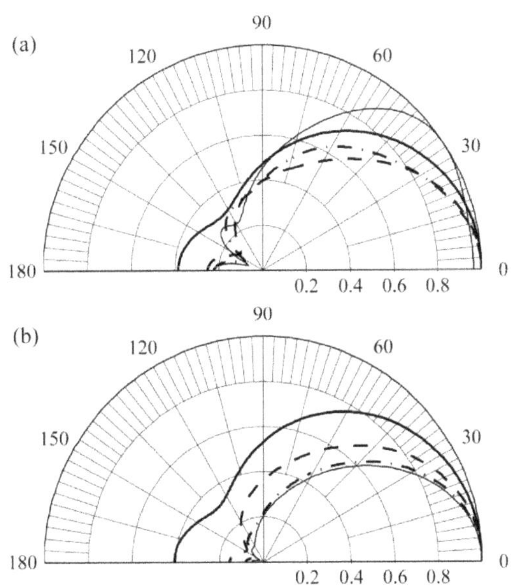

Figure 7.33: The modal directional factors of transducers at $\alpha=\pi/2$ baffle coverage for $ka=1$ (solid line), $ka=2$ (dashed line), $ka=3$ (dash-dotted line), $ka=4$ (thin solid line): (a) zero mode, (b) the first mode (see Section 7.2.3.3).

A remarkable property of the directional factors is that they almost do not change in a wide range of transducer wave sizes. It is also of note that the directional factors in this range are practically the same for the transducers with compliant baffles.

7.2.3 The First Order Transducer

The first order cylindrical transducer will be called a transducer comprised of the rings that have design shown in Figure 7.2(b), i.e., with electrodes split in two equal parts and connected in phase opposition. As follows from Eq. (7.8), the ring has the first order mode of vibration $\theta_1(\varphi) = \cos\varphi$ being the lowest electromechanically active mode. Though, unlike the case of the zero mode transducer, the first electromechanical active mode is not isolated. With full size electrodes all the odd order modes are also active. Thus, strictly speaking, the first order transducer has multiple degrees of freedom, and multi contour equivalent circuit shown in Figure 7.34 must be used for its calculation. The equivalent electromechanical parameters of the circuit are determined by the expressions (7.4), (7.5), (7.8), (7.11) and (7.12). The acoustic field related parameters are presented in Section 6.3.1.1. However, for practical calculations of the

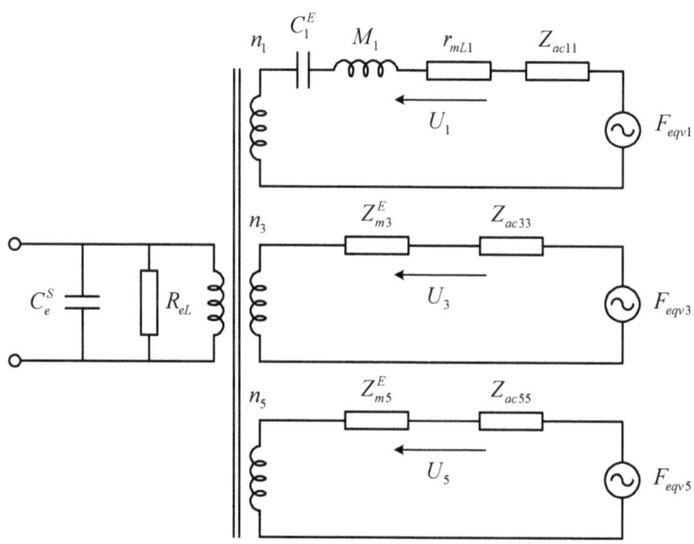

Figure 7.34: Multi contour equivalent circuit of the first order cylindrical transducer.

first order transducers operating in the frequency range around and below the first resonance frequency, which is usually the case, the equivalent circuit can be greatly simplified. The resonance frequencies that correspond to the successive modes of vibration according to formula (7.13) are: $\omega_1 = \omega_0\sqrt{2}$, $\omega_3 = \omega_0\sqrt{10}$, $\omega_5 = \omega_0\sqrt{26}$. Thus, the closest to the first mode frequency is 2.2 times higher. Besides the diffraction coefficients for higher modes drop significantly at small *ka*, as it follows from expression (6.119) and is illustrated in Figure 6.6 by the plot at $i = 3$. Therefore, contributions of the higher modes to results obtained using the first mode approximation (using the single contour of equivalent circuit that correspond to the mode shape $\theta_1(\varphi) = \cos\varphi$) can be neglected. The frequency range, in which the first mode of vibration dominates, can be broaden even up to $\omega \approx \omega_3$, as it may be needed for some applications discussed below, by reducing the size of electrodes down to $\Delta\varphi = 2\pi/3$. At this electrode size the third mode of vibration becomes inactive (according to expression (7.5) $\Omega_3 = 0$ and hence $n_3 = 0$), and the effective coupling coefficient for the first mode even increases (see plots in Figure 7.4). Effect of such changing the electrodes size on the directionality of the transducer is illustrated with plots in Figure 7.35, where directional factors of a transducer are presented at *ka* = 7 (in vicinity of the third mode resonance for PZT-4). calculated with full size and 120° degree electrodes.

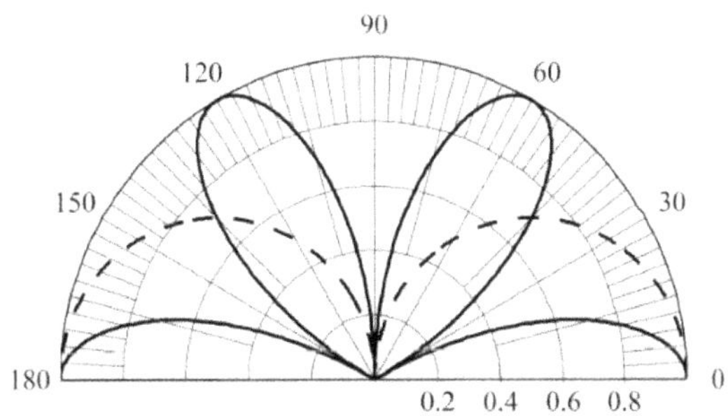

Figure 7.35: Directivity patterns of the first order transducer at *ka* = 7: full size electrode (solid line), 120° electrode (dashed line).

Taking into account all the above considerations we will further represent the equivalent circuit of the first order transducer without the acoustic baffles as is shown in Figure 7.36. Parameters of the circuit are:

$$M_1 = M = 2\pi aht\rho, \; K_{1l}^E = 4\pi th / as_{ll}^E, \; l = 1, 3 \tag{7.143}$$

for the radial and circumferential polarization;

$$n_1 = 8d_{31}h / s_{11}^E \tag{7.144}$$

for radial and

$$n_1 = 8d_{33}th / s_{33}^E\delta \tag{7.145}$$

for circumferential polarizations. Acoustic field related parameters, Z_{ac11} and k_{dif1}, are presented in Section 6.3.1.1.1. (Figures 6.6 and 6.7). The directional factor is $H(\varphi) = \cos\varphi$.

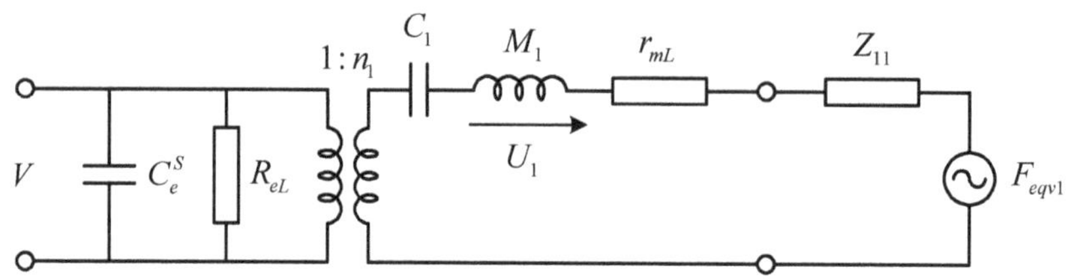

Figure 7.36: The equivalent electromechanical circuit of the first order transducer for an operational frequency range.

7.2.3.1 Dipole Directionality of the First Order Transducers

Many applications of the first order transducers are due to their dipole type directionality. In this capacity the first order transducers were mentioned in Ref. 14. Theoretically, under the assumption of ideal uniformity of electromechanical properties of the rings directionality of this type can be achieved in a broad range of frequencies. But the real piezoceramic rings, especially made as solid piezoelements, may be not sufficiently uniform, and this may result in significant corruption of the dipole directionality at frequencies below transducer resonance. In order to qualitatively estimate, what does it mean "not sufficiently uniform", consider a transducer operating in the receive mode assuming that electromechanical transformation coefficients of the halves of electrodes differ only by 2%, which can be considered as quite a precise match. Denote transformation coefficients of the halves that have opposite signs as n_a and n_b (according the note above Figure 7.2) and let it be $n_b = 1.02\, n_a$. Given that for the first mode n_a and n_b have opposite signs, at parallel in antiphase connection of the electrodes will be obtained that transformation coefficient for the first mode of vibration $n_1 = n_b + n_a = 2.02 n_a \approx 2n_a$, and for the zero mode $n_0 = n_b - n_a = 0.02\, n_a$. Thus, the zero mode of vibration becomes unwanted active, and the equivalent circuit of Figure 7.36 must be modified by including corresponding contour, as shown in Figure 7.37 for the receive mode of operation.

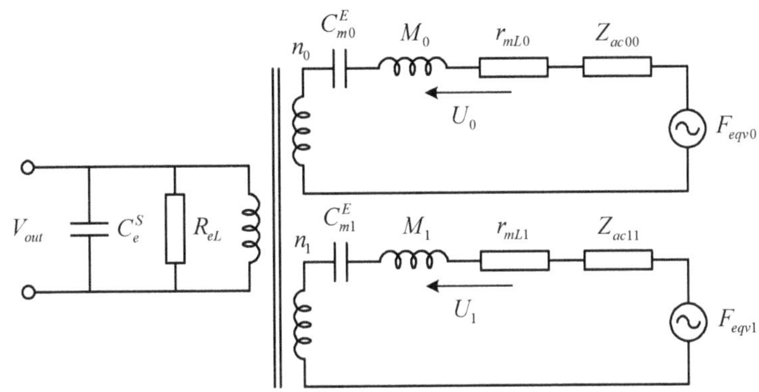

Figure 7.37: Equivalent circuit of the first order transducer with additional contour that takes into consideration unwanted contribution of the zero mode of vibration.

The output voltage of the transducer, V_{out}, will be found as superposition of effects of the first, V_1, and zero, V_0, modes. The output voltage dependence from the angle is

$$V_{out} = V_1 \cos\varphi + V_0 = V_1 \left(\cos\varphi + V_0 / V_1\right). \tag{7.146}$$

The second term in the parenthesis corrupts the dipole directivity. Especially strong contribution of this term can be expected in the frequency range around the zero-mode resonance due to difference in the mechanical impedances of the zero and first modes, and at low frequencies (at small *ka*) due to dominance of diffraction coefficient of zero mode over those of the first mode. The approximate quantitative estimations can be made as follows.

At the resonance frequency of the zero mode

$$V_0 = pS_\Sigma k_{dif0} \frac{n_0}{(r_{ac} + r_{mL})_0} j \frac{1}{\omega_{r0} C^S_{el}}, \; V_1 \approx pS_\Sigma k_{dif1} \frac{n_1 \omega_{ro}}{jK^E_{m1}} j \frac{1}{\omega_{ro} C^S_{el}}. \tag{7.147}$$

Taken into account that at ω_{r0} ($ka \approx 2.2$ for PZT-4) $(r_{ac0} + r_{mL})_0 \approx (\rho c)_w S_\Sigma$ (see Figure 6.7) and $k_{dif0} \approx k_{dif1}$ (see Figure 6.6), after some manipulations will be obtained that

$$\frac{V_0}{V_1} \approx j4 \frac{t}{a} \cdot \frac{\rho_c}{\rho_w} \cdot \frac{c_c}{c_w} \cdot \frac{n_0}{n_1}. \tag{7.148}$$

Thus, for the particular case of a cylinder with more or less typical for receivers ratio $(t/a) \approx 0.2$ made of PZT-4 under condition that $n_b = 1.02 n_a$ $ka < 1.0$

$$\frac{V_0}{V_1} \approx j0.13. \tag{7.149}$$

At low frequencies (at $ka < 1.0$)

$$V_0 = pS_\Sigma k_{dif\,0} \frac{n_0}{K_{m0}^E} \frac{1}{C_{el}^S}, \quad V_1 = pS_\Sigma k_{dif\,1} \frac{n_1}{K_{m1}^E} \frac{1}{C_{el}^S}. \tag{7.150}$$

As it follows from plots in Figure 6.6, at $|k_{dif\,0} / k_{dif\,1}| > 1.5 / ka$. Thus,

$$\frac{V_0}{V_1} = \frac{k_{dif\,0}}{k_{dif\,1}} \cdot \frac{K_{m1}^E}{K_{m0}^E} \cdot \frac{n_0}{n_1} \approx \frac{0.06}{ka} \cdot e^{(\arg k_{dif\,0} - \arg k_{dif\,1})}, \tag{7.151}$$

and distortions of the dipole directionality gradually increase with lowering the frequency. Distortions of the discussed type can be visualized with help of plots in Figure 7.38, where contributions of the first mode and unwanted contribution of the zero mode are presented qualitatively. It has to be remembered also that in general there exists shift of phase between these contributions, as is seen, for example, from formula (7.148).

To some extent the effect of mismatch of the transformation coefficients can be reduced by electrical equalizing the output voltages of transducer halves. This procedure will be considered in Ch. 14.

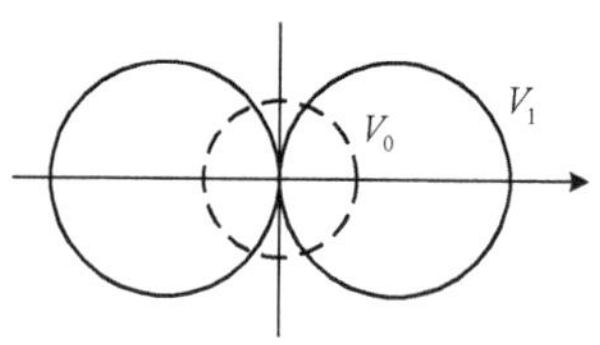

Figure 7.38: The main contribution to output voltage of a receiver of the first mode, V_1, and unwanted zero mode, V_0.

Another reason for distortions of the dipole directionality may be due to a possible nonuniformity of the mechanical properties of a ring (the most probable being in the density). In this case the mechanical interaction between zero and first modes introduces the unwanted contribution. Much better results in terms of uniformity of properties and thus broadening operational range of the first order transducer can be achieved with segmented ring design, in which case properties of the comprised prisms can be matched before cementing. Though, this improvement comes at the expense of the cost and much more complicated manufacturing of the transducers.

7.2.3.2 About Some Applications of the First Order Transducers.

One of the possible applications of first-order transducers is their use as pressure gradient receivers. Given that corresponding requirements for the frequency response of a receiver can be corrected electronically, the most challenging is requirement for exactness of $\cos\varphi$ directionality including deepness of the nulls. This presupposes uniformity of the mechanical and electromechanical properties of the rings, or in other words, immunity to action of the sound pressure, as was illustrated above. Such application will be considered in more detail in Ch. 14.

Some applications of the first order transducers require involvement of the zero order omnidirectional reference transducer for resolving an ambiguity in detecting direction to a source of sound with help of the first order transducer.

7.2.3.2.1 Cardioid Type Characteristic

One of examples of this kind is obtaining unidirectional characteristic of the cardioid type

$$H(\varphi) = 0.5(\cos\varphi + 1) \tag{7.152}$$

in a way that is shown in Figure 7.39.

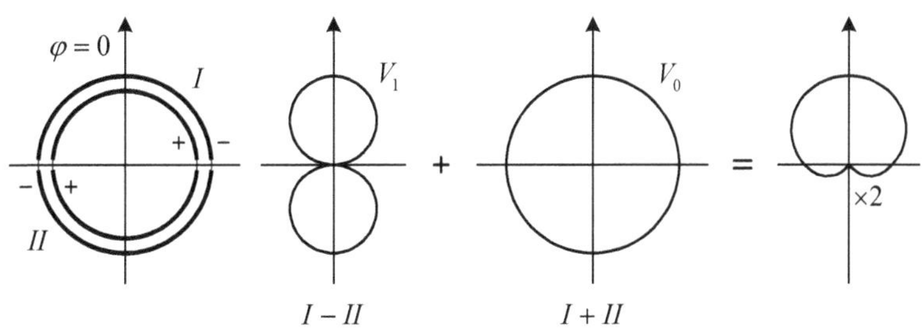

Figure 7.39: Forming the cardioid directional pattern by combining the first and zero modes of vibration of a cylindrical transducer.

For obtaining ideal cardioid pattern ratio of output voltages of the zero and first mode should be

$$\frac{V_1}{V_0} = \left|\frac{V_1}{V_0}\right| e^{\arg V_1 - \arg V_0} = 1, \tag{7.153}$$

i.e., the voltages should be equal by magnitude and phase. The phase difference between the output voltages can be represented as

$$\arg V_1 - \arg V_0 = \left(\arg Z_{mw1} - \arg Z_{mw0}\right) + \left(\arg k_{dif1} - \arg k_{dif0}\right), \qquad (7.154)$$

where $\arg Z_{mwi}$ is the argument of the mechanical impedance of a transducer in water. At frequencies significantly below f_{r0} the first term in parenthesis can be neglected (see formula (7.151)). At frequencies above f_{r1} the velocities are in phase (remember that for getting cosine directional pattern the size of electrodes must be reduced). In the range that includes the resonance frequencies the phase difference depends on the frequency, as

$$\arg Z_{mw1} - \arg Z_{mw0} \approx \arctan \Omega_1 Q_{mw1} - \arctan \Omega_0 Q_{mw0}, \qquad (7.155)$$

where according to expression (3.15) $\Omega_i = (f / f_{ri}) - (f_{ri} / f)$ and Q_{mwi} is the mechanical quality factor of a transducer in water. Example of the phase difference frequency response is presented in Ref. 15).

By splitting the electrodes in four equal parts and then combining the opposite parts in antiphase, as shown in Figure 7.40, two independent (except for a possible capacitive coupling between the quarters of the electrodes) electrical inputs will be obtained, which can be used for generating two orthogonal ($\cos\varphi$ and $\sin\varphi$) distributions of displacement.

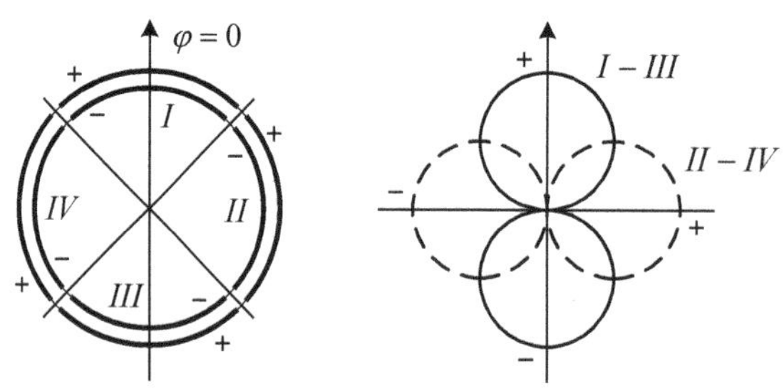

Figure 7.40: The electrodes configuration for forming two orthogonal first order ($\cos\varphi$ and $\sin\varphi$) displacement distributions.

For calculating thus formed identical transducers the equivalent circuit of Figure 7.36 is valid with electromechanical transformation coefficients that are determined using formula (7.10) at $m = 4$. Thus, at parallel antiphase connection of the electrode segments in pairs it will be $n_1 = 4\sqrt{2}d_{31}h / s_{11}^E$. Combination of such orthogonal displacement distributions can be used for designing a portable acoustic system for a direction finder. This can be done in different ways that involve different procedures of processing the output signals of the first order and zero order channels.

7.2.3.2.2 Double Dipole Transducer for Detecting Direction on a Sound Source

The principle of operation of one of the direction finder that employs vector addition of output signals of the channels having $|\cos\varphi|$ and $|\sin\varphi|$ directional characteristics on the display screen of a cathode-ray tube (CRT) is illustrated in Figure 7.41.

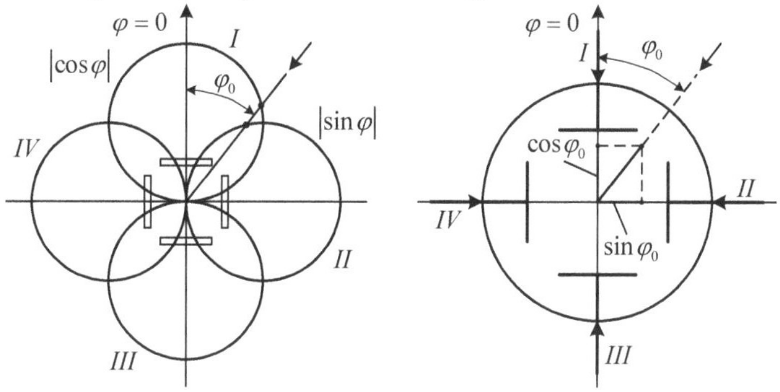

Figure 7.41: Illustration of principle of operating the direction finder with acoustic system that consist of four separate channels having $|\cos\varphi|$ and $|\sin\varphi|$ directional characteristics.

Accuracy of the bearing detection depends on how close the actual directional factors are to the ideal $|\cos\varphi|$ and $|\sin\varphi|$ configurations in a sufficiently broad operational frequency range. As it was shown, the first order transducer under consideration is capable of producing exact $\cos\varphi$ and $\sin\varphi$ characteristics. The problem is that output signals from their lobs that have opposite phases belong to the same channel. Being applied to the opposite deflecting plates of a CRT these signals will produce an ambiguous bearing. For resolving the ambiguity, the signals from the lobs having different phase must be separated and directed to their corresponding deflecting plates.

In principle this can be done by comparing the signals with reference output of omnidirectional transducer by phase with help of a phase detector and switch that directs signal, which is "in phase" with omnidirectional, to one of the deflecting plates, and those in "antiphase"- to the opposite plate. This is shown schematically in Figure 7.42.

The quotation signs are used because the signals are not exactly in phase or in antiphase due to the above estimated inherent phase difference between the first order and zero order outputs. All the needed signals theoretically can be obtained by combining the quarters of electrodes of the same transducer. Namely, the combinations $I - III$ and $II - IV$ give the orthogonal dipole outputs, and $I + II + III + IV$ gives omnidirectional output. In doing the comparison by phase the inherent phase difference between the first and zero order outputs must be remembered. Note that CRT in this example was used for illustrating the underlying principle and describing peculiarities of the acoustic system in Ref. 15. The real procedures of processing the signals can be different, including those realized in design of the Sonobuoy DIFAR (Ref. 13) for the same purpose.

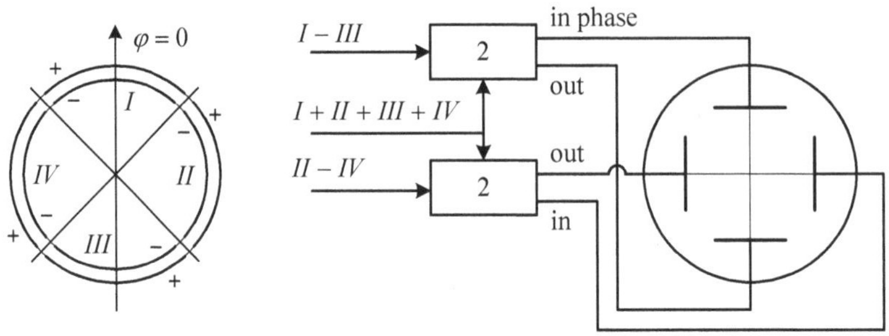

Figure 7.42: Schematic illustration of using outputs of orthogonal dipole cylindrical transducer (1) in combination with its omnidirectional output for detecting direction to a source of sound. Blocks 2 contain a phase detector and switch that separates close to in phase signals from those close to antiphase.

7.2.3.2.3 "Spiral Wave" Transducer

One more application of the two orthogonal first order distributions of radial displacement on the surface of a ring for developing a direction finder may be based on a different signals processing. Consider the combination of two identical orthogonal displacement distributions that are phase biased by 90 degrees. The resulting radial vibration of the surface will be denoted

$$\xi_r(\varphi,t) = \xi_1(\cos\varphi + j\sin\varphi)e^{j\omega t} = \xi_1 e^{j\varphi}e^{j\omega t}, \tag{7.156}$$

that is uniform in magnitude but has a phase that depends on azimuthal angle. The sound pressure generated in far field by this combination of displacements, P_{sp}, may be expressed as

$$P_{sp} = A(\omega,r)\cdot U_1 k_{dif1} e^{j\varphi_s} e^{j(\omega t - kr)}, \tag{7.157}$$

where $U_1 = |\dot{\xi}_r| = \dot{\xi}_1$ and $A(\omega,r)$ is a common function of propagating the spherical wave.

With addition of a reference omnidirectional transducer the combination of two first order transducers with orthogonal directional characteristics can be used as acoustic system of a beacon for underwater navigation that is described in Ref. 17. Results of investigating such acoustic system that can be called "spiral-wave" transducer are presented in Ref. 18. Direction to a source of sound, bearing φ_s, in this application is determined as phase difference between the received signals that are generated by the above combination of the first order transducers and by the omnidirectional transducer that produces sound pressure

$$P_0 = A(\omega, r) \cdot U_0 k_{dif\,0} e^{j(\omega t - kr)} \tag{7.158}$$

at the same point of the acoustic field. A systematic error due to the phase difference between the factors $U_1 k_{dif\,1}$ and $U_0 k_{dif\,0}$, which is the same as determined by Eq. (7.154), can be eliminated in course of calibrating the acoustic system. Obviously, the same acoustic system can be used in the receive mode.

In practical realization of the described acoustic system all the required directional characteristics can be obtained by using separate rings of small height that produce orthogonal dipole radiation and omnidirectional radiation. This configuration is shown in Figure 7.43.

It must be noted that in this acoustic system the comprising transducers have different acoustic centers, and an additional error in determining bearing may occur that depends on direction of sound wave propagation in the vertical plane. Minimizing this error requires reducing the heights of the rings. The concerns regarding phase difference between the signals due to difference in frequency dependencies of the impedances and diffraction coefficients of the zero and first mode transducers remain the same.

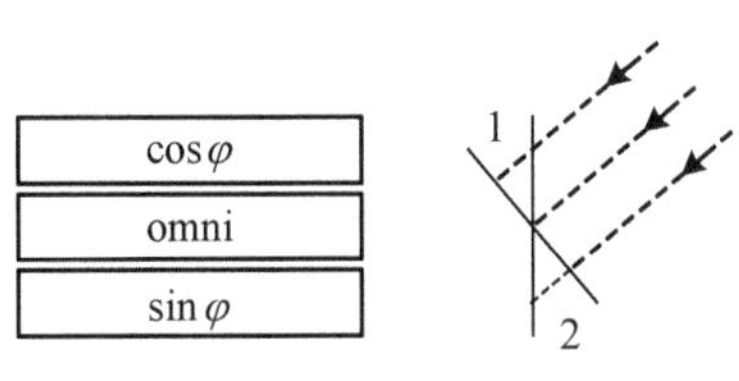

Figure 7.43: Acoustic system formed by the separate rings that produce $\cos\varphi$, $\sin\varphi$ and omnidirectional patterns in the horizontal plane. The segments 1 and 2 characterize unwanted phase shifts between the transducer outputs. Diameters of the zero and first order transducers at resonance frequencies are supposed to be equalized for eliminating an additional phase shift between the transducers in the horizontal plane.

But in this case an opportunity exists to significantly reduce this difference by way of equalizing the resonance frequencies of the rings. This can be achieved by employing the active-passive ring design, as was shown in Section 7.2.2.1.5, and by using different ceramic compositions for the first and zero order transducers. In order to get equal resonance frequencies at the same diameters, the sound speed in the zero-order transducer must be greater in factor of 1.4 (the ratio of the resonance frequencies f_{r1} / f_{r0}) in regard to the sound speed in the first order rings. Thus, for example, if the radially polarized rings for the first order transducers are made of PZT-5 ($c_1^E = 2.8 \cdot 10^3$ m/s), and for the zero order transducer an active–passive ring is used made of radially poled PZT-4 segments combined with aluminum segments of the same size (at $\beta = 1$), the sound speed c_{ap} will be $c_{ap} \approx 1.4 c_1^E$ according to formula (7.123). In this case the frequency dependent factors Ω_i in formula (7.155) are equal, and independent of frequency remaining phase shift can be eliminated in course of calibration of the acoustic system.

7.2.3.2 Unidirectional (Baffled) First Order Transducer

Unidirectionality of a transducer that has the first mode of vibration as the lowest active mode may be achieved in the frequency range around its resonance frequency by baffling a part of radiating surface that vibrates in antiphase. Assuming that the electrodes configuration meets condition (7.8) at $N = 1$, the most common is the 180° coverage of the surface, as shown in Figure 7.30. The next closest active mode of vibration has the resonance frequency $f_3 \approx 2.2 f_1$ and its contribution to vibration in the operating frequency range can be neglected. But due to non-symmetrical acoustic loading the acoustic coupling with electromechanically passive zero mode must be taken into consideration exactly in the same way, as it was done in Section 7.2.2.2 for the case that the zero mode was active, and acoustically coupled with the passive first mode. Thus, the equivalent electromechanical circuit of the baffled transducer with the first mode of vibration active looks like it is shown in Figure 7.44 with all the electromechanical parameters the same.

The acoustic field related characteristics of the transducer are considered in Section 6.3.1.1.2, where the general analytical expressions for the characteristics are given and plots are presented of their dependences on the wave size. For the modal diffraction coefficient and nondimensional coefficients of the radiation impedance they are presented in Figures 6.8 and 6.10, respectively. Coefficients of the mutual impedances between the first and zero modes are given in Figure 7.31. Plots of the directional factors are presented in Figure 7.33. They are similar to those for the baffled zero mode transducer, as is illustrated by the plots. The directionality of the baffled transducers does not change significantly at values of *ka* from about 2.0 to 3.5 that correspond to the range around resonance frequencies of the zero and first modes of the rings made of PZT ceramics.

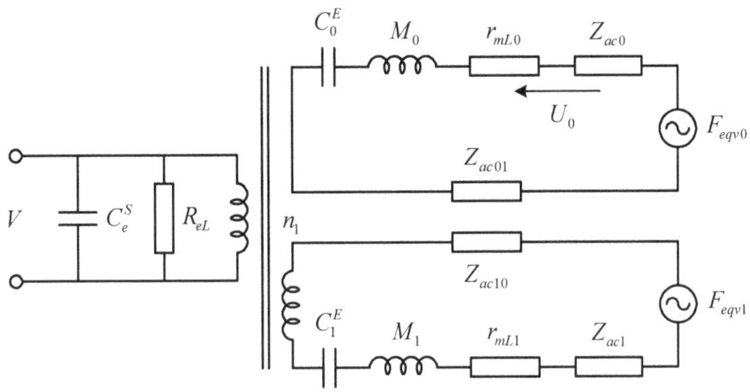

Figure 7.44: The equivalent electromechanical circuit of the baffled cylindrical transducer with the first mode of vibration active.

The frequency response of the transducer calculated with help of the equivalent circuit in Figure 7.44 has peculiarities due to contribution of the zero mode at frequencies around its resonance, because the ratio U_0 / U_1 has significant value at these frequencies. This effect is analogous to those produced on the frequency response of the baffled zero mode transducer by interference of the first mode in vicinity of its resonance frequency. These peculiarities can be seen on the plots of corresponding frequency responses.

7.2.3.3 Dual (0 ± 1) Active Modes Transducer

The operational bandwidth of a cylindrical transducer may be extended by simultaneous using both zero and first modes of vibration. This can be achieved at various conditions of energizing separate halves of electrodes of a baffled transducer.

When only one half of the electrodes is activated, these modes as well as the higher order odd modes are generated, as it follows from relations (7.10). Contribution of the higher modes to overall vibration in the operational range around and below the resonance frequency of the first mode can de neglected. Hence, the equivalent circuit of the baffled cylindrical transducer includes two active contours, as shown in Figure 7.45.

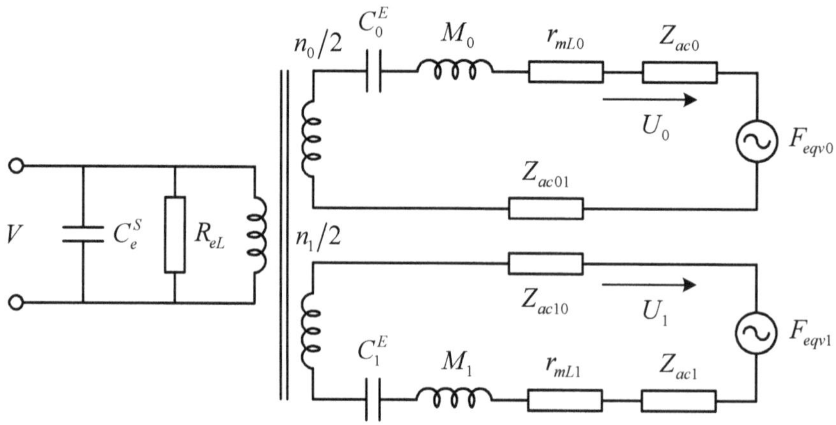

Figure 7.45: Two-contour equivalent electromechanical circuit of a transducer operating in the zero and first modes.

The operational characteristics (frequency responses, directional factors) of the transducer depend on whether the working electrode is outside or inside of the baffle (number *I*, or number *II* in Figure 7.46, respectively).

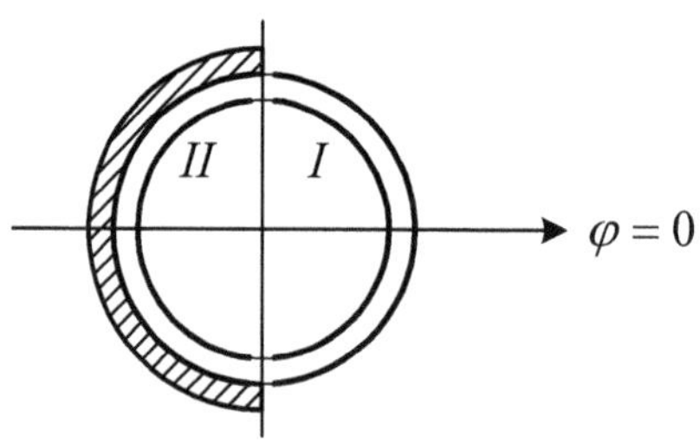

Figure 7.46: Baffled transducer with two pairs of electrodes.

These two cases differ by sign of the electromechanical transformation coefficients $n_1/2$, as this follows from expression (7.10): sign (+) for half *I* at $-\pi/2 < \varphi < \pi/2$, and sign (-) for half *II* at $\pi/2 < \varphi < 3\pi/2$. This change of the sign produces different results of superposition of the modal characteristics of the zero and first modes. The modal frequency responses typical for velocities of vibration U_0 and U_1 are qualitatively illustrated in Figure 7.47.

It follows from the Figure that with half *II* of the electrode active, the constructive superposition of the modal responses takes place in the frequency range between the resonance frequencies, and deconstructive at frequencies below and above the resonance frequencies.

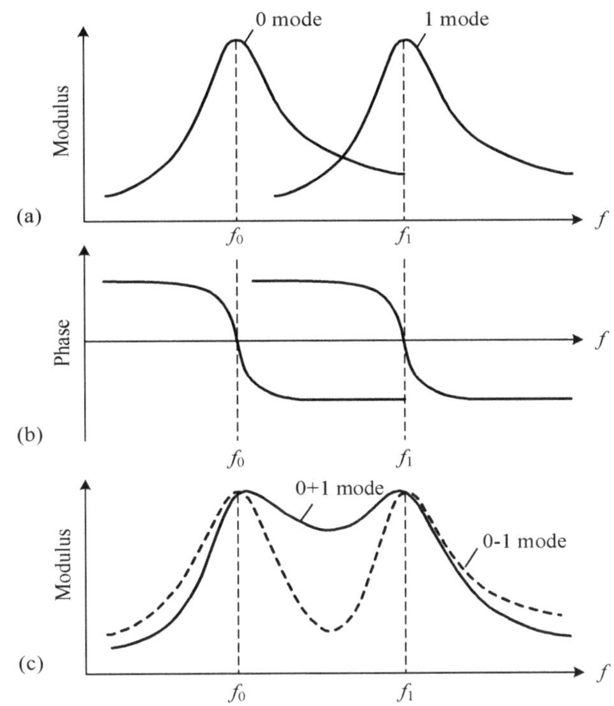

Figure 7.47: Qualitative view of the modulus (a) and phase (b) of the frequency responses of the modal velocities of vibration in the zero and first modes (U_0 and U_1), and of modulus of their superposition under condition that only a half of electrode is active (c). Mode 0+1 corresponds to the inner (#I*I*) half of the electrodes, mode 0-1 to the outer (# *I*) half.

In opposite, with half *I* active superposition is deconstructive between, and constructive outside the resonance frequencies. We will designate these cases for brevity as "0+1" and "0−1" modes of operation, where the plus and minus signs indicate the constructive and deconstructive superposition in the range between the resonance frequencies f_0 and f_1. Directional factor of the transducer in the horizontal plane and sound pressure on the axis may be calculated using the equivalent circuit of Figure 7.45 by formulas (7.141) and (7.142). The directional factors are shown in Figure 7.48.

The (0 ± 1) modes may be used in different applications. The 0-1 mode may be beneficial for operation in the receive mode at frequencies below f_0, or/and in the systems that simultaneously employ two individual resonances that need to be effectively separated. The 0+1 mode provides approximately half an octave operational bandwidth and appropriate directional factors in this range, as is illustrated in Figure 7.48. The directional factors correspond to excitation of "0+1" mode, and of the zero and first modes for comparison. It follows from the figure that in the frequency range around and between frequencies f_0 and f_1 the directional factors change not significantly.

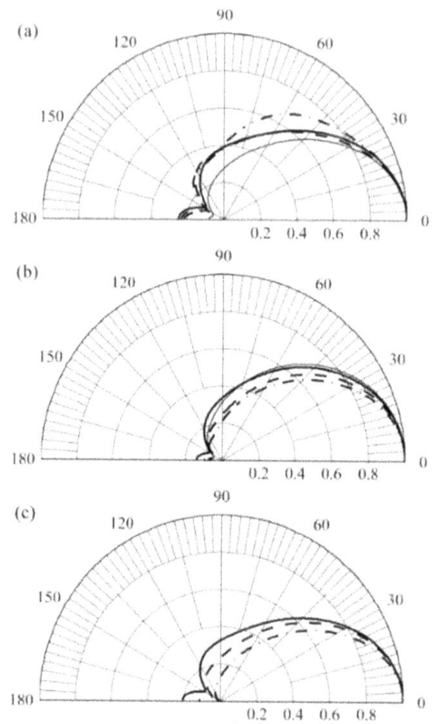

Figure 7.48: Calculated directional factors of the baffled cylindrical transducers for different modes of vibration[19]: (a) zero mode, (b) first mode, and (c) "0+1" mode for *ka* = 2.1 (solid line), *ka* = 2.6 (dashed line), and *ka* = 3.1 (dash-dotted line). The directional factors for a compliant baffled cylindrical transducer at *ka* = 2.1 (thin solid line) are also plotted in (a) and (b).

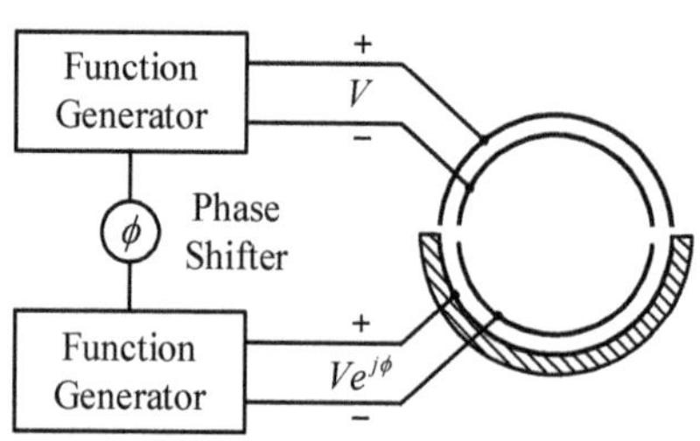

Figure 7.49: Circuit for generating the combined 0+1 and 0-1 mode of a baffled cylindrical transducer.

The unidirectional radiation of the "0+1" mode suggests its application for the broadband directional projectors. The disadvantage of the 0+1 mode, however, is the use of only a half of the piezoelectric material available, which results in a reduction of the sound pressure generated by the projector in comparison with the single mode projectors utilizing all the material. This problem may be reduced to some extent by using both 0+1 and 0-1 modes simultaneously, as it was described in Ref. 19. The idea behind this solution is in applying to both halves of the electrodes simultaneously voltages from electrical sources, whose relative phase can be changed. This can be implemented by employing two-channel transmit system that is schematically shown in Figure 7.49.

Applying the voltages V and $Ve^{j\phi}$ to separated electrode halves of the transducer results in the phase biased superposition of electromechanical forces acting in the mechanical system of the transducer.

For calculating the frequency responses of the system, the equivalent circuit of Figure 7.45 must be used once under action of voltage V and then under action of voltage $Ve^{j\varphi}$ and with sign of the electromechanical transformation coefficient n_1 changed, and the results must be added. Obviously, the phase shift $\varphi = 0$ results in excitation of the zero mode, and the phase shift $\varphi = 180^o$ results in excitation of the first mode. In general, by changing the phase shift between the two channels the frequency response of a projector can be controlled in favor of a certain frequency band in the range between the zero and first modes resonance frequencies. Analysis of the effects of various phase shifts produced in Ref. 19 showed that the flattest frequency response is achieved at $\varphi = 90^o$.

7.2.4 The Second Mode Transducers

The higher the mode number, the more circumferential becomes vibration of a ring, as follows from formula (4.266). Besides, the radial displacements in high modes have multiple nodes, and the smaller portions of radiating surface vibrate interchangeably in phase opposition. Altogether this makes high mode transducers poor acoustic sources. Probably, the last vibrational mode that may have practical applications is the second mode, $\theta_2(\varphi) = \cos 2\varphi$, with corresponding resonance frequency $f_2 = f_0\sqrt{5}$. For excitation of this mode of vibration the electrodes must be arranged, as it is shown in Figure 7.2(c). Namely, they must be divided into four parts, and the neighboring parts must be connected in antiphase. According to formulas (7.4) and (7.8) at $N = 2$, the lowest active mode has electromechanical transformation coefficient,

$$n_2 = 20\frac{d_{31}h}{s_{11}^E}, \tag{7.159}$$

if the parts of electrodes are connected in parallel. The next active mode is $\theta_6(\varphi) = \cos 6\varphi$ with resonance frequency $f_6 = f_0\sqrt{37}$. Thus, the second mode can be considered as isolated in the frequency range around and below its resonance frequency, and in terms of calculating parameters of a transducer that employs this mode of vibration the electromechanical equivalent circuit can be used having one active mechanical contour.

The distribution of displacements in the ring at the second mode is $\xi_r = \xi_o \cos 2\varphi$, $\xi_\varphi = 2\xi_o \sin 2\varphi$, and the equivalent mass and rigidity are according to formulas (4.262) and (4.263),

$$M_2 = 2.5M \text{ and } K_2^E = 25\pi th / as_{11}^E. \tag{7.160}$$

The electrical capacitance of the clamped ring is

$$C_{el}^S = 2\pi ah\left(1 - k_{31}^2\right)\varepsilon_{33}^T / t. \tag{7.161}$$

The effective coupling coefficient for the second mode transducer may be found with help of formulas (5.178) and (5.177),

$$\frac{k_{eff}^2}{1 - k_{eff}^2} = \alpha_c, \text{ where } \alpha_c = \frac{n_2^2}{K_2^E C_{el}^S}. \tag{7.162}$$

After substituting expressions (7.159)- (7.161) for the parameters involved it will be

$$k_{eff}^2 = \frac{8}{\pi^2} \cdot \frac{k_{31}^2}{1 - k_{31}^2\left[1 - \left(8/\pi^2\right)\right]}. \tag{7.163}$$

Note that this expression is the same as for a longitudinally vibrating bar for all the rings vibrating in high modes (at $i \geq 1$), because the distribution of stress in circumferential direction is the same. For a ring made of PZT-4 (with $k_{31} = 0.33$) $k_{eff} = 0.3$.

Acoustic field related parameters of a second mode transducer differ significantly depending on whether a part of its surface is covered by a baffle or not. We will assume that the overall height of the transducer is comparable with wavelength in the frequency range around its resonance frequency. Therefore, the modal acoustic characteristics of the transducer without baffles may be calculated by formulas presented in Section 6.3.1.1.1. The directional factor in the horizontal plane is $H_2(\varphi) = \cos 2\varphi$, i.e., those of the quadrupole. It is shown in Figure 7.50 by the solid line.

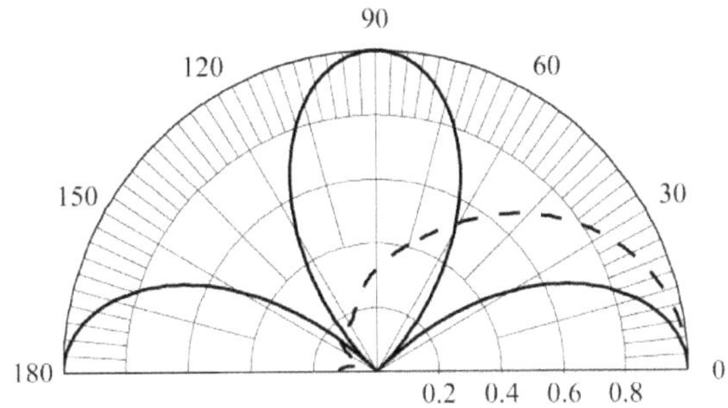

Figure 7.50: Directional factors of the second mode transducer: without baffles (solid line), and with rigid baffle on 3/4 of its surface at *ka* = 3.5 that corresponds to the resonance frequency of the ring made of PZT-4 ceramics (dashed line).

The modal radiation impedance is determined by formula (6.121) and (6.123) at $i = 2$, and can be represented as

$$Z_{ac2} = \rho c \cdot \pi a h\left[\alpha_2(ka) + j\beta_2(ka)\right]. \tag{7.164}$$

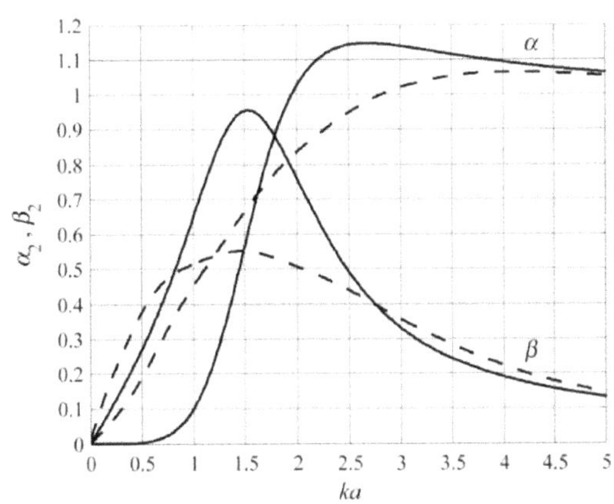

Figure 7.51: The nondimensional coefficients of radiation impedances of the second order transducer without baffles (solid lines) and with rigid baffle on 3/4 of its surface (dashed lines).

Plots of the nondimensional coefficients $\alpha_2(ka)$ and $\beta_2(ka)$ are depicted in Figure 7.51

for transducer without a baffle (solid lines) and with rigid baffle on 3/4 of surface (dashed lines).

It is noteworthy that the lowest mode of the flexural vibration of a ring has the same distribution of the radial displacement $\xi_{rfl}(\varphi) = \xi_{ofl}\cos 2\varphi$, as the second extensional mode (see Section 4.5.4). Therefore, acoustic coupling exists between these modes. It can be concluded from expression (6.41) that the corresponding mutual impedance is $z_{ac2,fl} = Z_{ac2}$ (as the modes of vibration are the same). Thus, the additional electromechanically passive mechanical contour must be included into equivalent circuit of the second mode transducer, which is responsible for the acoustically coupled flexural mode of vibration. Equivalent parameters of this contour according to formulas (4.287) and (4.289) are

$$M_{fl} = \frac{5}{4}M\,,\; K_{fl} = \frac{3\pi t^3 h}{4a^3 s_{11}^F}\,,\; f_{rfl} = 0{,}12\frac{t}{a^2}\sqrt{\frac{1}{\rho s_{11}^F}}\,. \tag{7.165}$$

The sound pressure generated by the transducer may be determined following formula (6.120) as

$$P_2(r,0) = \frac{k\rho c}{4}\sqrt{\frac{2}{\pi kr}}\,2\pi a U_2\, k_{dif\,2}(ka)e^{-j(kr-\pi/4)}\,, \tag{7.166}$$

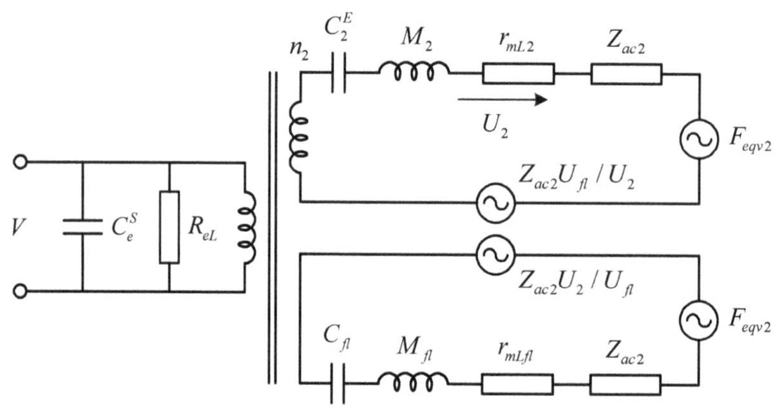

Figure 7.52: Equivalent electromechanical circuit of the second mode transducer. The inactive contour is due to acoustic coupling between the extensional and flexural modes that have the same velocity distribution on the transducer surface.

where $k_{dif\,2}(ka)$ is the diffraction coefficient for a transducer vibrating in the second mode without baffles that is determined by formula (6.119). Velocity of vibration, U_2, must be determined from the equivalent circuit of the transducer that is presented in Figure 7.52.

The equivalent force for determining sensitivity in the receive mode of operation is $F_2 = P_0 \cdot 2\pi ahk_{dif2}$. Plots of the diffraction coefficients for the transducer are shown in Figure 7.53.

In principle, the second mode transducer can be used, when the quadrupole directional pattern is needed. For example, such application is described in Ref. 21. The behavior of the diffraction coefficient shows that the second mode transducer is highly inefficient at frequencies below its resonance. Besides, distortions of the quadrupole patterns take place in vicinity of the frequency f_{rfl} due to coupling with flexural mode, and they can be corrupted due to unwanted contributions from the side of the zero and first modes.

This may occur by the same reasons that were noted regarding the dipole patterns in Section 7.2.3.1, as the quarters of a ring never can be ideally identical in electromechanical sense. The corruptions can be more pronounced, because of inherently smaller sensitivity of the second mode transducer.

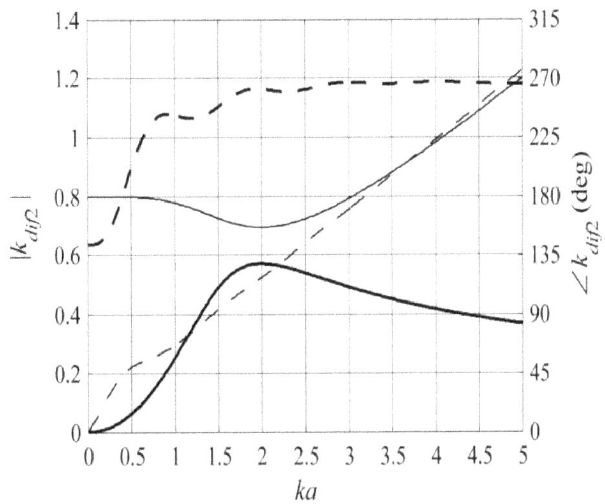

Figure 7.53: The diffraction coefficients for the second mode transducer: magnitude $|k_{dif2}|$ (thick lines) and phase $\angle k_{dif2}$ (thin lines) without baffles (solid lines) and with rigid baffle that covers 3/4 of the radiating surface (dashed lines).

Much more acoustically efficient and suitable for application in arrays is the transducer supplied with baffles that cover 3/4 of its surface, as shown in Figure 7.54.

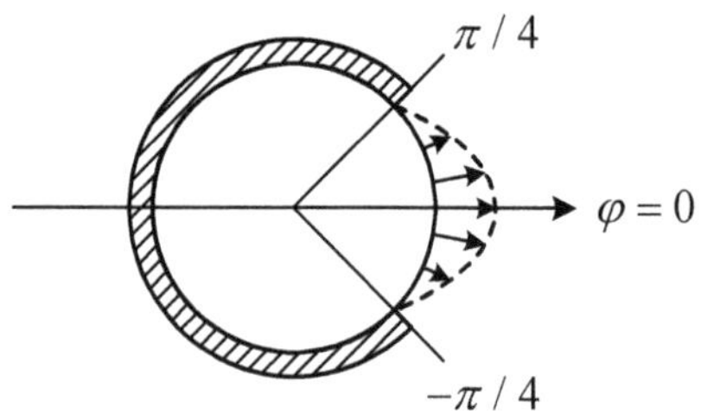

Figure 7.54: View of transducer with baffle that covers 3/4 of its surface.

The acoustic field related parameters of the second mode transducer with the baffle are determined as follows. The diffraction coefficient $k_{dif2}(ka, \pi/4)$ that is calculated with help of expression (6.146) with $\alpha = \pi/2$ replaced by $\alpha = \pi/4$ is plotted in Figure 7.53 by the dashed lines. The directional factors $H_2(ka, \pi/4)$ at values ka in vicinity of the resonance frequency calculated from expression (6.148) are presented in Figure 7.50. The radiation impedance

$$Z_{ac2}(ka, \pi/4) = \rho c \cdot 0.25\pi\left[\alpha_2(\pi/4) + j\beta_2(\pi/4)\right], \tag{7.167}$$

may be determined from expression (6.113) with coefficients a_{2i} calculated by formula (6.103) at $\alpha = \pi/4$. Thus, obtained nondimensional coefficients α_2 and β_2 are plotted in Figure 7.51 by the dashed lines.

When calculating the frequency response of the transducer in a broad range below the resonance frequency it must be taken in consideration that acoustic coupling of the active second mode exists not only with the flexural mode, as is mentioned above, but also with electromechanically passive zero and first modes. Therefore, the corresponding mechanical contours, strictly speaking, must be considered in the calculations. This is reflected in the equivalent circuit of the baffled second mode transducer that is shown in Figure 7.55.

The mutual acoustic impedances z_{20} and z_{21} that are responsible for the coupling can be calculated with help of expression (6.114) at $\alpha = \pi/4$. Judging by resemblance of the modes of vibration within interval $-\pi/4 < \varphi < \pi/4$, their values should be close to values of the self-impedances for the zero, first and second modes. The peculiarities in the frequency response of a transducer that are due to the acoustic coupling take place in vicinities of frequencies f_{fl}, f_0 and f_1. This is shown with a qualitative example of the frequency response in Figure 7.56. Quantitative analysis of these peculiarities hardly makes sense, though this can be done after the mutual impedances are determined. It is rather a matter of awareness of their existence and of avoiding the corresponding frequency ranges in operation, so far as a flat response of the transducer is required.

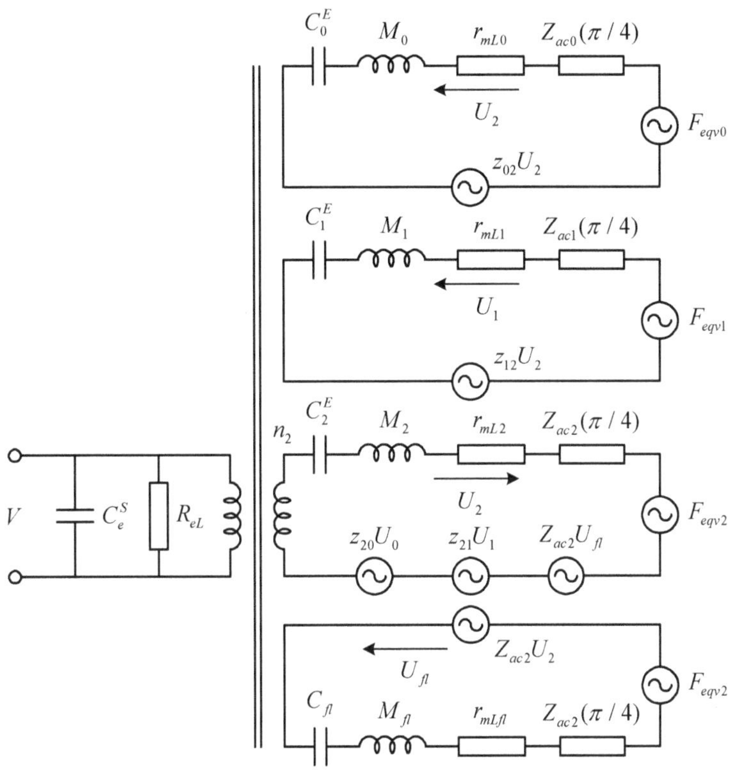

Figure 7.55: Equivalent electromechanical circuit of the baffled second mode transducer. The electromechanically passive contours are responsible for corrections due to acoustic coupling with the active contour.

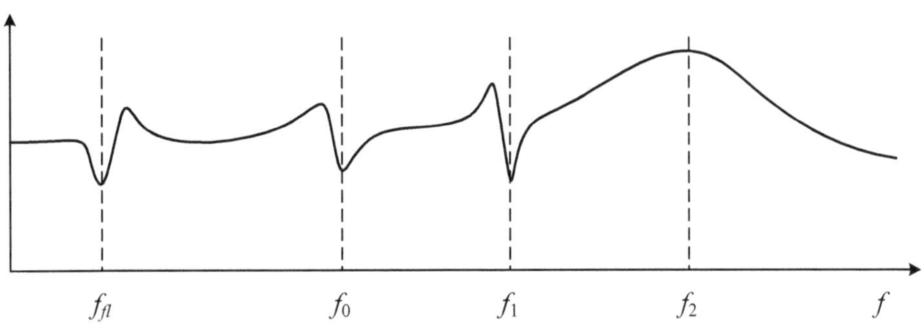

Figure 7.56: Qualitative example of frequency receive response of the second order transducer that illustrates existing of peculiarities in vicinity of the frequencies f_{fl}, f_0 and f_1.

It is noteworthy that the 3/4 baffled cylindrical transducer can be used as a unidirectional projector/receiver that is operating in the frequency bands around the zero, first and second mode resonance frequencies separately or simultaneously. Such in concert operation may be achieved by a proper for generating these modes of vibration combining the quarters of electrodes/

7.2.5 Incomplete Ring Transducer

Extensional vibrations of the incomplete rings (segments of a ring) made of passive materials were examined in Section 4.5.3. Consider now the extensional vibrating incomplete rings that are made from piezoelectric ceramics for designing transducers that may employ vibrations of such mechanical systems. Geometry of the ring is shown in Figure 7.57. The ring is assumed to be thin and short, therefore $T_r \approx 0$ and $T_h \approx 0$. In the crystallographic coordinates $T_r \to T_3$, $T_h \to T_2$, $S_\varphi \to S_1$ for the solid, and $T_r \to T_1$, $T_h \to T_2$, $S_\varphi \to S_3$ for the segmented rings. We will consider in detail a variant of the solid ring. The results obtained can be specified for the variant of the segmented ring in a straightforward way.

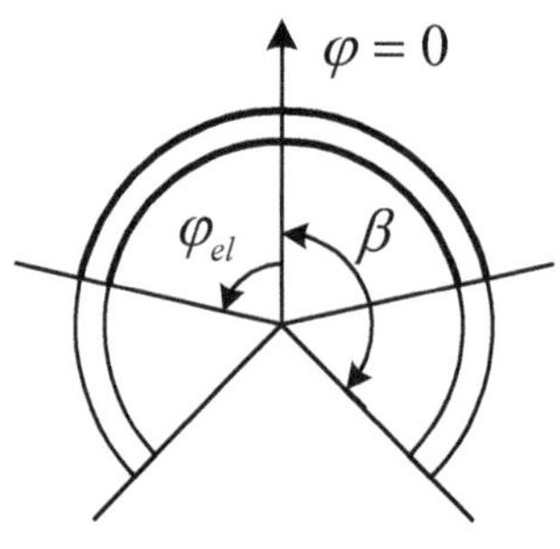

Figure 7.57: Geometry of the incomplete piezoceramic ring.

Status of a radial poled ring is governed by the piezoelectric equations

$$S_1 = s_{11}^E T_1 + d_{31} E_3, \quad D_3 = d_{31} T_1 + \varepsilon_{33}^T E_3, \tag{7.168}$$

where from follows that

$$T_1^E = \left(1/s_{11}^E\right) S_1 = Y_1^E S_1, \quad D_3^E = \left(d_{31}/s_{11}^E\right) S_1, \quad \varepsilon_{33}^{S_1} = \varepsilon_{33}^T \left(1 - k_{31}^2\right). \tag{7.169}$$

Thus, in all the expressions for the equivalent mechanical parameters of the passive rings by formulas that include Young's modulus Y it should be replaced by $Y_1^E = \left(1/s_{11}^E\right)$. The summary of the parameters is:

$$K_{eqvi}(\beta)=\frac{thY_1^E}{a}\beta\left\{1+\left[\frac{\pi}{2\beta}(2i+1)\right]^2\right\}^2, M_{eqvi}(\beta)=tha\rho\beta\left\{1+\left[\frac{\pi}{2\beta}(2i+1)\right]^2\right\}, \quad (7.170)$$

$$f_i(\beta)=\frac{1}{2\pi a}\sqrt{\frac{Y_1^F}{\rho}}\sqrt{1+\left[\frac{\pi}{2\beta}(2i+1)\right]^2}=f_{r0}\sqrt{1+\left[\frac{\pi}{2\beta}(2i+1)\right]^2}. \quad (7.171)$$

Note that all the parameters related to the incomplete ring transducers having opening angle β will be distinguished further by this value, as for example $Z_{ac}(\beta)$. In the expressions (7.170), (7.171) i=0,1,2,... resonance frequency of the lowest mode is

$$f_0(\beta)=f_{r0}\sqrt{1+(\pi/2\beta)^2}. \quad (7.172)$$

Thus, the lowest resonance frequencies at $\beta=\pi,(\pi/2),(\pi/4)$ are $f_{r0}\sqrt{1.25}$, $f_{r0}\sqrt{2}$, $f_{r0}\sqrt{5}$, respectively. It is noteworthy that the resonance frequencies for the half ring and for the quarter of ring are the same, as for the first and second modes of vibration of the complete ring. At $\beta=\pi$ the resalt corresponds to resonance frequency of a ring with a thin slot or with a crack.

The electromechanical conversion status of the ring is characterized by the electromechanical energy

$$W_{em}=\frac{1}{2}tha\int_{-\beta}^{\beta}D_3^E E_3 d\varphi, \quad (7.173)$$

where $D_3^E=\left(d_{31}/s_{11}^E\right)S_1$ and $E_3=(V/t)\theta_E(\varphi,\varphi_{el})$. The function $\theta_E(\varphi,\varphi_{el})=1$ for the parts of a ring covered by electrodes (at $|\varphi|<|\varphi_{el}|$ in Figure 7.57), and $\theta_E(\varphi,\varphi_{el})=0$ otherwise. Expression for the strain $S_1=S_\varphi$ obtained by formula (4.2687) after substituting expressions for displacements (4.272) and (4.273) is

$$S_1=S_\varphi=\frac{1}{a}\sum_{i=0}^{\infty}\xi_i\left\{1+\left[(2i+1)\frac{\pi}{2\beta}\right]^2\right\}\cos\left[(2i+1)\frac{\pi}{2\beta}\varphi\right] \quad (7.174)$$

After substituting E_3 and D_3^E under the integral (7.173) and performing integration we will obtain

$$W_{em} = \frac{1}{2} V \frac{hd_{31}}{s_{11}^E} \sum_{i=0}^{\infty} \xi_i \left\{ 1 + \left[\frac{\pi}{2\beta}(2i+1) \right]^2 \right\} \int_{-\varphi_{el}}^{\varphi_{el}} \theta_E(\varphi, \varphi_{el}) \cos\left[(2i+1)\frac{\pi}{2\beta}\varphi \right] d\varphi = \\ = \sum_{i=0}^{\infty} V \xi_i n_i, \tag{7.175}$$

where from follows that the coefficient of electromechanical transformation, n_i, is

$$n_i(\varphi_{el}) = \frac{4hd_{31}}{s_{11}^E} \frac{\beta}{\pi(2i+1)_i} \left\{ 1 + \left[\frac{(2i+1)\pi}{2\beta} \right]^2 \right\} \sin\frac{(2i+1)\pi}{2\beta}\varphi_{el}. \tag{7.176}$$

With the full size electrode, at $\varphi_{el} = \beta$,

$$n_i(\beta) = (-1)^i \frac{4hd_{31}}{s_{11}^E} \frac{\beta}{\pi(2i+1)_i} \left\{ 1 + \left[\frac{(2i+1)\pi}{2\beta} \right]^2 \right\}, \tag{7.177}$$

and for the case of the ring with a slot (at $\beta \approx \pi$)

$$n_i = (-1)^i \frac{4hd_{31}}{s_{11}^E} \frac{1}{(2i+1)_i} \left\{ 1 + \left[\frac{(2i+1)}{2} \right]^2 \right\}. \tag{7.178}$$

The effective coupling coefficients of the rings may be calculated through the coefficient $\alpha_c = n_i^2 / C_{el}^{S_1} K_i^E$, where parameters K_i^E and n_i must be determined in the general case from formulas (7.170) and (7.176), and $C_{el}^{S_1} = 2ha\varphi_{el}\varepsilon_{33}^T\left(1-k_{31}^2\right)/t$. Calculating coefficient α_c for the general case results in

$$\alpha_{ci}(\varphi_{el}) = \frac{8}{\pi^2} \frac{k_{31}^2}{1-k_{31}^2} \frac{\beta}{(2i+1)^2 \varphi_{el}} \sin^2\frac{(2i+1)\pi}{2\beta}\varphi_{el}. \tag{7.179}$$

For the case that $\varphi_{el} = \beta$

$$\alpha_{ci}(\beta)) = \frac{8}{\pi^2} \frac{k_{31}^2}{1-k_{31}^2} \frac{1}{(2i+1)^2}. \tag{7.180}$$

Thus, the effective coupling coefficients in this case do not depend on the opening angle of the ring. It is noteworthy that this value of α_c is the same as for the modes of extensional vibrations of a side electroded straight bar, and so are the effective coupling coefficients. The most usable

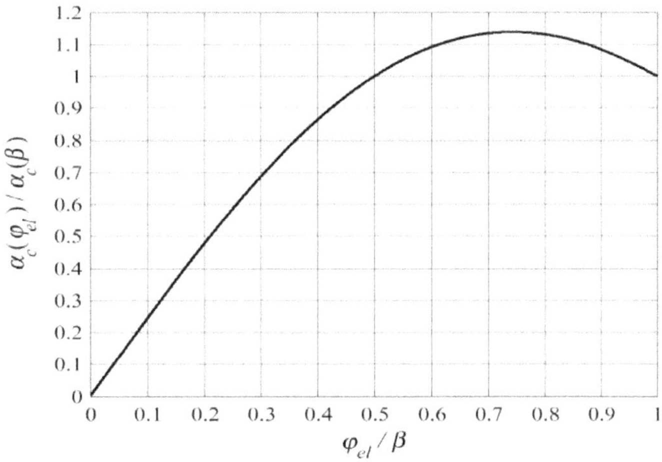

Figure 7.58; Plot that shows dependence of function $\alpha_c(\varphi_{el})/\alpha_c(\beta)$ on the angle of the electrode coverage.

for transducer applications is the lowest mode of vibration at $i = 0$. For this mode

$$k_{eff}^2 = \frac{8}{\pi^2} k_{31}^2 \frac{1}{1 - 0.19 k_{31}^2}, \tag{7.181}$$

and for all the piezoceramic compositions (with $k_{31} \le 0.35$) $k_{eff} = k_{31}\sqrt{8/\pi^2}$ within 2% accuracy. Expression (7.179) for the function $\alpha_c(\varphi_{el})$ shows that it has maximum at some value of φ_{el} (and so does the effective coupling coefficient). This function normalized to its value at $\varphi_{el} = \beta$,

$$\frac{\alpha_c(\varphi_{el})}{\alpha_c(\beta)} = \frac{\beta}{\varphi_{el}} \sin^2 \frac{(2i+1)\pi}{2\beta} \varphi_{el}, \tag{7.182}$$

is shown in Figure 7.58, where from it follows that function $\alpha_c(\varphi_{el})$ achieves its maximum at $\varphi_{el} \approx 3\beta/4$, and at $\varphi_{el} = \beta/2$ it has the same value, as for the full size electrode.

It must be remembered that all the calculations and conclusions regarding the electromechanical parameters of the incomplete rings are valid under the assumption that they are short enough for neglecting effects of coupled vibration between vibration in the circumferential and axial directions. Qualitative estimations in this regard can be made by comparison with situation that takes place for the complete rings, so far as angle β is relatively large, or with thickness poled plates at smaller β, in which case the segment of a ring becomes closer to the curved plate. As to the first case, a note is made in the end of Section 7.2.2.1.3. Analogous estimations for the second case can be made based on results presented in Section 4.6.2.2, where it is shown that the strongest coupling takes place at $(h/2a\beta)=1$ in our notations.

Table 7.5. Comparison of the calculated and measured at different aspect ratios effective coupling coefficients of the incomplete rings.

	Solid ring $\beta=\pi$	Ring with a slot $\beta=\pi$	$\beta=2\pi/3$	$\beta=\pi/2$	$\beta=\pi/3$
Calculated	0.33	0.30	0.30	0.30	0.30
Measured $h/2a=0.50$	0.32	0.29	0.24	0.21	0.2
Measured $h/2a=0.25$	0.33	0.30	0.30	0.30	0.3

With reduction of angle β the aspect ratio of the original solid ring, at which a harmful effect of coupled vibration on the effective coupling coefficient can be neglected, should be reduced. This is illustrated by comparison that is made in Table 7.5 between the calculated by above presented formulas and values of the effective coupling coefficients of incomplete rings measured with samples cut from the rings having different aspect ratios.

The acoustic field related parameters of incomplete rings depend on distribution of displacements on their surface in the radial direction. According to (4.272),

$$\xi_r(\varphi)=\sum_{i=0}^{\infty}\xi_i\cos\left[(2i+1)\frac{\pi}{2\beta}\varphi\right]. \tag{7.183}$$

Thus, the modal distributions (mode shapes) of the radial vibration are

$$\theta_i(\varphi,\beta)=\cos(2i+1)\frac{\pi}{2\beta}\varphi \tag{7.184}$$

The most usable may be the lowest modes (at $i=0$). For the opening angles $\beta=\pi$, $(\pi/2)$ and $(\pi/4)$ they are: $\theta_0(\varphi,\pi)=\cos(\varphi/2)$, $\theta_0(\varphi,\pi/2)=\cos\varphi$ and $\theta_0(\varphi,\pi/4)=\cos 2\varphi$, respectively. Note that $\theta_0(\varphi,\pi)=\cos(\varphi/2)=\sqrt{(1+\cos\varphi)/2}$ is the square root of the cardioid distribution. The mode shape at $\beta\approx\pi$ (for a ring with a slot) is shown in Figure 7.59 (a).

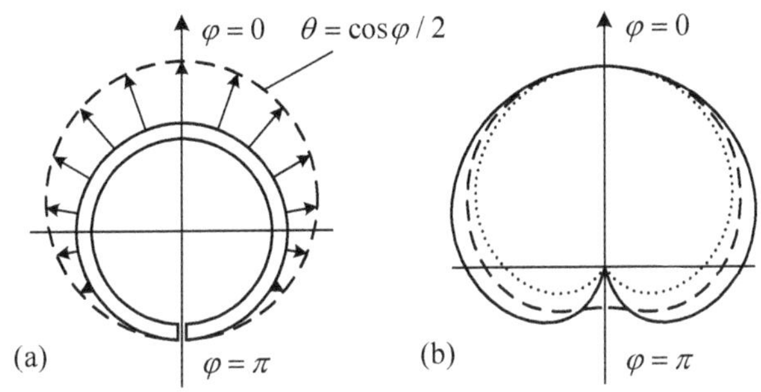

Figure 7.59: (a) The mode shape of radial vibration of a ring with a slot (at $\beta \approx \pi$), and (b) the corresponding directional factors in the horizontal plane at ka =2.4 (solid line) and ka =1 (dashed line). Plot of cardioid is presented by the dotted line for comparison.

It is noteworthy that a half ring (at $\beta = \pi / 2$) and a quarter of ring (at $\beta = \pi / 4$) have the same distributions of displacement and the same resonance frequencies (see (7.172)), as a complete ring vibrating in the first and second extensional modes. This is because the boundary conditions on the free ends of the segments that are cut out of a complete ring along the nodal lines remain the same, as in the complete ring.

Consider calculating the acoustic characteristics of variants of the transducers at different values of the opening angle.

7.2.5.1 Ring with a Slot

The closest resonance frequencies of the active modes of vibration are $f_0(\pi) = f_{r0}\sqrt{1.25}$ and $f_1(\pi) = f_{r0}\sqrt{3.25} = 1.6 f_0(\pi)$. Thus, in vicinity of the lowest resonance frequency the transducer made of rings with a slot may be considered as one mechanical degree of freedom system and may be represented by the single contour equivalent circuit of Figure 7.60.

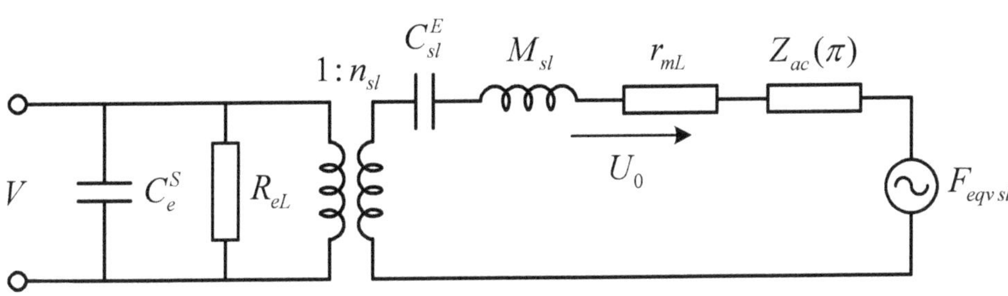

Figure 7.60: Equivalent electromechanical circuit of a transducer that employs piezoceramic ring with a slot.

The summary of equivalent parameters of the circuit is as follows:

$$M_{sl} = \frac{5}{8}M,\ K_{sl}^{E} = \frac{5\pi}{4}\cdot\frac{thY_1^E}{a},\ n_{sl} = \frac{5hd_{31}}{s_{11}^{E}};\tag{7.185}$$

$$C_{el}^{S_1} = 2\pi ah\varepsilon_{33}^{T}\left(1-k_{31}^{2}\right)/t,\ S_{av\,sl} = (2/\pi)S_{\Sigma},\ S_{eff\,sl} = 0.5S_{\Sigma}.\tag{7.186}$$

The acoustic field related quantities must be determined in the following way. The distribution of radial displacements can be represented in the form of series $\xi_r(\varphi,\pi) = \xi_o\cos(\varphi/2) = \xi_o(a_o + a_1\cos\varphi + a_2\cos 2\varphi + ...)$, where the Fourier coefficients are:

$$a_0 = \frac{2}{\pi},\ a_1 = \frac{4}{1\cdot 3\pi},\ a_2 = -\frac{4}{3\cdot 5\pi},\ a_3 = \frac{4}{5\cdot 7\pi},\ \ldots\tag{7.187}$$

The terms with a_i at $i \geq 3$ may be neglected, and

$$\xi_r = \xi_o\cos(\varphi/2) \approx \xi_o\left(\frac{2}{\pi} + \frac{4}{1\cdot 3\pi}\cos\varphi - \frac{4}{3\cdot 5\pi}\cos 2\varphi\right).\tag{7.188}$$

The sound field radiated by the transducer can be calculated as superposition of the sound fields radiated by the partial distributions of displacements that correspond to the terms of the series (7.188) with help of expression (6.117) and (6.119). Thus, the sound pressure generated may be presented as

$$P(\varphi,\pi) = A(r,\omega)2\pi ah\dot{\xi}_o\cdot\left(\frac{2}{\pi}k_{dif\,0} + \frac{4}{3\pi}k_{dif\,1}\cos\varphi - \frac{4}{3\cdot 5\pi}k_{dif\,2}\cos 2\varphi\right).\tag{7.189}$$

The velocity $\dot{\xi}_o = U_o$ must be determined from the equivalent circuit in Figure 7.60. It can be concluded from expression (7.189) that diffraction coefficient in this case is

$$k_{dif}(0,\pi) = \frac{2}{\pi}k_{dif\,0} + \frac{4}{3\pi}k_{dif\,1} - \frac{4}{3\cdot 5\pi}k_{dif\,2}.\tag{7.190}$$

The directional factor of the transducer is

$$H(\varphi,\pi) = \frac{P(\varphi,\pi)}{P(0,\pi)} = \frac{1 + \frac{2}{3}\frac{k_{dif\,1}}{k_{dif\,0}}\cos\varphi - \frac{2}{15}\frac{k_{dif\,2}}{k_{dif\,0}}\cos 2\varphi}{1 + \frac{2}{3}\frac{k_{dif\,1}}{k_{dif\,0}} - \frac{2}{15}\frac{k_{dif\,2}}{k_{dif\,0}}}.\tag{7.191}$$

As it follows from plots in Figure 6.6, at the wave sizes $ka \geq 2.4$ the diffraction coefficients up to $k_{dif\,3}$ are equal approximately to $k_{dif\,0}$. ($ka \approx 2.4$ corresponds to the resonance frequency of a solid ring made of PZT-4.). Therefore, in this range the expressions for the sound pressure, diffraction coefficient and directional factor become

$$P(\varphi,\pi) \approx A(r,\omega)\cdot 2\pi ah\dot{\xi}_o k_{dif\,0}\cos(\varphi/2), \tag{7.192}$$

$$k_{dif}(0,\pi) \approx k_{dif\,0}, \tag{7.193}$$

$$H(\varphi,\pi) \approx \cos(\varphi/2) = \sqrt{(1+\cos\varphi)/2}\,. \tag{7.194}$$

Plot of the directional factor is shown in Figure 7.59 (b) by the solid line. With decrease of ka the terms in the general expressions (7.189) and (7.191) that contain the diffraction coefficients at $i \geq 1$ gradually die out. At $ka \approx 1$ the expressions for the diffraction coefficient and directional factor reduce to

$$k_{dif}(0,\pi) \approx 1.06 k_{dif\,0}, \tag{7.195}$$

$$H(\varphi,\pi) \approx \frac{3}{5}\left(1+\frac{2}{3}\cos\varphi\right). \tag{7.196}$$

Plot of this function is shown in Figure 7.59 (b) by the dashed line. The conclusion can be made that the directional factor of the transducer almost does not change in the range of frequencies at least from $ka \approx 1.0$ up to $ka \approx 3.0$, and the diffraction coefficient is equal to those of the zero mode transducer. The transducer can be used for the same purposes as transducer having cardioid characteristic that is described in Section 7.2.3.2.1 in the much broader range of frequencies and without involvement of complications related to excitation of different modes of vibration of a solid ring.

Radiation impedance of the transducer may be determined from expression for power radiated by superposition of modes, through which the velocity of vibration is represented. According to (7.188)

$$U_r(\varphi,\pi) \approx U_o\left(\frac{2}{\pi}+\frac{4}{1\cdot 3\pi}\cos\varphi-\frac{4}{3\cdot 5\pi}\cos 2\varphi\right), \tag{7.197}$$

and following Eq. (6.123)

$$\bar{\dot{W}}_{ac}(\pi) = |U_o|^2\left[\left(\frac{2}{\pi}\right)^2 Z_{ac00}+\left(\frac{4}{3\cdot\pi}\right)^2 Z_{ac11}+\left(\frac{4}{3\cdot 5\pi}\right)^2 Z_{ac22}\right], \tag{7.198}$$

where Z_{acii} are the modal impedances of the cylindrical transducer without baffles,

$$Z_{acoo} = \rho c \cdot 2\pi ah\left[\alpha_{00}(ka) + j\beta_{00}(ka)\right],\ Z_{acii} = \rho c \cdot \pi ah\left[\alpha_{ii}(ka) + j\beta_{ii}(ka)\right]. \quad (7.199)$$

Plots of the nondimensional coefficients for the modal impedances are presented in Figure 6.7. The last term in (7.198) can be neglected. Thus, $\bar{\dot{W}}_{ac}(\pi) \approx Z_{ac}(\pi)|U_o|^2$, where

$$Z_{ac}(\pi) \approx \left[\left(\frac{2}{\pi}\right)^2 Z_{ac00} + \left(\frac{4}{3\cdot\pi}\right)^2 Z_{ac11}\right]. \quad (7.200)$$

After substituting expressions (7.199) for the impedances we obtain

$$Z_{ac}(\pi) = \rho c \cdot \pi ah\left[\alpha(\pi) + j\beta(\pi)\right], \quad (7.201)$$

where

$$\alpha(\pi) = \left[2\left(\frac{2}{\pi}\right)^2 \alpha_{00} + \left(\frac{4}{3\cdot\pi}\right)^2 \alpha_{11}\right] \text{ and } \beta(\pi) = \left[2\left(\frac{2}{\pi}\right)^2 \beta_{00} + \left(\frac{4}{3\cdot\pi}\right)^2 \beta_{11}\right]. \quad (7.202)$$

The wave sizes of rings with a slot made of PZT-4, PZT-5, and PZT-8 ceramics at frequencies around their resonances are within the range $2.0 < ka < 3.0$ (both for variants of radial and circumferential polarizations). At these values of ka $\alpha_{11} \approx \alpha_{00}$ and $\beta_{11} \approx \beta_{00}$. Hence,

$$\alpha(\pi) \approx \alpha_{11},\ \beta(\pi) \approx \beta_{11}, \text{ and } Z_{ac}(\pi) \approx \rho c \cdot \pi ah(\alpha_{11} + j\beta_{11}) = Z_{ac11}. \quad (7.203)$$

Note that $\pi ah = S_{eff}(\pi)$, where $S_{eff}(\pi)$ is the effective surface area of transducer with a slot determined by formula (6.140).

The equivalent force acting on the transducer in the receive mode is according to (6.34)

$$F_{eqv}(\pi) = P_o \cdot 2\pi ah \cdot k_{dif}(\pi). \quad (7.204)$$

At $ka > 1.0$ $k_{dif}(\pi) \approx k_{dif\,0}$.

At low frequencies (at $ka \to 0$) $k_{dif}(\pi) \to 2/\pi$, and

$$F_{eqv}(\pi) = P_o \cdot 4ah = P_o S_{av}(\pi). \quad (7.205)$$

Here $S_{av}(\pi) = (2/\pi)S_{\Sigma}$ is the average surface area of transducer determined by formula (6.8).

It is noteworthy that a zero mode ring that has a crack can be considered as unintended ring with a slot and its electromechanical parameters change accordingly. Thus, one of immediate indications of appearance a crack in the ring is raise of its resonance frequency in factor of $\sqrt{1.25} \approx 1.12$.

7.2.5.2 Half Ring Transducer

We will consider variants of the transducer made of a single half ring furnished with a rigid cylindrical baffle, as shown in Figure 7.61, and transducer made of two half rings mechanically isolated, as shown in Figure 7.62.

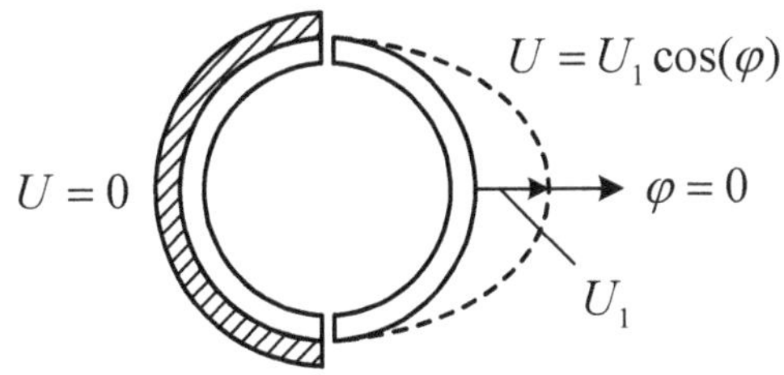

Figure 7.61: Transducer configuration that includes a half of active ring and cylindrical rigid baffle.

The lowest and the next closest resonance frequencies of the half ring transducer according to formula (7.171) are $f_0(\pi/2) = f_{r0}\sqrt{2}$ $f_0(\pi/2) = f_{r0}\sqrt{2}$ and $f_1(\pi/2) = f_{r0}\sqrt{10}$, therefore in the vicinity and below of the resonance frequency $f_0(\pi/2)$ the transducer can be considered as having single mechanical degree of freedom, and the equivalent circuit of Figure 7.60 can be used for its calculations. The equivalent parameters of the circuit are (see (7.170), (7.177)) $M_0(\pi/2) = \pi tha\rho$, $K_0^E(\pi/2) = 2\pi thY_1^E/a$, $n_0 = 4hd_{31}/s_{11}^E$,

$$C_{el}^{S_1} = \pi ah\varepsilon_{33}^T\left(1-k_{31}^2\right)/t, \qquad (7.206)$$

The effective coupling coefficient of the transducer by formula (7.181) is the same as for the first order transducer ($k_{eff} = 0.3$ for PZT-4 ceramics).

The acoustic field related characteristics of the transducer depend on the entire transducer design configuration. Single half ring transducer with rigid half cylindrical baffle has all the same acoustic characteristics as the baffled first order transducer described in Section 7.2.3.2, as the mode of vibration on the radiating surface is the same, $U(\varphi) = U_0\cos\varphi$. The advantage of the half ring transducer is that it has twice broader band of the transmit frequency response at -3 dB level around the resonance frequency than the first order transducer, because at the same value of radiation resistance it has twice smaller equivalent mass, and thus twice smaller $Q_{ac} = \omega_0 M_0(\pi/2)/r_{ac}$.

In the transducer that is composed of two half rings the rings can be used in the transmit and/or receive modes separately or simultaneously. The boundary conditions on the transducer surface can be presented as superposition of the two modes of operation, as shown in Figure 7.62. In the first mode ring *I* vibrates, surface of ring *II* is clamped; in the second mode ring *II* vibrates, surface of ring *I* is clamped.

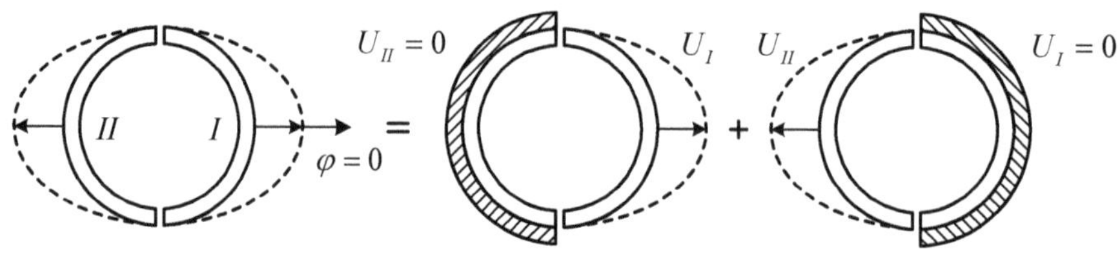

Figure 7.62. Operation of the two half ring transducer, as superposition of modes of operation of single half ring transducers with baffles.

Consider the first mode of operation. Acoustic characteristics of the transducer (Z_{acI} , k_{difI} , $H_I(\varphi)$) are the same as for the baffled first order transducer. Vibration of transducer *I* generates sound pressure on the clamped surface of transducer *II* that can be calculated by formula (6.106) for $i = 1$. The coefficients $a_{ni}(\alpha)$, which for our case should be specified as $a_{n1}(\pi/2)$, are the Fourier coefficients that are determined in (6.144) as $a_{10} = 2$, $a_{11} = \pi/2$, $a_{12} = 2/3$, $a_{13} = 0$. As the result, the distribution of the sound pressure on the clamped surface of transducer *II* is

$$P_{II}^{U}(a,\varphi) \approx -j\rho c U_{oI}\left[2\frac{H_0^{(2)}(ka)}{H_0^{(2)'}(ka)} + \pi\frac{H_1^{(2)}(ka)}{H_1^{(2)'}(ka)}\cos\varphi + \frac{4}{3}\frac{H_2^{(2)}(ka)}{H_2^{(2)'}(ka)}\cos 2\varphi\right]. \quad (7.207)$$

(Values of the higher order terms in series (7.207) are negligible.) In other words, the vibrating transducer *I* generates on the surface of transducer *II* the equivalent "acoustomotive" force,

$$\begin{aligned} F_{acII,I} &= ah\int_{\frac{\pi}{2}}^{\frac{3\pi}{2}} P_{II}^{U}(a,\varphi)\cos\varphi d\varphi = \\ &= -j\rho cah\frac{U_{oI}}{2\pi}\left[-4\frac{H_0^{(2)}(ka)}{H_0^{(2)'}(ka)} + \frac{\pi^2}{2}\frac{H_1^{(2)}(ka)}{H_1^{(2)'}(ka)} - \frac{8}{9}\frac{H_2^{(2)}(ka)}{H_2^{(2)'}(ka)}\right] \end{aligned}, \quad (7.208)$$

The equivalent force can be presented alternatively as

$$F_{acII,I} = z_{acII,I}\cdot U_{oI}, \quad (7.209)$$

where $z_{acII,I}$ is the mutual impedance between transducers *I* and *II*. Taking into consideration expressions (6.113) for the modal radiation impedances, and (7.208) for the force, we arrive at relation for the mutual impedance

$$z_{acII,I} = -\frac{1}{\pi^2} Z_{ac00} + \frac{1}{4} Z_{ac11} - \frac{4}{9\pi^2} Z_{ac22}, \qquad (7.210)$$

where Z_{acii} are the modal radiation impedances of cylindrical shell. Their plots are shown in Figure 6.7. In vicinity of the resonance frequency of the half ring $ka \approx 3$, and nondimensional coefficients of the radiation impedances in expression (7.210) are: $\alpha_{ii} \approx 1.0$, $\beta_{ii} \approx 0.2$. Accordingly, $Z_{ac00} \approx 2\pi ah\rho c(1+j0.2)$, $Z_{ac11} \approx Z_{ac22} \approx \pi ah\rho c(1+j0.2)$, and

$$z_{acII,I} \approx 2\pi ah\rho c(1+j0.2)\cdot 10^{-3}. \qquad (7.211)$$

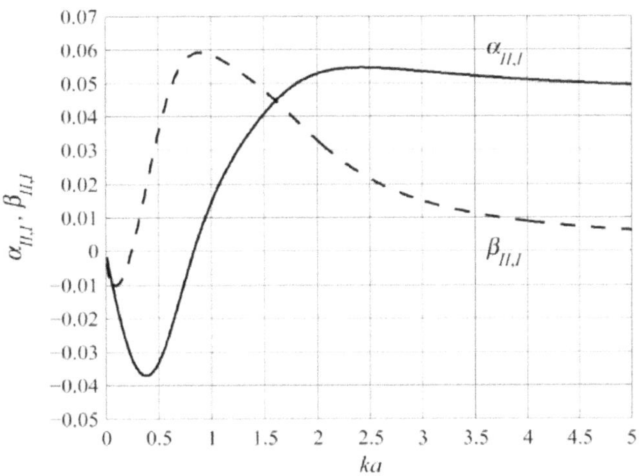

Figure 7.63: Nondimensional coefficients of the mutual radiation impedance between halves of the rings that comprise a transducer. Active component α (solid line), reactive component β (dashed line).

Thus, the mutual impedance between the halves of the transducer in vicinity of the resonance frequency is negligible in comparison with the self-impedance of the baffled first order transducer, for which the nondimensional coefficients are plotted in Figure 6.10. The halves of the cylinder are practically uncoupled in the vicinity and above the resonance frequency, because the wave size $ka \approx 3$ is large enough for the back side of the cylinder being in the shadow zone. The situation changes with the reduction of the wave size (at lower frequencies).

Dependence of the nondimensional coefficients of the mutual radiation impedance from the wave size is presented in Figure 7.63. This information will be useful for calculating the transducers in the receive mode of operation.

The sound pressure generated by the transducer I in the far field is according to formula (6. 152),

$$P_I(r,\omega,\varphi) \approx -j\rho c\sqrt{\frac{2}{\pi kr}}e^{-j(kr-\pi/4)}\frac{hU_I}{\pi}$$
$$\times\left[\frac{1}{H_0^{(2)'}(ka)} + j\frac{\pi}{2}\frac{1}{H_1^{(2)'}(ka)}\cos\varphi - \frac{2}{3}\frac{1}{H_2^{(2)'}(ka)}\cos 2\varphi\right]. \tag{7.212}$$

(Values of the higher mode terms are negligible.) The directional factor is

$$H_I(\varphi) = \frac{P_I(r,\omega,\varphi)}{P_I(r,\omega,0)} = \frac{\dfrac{1}{H_0^{(2)'}(ka)} + j\dfrac{\pi}{2}\dfrac{1}{H_1^{(2)'}(ka)}\cos\varphi - \dfrac{2}{3}\dfrac{1}{H_2^{(2)'}(ka)}\cos 2\varphi}{\dfrac{1}{H_0^{(2)'}(ka)} + j\dfrac{\pi}{2}\dfrac{1}{H_1^{(2)'}(ka)} - \dfrac{2}{3}\dfrac{1}{H_2^{(2)'}(ka)}}. \tag{7.213}$$

The sound pressure on the acoustic axis can be represented according to formula (6.120), as

$$P_I(r,\omega,0) = A(r,\omega)\cdot\pi ahU_I\cdot k_{difI}, \tag{7.214}$$

where it is denoted for brevity

$$A(r,\omega) = \frac{k\rho c}{4}\sqrt{\frac{2}{\pi kr}}e^{-j(kr-\pi/4)}. \tag{7.214a}$$

After comparing expressions (7.212) and (7.214) we obtain

$$k_{difI} = -j\frac{4}{\pi^2 ka}\left[\frac{1}{H_0^{(2)'}(ka)} + j\frac{\pi}{2}\frac{1}{H_1^{(2)'}(ka)} - \frac{2}{3}\frac{1}{H_2^{(2)'}(ka)}\right], \tag{7.215}$$

or, after using relations (6.119),

$$k_{difI} = \frac{2}{\pi}\left[k_{dif0} + \frac{\pi}{2}k_{dif1} - \frac{2}{3}k_{dif2}\right]. \tag{7.216}$$

Thus, the sound pressure generated by the transducer may be presented, as

$$P_I(r,\omega,\varphi) = A(r,\omega)\cdot\pi ahU_I\cdot k_{difI}H_I(\varphi). \tag{7.217}$$

Consider now the mode of operation, in which case transducer *II* vibrates, and the surface of transducer *I* is clamped. So far as the mode shape of vibration is the same as in the previous case, expressions for all the acoustic field related characteristics that do not depend on the magnitude of velocity remain the same. Just when superposing the sound fields of both modes of operation it must be taken into consideration that radiation of transducer *II* occurs in the opposite direction. Therefore, $H_{II}(\varphi) = H_I(\varphi + \pi)$, and

$$P_{II}(r,\omega,\varphi) = A(r,\omega)\cdot \pi ahU_{II} \cdot k_{difI} H_I(\varphi + \pi). \tag{7.218}$$

The resulting sound pressure that is generated by the transducer is

$$P_\Sigma(r,\omega,\varphi) = P_I + P_{II} = A(r,\omega)\pi ahk_{difI}\left[U_I H_1(\varphi) + U_{II} H_I(\varphi+\pi)\right]. \tag{7.219}$$

And the directional factor is

$$H_\Sigma(\varphi) = \frac{U_I H_1(\varphi) + U_{II} H_I(\varphi + \pi)}{U_I + U_{II} H_I(\pi)}. \tag{7.220}$$

The final result for the operational characteristics of the two half ring transducer depends on correlation between velocities U_I and U_{II} that can differ for various transducer applications. The following applications of the transducer may have practical sense.

One ring (for certainty #*I*) radiates, another is in the passive mode. Voltage V_I is applied to input of transducer *I*, and output of transducer *II* is or open, or short circuited.

Both the rings are in the receive mode. Outputs of the rings are open circuited.

Both the rings radiate. Voltage is applied to both transducer inputs. The most interesting are the variants, in which $V_I = V_{II}$ and $V_I = -V_{II}$.

Results of calculating the velocities and thus the related characteristics of the half ring transducers for all of these variants of operation can be obtained from combination of two acoustically coupled equivalent circuits that are presented in Figure 7.64 at different positions of the switches.

Two comments must be made regarding otherwise self-explanatory circuits. Firstly, we will assume that in the receive mode the sound wave is coming from the side of transducer *I* in direction of the acoustic axis ($\varphi = 0$). In this case $F_{eqvI} = P_o S_{av} k_{difI}$ and $F_{eqvII} = P_o S_{av} k_{difI} H_I(\pi)$. Secondly, if the rings are electromechanically identical, then $n_I = n_{II}$, $Z^E_{mI} = Z^E_{mII}$, $C^{S_1}_{eI} = C^{S_1}_{eII}$.

In general, they can be different. The acoustic field related parameters remain intact, so far as the mode shapes of the rings vibration do not change.

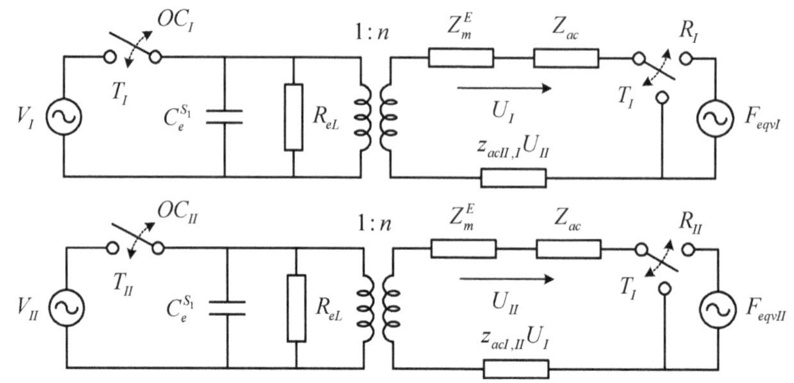

Figure 7.64: Acoustically coupled equivalent circuits of the half ring transducers that describe their combined electromechanical operation. The positions of switches *T, R, OC* correspond to operation in transmit or receive modes, and to the open circuited output.

In the case that one half ring transducer radiates the unidirectionality is achieved in the frequency range around the resonance frequency without using a baffle. The directional factors in this range are close to those of the first order transducers with baffles, because a contribution to acoustic field of vibration of the rear half ring turned to be negligible (the mutual impedance between the half rings is close to zero). One more advantage of the half ring transducer, which is mentioned above, is that it has twice the broader band of the transmit frequency response at -3 dB level around the resonance frequency than the first order transducer. The directional factors of the transducer in some frequency range around the resonance frequency are close to cosine in the half space. This feature makes possible using transducers of this kind in the receive mode in acoustic system of the direction finder, operational principle of which is illustrated in Figure 7.41. Employing the half ring transducer does not require a process of eliminating ambiguity in detecting the bearing that is necessary in case that the first order transducer is used for the same purpose. Though the simplicity is achieved for expense of the narrower operational range and lesser accuracy. Such a simple system may find applications in the case that these features are not critical (e.g., for communication between divers).

The mode of simultaneous radiating of both half rings under equal voltages applied in phase is equivalent to operation of a single half ring installed on the surface of rigid flat baffle, due to symmetry. And radiating under the voltages applied in antiphase is equivalent to operation of the half ring transducer installed on the compliant surface. These situations are illustrated in Figure 7.65.

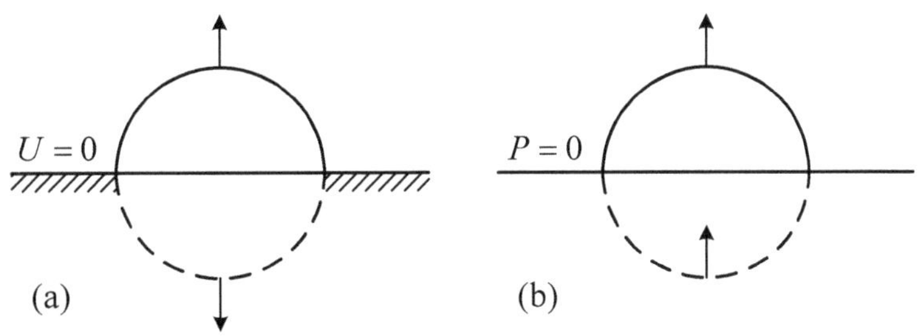

Figure 7.65: Half rings radiating in the (a) rigid and (b) compliant flat baffles. By solid lines are shown the rings, by the dashed lines are shown their images.

The directional factors of the transducers that operate under these conditions are presented for a range of the rings wave sizes in Figure 7.66. These characteristics and all the parameters of the transducers remain practically the same in case the flat baffles of about a wavelength size are used.

It has to be noted that transducers made of incomplete rings have peculiarities that must be considered in their practical design for underwater applications. First of all, the air-backed variant of the transducer design cannot withstand significant hydrostatic pressure, and the variants of the liquid (PU) filled designs must be considered in this respect. On the positive side is that in terms of influencing the liquid filled internal volume on parameters of a transducer the extensionally vibrating incomplete rings behave in the same way as the complete rings.

Other peculiarities are related to the frequency responses of the transducers. Except for the modes of vibration, which were considered so far for operating in the frequency ranges around the extensional resonance frequencies, the incomplete rings may vibrate as a whole (that formally corresponds to the resonance frequency $\omega = 0$) and in the flexural modes (that have resonance frequencies much below the extensional ones).

These "parasitic" modes of vibration do not have electromechanical coupling with the working extensional modes but may influence the frequency response of transducer through the acoustic interactions with the working modes. These interactions may be especially pronounced in vicinity of the "parasitic" resonances, and they result in a drop of sensitivity of transducer at low frequencies and in the notches in the transducer frequency response in vicinity of the flexural resonances.

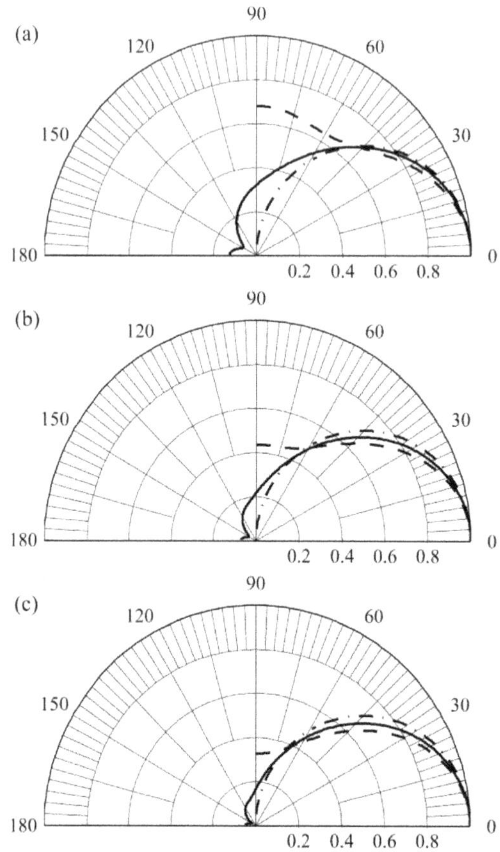

Figure 7.66: The directional factors of the half rings radiating in a cylindrical rigid baffle (solid lines), in a rigid flat baffle (dashed lines), and in a compliant flat baffle (dash-dotted lines) at (a) $ka = 2.2$, (b) $ka = 3.0$, (c) $ka = 4.0$.

Example of analytical approach to estimating such effects is illustrated with the equivalent circuit in Figure 7.55, and resulting qualitative frequency response in Figure 7.56. Here we assume that operational range of the incomplete ring transducers can be recommended in the more or less broad band around their resonance frequencies. Therefore, the contours that are accounting for these "parasitic" effects where not included in the equivalent circuits of the transducers for simplicity.

7.3 Transducers Operating in the Flexural Modes

When considering cylindrical transducers of this type the results will be used that are presented in Section 4.5.4 and 4.5.10, where flexural vibrations of the complete and slotted passive rings are discussed. The same assumption will be in place that the rings are thin and short, unless it will be noted otherwise. Therefore, the correspondence between strain and stress in the crystallographic and geometry coordinates, and governing piezoelectric equations are the same, as for the incomplete rings in Section 7.2.5. Thus, $T_r \to T_3$, $T_h \to T_2$, $S_\varphi \to S_1$ for the solid, and $T_r \to T_1$, $T_h \to T_2$, $S_\varphi \to S_3$ for the segmented rings. The relations (7.169) modified for both radial (at $i = 1$) and circumferential (at $i = 3$) polarizations that follow from the piezoelectric equations,

$$T_i^E = \left(1/s_{ii}^E\right)S_i = Y_i^E S_i,\ D_3^E = \left(d_{3i}/s_{33}^E\right)S_i\ \ \varepsilon_{33}^{S_i} = \varepsilon_{33}^T\left(1-k_{3i}^2\right), \qquad (7.221)$$

can be used in this case as well. The main analysis will be made for the radially polarized rings (transverse piezoeffect). Results for the segmented ring design will be presented based on this analysis.

7.3.1 Complete Rings Transducers

The main content of this section is after Ref. 21.

7.3.1.1 Equivalent Electromechanical Parameters of the Rings

Consider the design features of a ring made of piezoceramic intended to generate the lowest mode of flexural vibrations. Geometry of the ring and configuration of electrodes and electric fields in its body are presented in Figure 7.67 with example of a solid ring for simplicity. In reality transducers of this type are usually cemented out of prisms shown in Figure 7.1(c)

As it was shown by analyzing the general expression (7.1) for the electromechanical energy associated with vibration of piezoceramic rings, excitation of the flexural deformations only is possible under the condition that function $\Omega_1(z)$ in the distribution of electric field in the volume of a ring represented by formula,

$$E_3(z,\varphi) = |E|\cdot\Omega_1(z)\Omega_2(\varphi), \qquad (7.222)$$

is odd, i.e., $\Omega_1(z) = -\Omega_1(-z)$. Under this condition, which is realized at configuration of electrodes and relative directions of the operational electric field and field of polarization that are shown in Figure 7.1(c) and 7.67, the bending moment is generated in the piezoelement. In this case the electromechanical conversion under the flexural deformation is characterized by the energy

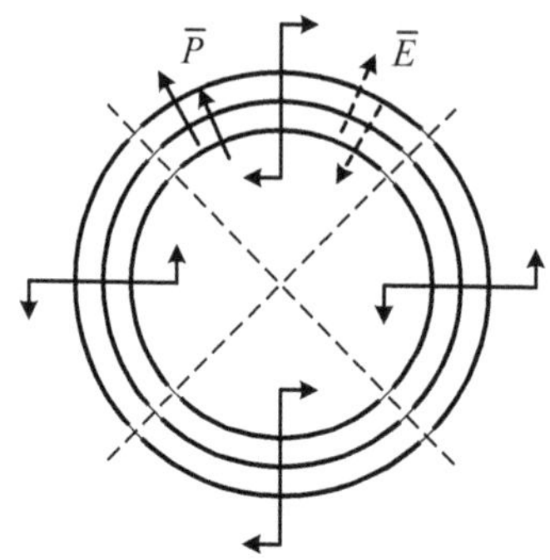

Figure 7.67: Geometry of the piezoceramic ring intended for excitation of the flexural vibration, and configuration of electrodes in the body of the ring. The electrodes are shown by solid lines); mutual directions of the electric field of polarization ($\bar{\boldsymbol{P}}$) and operational electric field ($\bar{\boldsymbol{E}}$) in the layers above and below the neutral surface are shown by solid and dashed arrows. For generating the flexural deformation of a ring, the bending moments in neighboring quarters must change signs (operational fields, or fields of polarization must change signs).

$$W_{em} = \frac{1}{2}\frac{d_{31}hE}{s_{11}^E a}\int_0^{2\pi}\left(\int_{-t/2}^{t/2}\Omega_1(z)zdz\right)\Omega_2(\varphi)\left(\frac{\partial^2\xi_r}{\partial\varphi^2}+\xi_r\right)d\varphi\,. \tag{7.223}$$

Following the results presented in Section 4.5.4, the radial modes of flexural vibration,

$$\xi_r(\varphi,t) = \xi_i(t)\cos i\varphi\,, \tag{7.224}$$

and expressions for the equivalent mechanical parameters and resonance frequencies (with elastic modulus Y replaced by Y_1^E) are

$$M_{eqvi} = \frac{M(1+i^2)}{2i^2}\,,\; K_{eqvi} = \frac{1}{C_{eqvi}} = \frac{\pi Y_1^E t^3 h(i^2-1)^2}{12a^3}\,,\; i=2,3\,, \tag{7.225}$$

$$f_i = \frac{1}{2\pi\sqrt{M_{eqvi}C_{eqvi}}} = \frac{t}{2\pi\sqrt{12}\,a^2}\sqrt{Y_1^E/\rho}\,\frac{i(i^2-1)}{\sqrt{i^2+1}}\,. \tag{7.226}$$

After substituting expression (7.224) for displacements ξ_r under the integral in (7.223) and taking into consideration definition for the electromechanical transformation coefficient, n, we arrive at the expression

$$W_{em} = \frac{1}{2}\frac{2d_{31}h|E|}{s_{11}^E a}\int_0^{\frac{t}{2}}\Omega_1(z)zdz\int_0^{2\pi}\left[\Omega_2(\varphi)\sum_{i=2}^{\infty}(i^2-1)\xi_i\cos i\varphi\right]d\varphi = \\ = \frac{1}{2}V\sum_{i=2}^{\infty}n_i\xi_i \tag{7.227}$$

It follows from this expression that the flexural vibration modes can be electromechanically active only, if function $\Omega_2(\varphi)$ has a form that insures nonzero value of integral

$$\Omega_{2i} = (i^2-1)\int_0^{2\pi}\Omega_2(\varphi)\cos i\varphi d\varphi. \tag{7.228}$$

Obviously, this takes place at $i \geq 2$, and in the case that function $\Omega_2(\varphi)$ changes sign simultaneously with $\cos i\varphi$. Thus, when the electrodes are divided into $2N$ equal parts along the circumference and the adjacent parts are connected electrically in anti-phase (the bending moments of the opposite sign are generated in these parts), then

$$\Omega_{2i} = (-1)^{l+1}\cdot 4(i^2-1)/i \text{ at } i=(2l-1)N,\ l=1,2,\ldots \tag{7.229}$$

and $\Omega_{2i}=0$ otherwise. The lowest active mode of vibration $\theta_2(\varphi)=\cos 2\varphi$ is generated at N = 2, when the electrodes are divided in 4 parts, as shown in Figure 7.67. This mode of vibration is of the greatest practical interest. The corresponding coefficient of electromechanical transformation will be obtained at $\Omega_{22}=6$, and the resonance frequency of the ring is

$$f_2 = (0.12t/a^2)\sqrt{Y_1^E/\rho}. \tag{7.230}$$

The next closest active vibration mode is $\theta_6(\varphi)=\cos 6\varphi$ (at $N=2,\ l=2$). The resonance frequency that corresponds to this mode is $f_6=12.8f_2$ according to formula (7.226). Thus, the vibration mode $\theta_2(\varphi)$ can be considered as dominant within a broad frequency range. Therefore, the transducers employing rings vibrating in this mode can be treated as having a single mechanical degree of freedom. For calculating transducer properties, the regular single degree of freedom equivalent electromechanical circuit can be used with equivalent mechanical parameters,

$$M_{eqv} = \frac{5}{8}M,\ K_{eqv}^E = \frac{1}{C_{eqv}^E} = \frac{3\pi Y_1^E t^3 h}{4a^3}, \tag{7.231}$$

according to expressions (7.225).

For specifying values of the transformation coefficients and capacitances, the variants of electrical connection of parts of electrodes in piezoelements must be considered. They are illustrated in Figure 7.68.

We assume that the thickness of a ring is small, and dependence of electric field from z can be neglected. In the variant of parallel connection of the concentric parts of a ring (Figure 7.68 (a)) $|E| = 2V/t$ and

$$n = 3d_{31}th/s_{11}^{E}a, \; C_{el}^{S_1} = 8\pi ah\varepsilon_{33}^{T}\left(1-k_{31}^{2}\right)/t. \tag{7.232}$$

In the variant of series connection of the electrodes (Figure 7.68 (b)) $|E| = V/t$ and

$$n = 1.5d_{31}th/s_{11}^{E}a, \; C_{el}^{S_1} = 2\pi ah\varepsilon_{33}^{T}\left(1-k_{31}^{2}\right)/t. \tag{7.233}$$

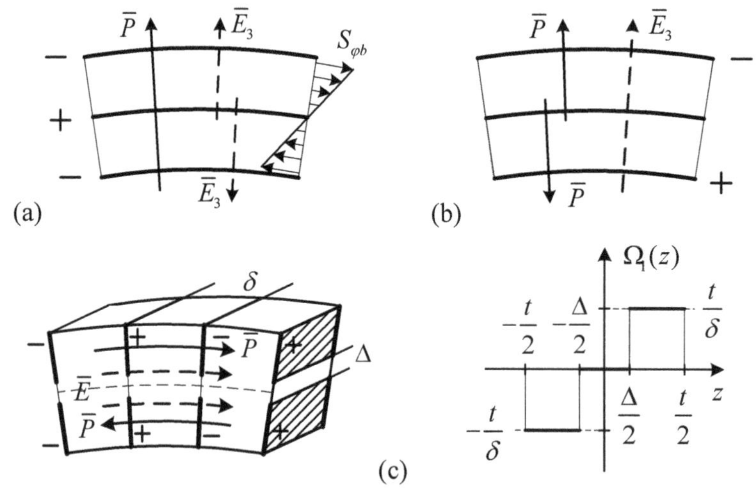

Figure 7.68: Configuration of piezoelements that produce the bending deformations, and variants of electrodes connections: (a) transverse piezoeffect, parallel connection of electrodes; (b) transverse piezoeffect, series connection of electrodes; (c) longitudinal piezoeffect and distribution of the electric fields through the thickness of the prisms.

The piezoelements that realize the longitudinal piezoelectric effect can be made up of the prisms (wedges) with electrodes on their sides separated into two parts, and the electric fields of polarization and operation directed as shown in Figure 7.68 (c). Under the assumption that ratio δ/t is small enough for the electric field within the stripe Δ to be neglected, the operational electric field is $|E| = V/\delta$. It may be obtained in the same way, as it was done for the case of the transverse piezoeffect, that for the case of the parallel connection of the prisms (which is the most practical),

$$n_2 = 3d_{33}h\left(t^2 - \Delta^2\right)/ s_{33}^E a\delta\,,\; C_{el}^{S_3} = 2\pi a\left(t - \Delta\right)h\varepsilon_{33}^T\left(1 - k_{33}^2\right)/\delta^2\,. \tag{7.234}$$

Besides, in formulas for the rigidities the elastic modulus, Y_1^E must be replaced by $Y_3^E = 1/s_{33}^E$. It is noteworthy that the segmented transducer design is the most practical, because the transducer piezoelement made of the concentric solid rings is hard to manufacture, not to speak about lesser effective coupling coefficient that is critical for this transducer type.

Knowing the equivalent parameters allows determining the effective coupling coefficient of the transducer. This will be done for the segmented design as the most usable for the projector application. At first determine the coefficient $\alpha_c = n^2 / K_{eqv}^E C_{el}^{S_3}$. After substituting expressions (7.234) and (7.231) for the parameters involved (the latter with replacement Y_1^E by Y_3^E), we will arrive at expression for α_c

$$\alpha_c = \frac{3}{\pi^2}\left(1 + \frac{\Delta}{t}\right)\left[1 - \left(\frac{\Delta}{t}\right)^2\right]\frac{k_{33}^2}{1 - k_{33}^2}\,. \tag{7.235}$$

The coefficient α_c as a function of the ratio Δ/t has maximum. Plot of the function $\alpha_c(\Delta/t)/\alpha_c(0)$ vs. Δ/t is presented in Figure 7.69. It is seen that the maximum value $\alpha_{c\,max}$ is achieved at $\Delta/t = 1/3$. The effective coupling coefficient at $\alpha_c = \alpha_{c\,max}$ is

$$k_{eff}^2 = \frac{\alpha_{c\,max}}{1 + \alpha_{c\,max}} = \frac{k_{33}^2}{2.8 - 1.8k_{33}^2}\,. \tag{7.236}$$

In case that PZT-4 ceramics is used $k_{eff} \approx 0.72k_{33}$.

It must be kept in mind that increase of the ratio Δ/t is followed by reduction of the electromechanical transformation coefficient according to formula (7.234). Thus, the equivalent electromechanical force $F_{em} = nV$ that drives the flexural vibration reduces. On the other hand, the gap Δ between the electrodes cannot be small because this gap must withstand polarizing voltage that is applied in the course of poling process.

The effective coupling coefficient can be further optimized by changing the size of the electrodes (the active segment of a ring) in circumferential direction. Consider the general expressions for the capacitance and transformation coefficient under condition that only a part of each quarter of the volume of the ring within sector $\pm 45^o$ around its axis is active. The electromechanical transformation coefficient is proportional to the coefficient calculated by formula (7.228). Ratio of values of this coefficient determined by integrating over shortened and full-size electrodes, and of respective transformation coefficients is,

$$n(\varphi_{el}) = n(\pi/2)\sin\varphi_{el}. \tag{7.237}$$

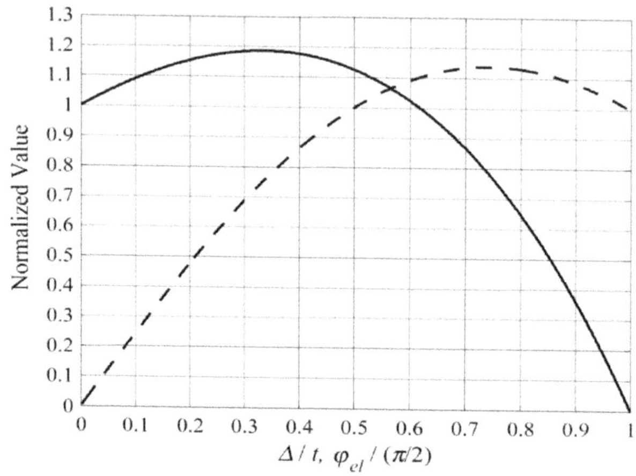

Figure 7.69: Normalized functions that characterize dependence of the effective coupling coefficient from separation between electrodes through the thickness, $\alpha_c(\Delta/t)/\alpha_c(0)$ (solid line), and from the angular size of the electrodes, $\alpha_c(\varphi_{el})/\alpha_c(\pi/2)$ (dashed line).

The capacitance of the active part of the ring may be represented as

$$C_{el}^S(\varphi_{el}) = C_{el}^S(\pi/2)2\varphi_{el}/\pi. \tag{7.238}$$

After substituting these expressions for $n(\varphi_{el})$ and $C_{el}^S(\varphi_{el})$ into formula for α_c we obtain,

$$\alpha_c(\varphi_{el}) = \alpha_c(\pi/2)\frac{\pi\sin^2\varphi_{el}}{2\varphi_{el}}. \tag{7.239}$$

Dependence $\alpha_c(\varphi_{el})/\alpha_c(\pi/2)$ from angle φ_{el} is depicted in Figure 7.69 by the dashed line. This function has maximum at $\varphi_{el} = 3\pi/8$, and $\alpha_c(3\pi/8) = 1.23\alpha_c(\pi/2)$. The maximum value of the effective coupling coefficient under the combined effect of optimizing the electrodes configuration through the thickness and over circumference is

$$k^2_{effmax} = \frac{k^2_{33}}{2.4 - 1.4k^2_{33}}. \tag{7.240}$$

With PZT-4 ceramics used this results in $k_{effmax} = 0.94k_{33}$.

The parts of the ring that are cleared out of electrodes (piezoelectrically inactive parts) are located around the nodal lines. They can be replaced by a passive material, or alternatively by piezoelements that excite shear deformations. The reason for this is that while the bending deformations are minimal near the nodal lines, the shear deformations are maximal. The situation is similar to those described in Ref. 22 in regard to a piezoceramic beam vibrating in flexure, and calculation of contribution of electromechanical transformation due to shear deformations can be made in the same way, as it is done therein. Some detail on this issue will be discussed in Section 9.2.2.2.

7.3.1.2 Acoustic Field Related Characteristics of the Transducers

In order to complete calculation of an electromechanical transducer as electroacoustic device, the acoustic field related parameters that are presented in the equivalent circuit by the radiation impedance, Z_{ac}, and the equivalent force, F_{eqv}, must be determined. Peculiarities of the acoustic field related characteristics of a cylindrical transducer of the flexural type arise from the fact that the neighboring quarters of its surface vibrate in antiphase and its diameter is small compared to wavelength at resonance frequency. Namely, using formula (7.230) for the resonance frequency we obtain

$$(ka)_{res} = 0.75\frac{c_c}{c_w}\frac{t}{a}. \tag{7.241}$$

In the case that PZT-4 is used $(c_c / c_w) \approx 2$, and for the thin rings $(t/a) < (0.2 \text{ to } 0.25)$. Thus, the wave size of the rings can be estimated as $(ka)_{res} \approx 0.3$ or less.

In order to address the problems related to using transducers as low frequency projectors, at first, they must be built from a number of mechanically isolated rings stacked axially so that the overall height (h_Σ) of the transducer is comparable with wavelength at the resonance frequency. Acoustic field generated by such transducer in the horizontal plane can be approximately considered, as radiated by the infinitely long cylinder having the same distribution of velocity. Radiation of tall cylindrical shells having modal distributions of velocity is considered in Section 6.3.1.1.1, where expressions for the directional factor, diffraction coefficient and radiation impedances are presented for different modes of vibration including $\cos 2\varphi$. Plots of these quantities as functions of *ka* are shown in Figs. 6.6 through 6.8. Thus, the expression for the radiation impedance is

$$Z_{ac} = -j(\rho c)_w h_\Sigma \pi a \frac{H_2^{(2)}(ka)}{H_2^{(2)'}(x)\big|_{x=ka}}. \tag{7.242}$$

For small *ka* ($ka < 0.3$), which is our case,

$$Z_{ac} = r_{ac} + jx_{ac} = (\rho c)_w h_\Sigma \pi a \left[\frac{\pi (ka)^5}{32} + j\frac{ka}{2}\right], \tag{7.242a}$$

$$r_{ac} = (\rho c)_w h_\Sigma \pi a \frac{\pi (ka)^5}{32}, \tag{7.243}$$

$$x_{ac} = (\rho c)_w h_\Sigma \pi a \frac{ka}{2} = \omega \rho \frac{h_\Sigma \pi a^2}{2} = \omega \frac{M_w}{2}, \tag{7.244}$$

where M_w is the mass of water in the volume of the cylinder per unit height.

The directional factor of the transducer ($H(\varphi) = \cos 2\varphi$) is not appropriate for underwater applications in the first place. Besides, the radiation impedance of the transducer at frequencies around its resonance is almost purely reactive. The ratio of its active component to reactive is $\left(r_{ac} / x_{ac}\right) = \pi (ka)^4 / 16$, where $ka < 0.3$. This could be expected, as the volume velocity of the surface vibration for mode $\theta_2(\varphi) = \cos 2\varphi$ is zero. Obviously, transducer having such infinitesimal radiation resistance is not practical. In order to change the spatial distribution of energy radiated and to increase the radiation resistance, the parts of transducer surface that vibrate in antiphase should be covered by acoustic baffles, as shown in Figure 7.70.

The variant with two opposite quarters of a ring covered with baffles, as shown in Figure 7.70 (a), is the most suitable for designing the column like transducers, and the variant with three quarters covered is more appropriate for using in array, as shown in Figure 7.70(b). If to assume that the baffles are absolutely rigid, then the distribution of velocity over surface of transducer with two quarters of surface baffled will be $\theta(\varphi) = |\cos 2\varphi|$ at $-\pi/4 \le \varphi \le \pi/4$ and $3\pi/4 \le \varphi \le 5\pi/4$, and $\theta(\varphi) = 0$ otherwise. The directional factor and radiation impedance of the transducer can be calculated following the procedure described in Section 6.3.1.1.2.

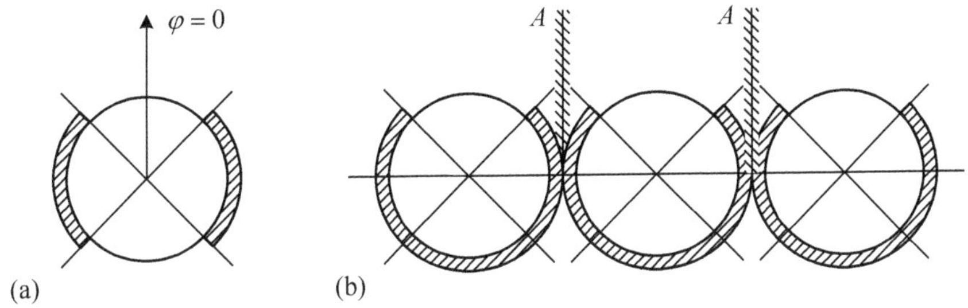

Figure 7.70: Variants of baffling the transducer surface: (a) baffles cover the opposite quarters of a ring, (b) baffles cover three quarters of a ring, when it is used in an array of a large size.

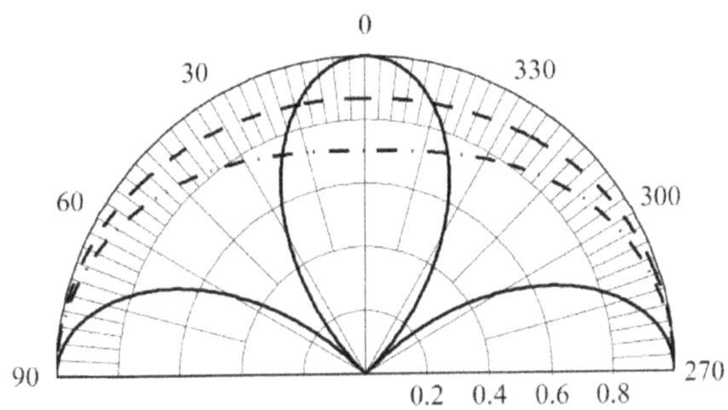

Figure 7.71: Directional factors of a cylindrical transducer vibrating in the mode $\theta(\varphi) = \cos 2\varphi$ without baffles (solid line); with the baffles on the opposite quarters of surface at $ka = 0.4$ (dashed line) and $ka = 0.6$ (dash-dotted line).

Results of calculation performed at $ka = 0.4$ and $ka = 0.6$ are presented in Figure 7.71 by the dashed and dash-dotted lines, respectively. For obtaining even closer to omnidirectional response at these values of *ka*, the baffled rings that are adjacent by height must be turned ninety degrees in respect to each other. Components of the radiation impedance per unit length of a long column like baffled transducer can be presented according to expression (6.358) for a simple cylindrical source as

$$r_{ac} = (\rho c)_w ka^2 = \frac{\pi}{2}(\rho c)_w \frac{S_{av}^2}{\lambda}, \tag{7.245}$$

($S_{av} = 2a$ per unit length) and

$$x_{ac} \approx 0.9\pi a^2 \rho\omega . \tag{7.246}$$

For a tall transducer of finite height h_z the radiation impedance must be multiplied by the height.

It is convenient for practical applications to use completed flexural ring transducer units that can operate as single projectors of small wave size or can populate projector systems of a large size. The height of such units is usually comparable to their diameter for technological reasons. Transducers of this small wave size ($ka < 0.3$) may be considered as the equivalent spherical simple sources. The radiation impedance of an equivalent spherical simple source can be determined by modifying expressions (2.24) and (2.25) for components of radiation impedance of the pulsating sphere of a small wave size. The radiation resistance is

$$r_{ac} = \pi(\rho c)_w S_\Sigma^2 / \lambda^2 . \tag{7.247}$$

In this case $S_\Sigma = S_{av} \cdot h_{unit}$, where $S_{av} = 2a$ per unit length. The radiation reactance will be estimated as for a pulsating sphere having the same volume velocity, and the result is

$$x_{ac} = 4\pi a_{eqv}^3 \rho\omega , \tag{7.248}$$

where a_{eqv} must be determined from the condition $4\pi a_{eqv}^2 = S_\Sigma$, which is equivalent to the condition of equality of the volume velocities (remember that volume velocity is $U_{\tilde{V}} = S_\Sigma U_0$).

In the case that transducers with baffles on the three quarters of surface populate an array of a large size, the radiation impedance can be estimated as follows. A strip of an array having the width of transducer (shown between imaginary planes *A-A* in Figure 7.70(b)) being small compared to wave-length can be considered as vibrating uniformly with the same volume velocity per unit length, $U_{\tilde{V}} = aU_o$, that a single ring has. The average velocity of uniform vibration of the strip will be found from equality of the volume velocities, $U_{\tilde{V}} = aU_o = 2aU_{av}$, where from $U_{av} = U_o / 2$. The acoustic power radiated from a strip of width $2a$ per unit length of a large array, or per unit height of the single ring located within the strip, is $\dot{W}_{ac} = (\rho c)_w 2aU_{av}^2 = (\rho c)_w aU_o^2 / 2$. The radiation resistance per unit height of a single ring in this case is

$$r_{ac} = (\rho c)_w a / 2 . \tag{7.249}$$

The radiation reactance in array of a large size that vibrates with uniform average velocity can be neglected.

So far as the radiation impedances for the usable variants of the flexural type cylindrical transducers made of complete rings with baffles are determined, velocity of vibration U_o can be obtained from the equivalent circuit of the transducers operating in the transmit mode. After this the sound pressure that column like transducer generates may be calculated by formula (2.64), as for cylindrical simple source having volume velocity $U_{\tilde{V}} = S_{av}U_o = 2ah_{\Sigma}U_o$.

It follows from the above considerations that quality of the baffles is essential for determining the acoustic characteristics and effectiveness of the flexural ring transducers. The model of a rigid baffle used for estimating acoustic characteristics is idealized, but experience obtained with baffling cylindrical extensional transducers showed that a good agreement was obtained between characteristics calculated under the assumption that the baffles are ideally rigid, and results obtained experimentally at moderate operating depth with baffles made of corprene. More appropriate for transducers of this kind may be metal baffles that are described in Ref.23.

7.3.2 Slotted Ring Transducers

A short history of development of this transducer type in USA with related references and approximate equivalent circuit model of the transducer can be found in Ref. 24. More detailed analysis was later produced in Ref. 25. Some work on the slotted ring transducers was done in parallel in Russia with results reported in Ref. 26 and later significantly expended in Ref. 27. The main content of this Section is after Ref. 27.

7.3.2.1 Equivalent Electromechanical Parameters of the Transducer

Free flexural vibrations of a thin and short slotted ring are considered in Section 4.5.10, where the mode shape of vibration and equivalent mechanical parameters of the ring are determined. A qualitative comparison between the mode shapes of the radial displacements for the complete ring and slotted ring is made in Figure 7.72.

Comparison of the mode shapes illustrates the difference between these vibrating systems as sources of low frequency radiation. Whereas the complete ring constitutes a quadrupole source with $S_{av} = 0$, and baffling is required of the areas vibrating in anti-phase for its effective radiation, the slotted ring is a monopole source with $S_{av} \neq 0$, and may have a fairly decent value of the radiation resistance without additional measures taken. The disadvantage of the slotted ring vibrating in flexure is its relatively small hydrostatic pressure strength.

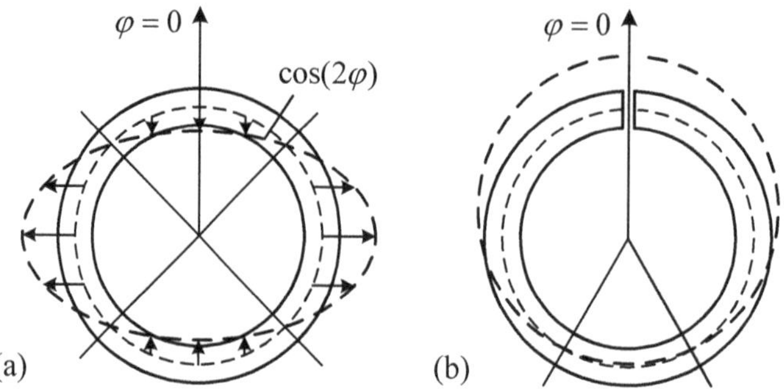

Figure 7.72: Qualitative comparing the mode shapes: (a) radial displacements for a complete ring and (b) for a slotted ring.

In case that the ring is made of piezoceramics the relations (7.221) that follow from the piezoelectric equations of state must be used, and in the mechanical quantities the elastic modulus Y must be replaced by Y_i^E. The summary of mechanical equivalent parameters that follows from expressions (4.517)-(4.518) is,

$$M_{eqv} = \frac{\pi}{2}M\,,\ K_m^E = \frac{1}{C_m^E} = \frac{\pi t^3 h}{20a^3 s_{ii}^E}\,,\ f_r = 0.02\frac{t}{a^2}\sqrt{\frac{Y_i^E}{\rho}}\,. \tag{7.250}$$

Configuration of piezoelements and variants of the electrodes connection for producing the bending moments in the ring are the same as shown in Figure 7.68. And general expressions (7.222), (7.223) for the distribution of the operating electric field and electromechanical energy are the same as well. Different is expression for distribution of the radial displacement, which in the case of slotted ring is according to formula (4.513)

$$\xi_r(\varphi) = \xi_{ro}\left(0.55\cos\varphi + 0.70\sin\varphi - 0.22\varphi\sin\varphi + 0.44\right). \tag{7.251}$$

After substituting expressions (7.222) for the electric field and (7.251) for the displacement distribution into the general expression for the electromechanical energy (7.223) we obtain

$$W_{em} = \frac{1}{2}0.44\xi_{ro}\frac{hd_{3i}\left|E_3\right|}{s_{ii}^E a}\int_{-t/2}^{t/2}\Omega_1(z)zdz\int_0^{2\pi}\Omega_2(\varphi)(1-\cos\varphi)d\varphi = \frac{1}{2}nV\xi_{ro}\,. \tag{7.252}$$

As the basic variant of transducer design, we will consider the parallel connection of electrodes shown in Figure 7.68 (a) that cover all the ring surface. The electromechanical transformation coefficient and capacitance in this case are

$$n = 0.44\pi d_{31}ht/s_{11}^E a\,,\ C_{el}^{S_i} = \varepsilon_{33}^T\left(1-k_{31}^2\right)8\pi ah/t\,. \tag{7.253}$$

The effective coupling coefficient for the basic transducer design being calculated by formula $k_{eff}^2 = \alpha_c / (1+\alpha_c)$, where $\alpha_c = n^2 / K_m^E C_{el}^{S_1}$, is

$$k_{eff}^2 = \frac{k_{31}^2}{2.1 - 1.1k_{31}^2}. \qquad (7.254)$$

If PZT-4 ceramics is used, $k_{eff} \approx 0.71k_{31}$.

For the variant of design that employs the longitudinal piezoeffect, in which case the ring is made up of the prisms shown in Figure 7.68 (c), it can be obtained analogous to expressions (7.234) that

$$n = 0.44\pi d_{33} h(t^2 - \Delta^2)/ s_{33}^E a\delta, \; C_{el}^{S_3} = \varepsilon_{33}^T (1 - k_{313}^2) 2\pi a h (t - \Delta)/\delta^2. \qquad (7.255)$$

Given that distribution of strain exists in the volume of the ring both through the thickness and over circumference, optimizing of the effective coupling coefficient is possible in the same way as it was done for the complete flexural ring. For the basic design optimizing can be achieved by changing the size of the electrodes over circumference. Assuming that the electrodes cover segment of the surface at $-\psi \le \varphi \le \psi$, from expression (7.252) will be obtained that

$$n(\psi) = n(0)(\pi - \psi + \sin\psi)/\pi. \qquad (7.256)$$

Besides,

$$C_{el}^{S_1}(\psi) = C_{el}^{S_1}(0)(\pi - \psi)/\pi. \qquad (7.257)$$

Here the factors $n(0)$ and $C_{el}^{S_1}(0)$ are the values of these quantities determined for the full size electrodes (at $\psi = 0$) in variants with transverse or longitudinal piezoeffect. After substituting (7.256) and (7.257) into expression for function α_c, we arrive at relation

$$\frac{\alpha_c(\psi)}{\alpha_c(0)} = \frac{(\pi - \psi + \sin\psi)^2}{\pi(\pi - \psi)}. \qquad (7.258).$$

Note that in the course of manipulations it was neglected insignificant changing of equivalent rigidity, which strictly speaking may occur due to difference between values of elastic moduli on the parts of volume before and after they are deprived of electrodes. In the case that the electrodes on these parts are electrically separated and short circuited, no changes of elastic modulus take place at all. The function (7.258) is plotted in Figure 7.73.

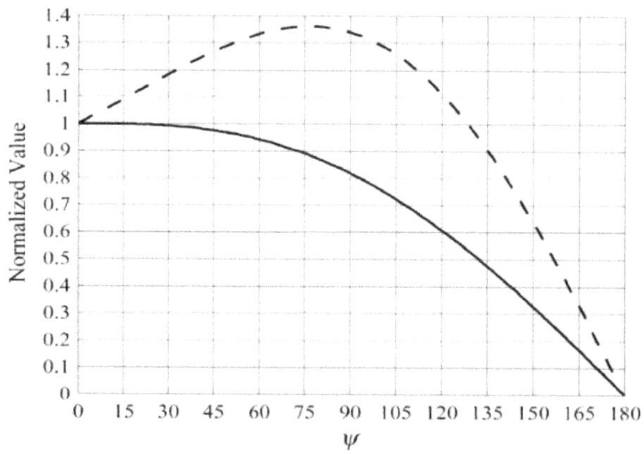

Figure 7.73: Plots of functions $n(\psi)/n(0)$ (solid line) and $\alpha_c(\psi)/\alpha_c(0)$ (dashed line) vs. angle, up to which the parts of ceramic volume become inactive.

The same result is valid for the longitudinal piezoeffect, when ring is composed of the prisms shown in Figure 7.68 (c). It follows from the plot that function $\alpha_c(\psi)/\alpha_c(0)$ has maximum at $\psi = 75°$. The effective coupling coefficient also have maximum at this value of $\alpha_c(75^o)$, and after some manipulations it can be shown that

$$k_{effmax}^2 = \frac{1.35k_{3i}^2}{2.1 - 0.75k_{3i}^2}. \tag{7.259}$$

With PZT-4 ceramics used $k_{effmax} = 0.82k_{31}$ at the transverse piezoelectric effect, and $k_{effmax} = 0.88k_{33}$ at longitudinal piezoeffect. In the variant of the segmented ring an additional optimization of the effective coupling coefficients may be achieved by changing ratio Δ/t, as it was shown regarding the complete ring vibrating in flexure and illustrated with plot in Figure 7.69. At $\Delta/t = 1/3$ $\alpha_c(\Delta/t) = 1.18\alpha_c(0)$. Under the combined effect of optimizing the electrodes configuration through the thickness and over circumference the maximum value of the effective coupling coefficient will be

$$k_{effmax}^2 = \frac{k_{33}^2}{1.24 - 0.24k_{33}^2}. \tag{7.260}$$

With PZT-4 ceramics $k_{effmax} = 0.94\text{k}_{33}$.

Optimizing the effective coupling coefficient is accompanied by reducing the electromechanical transformation coefficient, *n*, according to formula (7.256), and thus by reducing the effective electromechanical force $F_{em} = Vn$. Reducing the electrodes up to their optimal size results in 10% loss of the transformation coefficient, as follows from the plot in Figure 7.73. Note that the physical effect of bending is produced by bending moments distributed over the circumference of a ring. The "effective electromechanical force" is the imaginary force that being applied to the reference point in the radial direction produces the same effect as a real bending moment distribution.

One of the problems related to underwater applications of the slotted ring transducers is their limited strength under hydrostatic pressure. A partial solution for transducers intended for operating at moderate depth is in using half passive (bilaminar) design of their mechanical system that is considered below. (Analogous analysis is performed regarding vibration of a bilaminar beam in Section 4.5.6.4. and of the flexural type plate transducers in Chapter 9. This analysis is partially reproduced in the next section for convenience of reading.)

7.3.2.2 Bilaminar (Active-Passive) Slotted Ring Design

The tensile strength of ceramics is much less than compressive strength. Therefore, the replacement of the layer of ceramics located exterior to the neutral (free of stress) surface of a ring by metal would result in significant increase of the ring strength under action of hydrostatic pressure and in the corresponding increase of operating depth of transducer. Replacing the outer layer of a ring by metal leads to the bilaminar ring design, a fragment of which is shown in Figure 7.74.

Previously (see Figure 7.1 (c)) such design was mentioned as typical for excitation both the extensional and flexural modes of vibration. Indeed, the extensional vibration of the ring with a slot that is considered in Section 7.2.6.1 will be generated as well. But the resonance frequency of the extensional vibration is a way higher, and these types of vibration can be considered as independent. Although, if for some reason a simultaneous radiation of high and low frequencies by the same acoustic system is desirable, transducer of such design can be used for this purpose.

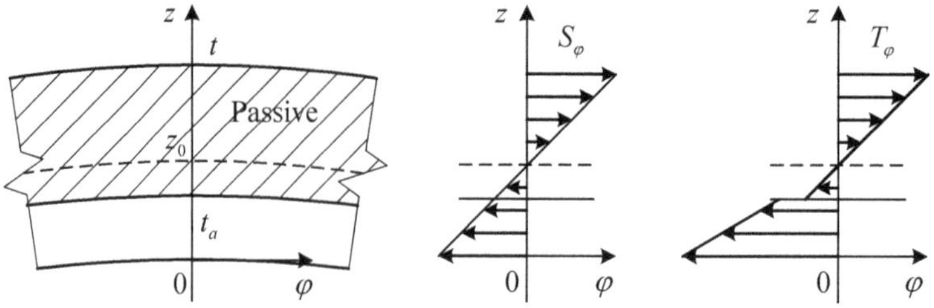

Figure 7.74: Fragment of the bilaminar ring, and distributions of strain and stress over thickness of the ring. The active layer is shown as solid (under the transverse piezoeffect), but it can be segmented (under the longitudinal piezoeffect) as well.

When analyzing properties of the bilaminar ring, the assumption will be kept that the thickness to diameter ratio, $t/2a$, is sufficiently small for elementary theory of bending being applicable to the first approximation. Elastic moduli and densities of the active and passive layers will be denoted $Y_a^E = 1/s_{ii}^E$, Y_p and ρ_a, ρ_p. And notations will be introduced for brevity

$$\frac{Y_p}{Y_a} = \gamma_Y, \quad \frac{\rho_p}{\rho_a} = \gamma_\rho. \tag{7.261}$$

Values of the equivalent parameters for the active-passive rings will be distinguished by subscript *ap*, and for the fully active - just *a*.

The peculiarity of this case is that the neutral surface under bending does not coincide with the middle surface as it was in the case of the fully active ring. Thus, as the first step the location of the neutral surface (coordinate z_0) must be determined. By definition the neutral surface should be free of stress, and therefore its coordinate z_0 may be found from the condition that

$$\int_0^{z_0} T_\varphi(z)\,dz + \int_{z_0}^{t} T_\varphi(z)\,dz = 0, \tag{7.262}$$

where

$$T_\varphi(z) = -\xi_{ro} Y(z)\frac{z - z_0}{a^2}\left(\frac{d^2\theta_r}{d\phi^2} + \theta_r\right). \tag{7.263}$$

If the thickness of the active layer is t_a, then condition (7.262) is equivalent to

$$\int_0^{t_a} Y_a^E (z - z_0)\,dz + \int_{t_a}^{t} Y_p (z - z_0)\,dz = 0, \tag{7.264}$$

where from

$$\frac{z_0}{t} = \frac{1}{2} \cdot \frac{\gamma_Y + [1-\gamma_Y](t_a/t)^2}{\gamma_Y + [(1-\gamma_Y)](t_a/t)}. \tag{7.265}$$

In a rational transducer design z_0 should be greater than t_a (otherwise the electromechanical effects in the piezoelectric element above and below the neutral plane would be in opposite phase). Denote the value of z_0 that is equal to t_a as z_{0m}, i.e., $z_{0m} = t_a$ (the subscript *m* stands for the minimum reasonable value of z_0). It follows from Eq. (7.265) that

$$\frac{z_{0m}}{t} = \frac{\sqrt{\gamma_Y}}{1+\sqrt{\gamma_Y}}. \tag{7.266}$$

Thus, for example, for combinations of solid piezoelement made of PZT-4 with aluminum and steel the values of ratio z_{0m}/t are 0.48 and 0.62, respectively.

Now we can determine the equivalent parameters of the active-passive slotted ring with the piezoelectric elements of different thickness. Values of the equivalent parameters for the active-passive rings will be distinguished by subscript *ap*, and for the fully active - just *a*. The expression for the mechanical energy of a bilaminar ring is

$$W_m^E = \frac{1}{2}\xi_{ro}^2 \frac{2h}{a^3}\left[\int_0^{t_a}(z-z_0)^2 Y_a dz + \int_{t_a}^{t}(z-z_0)^2 Y_p dz\right]\int_0^{\pi}\left(\frac{\partial^2\theta_r}{\partial\varphi^2}+\theta_r\right)d\varphi. \tag{7.267}$$

The term in the brackets is equal to

$$\frac{1}{3}\left[(t_a - z_0)^3 + z_0^3\right]Y_a + \left[(t-z_0)^3 + (z_0 - t_a)^3\right]Y_p. \tag{7.268}$$

For the fully active ring this term would be equal to $Y_a^E t^3/12$. Thus, if to denote

$$Y_{eqvap}(z_0) = 4\left\{\left[\left(\frac{t_a}{t}-\frac{z_0}{t}\right)^3 + \left(\frac{z_0}{t}\right)^3\right]Y_a + \left[\left(1-\frac{z_0}{t}\right)^3 + \left(\frac{z_0}{t}-\frac{t_a}{t}\right)^3\right]Y_p\right\}, \tag{7.269}$$

then expression (7.267) for the mechanical energy will look in the same way as for fully active ring with the only difference that Y_a^E will be replaced by Y_{eqvap}. Accordingly, formula (7.250) for the equivalent rigidity is valid, if to use value of Y_{eqvap} from expression (7.269) instead of $1/s_{ii}^E$, i.e.,

$$K^E_{eqvap} = K^E_{ma} \cdot \left(Y_{eqvap} / Y^E_a\right). \tag{7.270}$$

In the particular case that $z_0 = z_{0m} = t_a$ using relation (7.266) we obtain

$$Y_{eqvap}\left(z_{0m}\right) = Y^E_a \cdot \frac{4\gamma_Y}{\left(1+\sqrt{\gamma_Y}\right)^2}. \tag{7.271}$$

The kinetic energy of the ring is

$$W_{kin} = \frac{\dot{\xi}^2_{ro}}{2} 2ha\int_0^\pi \left[\rho_a t_a + \rho_p\left(t - t_a\right)\right]\theta_r^2 d\varphi = \frac{\dot{\xi}^2_{ro}}{2}\rho_{eqvap} t S_{eff} = \frac{\dot{\xi}^2_{ro}}{2} M_{eqvap}, \tag{7.272}$$

where

$$\rho_{eqvap} = \rho_a\left\{\left(\frac{t_a}{t}\right) + \gamma_\rho\left[1 - \left(\frac{t_a}{t}\right)\right]\right\} \tag{7.273}$$

is the equivalent density of the ring, and

$$S_{eff} = 2ha\int_0^\pi \theta_r^2 d\varphi \tag{7.274}$$

is the effective mean surface of the ring. Thus

$$M_{eqvap} = \rho_{eqvap} t S_{eff} = \left(\rho_{eqvap} / \rho_a\right) M_{eqva}. \tag{7.275}$$

In the case that $z_0 = z_{0m} = t_a$

$$\rho_{eqvap} = \rho_a \frac{1}{1+\sqrt{\gamma_Y}}\left(\sqrt{\gamma_Y} + \gamma_\rho\right). \tag{7.276}$$

Using expressions (7.270) and (7.275) for the equivalent rigidity and mass we arrive at the formula for the resonance frequency of the bilaminar ring

$$f_{rap}\left(z_0\right) = f_{ra}\sqrt{\frac{\rho_a}{Y^E_a}} \cdot \sqrt{\frac{Y^E_{eqvap}\left(z_0\right)}{\rho_{eqvap}\left(z_0\right)}}, \tag{7.277}$$

where f_{ra} is the resonance frequency of the fully active ring, and Y^E_{eqvap} and ρ_{eqvap} are given by expressions (7.269) and (7.273). In the case that $z_0 = z_{0m}$

$$f_{rap}\left(z_{0m}\right) = f_{ra}\frac{2\sqrt{\gamma_Y}}{\sqrt{\left(1+\sqrt{\gamma_Y}\right)\left[\sqrt{\gamma_Y} + \gamma_\rho\right]}}. \tag{7.278}$$

The electromechanical energy is

$$W_{em} = \frac{\xi_{ro}}{2} E_3 \frac{hd_{3i}}{as_{ii}^E} \int_0^{t_a} (z_0 - z)\,dz \int_0^{2\pi} \left(\frac{\partial^2 \theta_r}{\partial \varphi^2} + \theta_r \right) d\varphi = \frac{1}{2} \xi_{ro} V n_{ap} . \tag{7.279}$$

Comparing expressions for the transformation coefficients of bilaminar and fully active rings results in relations:

$$n_{ap}(z_0) = n_a \left(\frac{2z_0}{t} - \frac{t_a}{t} \right) \tag{7.280}$$

for the general case, and

$$n_{ap}(z_{0m}) = n_a \cdot \frac{t_a}{t} = n_a \cdot \frac{\sqrt{\gamma_Y}}{1+\sqrt{\gamma_Y}} \tag{7.281}$$

for the case that $z_0 = z_{0m}$.

Capacitances of the ring in these cases are

$$C_{el\,ap}^{S_\varphi}(z_0) = \varepsilon_{33}^{S_1} \frac{2\pi a h}{t_a} = C_{el\,a}^{S_1} \cdot \frac{t}{4t_a}, \tag{7.282}$$

and

$$C_{el\,ap}^{S_1}(z_{0m}) = C_{el\,a}^{S_1} \frac{1+\sqrt{\gamma_Y}}{4\sqrt{\gamma_Y}}, \tag{7.283}$$

respectively.

Thus, the nondimensional coefficient α_c, through which we express the effective coupling coefficient, is for the bilaminar ring

$$\alpha_{c\,ap}(z_0) = \frac{C_{el\,ap}^{S_1}(z_0)}{n_{ap}^2(z_0) C_{eqv\,ap}^E(z_0)} = \alpha_{ca} \frac{4t_a(2z_0 - t_a)^2}{t^3 \left(Y_{eqvap}^E / Y_a \right)} . \tag{7.284}$$

In the case that $z_0 = z_{0m}$

$$\alpha_{cap}(z_{0m}) = \alpha_{ca} \cdot \frac{\sqrt{\gamma_Y}}{1+\sqrt{\gamma_Y}} . \tag{7.285}$$

and the effective coupling coefficient is

$$k^2_{eff\,ap} = \frac{k^2_{eff\,a}\sqrt{\gamma_Y}}{1+\sqrt{\gamma_Y} - k^2_{eff\,a}}. \tag{7.286}$$

Thus, for combination of aluminum and PZT-4 ($Y_p = 70 \cdot 10^9\,Pa$, $Y_a = 81 \cdot 10^9\,Pa$ for transverse piezoeffect) $\gamma_Y = 0.86$, and $k_{eff\,ap} = 0.5k_{31}$ instead of $k_{eff\,a} = 0.71k_{31}$.

7.3.2.3 Acoustic Field Related Properties of the Transducers

Peculiarities of the acoustic properties of the slotted rings and their combinations are due to small wave sizes of their elementary units (technologically completed units that can be used as the single transducers). The situation is similar to those considered in Section 7.3.1.2 regarding the complete rings vibrating in flexure. The difference is that the slotted ring transducers have significantly (six times) lower resonance frequency at the same geometry as the complete ring vibrating in flexure has. Therefore, it is hard to imagine a column like mechanically solid slotted ring transducer that has the height comparable with operational wavelength, and more realistic it is to consider all the radiating systems configurations as combination of acoustically interacting simple sources. Assuming that the elementary units have comparable height and diameter they can be considered as simple three-dimensional sources, for which the radiation resistance and reactance can be represented by the relations (7.247) and (7.248), namely,

$$r_{ac} = \pi(\rho c)_w S^2_{av} / \lambda^2, \tag{7.287}$$

$$x_{ac} = 4\pi a^3_{eqv} \rho\omega, \tag{7.288}$$

where a_{eqv} may be determined from the condition $4\pi a^2_{eqv} = S_{av}$. The average surface area is by definition

$$S_{av} = 2ah_{unit}\int_0^{\pi}\theta_r(\varphi)\,d\varphi. \tag{7.289}$$

After substituting $\theta_r(\varphi)$ from expression (7.251), it will be obtained that $S_{av} = 0.62S_\Sigma$, where S_Σ is the total outer surface of a transducer. It is noteworthy that the average surface determined for the part of ring within segments $|\varphi| \le 150°$, i.e., up to the nodal lines, where displacement changes sign, is $S_{av}(\pm 150°) = 0.65S_\Sigma$. The 5% difference characterizes distractive contribution of the parts of the surface vibrating in anti-phase. The equivalent force for a transducer having small wave size is determined by the formula

$$F_{eqv} \approx S_{av} P_o \,. \tag{7.290}$$

In practice the low frequency transducers are composed of a number of elementary slotted rings. The most natural is the column like transducer design shown in Figure 7.75(a).

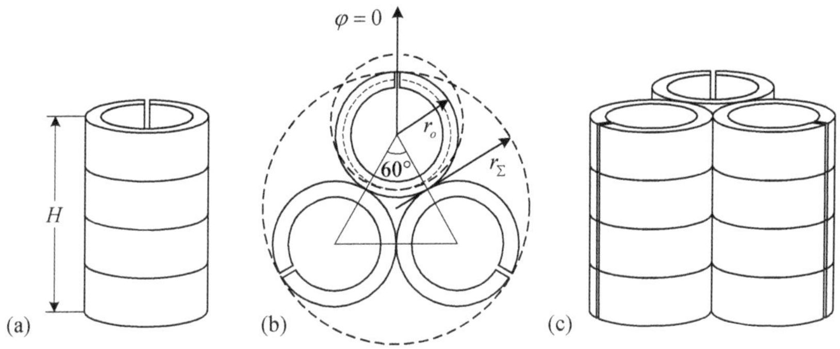

Figure 7.75: Variants of transducers composed of elementary slotted rings: (a) column like transducer, (b) three-ring assembly and (c) assembly of three column like transducers.

The radiation impedance of a combination of elementary slotted ring transducers can be determined considering acoustic interaction between them as between the simple sources by formula (6.356)

$$Z_{12}(d) = r_{11}(\sin kd + j\cos kd)/kd \,, \tag{7.291}$$

where $Z_{12}(d)$ is the mutual impedance between two sources separated by distance d, and r_{11} is the radiation resistance by formula (7.287). Thus, in a linear transducer made of three ring units (№ 1 in the center, 2 and 3 on the edges, separation between centers of the rings d_1, $Z_{12} = Z_{13}$) the radiation impedances of the rings can be calculated as

$$Z_1 = Z_{11} + 2Z_{12}(d_1),\ Z_2 = Z_3 = Z_{11} + Z_{12}(d_1) + Z_{13}(2d_1)\,. \tag{7.292}$$

Here $Z_{11} = r_{ac} + jx_{ac}$ is the self-impedance of a single ring unit with r_{ac} and x_{ac} according expressions (7.287) and (7.288). The radiation resistance per unit ring can be increased in the composition of slotted rings shown in Figure 7.75 (b). The three rings are tangential to each other at the nodal lines of the modes of vibration. The internal 60° segments of the ring that vibrate in anti-phase with the other parts are self-baffled and do not participate in outside radiation. Therefore, when determining average surface area by formula (7.196) integration should be performed over the sectors $0 \pm 150°$. This results in $S_{av}(150°) = 0.65 S_\Sigma$ as noted before.

From the geometry considerations follows that radius r_Σ of the circle inclosing the rings and tangential to their surfaces is $r_\Sigma = r_o\left(1+2/\sqrt{3}\right) \approx 2.15 r_o$, and this assembly still remains small enough for considering the transducer as a simple source. The acoustic power radiated by a simple source can be presented as $\dot{W}_{ac} = r_{ac}U_o^2 = \pi(\rho c)_w U_{\tilde{V}}^2 / \lambda^2$, where from the radiation resistance is $r_{ac} \sim U_{\tilde{V}}^2$. The total volume velocity generated by the three rings is $U_{\tilde{V}3} = 3 \cdot 0.65 S_\Sigma \cdot U_o$, whereas generated by a single ring is $U_{\tilde{V}1} = 0.62 S_\Sigma \cdot U_o$. Thus, the ratio of the radiation resistances of the three-ring assembly and of a single ring is

$$\frac{3r_{ac3}}{r_{ac1}} = \frac{U_{\tilde{V}3}^2}{U_{\tilde{V}1}^2} \approx 9.9, \tag{7.293}$$

i. e., the radiation resistance of a ring in assembly of three is $r_{ac3} = 3.3 r_{ac1}$ compared with radiation resistance of the same ring operating as a single. Calculating the radiation impedances in the column like transducer shown in Figure 7.75(c) can be performed applying formula (7.291) to a number of the three-ring simple sources.

Results of this section allow calculating operating characteristics of the transducers that employ flexural vibration of the slotted rings. Estimation of their operating properties as projectors will be made in Ch. 13 in comparison with properties of transducers of different types.

7.4 References

1. S. Sarangapani, D. A. Brown and B. Aronov, "Analysis of electromechanical parameters of thick rings under radial, axial and circumferential modes of polarization (A)," J. Acoust. Soc. Am., **132**(3), 2067 (2012).
2. C. V. Stephenson, "Radial vibrations in short, hollow cylinders of barium titanate," J. Acoust. Soc. Am., **28**(1), 51-56 (1956).
3. B. S. Aronov, C. Bachand, and D. A. Brown, "Analysis of parameters of stripe-electroded piezoelectric elements," J. Acoust. Soc. Am., **129**(5), 2960-2967, (2011).
4. S. Sarangapani and D. A. Brown, "Improved calculations of the electromechanical properties of tangentially poled stripe-electroded piezoelectric bars and cylinders with nonuniform electric fields," J. Acoust. Soc. of Am., **132**(5), 3068-3075 (2012).
5. C. R. Bowen, A. Bowles, S. Drake, N. Johnson, and S. Mahon, "Fabrication and finite element modelling of interdigitated electrodes," Ferroelectrics, **228**, 257-269 (1999).
6. C. R. Bowen, L. J. Nelson, R. Stevens, M. G. Cain, and M. Stewart, "Optimization of interdigitated electrodes for piezoelectric actuators and active fiber composites," J. Electroceram., **16**, 263-260 (2006).
7. B. Aronov, "Coupled vibration analysis of the thin-walled cylindrical piezoelectric ceramic transducers," J. Acoust. Soc. Am., **125**(2), 803-818 (2009).
8. B. S. Aronov, D. A. Brown, and S. Regmi, "Experimental investigation of coupled vibrations in piezoelectric cylindrical shells," J. Acoust. Soc. Am., **120**(3), 1374-1380 (2006).
9. D. A. Berlincourt, D. R. Curran, and H. Jaffe, *Piezoelectric and Piezomagnetic Materials and their Function in Transducers*, in Physical Acoustics, Vol. I, Part A, edited by W. P. Mason (Academic, New York, 1964).
10. W. J. Trott, "Sensitivity of piezoceramic tubes with capped or shielded ends above the omnidirectional frequency range," J. Acoust. Soc. Am., **62**, 565-568 (1977).
11. B. Aronov, D. A. Brown, and C. L. Bachand, "Effects of coupled vibrations on the acoustic performance of underwater cylindrical shell transducers," J. Acoust. Soc. Am., **129**(2), 582-584 (2011).
12. B. Aronov, D. A. Brown, and C. L. Bachand, "Effects of coupled vibrations on the parameters of stripe-electroded piezoelectric cylinders (L)," J. Acoust. Soc. Am., **129**(2), 582-584 (2011).
13. J. L. Butler, "Model for a ring transducer with inactive segments," J. Acoust. Soc. Am., **59**, 480-482 (1976).
14. R. S. Gordon, J. L. Butler, and L. Prasad, "Equivalent circuit of a ceramic ring transducer operated in the dipole mode," J. Acoust. Soc. Am., **58**, 1311-1314 (1975).
15. B. S. Aronov, "Calculation of first order cylindrical piezoceramic receivers," Sov. Phys. Acoust., **34**(5), (1988)

16. M. E. Higgins, "DIFAR system overview," in Proc. of the Workshop on Directional Acoustic Sensors, NUWC/ONR, April 2001; S. L. Erlich and P. D. Frelich, U. S. Patent #3,176,262: Directional Sonar Systems (March 30, 1965).
17. B. Hefner and B. Dzikowicz, "A spiral wave front beacon for underwater navigation: Basic concept and modeling," J. Acoust. Soc. Am. **129**(6), 3630–3639 (2011).
18. D. A. Brown, B. Aronov, and C. Bachand, "Cylindrical transducer for producing an acoustic spiral wave for underwater navigation (L)," J. Acoust. Soc. Am., **132**, 3611 (2012).
19. T. Oishi, B. Aronov, and D. A. Brown, "Broadband multimode baffled piezoelectric cylindrical shell transducer," J. Acoust. Soc. Am., **121**(6), 3465–3471 (2007).
20. J. L. Butler, A. L. Butler, and J. A. Rice, "A tri-modal directional transducer," J. Acoust. Soc. Am., **115**(2), 658-665 (2004).
21. B. S. Aronov, "Piezoelectric circular ring flexural transducers," J. Acoust. Soc. Am., **134**(2), 1021-1030 (2013).
22. B. S. Aronov and L. B. Nikitin, "Electromechanical waver transducer using shear strain," Sov. Phys. Acost., **35**(2), (1989).
23. V. E. Glazanov and A. V. Michailov, *Baffling of hydroacoustic transducers*, (Elmore Publishing House, St. Petersburg, Russia, 2004) (in Russian).
24. C. H. Sherman and J. L. Butler, *Transducers and Arrays for Underwater Sound* (Springer, New York, 2007), pp. 142–145.
25. J. C. Cochran, *A General Theory of the Slotted Cylinder Projector (SCP) Transducer*, Ph. D. Dissertation, Electrical Engineering, Univ. of Massachusetts, Dartmouth (2018).
26. L. D. Lubavin, "Acoustic transducer in the shape of a cylinder cut along its element," Voprosy Electroniki, Series XVII, Hydroacoustics (Sudostroenie, Leningrad, 1961) (in Russian), pp. 30–41.
27. B. S. Aronov, "Piezoelectric slotted ring transducer," J. Acoust. Soc. Am., **133**(6), 3875-3884 (2013).

CHAPTER 8

SPHERICAL TRANSDUCERS

8.1 Introduction

Piezoceramic spherical shell transducers are common in underwater acoustics predominantly as omnidirectional (zero mode) projectors and hydrophones, and less frequently as bidirectional (first order) hydrophones and multimode hydrophones. Spherical shell transducers have also been suggested in Refs. 1-3 as directional projectors by combining different modes of vibration, as discussed theoretically in Ref. 3, however the directionality is achieved in this case at the expense of significant reduction in bandwidth and efficiency of the transducer. Demand for moderately directional broadband transducers of small size is increasing with the growth of underwater acoustic communications, networks, and unmanned underwater vehicles and gliders.

A practical way of achieving the goal of simultaneously directional and broadband radiation with spherical shells is in employing the conformal baffles in the same manner, as it was illustrated with examples of cylindrical transducers in the previous Chapter. This approach was demonstrated in Ref. 8. Another possibility for achieving this goal is in the use of incomplete spherical shells, e.g., hemispheres and open spheres of various shapes for the transducers designing. The practical designing directional multimode spherical shell transducers for a broad range of applications requires a general treatment of such related problems as vibration of spherical shells, electromechanical excitation of their vibration and acoustic radiation by the vibrating spherical shells. The modal analysis of vibration of the complete and incomplete passive spherical shells was performed in Ref. 5 and in Section 4.5.5, and the modal characteristics of acoustic radiation by spheres including those with baffles were considered in Section 6.3.2. Objectives of this Chapter are in considering the electromechanical excitation of different modes of vibration of spherical shells made of piezoelectric ceramics, and in completing analytical analysis of the spherical transducers that may be intended for various applications.

8.2 Modes of Spherical Shell Vibrations

Consider an element of the spherical piezoceramic shell in the crystallographic coordinate system shown in Figure 8.1 (b). Following the common notations for the crystallographic axes with poling axis denoted as 3, we have correlations $S_{\varphi\varphi} = S_1$, $S_{\theta\theta} = S_2$; $T_{\varphi\varphi} = T_1$, $T_{\theta\theta} = T_2$, $T_r = T_3$. For a thin shell it is assumed that $T_3 = 0$, and the piezoelectric equations for element of the shell volume simplify to the following form

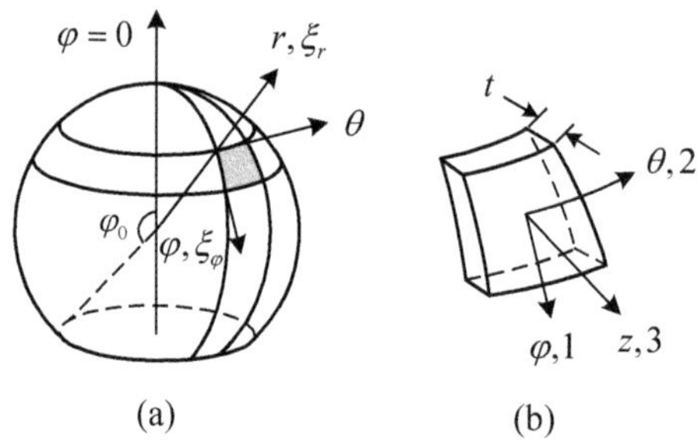

Figure 8.1: (a) Geometry of the spherical shell and (b) differential element of the shell and coordinate systems (geometrical and crystallographic) used.

$$S_1 = s_{11}^E T_1 + s_{12}^E T_2 + d_{31} E_3 , \tag{8.1}$$

$$S_2 = s_{12}^E T_1 + s_{22}^E T_2 + d_{31} E_3 , \tag{8.2}$$

$$D_3 = d_{31}(T_1 + T_2) + \varepsilon_{33}^T E_3 . \tag{8.3}$$

Substituting the stresses T_1 and T_2 from Eqs. (8.1) and (8.2) into Eq. (8.3) yields

$$D_3 = D_e^{S_{1,2}} + D_3^E . \tag{8.4}$$

Components of the charge density are introduced here as follows:

$$D_e^{S_{1,2}} = \varepsilon_{33}^{S_{1,2}} E_3 , \tag{8.5}$$

where $\varepsilon_{33}^{S_{1,2}} = \varepsilon_{33}^T(1 - k_p^2)$ and $k_p^2 = 2d_{31}^2 / [\varepsilon_{33}^T s_{11}^E (1 + s_{12}^E / s_{11}^E)]$ are the dielectric constant and the square of the planar coupling coefficient of a piezoceramic material in the clamped sphere (superscript $S_{1,2}$ indicates that the deformations S_1 and S_2 are set to zero). The term

$$D_3^E = D_{em}(S_1, S_2) = \frac{d_{31}}{s_{11}^E + s_{12}^E}(S_1 + S_2) \tag{8.6}$$

is the charge density induced by the deformations at $E_3 = 0$.

The stresses in a sphere at $E_3 = 0$ are found from Eqs. (8.1) and (8.2) as

$$T_1^E = \frac{Y_1^E}{1-(\sigma_1^E)^2}(S_1 + \sigma_1^E S_2)\,, \tag{8.7}$$

$$T_2^E = \frac{Y_1^E}{1-(\sigma_1^E)^2}(\sigma_1^E S_1 + S_2)\,. \tag{8.8}$$

Eqs. (8.7) and (8.8) are identical with Eqs. (4.295), if Y and σ are replaced with Y_1^E and σ_1^E. Therefore, all the results obtained for the isotropic passive elastic shells in terms of their modal equivalent parameters that characterize mechanical behavior of the shells are completely applicable to the thin-walled thickness poled piezoelectric shells. It is noteworthy that in the numerical examples the values of resonance frequency related factor, Ω, may slightly deviate from those presented in Table 4.1 due to difference between values of σ_1^E for a particular ceramic composition and $\sigma = 0.3$ that is used in calculations for the Table, and for the bending modes due to change of thickness to radius ratio compared with $t/a = 0.1$.

Consider now parameters related to the electromechanical conversion in the vibrating spherical shells. The electric energy supplied to the piezoelectric shell under the condition that it is clamped is

$$W_{el}^{S_{1,2}} = \frac{1}{2}\int_{\tilde{V}} D_e^{S_{1,2}} E_3 d\tilde{V} = \frac{1}{2}\int_{\tilde{V}} \varepsilon_{33}^{S_{1,2}} E_3^2 d\tilde{V} = \frac{1}{2} C_e^{S_{1,2}} V^2\,, \tag{8.9}$$

where $C_e^{S_{1,2}}$ is the electric capacitance of a clamped sphere. Its value depends on configuration of electrodes.

The electromechanical energy is

$$W_{em} = \frac{1}{2}\int_{\tilde{V}} D_{em}(S_1, S_2)\, E_3\, d\tilde{V} = \frac{1}{2}\frac{d_{31}}{s_{11}^E + s_{12}^E}\int_{\tilde{V}} (S_1 + S_2)\, E_3\, d\tilde{V}\,. \tag{8.10}$$

Note that elementary volume $d\tilde{V}$ of the spherical shell is $d\tilde{V} = r^2 \sin\varphi dr d\varphi\theta$. Let us place the origin of the axis z of the crystallographic coordinate system on the middle surface of the shell. Given that radius of the middle surface is a and the thickness of a shell is small compared with its radius, the following manipulations can be made: $r = a + z$, $dr = dz$, $r^2 \approx a^2 + 2az$, and as the result

$$d\tilde{V} \approx (a^2 + 2az)\sin\varphi dz d\varphi d\theta\,. \tag{8.11}$$

Integration by φ and z has to be performed within intervals $0 \le \varphi \le \varphi_0$, where φ_0 is the opening angle of the sphere, and $-t/2 \le z \le t/2$. When representing expression for S_1+S_2 under the integral, it must be remembered that according to formulas (4.294)

$$S_1 = S_\varphi + z\chi_\varphi, \; S_2 = S_\theta + z\chi_\theta, \tag{8.12}$$

where S_φ and S_θ are the strains of the middle surface of the sphere ("membrane" strains), and χ_φ, χ_θ are the changing of curvature of the middle surface in directions of meridian and azimuth. Thus,

$$S_1 + S_2 = (S_\varphi + S_\theta) + z(\chi_\theta + \chi_\varphi). \tag{8.13}$$

The strains and curvatures can be expressed through displacements $\xi_r(\varphi)$ and $\xi_\varphi(\varphi)$ of the middle surface according to relations (4.292) and (4.293) as

$$S_\varphi = \frac{1}{a}\left[\xi'_\varphi(\varphi) + \xi_r(\varphi)\right], \quad S_\theta = \frac{1}{a}\left[\xi_\varphi(\varphi)\cot\varphi + \xi_r(\varphi)\right], \tag{8.14}$$

$$\chi_\varphi = \frac{1}{a^2}\left[\xi'_\varphi(\varphi) - \xi''_r(\varphi)\right], \quad \chi_\theta = \frac{1}{a^2}\cot\varphi\left[\xi_\varphi(\varphi) - \xi'_r(\varphi)\right]. \tag{8.15}$$

We will denote for brevity

$$S_\varphi + S_\theta = \frac{1}{a}[2\xi_r(\varphi) + \xi_r{}'(\varphi) + \xi_\varphi(\varphi)\cot\varphi] = \frac{1}{a}A(\varphi), \tag{8.16}$$

$$\chi_\varphi + \chi_\theta = \frac{1}{a^2}\{[\xi_\varphi(\varphi) - \xi_r{}'(\varphi)]\cot\varphi + [\xi_\varphi{}'(\varphi) - \xi_r{}''(\varphi)]\} = \frac{1}{a^2}B(\varphi). \tag{8.17}$$

The terms $A(\varphi)$ and $B(\varphi)$ will be marked as $A_i(\varphi)$ and $B_i(\varphi)$ for the ith mode of vibration. The displacements $\xi_r(\varphi)$ and $\xi_\varphi(\varphi)$ are represented by series (4.299), namely,

$$\xi_r(\varphi) = \sum_{i=0}^{\infty}\xi_{ri}P_i(\cos\varphi), \quad \xi_\varphi(\varphi) = \sum_{i=1}^{\infty}\xi_{\varphi i}P_i'(\cos\varphi). \tag{8.18}$$

The modal displacements ξ_{ri} and $\xi_{\varphi i}$ are related according to (4.300) as

$$\xi_{ri}(\varphi) = \xi_{ri}P_i(\cos\varphi), \quad \xi_{\varphi i}(\varphi) = -\xi_{ri}(1+\sigma)C_iP_i{}'(\cos\varphi), \quad i = 1,2,3,\ldots, \tag{8.19}$$

where coefficients C_i are presented in Table 4.1. Thus, the radial displacements ξ_{ri} only can be taken for the generalized coordinates. Subscript r will be further omitted for brevity.

The electric field in a thin shell can be represented as

$$E_3 = \frac{V}{t}\Omega(\varphi), \tag{8.20}$$

where $\Omega(\varphi)$ is a function of the geometry of electrodes (a small change of electric field through the thickness in a thin-walled shell is neglected). Function $\Omega(\varphi)=0$ on the parts of surface deprived of electrodes. Otherwise, $\Omega(\varphi)=1$, if the pats of electrodes are unipolar, or it has to be changed accordingly, if the parts are connected in different polarities or/and in series.

Upon substituting expressions (8.13)-(8.17) under the integral (8.10) and integration over the volume of the shell, we arrive at the following expression for the modal electromechanical energy of a spherical shell

$$W_{em\,i}=\frac{1}{2}\frac{2\pi a d_{31}}{s_{11}^E+s_{12}^E}V\int_0^{\varphi_0}\left[A_i(\varphi)+\frac{t^2}{6a^2}B_i(\varphi)\right]\Omega(\varphi)\sin\varphi\,d\varphi=\frac{1}{2}Vn_i\xi_i\,, \tag{8.21}$$

where n_i is the modal coefficient of electromechanical transformation. The first term in the brackets is due to the extensional deformations of the shell, and it constitutes the main part of the electromechanical energy. The second term is due to flexural deformations of the shell. Under the membrane theory approach (at $t/a\to 0$) this flexural term vanishes. For the relative thicknesses that are typical of transducer applications ($t/a<0.2$) this term accounts for a small contribution to the electromechanical conversion of the bending modes of the shell vibration. The total electromechanical energy for an arbitrary vibrating spherical shell can be represented as

$$W_{em}=\sum_i W_{em\,i}=\frac{V}{2}\sum_i n_i\xi_i\,. \tag{8.22}$$

The electromechanical conversion in the complete and incomplete spherical shells will be considered separately.

8.2.1 Electromechanical Conversion in the Complete Spherical Shells

Upon substituting the displacements (8.19) into expressions (8.16) and (8.17) the terms $A_i(\varphi)$ and $B_i(\varphi)$ that correspond to i^{th} mode of vibration for the complete spherical shell become

$$A_i(\varphi)=\xi_i\{2P_i(\cos\varphi)-(1+\sigma)C_i[\cot\varphi\,P_i'(\cos\varphi)+P_i''(\cos\varphi)]\}\,, \tag{8.23}$$

$$B_i(\varphi)=-\xi_i[1+(1+\sigma)C_i][\cot\varphi\,P_i'(\cos\varphi)+P_i''(cos\varphi)]\,, \tag{8.24}$$

where i = 0, 1; $2m$, $2b$; $3m$, $3b$; after differentiating with respect to φ, denoting $\cos\varphi=x$ and using Eq. (4.298) the expression in brackets may be transformed as follows

$$\cot\varphi P_i'(\cos\varphi) + P_i''(\cos\varphi) = (1-x^2)\frac{d^2P_i(x)}{d^2x} - 2x\frac{dP_i(x)}{dx} = -i(i+1)P_i(x). \tag{8.25}$$

Taking into account relation (8.25), the expressions (8.23) and (8.24) become

$$A_i(x) = \xi_i[2 + i(i+1)(1+\sigma)C_i]P_i(x), \tag{8.26}$$

$$B_i(x) = \xi_i\, i(i+1)[1 + (1+\sigma)C_i]P_i(x). \tag{8.27}$$

Thus, the Eq. (8.21) for the modal electromechanical energy W_{emi} in the case of the complete spherical shell (at $\varphi_0 = \pi$) may be represented as

$$W_{emi} = \frac{1}{2}\frac{2\pi a d_{31}}{s_{11}^E + s_{12}^E}V\int_1^{-1}\left[A_i(x) + \frac{t^2}{6a^2}B_i(x)\right]\Omega(x)\,dx. \tag{8.28}$$

Taking into account expressions for $A_i(x)$ and $B_i(x)$, after referring to Eq. (8.21) we arrive at the following expression for the modal coefficients of electromechanical transformation, n_i,

$$n_i = \frac{2\pi a d_{31}}{s_{11}^E + s_{12}^E}\left\{[2 + i(i+1)(1+\sigma)C_i] + \frac{t^2}{6a^2}i(i+1)[1 + (1+\sigma)C_i]\right\}\times \\ \times\int_1^{-1}P_i(x)\Omega(x)dx \tag{8.29}$$

For the zero mode of vibration (at $i = 0$) the electromechanical transformation coefficient does not depend on coefficient C (as well as other equivalent parameters, which is noted in the Table 4.1). As it follows from Table 4.1, for the first mode $C_1 = 1/2(1+\sigma)$ and for the higher modes at $i \geq 2$ the coefficient C_i has two values: larger C_{im} and smaller C_{ib}. The large value corresponds to the membrane mode of vibration having higher resonance frequency, and the smaller corresponds to the bending mode having significantly lower resonance frequency. Respectively, the electromechanical transformation coefficients for these modes, n_{im} and n_{ib}, have different values. (Remember that the membrane and bending modes that have the same number *i* have the same distribution of radial displacements).

Note that the bending modes do not have useful applications. In opposite, they produce corruption of the frequency responses that correspond to active membrane modes, because their resonances appear within operating frequency ranges (predominantly at their lower parts). Therefore, it is desirable to take measures, if possible, for suppressing or diminishing effect of electromechanical conversion of the bending modes by rational designing the electrodes.

Numerical values of the modal transformation coefficients critically depend on the form of the function $\Omega(x)$, i.e., on the geometry and electrical connection of the parts of electrodes applied to the shell. In the variant that the unipolar electrodes cover the entire outer and inner surfaces of the shell, as shown in Figure 8.2 (a), $\Omega(x)=1$. Since $P_0(x)=1$ and for $i\geq 1$

$$\int_{1}^{-1} P_i(x)\,dx = 0\,, \tag{8.30}$$

we conclude that the only active mode of vibration is zero or "breathing" mode with $\xi_r(\varphi)=\xi_{r0}$ and the corresponding electromechanical transformation coefficient is

$$n_0 = \frac{8\pi a d_{31}}{s_{11}^E + s_{12}^E}\,. \tag{8.31}$$

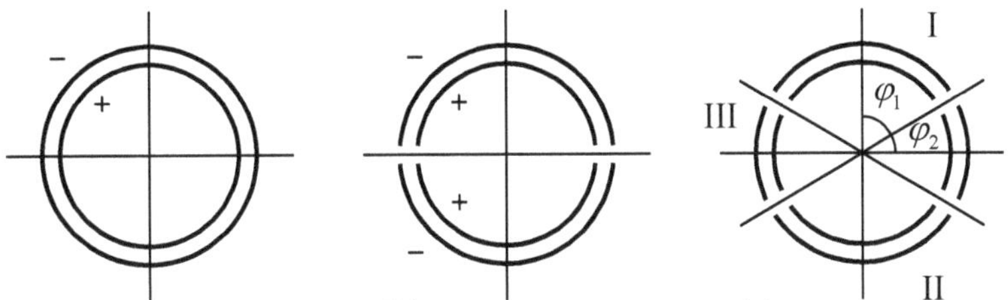

Figure 8.2: Illustration of different electrode configurations: (a) unipolar electrodes, (b) bipolar electrodes, (c) variable angle φ_1 electrode division.

If the electrodes are split into two halves and the halves are connected in opposite phase ("bipolar" electrodes connection) as is shown in Figure 8.2 (b), then $\Omega(x)=1$ at $0<x\leq 1$ and $\Omega(x)=-1$ at $0>x\geq -1$. According to the properties of Legendre polynomials (see Appendix C.3)

$$\begin{gathered}\int_0^1 P_{2l+2}(x)dx = 0,\\ \int_0^1 P_{2l+1}(x)\,dx = \int_0^{-1} P_{2l+1}(x)\,dx = \frac{(-2l+1)(-2l+3)\cdots(-3)(-1)}{(2l+2)\cdot 2l\cdots 4\cdot 2},\end{gathered} \tag{8.32}$$

(here l = 0, 1, 2, …) all the even modes are inactive. For the odd modes the values of integral in (8.29), which we will denote as I_i, are

$$I_1 = 1,\quad I_3 = 1/4,\quad I_5 = 1/8,\ldots \tag{8.33}$$

Thus, for the electromechanical transformation coefficients of the first and $3m$ modes from Eq. (8.29) we arrive at

$$n_1 = -\frac{6\pi a d_{31}}{s_{11}^E + s_{12}^E}, \quad n_{3m} = \frac{4.2\pi a d_{31}}{s_{11}^E + s_{12}^E}. \tag{8.34}$$

The sign of electromechanical transformation coefficient must be taken into consideration only if several modes of vibration are employed simultaneously. The convention is that on the part of the electrode, where $\varphi > 0$ (at $0 < \varphi \le \pi/2$ in the variant of the "bipolar" connection), the direction of operating electric field should coincide with the direction of polarization. The next closest active is the membrane mode $3m$ with resonance frequency $f_{3m} \approx 1.9 f_1$, which is accompanied by the bending mode $3b$ with resonance frequency $f_{3b} \approx 0.44 f_1$. Contribution of the membrane mode $5m$ with resonance frequency $f_{5m} \approx 3.0 f_1$ that is far beyond the range of interest can be neglected. But the bending mode $5b$ having resonance frequency $f_{5b} \approx 0.6 f_1$ may produce unwanted effect on the operating characteristics of transducer at low frequencies and has to be considered as well as the mode $3b$. It follows from Eq. (8.29) that electromechanical transformation coefficients for the bending modes are

$$n_{3b} = \frac{0.25\pi a d_{31}}{s_{11}^E + s_{12}^E}, \quad n_{5b} = \frac{0.08\pi a d_{31}}{s_{11}^E + s_{12}^E}. \tag{8.35}$$

In the variant that only one half of the electrode is used, $\Omega(x) = 0$ at $0 < x \le 1$ or at $0 > x \ge -1$ and both the zero and first modes of vibration are active together with other odd modes, while the even modes remain inactive. The magnitudes of the corresponding transformation coefficients will be reduced by a factor of 2. The sign of n_1 changes depending on which half of the electrode is used. It is positive, if the electrodes are facing in the direction of $\varphi = 0$, and negative for the opposite half of the electrodes. The zero and the first modes of the spherical shell vibration (with resonance frequencies f_0 and f_1) and their possible combinations are the most practical and effective for transducer applications.

Although using the higher order modes of vibrations is generally less practical, employing the membrane modes of higher order for the unidirectional multi resonance transducer operation may have applications, when part of the shell surface is baffled accordingly. Whereas the third mode is automatically generated together with the first mode at the bipolar full-size electrode connection, generating the second mode requires a special electrode configuration.

Consider the configurations of electrodes shown in Figure 8.2 (c). If the electrodes I and II are used in the unipolar connection, then $\Omega(x)=1$ at $0<x\leq\cos\varphi_1$ and at $-\cos\varphi_1>x\geq-1$, and $\Omega(x)=0$ elsewhere. Given that $P_2(x)=0.5(3x^2-1)$, the integral in Eq. (8.29) becomes

$$I_2(\varphi_1)=-\cos\varphi_1\sin^2\varphi_1\,. \tag{8.36}$$

Thus, for the transformation coefficient of the second membrane mode we obtain from Eq. (8.29) (at $t/a<0.2$),

$$n_{2m}(\varphi_1)=-\frac{11.2\pi a d_{31}}{s_{11}^E+s_{12}^E}\cos\varphi_1\sin^2\varphi_1\,. \tag{8.37}$$

The function $|I_2(\varphi_1)|$ has maximum at $\varphi_1=54.7°$. At this angle $|I_{2\max}(\varphi_1)|=0.38$ and $n_{2m}(54.7°)=-4.3\pi a d_{31}/(s_{11}^E+s_{12}^E)$. This outcome could be predicted using results presented in Ref. 6 and qualitatively estimated by observing the plots of the mode shapes in Figure 8.3.

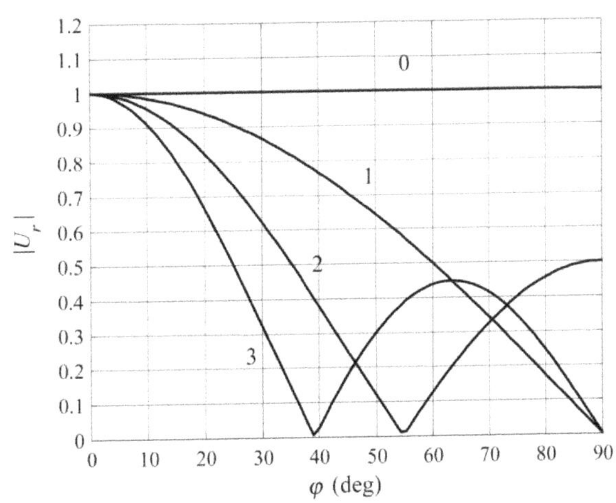

Figure 8.3: The modal radial velocity distributions for the complete spherical shells, $P_n(\cos\varphi)$, $n=0, 1, 2, 3$.

The value of integral $I_2(\varphi_1)$ is proportional to the area under the curve corresponding to the mode of interest within interval of angles $[0,\ \varphi_1]$ covered by the electrodes. (Note that $P_2(x)=0$ at $x=1/\sqrt{3}$, which corresponds to $\varphi_1=54.7°$.) If the pairs of electrodes I and III are connected in opposite phase ($\Omega(x)=1$ at $0<x\leq\cos\varphi_1$, and $\Omega(x)=-1$ at $-\cos\varphi_1>x\geq-1$), the integral $I_2(\varphi_1)=0$ and the second mode remains inactive.

If the electrodes have configuration III shown in Figure 8.2 (c), $\Omega(x)=1$ at $-\sin\varphi_2 \le x \le \sin\varphi_2$ and it is zero elsewhere. In this case

$$I_2(\varphi_2) = \sin\varphi_2 \cos^2\varphi_2 \tag{8.38}$$

and

$$n_{2m}(\varphi_2) = \frac{11.2\pi a d_{31}}{s_{11}^E + s_{12}^E} \sin\varphi_2 \cos\varphi_2^2 . \tag{8.39}$$

If $\varphi_2 = 90° - \varphi_1$, then $I_2(\varphi_2) = -I_2(\varphi_1)$. Thus, if the electrodes on the inner and outer surfaces of the spherical shell are split along the meridian lines at angles 54.7° and 126.7° and the pairs of electrodes of groups I and III are connected in phase and together connected to group II in opposite phase (all the connections being in parallel), then the maximum transformation coefficient for the second mode will reach the value

$$n_{2m}(\varphi_1, \varphi_2)_{\max} = -\frac{8.6\pi a d_{31}}{(s_{11}^E + s_{12}^E)} . \tag{8.40}$$

The notation $n_{2m}(\varphi_1, \varphi_2)$ is introduced to distinguish the electromechanical transformation coefficient for this combination of electrodes from those related to the cases that only electrodes I and II or electrodes III are used. The latter are denoted $n_{2m}(\varphi_1)$ and $n_{2m}(\varphi_2)$. For the bending mode, which accompanies the second membrane mode at the same electrodes configuration, from Eq. (8.29) follows that

$$n_{2b}(\varphi_1, \varphi_2)_{\max} = 0.29\pi a d_{31} / (s_{11}^E + s_{12}^E) . \tag{8.41}$$

As we can see, the three consecutive modes of vibration of the spherical shell may be generated by combining the parts of electrodes shown in Figure 8.42 (c): zero mode, if all the parts are connected in parallel in the same polarity (I+II+III); first mode, if the parts I and II are connected in the opposite phase (I-II); second mode, if the parts I and II are connected in phase and in the opposite phase to part III (I+II-III).

Determining the best position for dividing the electrodes is a matter of evaluating the desirable contribution of the modes of vibration. While being optimal for generating the second mode, this position is not favorable for generating the first mode. To find a possible compromise between the electromechanical activities of these modes, dependence of the transformation coefficient for the first mode from the electrode size must be considered. It is also important to investigate the effective coupling coefficients of the transducer for all the modes and their dependence on the electrode geometry to be able to make a balanced choice. With this goal we consider the coefficient

$$\alpha_{ci} = n_i^2 / K_{eqvi}^E C_e^{S_{1,2}} \tag{8.42}$$

that is related to the effective coupling coefficient as $k_{eff}^2 = \alpha_{ci} / (1+\alpha_{ci})$, with which they reach maximum values simultaneously. Expressions for the transformation coefficient n_1 and capacitance $C_e^{S_{1,2}}$ of the transducer in the case that the electrodes cover axially symmetric segments $0 \le \varphi \le \varphi_1$ around both poles and are connected in parallel in the opposite phase, can be obtained from expressions (8.29) and (8.9), in which integration has to be performed over the surface of the segments, in the form

$$n_1(\varphi_1) = \frac{6\pi a d_{31}}{s_{11}^E + s_{12}^E} \sin^2 \varphi_1, \tag{8.43}$$

$$C_e^{S_{1,2}}(\varphi_1) = (4\pi a^2 \varepsilon_{33}^{S_{1,2}} / t)(1-\cos\varphi_1). \tag{8.44}$$

The equivalent rigidities K_{eqvi}^E may be obtained from Table 4.1 after replacing Y and σ by Y_1^E and σ_1^E. These quantities do not change depending on the angle φ_1, if the remaining parts of the electrodes are retained and are electrically short-circuited. They change insignificantly if those electrodes are removed. After substituting K_{eqv1}^E and expressions (8.43), (8.44) for n_1 and $C_e^{S_{1,2}}$, we arrive at

$$\alpha_{c1}(\varphi_1) = \frac{3}{4}\frac{k_p^2}{1-k_p^2}\frac{\sin^4\varphi_1}{1-\cos\varphi_1}. \tag{8.45}$$

In the variant of the full-size electrodes in bipolar connection $\alpha_{c1}(90°) = 0.75k_p^2/(1-k_p^2)$

The corresponding effective coupling coefficient is $k_{eff1} \approx 0.90\,k_p$, if PZT-4 ceramics is used having $k_p = 0.58$. Function $\alpha_{c1}(\varphi_1)$ has maximum at the angle $\varphi_1 = 70.5°$, which can be found from the condition $[\alpha_{c1}(\varphi_1)]' = 0$. This results in $\alpha_{c1}(70.5°) = 0.89k_p^2/(1-k_p^2)$. The corresponding effective coupling coefficient is $k_{eff1\max} \approx 0.96\,k_p$. The function $\alpha_{c1}(\varphi_1)$ normalized to $\alpha_{c1}(90°)$ is depicted in Figure 8.4 (curve 1). It is of note that the increase of k_{eff1} occurs as a result of disproportional reduction of the electromechanical transformation coefficient and capacitance, $n_1(\varphi_1)$ and $C_e^{S_{1,2}}(\varphi_1)$, which is more beneficial for operating in receive rather than transmit mode. Dependencies of these functions from the angle φ_1 are shown in Figure 8.4 as well.

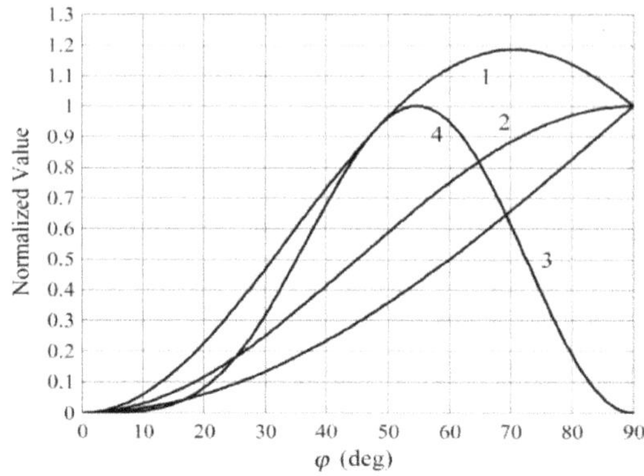

Figure 8.4: Plots of normalized functions related to optimizing the electromechanical conversion as a function of angle: (1) function $\alpha_{c1}(\varphi)$ normalized to $\alpha_{c1}(90°)$, (2) function $n_1(\varphi)$ normalized to $n_1(90°)$, (3) the capacitance $C_e^{S_{1,2}}(\varphi)$ normalized to $C_e^{S_{1,2}}(90°)$, (4) function $\alpha_{c2m}(\varphi_1, \varphi_2)$ normalized to its maximum value.

Considering that the electrode configuration I+II-III is used for generating the second mode, the integral in Eq. (8.29) for $n_{2m}(\varphi_1, \varphi_2)$ may be found as $I_2(\varphi_1) - I_2(90^o - \varphi_1)$. As the result of this manipulation, we arrive at the expression

$$n_{2m}(\varphi_1, \varphi_2) = -\frac{22.4\pi a d_{31}}{s_{11}^E + s_{12}^E} \cos\varphi_1 \sin^2\varphi_1 \,. \tag{8.46}$$

The capacitance for the full size electrodes is

$$C_e^{S_{1,2}}(\varphi_1) = 4\pi a^2 \varepsilon_{33}^{S_{1,2}} / t \,. \tag{8.47}$$

Note that for the assumed electrode connection $C_e^{S_{1,2}}$ and n_{2m} are determined under the condition that electrodes cover all the surface of a sphere. In practical transducer designs the separation between electrodes having different signs must be made of finite width to ensure the electrical strength of the transducer under applied operating voltage. Therefore, strictly speaking, these quantities must be calculated accordingly. With K_{eqv2m}^E taken from Table 4.1 we obtain

$$\alpha_{c2m}(\varphi_1,\varphi_2) = 3.5(1+\sigma_1^E)\frac{k_p^2}{1-k_p^2}\cos^2\varphi_1 \sin^4\varphi_1 \tag{8.48}$$

This function has maximum, $\alpha_{c2m}(\varphi_1,\varphi_2)_{\max} = 0.52(1+\sigma_1^E)k_p^2/(1-k_p^2)$, at angle $\varphi_1 = 54.7°$. For the sphere made of PZT-4 $k_{eff\,2m} \approx 0.79\,k_p$. The function (8.48) normalized to its maximum value is presented in Figure 8.4 (curve 4).

The resonance frequencies of the bending modes, which are labeled as modes $2b$, $3b$, $5b$ and so on, may be situated within the operating ranges of the transducers employing the membrane modes. The presence of the unwanted bending modes may corrupt the operating frequency responses and directional factors of the transducers. Therefore, an estimation of the electromechanical conversion related parameters for the bending modes, and their behavior in the course of optimizing parameters of transducers by altering the electrodes geometry is of interest. Thus, maximizing the coupling coefficient for a certain mode may be achieved by removing electrodes from those parts of the spherical shell, in which the strains are relatively small. For example, for the first mode the electrodes can be removed from some segment around the line $\varphi = \pi/2$, which brings us to the configuration of electrodes I-II (bipolar connection at the reduced electrodes) illustrated in Figure 8.2 (c) and to the expression (8.45) for $\alpha_1(\varphi_1)$.

The bending modes limit the operating range, in which the first mode of vibration dominates, at frequencies below its resonance. At frequencies above the resonance the operating range is limited by interference on the side of the high membrane modes, the most influential of which is the closest $3m$ mode. To avoid corruption of the operating characteristics in a broad frequency range, both the bending and $3m$ membrane modes of vibration must be suppressed. As it follows from Eq. (8.29), this can be achieved by determining the size of electrodes (angle φ_1) from condition that

$$\int_1^{x_1} P_3(x)\,dx = 0, \tag{8.49}$$

where $x_1 = \cos\varphi_1$. Given that $P_3(x) = (5x^3 - 3x)/2$, this condition is equivalent to

$$5\cos^4\varphi_1 - 6\cos^2\varphi_1 + 1 = 0, \tag{8.50}$$

where from $\varphi_1 = 63.5^o$. With electrodes of this size $n_{3m} = n_{3b} = 0$, whereas according to Eq. (8.43) $n_1 = 4.8\pi a d_{31}/(s_{11}^E + s_{12}^E)$. This feature is especially useful for the bipolar transducers intended for producing dipole directional characteristics in a broad frequency range. The effective coupling coefficient for the first mode at this electrode size slightly deviates from its maximum value at $\varphi_1 = 70.5^o$, but this reduction is negligible, as it follows from curve 1 in Figure 8.4.

The general expressions of transformation coefficients for the modes of vibration at electrodes size reduced to an angle φ_1 can be obtained following Eq. (8.29). In addition to formula (8.43) for $n_1(\varphi_1)$ they are:

$$n_{3m}(\varphi_1) = n_{3m}(\pi/2)\cdot(5\cos^4\varphi_1 - 6\cos^2\varphi_1 + 1)\,, \tag{8.51}$$

$$n_{3b}(\varphi_1) = n_{3b}(\pi/2)\cdot(5\cos^4\varphi_1 - 6\cos^2\varphi_1 + 1)\,, \tag{8.52}$$

$$n_{5b}(\varphi_1) = n_{5b}(\pi/2)\cdot(21\cos^6\varphi_1 - 35\cos^4\varphi_1 + 15\cos^2\varphi_1 + 1)\,. \tag{8.53}$$

Here $n(\pi/2)$ are the values that correspond to the full-size electrodes.

8.2.2 Electromechanical Conversion in the Incomplete Spherical Shells

The modal displacements of the incomplete spherical shells are given in general by expressions (4.306). The transformation coefficient can be obtained upon determining the functions $A(\varphi)$ and $B(\varphi)$ from expressions (8.17), (8.18) after their substituting under the integral in (8.28). The most usable out of incomplete spherical shells is the hemisphere with free boundary. It is shown in Section 4.5.5.3 that spectrum of resonance frequencies of the hemisphere is the same as of the complete sphere vibrating in the odd modes. Thus, the closest membrane frequencies are f_{1m} and $f_{3m} \approx 1.9 f_{1m}$. Frequency f_{1m} is preceded by the frequencies f_{3b} and f_{5b} that belong to bending modes of vibration that practically coincide with those for the complete sphere. And distributions of displacements for the first membrane and first bending (labeled as $3b$) modes are

$$\xi_{rm}(\varphi) \cong \xi_r P_1(\cos\varphi), \quad \xi_{\varphi m}(\varphi) \cong -0.39(1+\sigma)\xi_r P_1'(\cos\varphi)\,; \tag{8.54}$$

$$\xi_{rb}(\varphi) \cong \xi_r P_3(\cos\varphi), \quad \xi_{\varphi b}(\varphi) \cong 0.1(1+\sigma)\xi_r P_3'(\cos\varphi)\,. \tag{8.55}$$

Following the above procedure of calculating the electromechanical transformation coefficients for these modes, it will be obtained that they are equal to half of those for the complete sphere at the bipolar electrode connection, $n(\pi/2) = 0.5n(\pi)$. Taking into consideration that the equivalent rigidity of hemisphere is half of rigidity of complete sphere, $K_{eqv}^{E}(\pi/2)=0.5K_{eqv}^{E}(\pi)$, as shown in Section 4.5.5.3, and the capacitance is half the capacitance of complete sphere, $C_{el}^{S_{1,2}}(\pi/2)=0.5C_{el}^{S_{1,2}}(\pi)$, we will come to conclusion that the coefficient $\alpha_c = n^2/C_{el}^{S_{1,2}}K_{eqv}^{E}$ and hence the effective coupling coefficient remain the same as for the complete sphere,

$$\alpha_c(\pi/2)=[n(\pi/2)]^2/[C_{el}^{S_{1,2}}(\pi/2)]\cdot[K_{eqv}^{E}(\pi/2)]=\alpha_c(\pi),$$
$$k_{eff}^{2}(\pi/2)=k_{eff}^{2}(\pi). \tag{8.56}$$

The considerations for optimizing the coupling coefficient for the first mode of the complete sphere are also applicable to the hemisphere. Thus, it reaches maximum at $\varphi_1 = 70.5°$.

As one more example of incomplete spherical shell that may have application for unidirectional radiation, consider the spherical segment with opening angle $\varphi_0 = \pi/3$. The resonance frequency of the first membrane mode is $f(\pi/3)\approx 1.5f_0$, where f_0 is the resonance frequency of complete sphere of the same radius. Two bending modes exist that have resonance frequencies at 0.37 and 0.55 of $f(\pi/3)$. Following Section 4.5.5.3 the simplified expressions for displacements in the first membrane and the lowest bending modes are

$$\xi_{rm}(\varphi)=\xi_r P_{1.6}(\cos\varphi),\quad \xi_{\varphi m}(\varphi)=-0.45(1+\sigma)\xi_r P'_{1.6}(\cos\varphi); \tag{8.57}$$

$$\xi_{rb}(\varphi)=\xi_r P_{3.6}(\cos\varphi),\quad \xi_{\varphi b}(\varphi)=-0.07(1+\sigma)\xi_r P'_{3.6}(\cos\varphi). \tag{8.58}$$

The values of the electromechanical transformation and of the effective coupling coefficients for the first membrane mode of the hemisphere and of the segment at $\varphi_0 = \pi/3$ are as follows (the numerical values are for PZT-4 ceramics:

$$n_{1m}(\pi/2)=\frac{3\pi a d_{31}}{S_{11}^{E}(1-\sigma_1^{E})},\quad k_{eff\,1m}(\pi/2)=0.53; \tag{8.59}$$

$$n_{1m}(\pi/3)=\frac{5\pi a d_{31}}{S_{11}^{E}(1-\sigma_1^{E})},\quad k_{eff\,1m}(\pi/3)=0.51. \tag{8.60}$$

A detailed analysis of parameters of the bending modes does not make practical sense except for predicting a damaging effect on the frequency responses in vicinity of their resonance frequencies.

8.3 Spherical Transducer Types

8.3.1 Complete Spherical Shell Transducers without Baffles

8.3.1.1 Omnidirectional Transducer

With electrodes covering all the surfaces of the sphere the only mode of vibration generated is zero (pulsating) mode. In the broad frequency range (up to frequencies close to resonance frequency of vibration through the thickness of the shell) the transducer represents a classical example of the single degree of freedom system. Calculation of operating characteristics of the transducer can be performed using common equivalent electromechanical circuit with single mechanical contour shown in Figure 8.5. All the equivalent parameters of the circuit including the acoustic field related parameters are determined in Section 2.2.

Expressions for the parameters are summarized as follows:

$$K_{eqv}^{E}=\frac{1}{C_{eqw}^{E}}=\frac{8\pi t}{s_{11}^{E}+s_{12}^{E}},\ M_{eqv}=M=4\pi a^{2}t\rho,\ n=\frac{8\pi a d_{31}}{(s_{11}^{E}+s_{12}^{E})},\tag{8.61}$$

$$C_{e}^{S_{1,2}}=4\pi a^{2}\varepsilon_{33}^{T}(1-k_{p}^{2})/t,\ f_{0}=\frac{1}{2\pi a\sqrt{\rho s_{11}^{E}}}\sqrt{\frac{2}{1-\sigma_{1}^{E}}}.\tag{8.62}$$

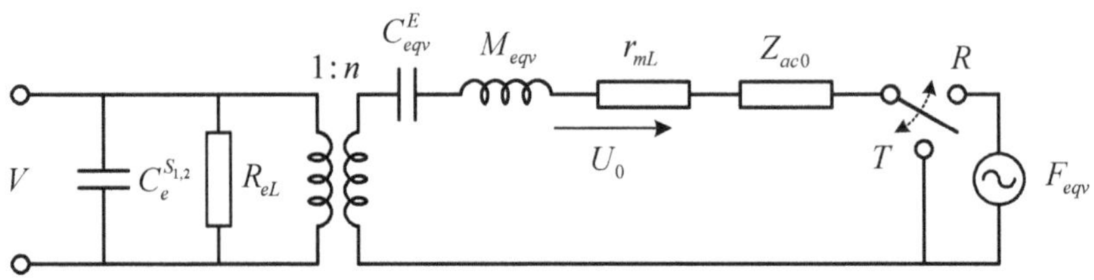

Figure 8.5: Equivalent electromechanical circuit of the zero-mode spherical transducer, Positions *T* and *R* of the switch correspond to the transmit and receive modes of operation.

The acoustic field related quantities, which in this case can be presented in the closed analytical form, following expressions (2.23), (2.27)-(2.29), and (2.33) are:

the radiation impedance

$$Z_{ac0}=4\pi a^{2}(\rho c)_{w}\left[\frac{(ka)^{2}}{1+(ka)^{2}}+j\frac{ka}{1+(ka)^{2}}\right]=r_{ac}+jx_{ac}=(\rho c)_{w}S_{\Sigma}(\alpha_{r}+j\beta_{r});\tag{8.63}$$

the sound pressure generated

$$P(\boldsymbol{r},\omega) = P_0(kr)k_{dif\,0}\,, \tag{8.64}$$

where

$$P_0(kr) = \frac{\rho c}{2\lambda r}U_{\tilde{V}}e^{-j(kr-\pi/2)}\,, \tag{8.65}$$

$U_{\tilde{V}} = U_o 4\pi a^2$ is the volume velocity (strength of the source), and

$$k_{dif\,0} = \frac{1}{1+jka}e^{jka} = \frac{1}{\sqrt{1+(ka)^2}}e^{j(ka-\arctan ka)}\,; \tag{8.66}$$

the equivalent force

$$F_{eqv0} = P_o\,k_{dif\,0}\,S_\Sigma, \tag{8.67}$$

where P_o is the sound pressure in the plane wave, and $S_\Sigma = 4\pi a^2$. The wave size of the transducer at resonance frequency is

$$(ka)_0 = \frac{1}{c_w}\sqrt{\frac{Y_1^E}{\rho}}\sqrt{\frac{2}{1-\sigma_1^E}}\,. \tag{8.68}$$

For a sphere made of PZT-4 $(ka)_0 \approx 3.8$.

8.3.1.2 Dipole Transducer

The main usable property of the transducer of this type is in providing the dipole like directional factor, $H(\omega,\varphi) = \cos\varphi$, in as broad frequency range, as is required by an intended application. In this case the electrodes are split into two halves and the halves are connected in opposite phase, as is shown in Figure 8.2 (b). The odd modes of vibration $\xi_{ri}(\varphi) = \xi_i P_i(\cos\varphi)$ at i = 1, 3, 5... that include the membrane (1, 3*m*, 5*m* ...) and bending (3*b*, 5*b*...) modes are generated. Equivalent mechanical parameters that correspond to these modes are presented in Table 4.1 (Section 4.5.5.2). The resonance frequencies of the modes that are close to the resonance frequency $f_1 = 1.22 f_0$ of the first (operating) mode and therefore have to be taken into consideration, when calculating operating characteristics of the transducer, are: $f_{3m} \approx 1.9 f_1$, $f_{3b} \approx 0.44 f_1$, $f_{5b} \approx 0.6 f_1$. Thus, the transducer cannot be treated as the single degree of freedom system, if an intended operating frequency range of the dipole transducer is from low frequencies and up to frequencies above the first resonance. Strictly speaking, the velocities $\dot{\xi}_{1m}$, $\dot{\xi}_{3m}$, $\dot{\xi}_{3b}$, and $\dot{\xi}_{5b}$ have to be considered as the generalized velocities.

It must be noted that spherical shell can vibrate as a rigid body with the same radial distribution of velocity as in the first membrane mode. This movement itself does not produce electromechanical conversion and cannot be generated electromechanically. But the sphere can vibrate as the whole under action of the acoustic field and can produce indirect effect on vibration in the first membrane mode through the acoustic interaction. This mechanism will be considered later. At least mechanically the resulting magnitude of vibration in the first mode must be presented as

$$\xi_1 = \xi_{1m} + \xi_{1t}, \tag{8.69}$$

where ξ_{1m} is responsible for deformation in the first membrane mode, and ξ_{1t} (t stands for "transition") is due to vibration of sphere as rigid body.

Thus, in general the resulting directional factor of the transducer may be represented as

$$H(\omega,\varphi) = \frac{\dot{\xi}_1 \cos\varphi + (\dot{\xi}_{3m} + \dot{\xi}_{3b})P_3(\cos\varphi) + \dot{\xi}_{5b}P_5(\cos\varphi)}{\dot{\xi}_1 + \dot{\xi}_{3m} + \dot{\xi}_{3b} + \dot{\xi}_{5b}}. \tag{8.70}$$

From this expression follows that the transducer has uncorrupted dipole directionality only in the frequency ranges, within which the additional terms in the nominator can be neglected. For numerical estimations, the equivalent circuit shown in Figure 8.6 may be used. The electromechanical equivalent parameters of the circuit that are taken from Table 4.1 (mechanical parameters) and electromechanical transformation coefficients are summarized below.

For the equivalent masses

$$M_{eqv1t} = M,\ M_{eqvi} = 4\pi a^2 t\rho\cdot\gamma_{Mi},$$

$$\gamma_{M1m} = 0.5\ \gamma_{M3m} = 0.62\ \gamma_{M3b} = 0.17,\ \gamma_{M5b} = 0.10. \tag{8.71}$$

For the equivalent rigidities,

$$K^E_{eqv\,i} = \frac{2\pi t Y_1^E}{1-\sigma_1^{E2}}\cdot\gamma_{Ki};$$

$$\gamma_{K1n} = 3(1+\sigma_1^E),\ \gamma_{K3m} = 8.9,\ \gamma_{K3b} = 0.23,\ \gamma_{K5b} = 0.16. \tag{8.72}$$

At $i \ge 2$ the mutual rigidities, K_{imb}, exist that characterize the elastic coupling between the membrane and bending modes of the same order. In our case

$$K_{3bm} = 16\cdot 10^{-4}\frac{2\pi t Y_1^E}{1-\sigma_1^{E2}}, \tag{8.73}$$

$$n = \frac{4\pi a d_{31}}{s_{11}^E(1-\sigma)}\cdot \gamma_{ni}, \tag{8.74}$$

$$\gamma_{n1} = 1.50,\ \gamma_{n3m} = -1.04,\ \gamma_{n3b} = -0.06,\ \gamma_{n5b} = 0.04. \tag{8.75}$$

Note that this is a typical situation, in which the signs of the modal electromechanical transformation coefficients matter, because the modes with transformation coefficients having different signs participate in operation simultaneously (see the remark under formula (8.34)).

The additional rigidities that are introduced due to the coupling to K_{3m}^E and K_{3b}^E are $K_{3mb}(U_{3b}/U_{3m})$ and $K_{3mb}(U_{3m}/U_{3b})$, respectively. Since the ratios of velocities of vibration in water are not large enough to make the values of introduced rigidities comparable with the self-rigidities, the effect of the elastic coupling between the modes can be neglected. Note that in case of vibration in air the situation may change in vicinity of the resonance frequencies of the modes due to high mechanical Q of the shells.

Acoustic field related parameters that correspond to the involved modes of vibration were considered in Section 6.3.2. The general expressions for the diffraction coefficients and radiation impedances are given by formulas (6.236), (6.240) and plotted in Figures 6.27, 6.28. For the first mode of vibration, they are

$$k_{dif1} = 1/(ka)^2 h_1^{(2)\prime}(ka), \tag{8.76}$$

$$Z_{ac1} = (\rho c)\frac{4\pi a^2}{3}e^{-jka}\frac{1+jka}{h_1^{(2)\prime}(ka)} = (\rho c)S_{eff1}(\alpha_{11}+j\beta_{11}). \tag{8.77}$$

Here $h_1^{(2)}$ is the spherical Hankel function of the second kind, and $S_{eff\,i} = 4\pi a^2/(2i+1)$.The Figures 6.27 and 6.28 are also reproduced here for the sake of convenience as Figure 8.6 and Figure 8.7. In case that $i \ge 2$ distribution of radial vibration in the membrane and bending modes are the same and correspond to the radial distribution of vibration $\xi = \xi_1\cos i\varphi$. Therefore $k_{dif\,im} = k_{dif\,ib} = k_{dif\,i}$, and $Z_{acim} = Z_{acib} = Z_{aci}$. In particular,

$$k_{dif\,3b} = k_{dif\,3m} = k_{dif\,3},\ k_{dif\,5b} = k_{dif\,5m} = k_{dif\,5} \text{ and } Z_{ac3b} = Z_{ac3m} = Z_{ac3}. \tag{8.78}$$

The equivalent forces, F_{eqvi}, that are generating vibration in different modes must be determined by formula analogous to (8.67) with corresponding diffraction coefficients.

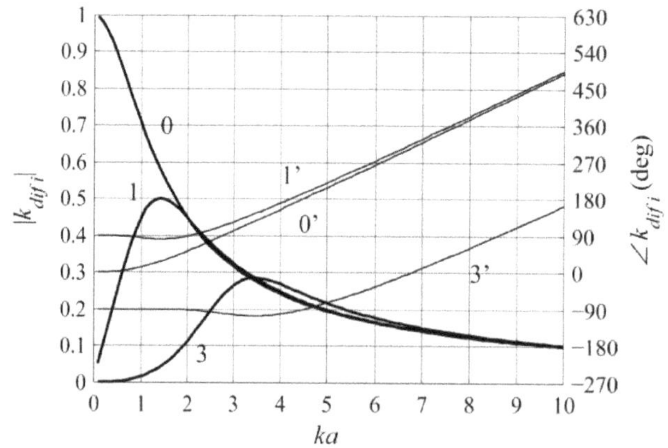

Figure 8.6: Diffraction coefficients $k_{dif\,i}$ for spheres without baffles: Magnitude (thick lines) and phase (thin lines, labeled with ') for $i = 0, 1, 3$.

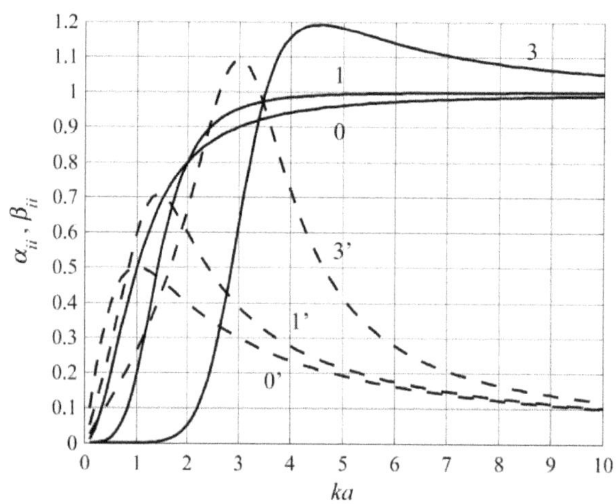

Figure 8.7: Nondimensional modal radiation impedance coefficients for spheres without baffles: α_{ii} (solid lines) and β_{ii} (dashed lines, labeled with ') for $i = 0, 1, 3$.

Since the magnitude U_i in the radial distribution of velocity $U_i(\varphi) = U_i \cos i\varphi$ at $i \geq 2$ is superposition of the generalized velocities that belong to the membrane and bending modes of vibration,

$$U_i = U_{im} + U_{ib}\,. \tag{8.79}$$

The acoustic energy produced by this distribution of velocity can be represented as

$$\bar{\bar{W}}_{ac\,i} = U_i U_i^* Z_{ac\,i} = \left[|U_{im}|^2 + (U_{im}^* U_{ib} + U_{im} U_{ib}^*) + |U_{ib}|^2 \right] Z_{ac\,i} \,. \tag{8.80}$$

This expression in its turn can be represented in the form

$$\begin{aligned} \bar{\bar{W}}_{ac\,i} &= |U_{im}|^2 \left[Z_{ac\,i} + Z_{ac\,i}(U_{ib} / U_{im}) \right] + |U_{ib}|^2 \left[Z_{ac\,i} + Z_{ac\,i}(U_{im} / U_{ib}) \right] = \\ &= |U_{im}|^2 \, Z_{ac\,im} + |U_{ib}|^2 \, Z_{ac\,ib}, \end{aligned} \tag{8.81}$$

where

$$Z_{ac\,im} = Z_{ac\,i} + Z_{ac\,i}(U_{ib} / U_{im}) \text{ and } Z_{ac\,ib} = Z_{ac\,i} + Z_{ac\,i}(U_{im} / U_{ib}) \tag{8.82}$$

are the radiation impedances that correspond to the membrane and bending modes of vibration.

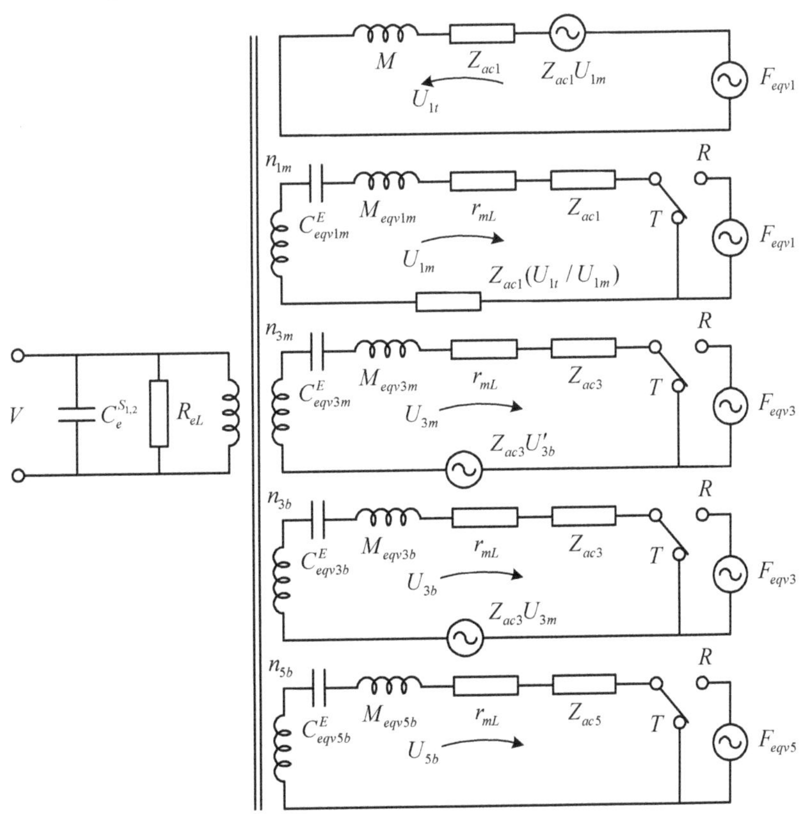

Figure 8.8: Equivalent electromechanical circuit of the spherical transducer with full size bipolar electrodes.

From relations (8.82) follows that acoustic interaction between the membrane and bending modes exists with mutual impedances $z_{ac\,imb} = Z_{ac\,i}$. Likewise, it can be concluded based on the expression (8.57) for the total radial displacement in the first mode of vibration that the acoustic interaction between the membrane component of vibration and vibration of the shell as a rigid body exists with mutual impedance Z_{ac1}, and the diffraction coefficient for the "transitional" mode of vibration is k_{dif1}.

After all the equivalent parameters are specified, the magnitudes of the generalized velocities can be determined with help of equivalent circuit in Figure 8.8, and hence the directional factor of the transducer may be obtained by formula (8.70).

A qualitative estimation can be made regarding the possible contribution of the acoustically induced passive mode of vibration to results of calculating transducer characteristics. It is appropriate to make such an estimation for this particular case because the acoustic coupling between the first active mode and passive ("transitional") mode of movement transducer body as a whole is the strongest in this case. It is reasonable to make the estimations separately for the transmit and receive modes of operation, i.e., in the range around the resonance frequency of transducer and at low frequencies. As it follows from considering the passive contour in Figure 8.8 in the transmit mode

$$U_{1t} = \frac{Z_{ac1}}{j\omega M + Z_{ac1}} U_{1m} = \frac{1}{1 + (j\omega M / Z_{ac1})} U_{1m}. \tag{8.83}$$

Value of the ratio $(j\omega M / Z_{ac1}) = k_t$ can be estimated as follows. At frequencies around the resonance frequency of transducer the wave number $(ka)_1 \approx 3.8$, i.e., $\omega \approx 3.8 c_w / a$; $Z_{ac1} \approx (\rho c)_w \cdot 4\pi a^2 / 3$. Thus, $k_t \approx j11.4 \cdot (t/a) \cdot (\rho_c / \rho_w)$. Assuming that $(t/a) \approx 0.2$ and ratio of the ceramics and water densities is $(\rho_c / \rho_w) \approx 7.5$, we obtain that $k_t \approx j17$. Thus, we arrive at the conclusion that an additional impedance introduced in the active contour due to coupling with the passive one is $|U_{1t}| \approx 0.05 |U_{1m}|$. Such addition is within accuracy of determining the radiation impedance and can be neglected to the first approximation.

In the receive mode of operation at frequencies much below the resonance frequency the magnitude of vibration in the passive contour can be estimated to the first approximation (without secondary influence of the first active mode) as

$$U_{1t} = \frac{F_{eqv1}}{j\omega M + Z_{ac1}}. \tag{8.84}$$

The force induced by vibration of transducer body as a whole in the first mode contour, which is $F_{in} = Z_{ac1} \cdot U_{1t}$, will be estimated as

$$F_{in} = F_{eqv1} \frac{1}{1 + j\omega M / Z_{ac1}} = F_{eqv1} \frac{1}{1 + k_t}. \tag{8.85}$$

Given that at low frequencies $Z_{ac1} \approx j\omega \cdot (2\pi a^3 / 3)\rho_w$ according to (6.257), $k_t \approx 6 \cdot (t/a) \cdot (\rho_c / \rho_w) = 9$. Thus, $F_{in} \approx 0.1 F_{eqv1}$. This may result in changing output of the transducer by about 10%.

Note that there is no acoustic coupling between contours of the equivalent circuit that belong to velocity distributions of different order ($\cos i\varphi$ at $i = 1,3,5$). Thus, the velocities U_1, U_3, U_5 that are determined by formula (8.79) can be calculated independently. The sound pressure radiated may be obtained using formula (8.64), which in this case can be modified as follows,

$$P(r,\omega) = \frac{\rho c}{2\lambda r} e^{-j(kr - \pi/2)} \cdot 4\pi a^2 [U_1 k_{dif1} + U_{3b} k_{dif3} + U_{5b} k_{dif5}]. \tag{8.86}$$

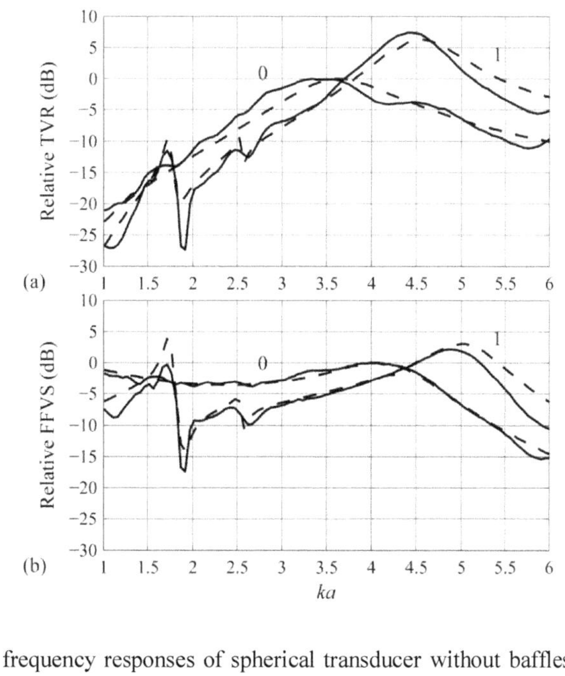

Figure 8.9: The frequency responses of spherical transducer without baffles with unipolar and bipolar electrode configurations: (a) transmit mode (TVR) and (b) receive mode (FFVS). Calculated responses are shown by the solid lines and measured by the dashed lines.

Calculated and measured frequency responses of the spherical transducer without baffles that operate with unipolar and bipolar full size electrodes configurations are presented in Figure 8.9. Specifics of the bipolar configuration in contribution of the bending modes of vibration is clearly illustrated.

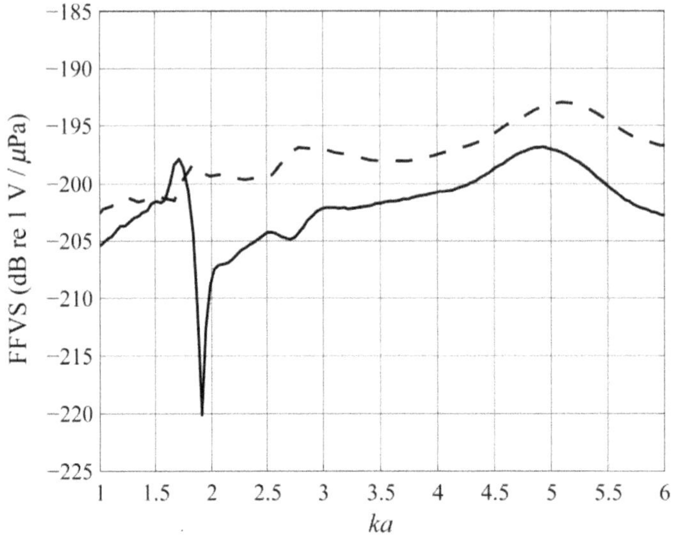

Figure 8.10: The frequency response (FFVS) of spherical transducer without baffles for bipolar (first mode) electrode configuration. With full size electrodes (solid line) and with optimized electrodes (dashed line).

In practical designing of the bipolar transducer the general formulas may simplify depending on the transducer application and hence on its operating range. Thus, in the transmit mode (one of such possible applications will be considered in the next section) the operating range is around the resonance frequency of the first mode. Contribution of all the other modes can be neglected, and the equivalent circuit in Figure 8.10 reduces to the two first contours. The wave size of the first order transducer at resonance is

$$(ka)_1 = \frac{1}{c_w}\sqrt{\frac{Y_1^E}{\rho}}\sqrt{\frac{3}{1-\sigma_1^E}} \,. \tag{8.87}$$

For transducer made of PZT-4 $(ka)_1 \approx 4.6$. At wave numbers close to this value the non-dimensional coefficients of radiation impedance are $\alpha_1 \approx 1$, $\beta_1 \approx 0.25$. The diffraction coefficient can be approximated with reference to plots in Figure 8.6 and formula (8.66) as

$|k_{dif1}| \approx 1/ka$.

The most challenging is the application of the bipolar transducer as a broadband dipole (pressure gradient) hydrophone. Without special measures taken the frequency response and directional characteristics of the hydrophone are significantly corrupted in the frequency ranges below and above the first membrane resonance frequency, as it was pointed out above. The characteristics of hydrophone in a broad band can be greatly improved, if to optimize the electrodes dimensions by reducing their size to $\varphi_1 = 70.5^o$. In this case the electromechanical transformation coefficients are determined by formulas (8.43). Thus, the third membrane mode of vibration is suppressed, and the bending modes are greatly weakened.

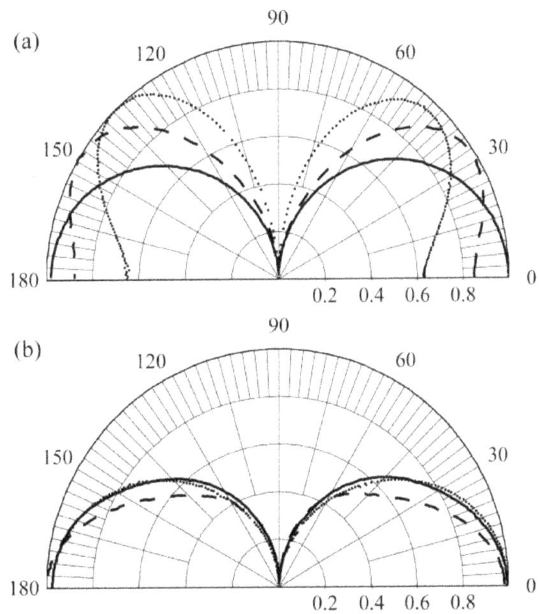

Figure 8.11: Directional factors of spherical transducer without baffles with bipolar electrodes. (a) Full size electrode configuration: *ka* = 4.3 (solid line), *ka* = 1.8 (dotted line), *ka* = 2.6 (dashed line). In Figure (b) are shown results of optimizing the electrodes configuration at the same frequencies and with the same line styles.

As the result the directional characteristics and frequency response in the broad frequency range are significantly improved. This is shown in Figure 8.10 and Figure 8.11 (only halves of the plots are presented in the figure due to symmetry). Equivalent circuit of the transducer in

Figure 8.9 can be reduced to the first two contours that are related to the only active first mode of vibration.

8.3.2 Complete Spherical Shell Transducers with Baffles

The complete spherical shells may have application as unidirectional transducers, in which case unidirectionality is achieved by employing the acoustic baffles, as shown in Figure 8.12.

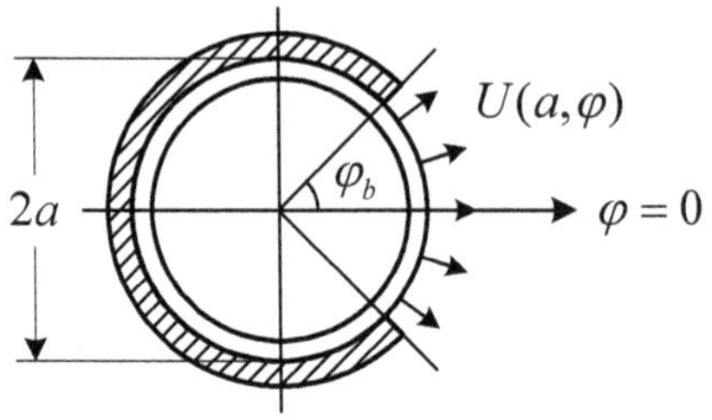

Figure 8.12: Illustration of a spherical transducer with the conformal baffle. Opening angle of the baffle is φ_b.

Radiation of the spherical shells with baffles related issues were considered in Section 6.3.2.3. The modal (pertaining to a single mode of vibration) directional characteristics, diffraction coefficients and radiation impedances as well as mutual intermodal radiation impedances that characterize acoustic coupling between the modes were determined therein. Due to the existence of the acoustic interaction between modes of vibration, not only radiation related parameters of the active modes change, but some additional passive modes of vibration of mechanical system can be generated acoustically. Therefore, peculiarities in calculating the baffled single active mode transducers appear that must be considered. Besides, existence of the baffles makes possible using multiple active modes of the spherical shell vibration, and thus broadening an operating frequency range of transducer. Calculating the multimode transducers will also be considered in this section.

8.3.2.1 Single Active Mode Transducers

All the peculiarities of calculating the baffled transducers with intentionally single active modes of vibration arise due to their possible acoustic interaction with passive modes. Therefore, first the mutual impedances between the modes must be determined. In general, they are different for different active mode transducers. Expression for the mutual intermodal impedances for a spherical shell with rigid baffle is given in Section 6.3.2.1 in the form of Eq. (6.241),

$$z_{acip}(\varphi_b) = -j\rho c\, 4\pi a^2 \sum_{l=0}^{\infty} \frac{1}{2l+1} a_{pl}(\varphi_b)\, a_{il}(\varphi_b) \frac{h_l^{(2)}(ka)}{h_l^{(2)\prime}(ka)} . \tag{8.88}$$

In this section we will consider baffles with opening angle $\varphi_b = \pi/2$. As it follows from the plots in Figures. 6.30-6.31 the radiation impedances (including mutual) and diffraction coefficients with compliant baffles have practically the same values as with the rigid baffles at least for this opening angle. The coefficients $a_{mn}(\varphi_b)$ according to formula (6.229) are

$$a_{ml}(\pi/2) = (l+1/2) \int_0^{\pi/2} P_m(a,\varphi)\; P_l(\cos\varphi)\sin\varphi d\varphi . \tag{8.89}$$

With opening angle $\varphi_b = \pi/2$ for the case that $m = 0$

$$a_{00} = 1/2,\; a_{01} = 3/4,\; a_{03} = -7/16,\; a_{05} = 11/32, \ldots, \text{ and } a_{0l} = 0 \text{ for } l \text{ even.} \tag{8.90}$$

For the case that $m = 1$,

$$a_{11} = 1/2,\; a_{10} = 1/4,\; a_{12} = 5/16,\; \ldots, \text{ and } a_{1l} = 0 \text{ for } l > 1 \text{ odd.} \tag{8.91}$$

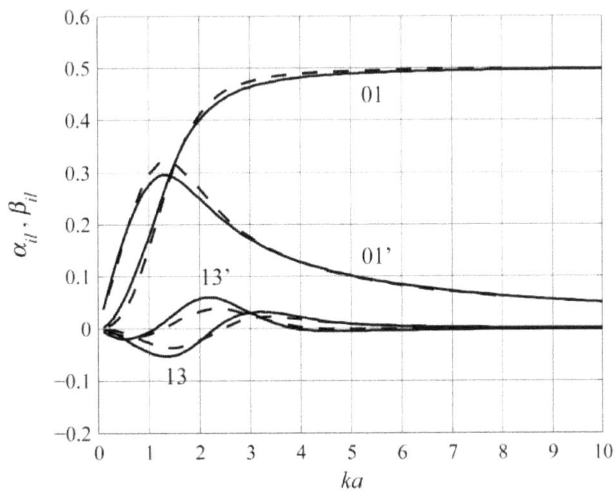

Figure 8.13: Nondimensional coefficients of the mutual radiation impedances of a baffled ($\varphi_b = \pi/2$) sphere: with rigid baffle (solid lines) and with compliant baffle (dashed lines). β_{il} is labeled with' The mutual radiation impedances are $z_{acil} = \rho c \cdot 2\pi a^2(\alpha_{il} + j\beta_{il})$.

Calculations show that magnitudes of the mutual impedances between modes drop abruptly, as separation between modes increases, and especially so the higher the orders of the modes are. Plots presented in Figure 8.13 show that only the mutual impedance z_{ac01} between the zero and first modes has significant value, and z_{ac13} can be already neglected. Therefore, only interaction between the zero and first modes will be further taken into consideration.

Expression for $z_{ac01} = z_{ac10}$ can be determined from formula (6.241) in the form

$$z_{ac01} = -j\rho c \cdot 2\pi a^2 \cdot \frac{1}{4}\left[\frac{h_0^{(2)}(ka)}{h_0^{(2)\prime}(ka)} + \frac{h_1^{(2)}(ka)}{h_1^{(2)\prime}(ka)}\right]. \quad .(8.92)$$

8.3.2.1.1 Unipolar (0th Active Mode) Transducer

Equivalent circuit of the transducer with baffle is presented in Figure 8.14. In comparison with the circuit in Figure 8.5 of the transducer without baffle it contains the passive (not having a direct coupling with the electrical side) mechanical contours related to vibrations in the modes that are generated due to acoustic coupling with the zero mode of vibration.

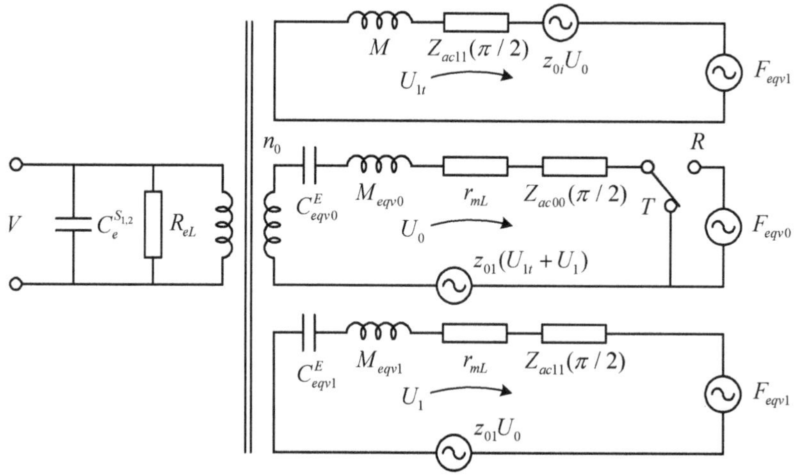

Figure 8.14: Equivalent electromechanical circuit of the baffled transducer with unipolar electrodes

All the mechanical equivalent parameters of the corresponding contours remain the same, as for the transducer without a baffle. Only the acoustic field related parameters change. They must be calculated using general formulas presented in Section 6.3.2.1. Thus, the radiation impedances $Z_{ac00}(\pi/2)$ and $Z_{ac11}(\pi/2)$ following formula (6.240) are

$$Z_{ac00}(\pi/2) = -j\rho c \cdot 4\pi a^2 \cdot \left[\frac{1}{2}\frac{h_0^{(2)}(ka)}{h_0^{(2)\prime}(ka)} + \frac{3}{8}\frac{h_1^{(2)}(ka)}{h_1^{(2)\prime}(ka)}\right], \quad (8.93)$$

$$Z_{ac11}(\pi/2) = -j\rho c \cdot 4\pi a^2 \cdot \left[\frac{1}{16} \frac{h_0^{(2)}(ka)}{h_0^{(2)\prime}(ka)} + \frac{1}{12} \frac{h_1^{(2)}(ka)}{h_1^{(2)\prime}(ka)} \right]. \tag{8.94}$$

Nondimensional coefficients of these impedances are plotted in Figure 8.15. The impedances introduced into the contours due to acoustic coupling are

$$Z_{10} = z_{ac01} \frac{U_0}{U_1} \text{ and } Z_{01} = z_{ac01} \frac{U_1}{U_0}. \tag{8.95}$$

Alternatively, the acoustic coupling can be accounted for by introducing the forces

$$F_{ac10} = z_{ac01} U_0 \text{ and } F_{ac01} = z_{ac01} U_1 \tag{8.96}$$

into contours that correspond to the zero and first modes of vibration, respectively.

The modal diffraction coefficients $k_{dif\,0}(\pi/2)$ and $k_{dif\,1}(\pi/2)$ following formula (6.236) are

$$k_{dif\,0}(\pi/2) = -j\frac{1}{2(ka)^2}\left[\frac{1}{h_0^{(2)\prime}(ka)} + j\frac{3}{2 \cdot h_1^{(2)\prime}(ka)} + \frac{7}{8 \cdot h_3^{(2)\prime}(ka)} \right], \tag{8.97}$$

$$k_{dif\,1}(\pi/2) \approx -j\frac{1}{2(ka)^2}\left[\frac{1}{2 \cdot h_0^{(2)\prime}(ka)} + j\frac{1}{h_1^{(2)\prime}(ka)} - \frac{5}{8 \cdot h_2^{(2)\prime}(ka)} \right]. \tag{8.98}$$

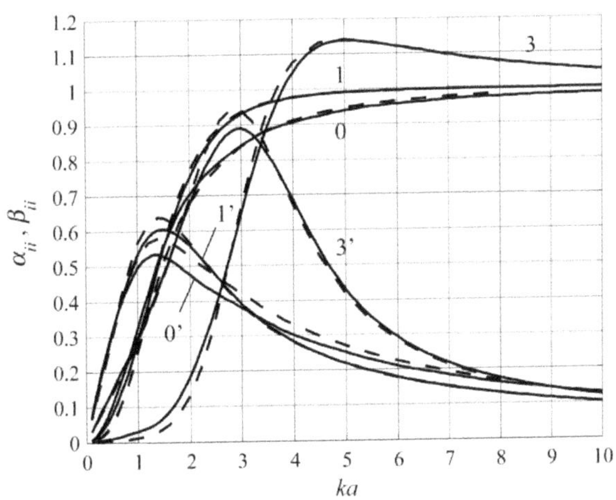

Figure 8.15: Nondimensional coefficients of the modal self-radiation impedances of a baffled ($\varphi_b = \pi/2$) sphere for the rigid baffle (solid lines) and for the compliant baffle (dashed lines). β_{ii} is labeled with′ The radiation impedances are $Z_{acii} = \rho c S_{eff\,i}(\alpha_{ii} + j\beta_{ii})$, where $S_{eff\,0} = 2\pi a^2$ and $S_{eff\,1} = 2\pi a^2/3$.

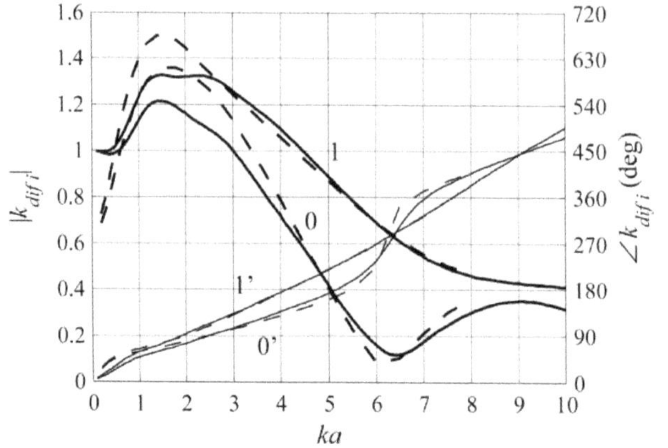

Figure 8.16: Modal diffraction coefficients of a baffled ($\varphi_b = \pi/2$) sphere: $|k_{dif\,i}|$ (thick lines) and phase (thin lines, labeled with ') for $i = 0$, 1 with rigid baffle (solid lines) and compliant baffle (dashed lines).

Plots of these functions are presented in Figure 8.16. The equivalent forces that are acting in the contours in the receive mode are by definition (6.34)

$$F_{eqvi} = 2\pi a^2 P_o k_{dif\,i}(\pi/2)\,, \tag{8.99}$$

where P_o is the sound pressure in the plane wave. Using the equivalent circuit of Figure 8.14 with all the parameters known, the velocities U_0, U_1, and frequency response in the receive mode can be determined.

The sound pressure generated by the transducer on the acoustic axis in the far field according to formula (6.236) is

$$P(r,0,\pi/2) = j\rho\omega\frac{e^{-jkr}}{4\pi r}\cdot 2\pi a^2[U_0 k_{dif\,0}(\pi/2) + U_1 k_{dif\,1}(\pi/2)]\,. \tag{8.100}$$

The directional factor of the transducer following expression (6.235) is

$$H_0(\varphi,\pi/2) = \frac{U_0 k_{dif\,0}(\pi/2)P_0(\cos\varphi) + U_1 k_{dif\,1}(\pi/2)P_1(\cos\varphi)}{U_0 k_{dif\,0}(\pi/2) + U_1 k_{dif\,1}(\pi/2)}\,. \tag{8.101}$$

8.3.2.1.2 Bipolar (1st Active Mode) Transducer

In distinction from the circuit in Figure 8.8 the equivalent electromechanical circuit of the baffled bipolar transducer includes additional passive contour that corresponds to the zero mode vibration due to its acoustic coupling with the first mode. The circuit is shown in Figure 8.17.

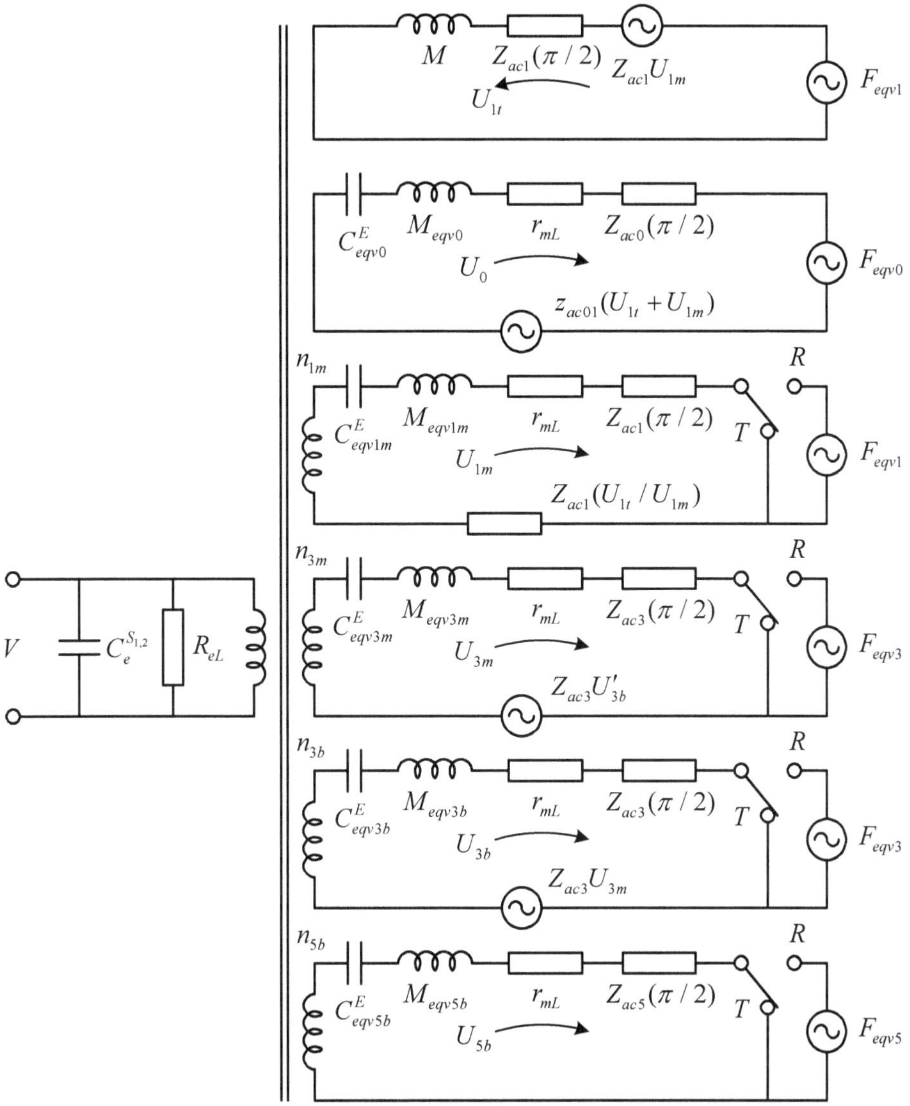

Figure 8.17: The electromechanical equivalent circuit of the baffled bipolar spherical transducer.

All the electromechanical and mechanical equivalent parameters of the contours remain the same as they were listed in Section 8.3.1.1 and 8.3.1.2. The radiation impedances and diffraction coefficients must be used that are determined in the previous section. Formulas (8.100) and (8.101) for calculating the sound pressure generated and the directional characteristics in general must be modified in order to include contributions of the bending modes. Thus, they will be

$$P(r,0,\pi/2) = j\rho\omega\frac{e^{-jkr}}{4\pi r}2\pi a^2 \times$$
$$\times[U_0 k_{dif0}(\pi/2) + U_1 k_{dif1}(\pi/2) + U_{3b} k_{dif3}(\pi/2) + U_{5b} k_{dif5}(\pi/2)], \tag{8.102}$$

$$H_0(\varphi,\pi/2) = \frac{U_0 k_{dif0}(\pi/2)P_0(\cos\varphi) + U_1 k_{dif1}(\pi/2)P_1(\cos\varphi)}{U_0 k_{dif0}(\pi/2) + U_1 k_{dif1}(\pi/2) + U_{3b} k_{dif3}(\pi/2) + U_{5b} k_{dif5}(\pi/2)} +$$
$$+\frac{U_{3b} k_{dif3}(\pi/2)P_3(\cos\varphi) + U_{5b} k_{dif5}(\pi/2)P_5(\cos\varphi)}{U_0 k_{dif0}(\pi/2) + U_1 k_{dif1}(\pi/2) + U_{3b} k_{dif3}(\pi/2) + U_{5b} k_{dif5}(\pi/2)}. \tag{8.103}$$

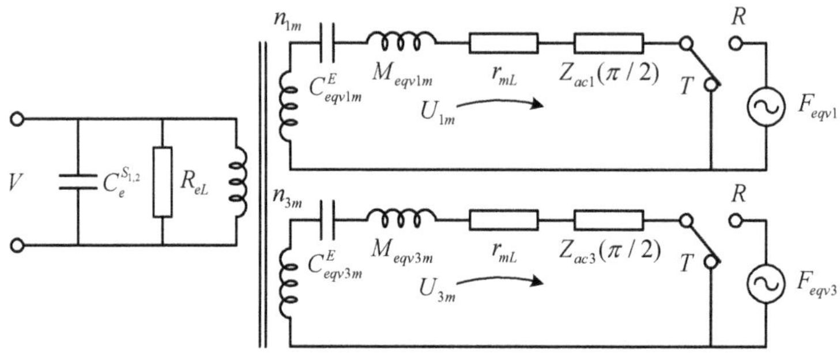

Figure 8.18: Simplified equivalent electromechanical circuit of the baffled bipolar spherical transducer for operating range around the resonance frequency.

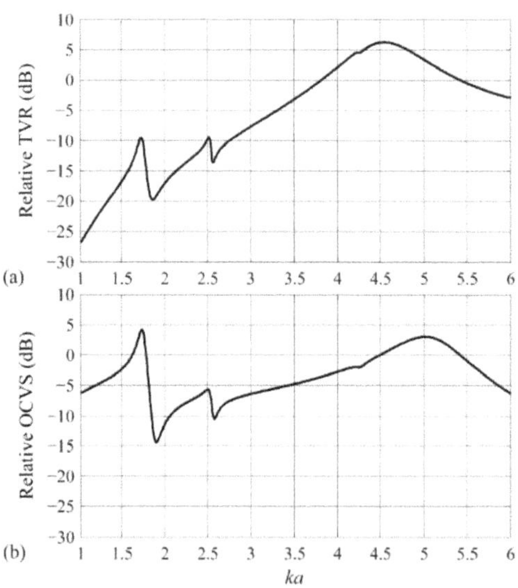

Figure 8.19: Frequency responses of the baffled bipolar spherical transducer: (a) transmit mode, (b) receive mode.

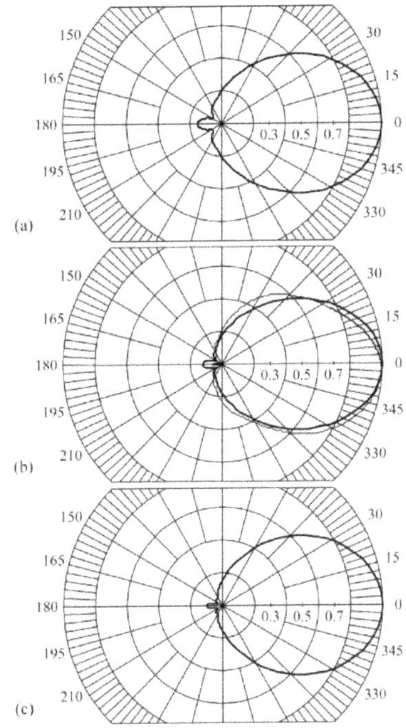

Figure 8.20: Directional factors of the rigidly baffled spherical bipolar transducer in the frequency band around its resonance frequency at different *ka*: (a) 3.5; (b) 4.5; (c) 5.5. Note that the characteristics are the same as in Figure 6.33. In plot (b) is also shown the measured directional factor of a hemispherical transducer (thin solid line).

It must be taken into consideration that the terms related to the bending modes as well as those induced due acoustic coupling with zero mode may have noticeable contribution in vicinity of resonance frequencies of these modes only, i.e., in the frequency range below the resonance frequency of the bipolar transducer ($f_{3b} \approx 0.37 f_1$, $f_{5b} \approx 0.50 f_1$). At corresponding wave sizes (with PZT-4 ceramics $(ka)_{3b} \approx 1.7$ and $(ka)_{5b} \approx 2.3$, whereas $(ka)_1 \approx 4.5$) the directional characteristics of the bipolar baffled transducer significantly widen (see Figure 8.11(a)) and become apparently not usable. Thus, the peculiarities of the frequency responses and directional characteristics that occur due to existence of the bending modes are beyond an operating frequency range. Therefore, for practical calculations the contours related to the bending modes and the corresponding terms in formulas (8.102), (8.103) may be neglected in this case. Thus, for practical calculations the equivalent circuit of the transducer can be simplified to those shown in Figure 8.18.

For calculating transmit frequency response and directional characteristics of the transducer the formulas (8.100) and (8.101) can be used. The frequency responses and directional characteristics of the baffled bipolar transducer calculated for the most practical operating range around its resonance frequency are presented in Figure 8.19 and Figure 8.20.

8.3.2.2 Multimode (0±1 Active Modes) Transducer

Baffled spherical transducer with electrodes divided into halves can be used in multimode operation at different electrodes connections that are shown in Figure 8.21: Variants of connection of half electrodes for multimode operation of the baffled spherical transducer: (a) unipolar (0 mode) operation, (b) bipolar (1 mode) operation, (c) operation with two simultaneously active modes (0+1 mode). The situation is analogous to those considered for the cylindrical transducers in Section 7.2.3.3.

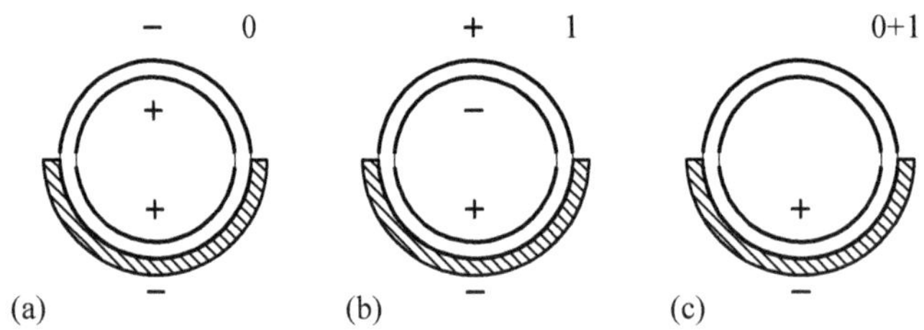

Figure 8.21: Variants of connection of half electrodes for multimode operation of the baffled spherical transducer: (a) unipolar (0 mode) operation, (b) bipolar (1 mode) operation, (c) operation with two simultaneously active modes (0+1 mode).

Transition between the zero and first mode operating resonance frequencies is achieved by switching the half electrodes. Calculating the operating characteristics can be performed as described above for unipolar and bipolar baffled transducers.

If only one half of electrodes is used, both zero and first modes of vibration are active simultaneously. In this case the magnitudes of the transformation coefficients must be reduced in factor of 2, and the sign of $n_1/2$ must be taken into account.

The sign of $n_1/2$ changes depending on which half of the electrodes is used. Thus, $n_1 < 0$ if electrodes are under the baffle, as it follows from analysis made in Section 8.2.1. By the considerations that are illustrated with Figure 7.47 for the analogous situation regarding the cylindrical transducers, in this case superposition of modal responses between their resonances is constructive and leads to a broadband operation of the transducer. The operating frequency range extends from below resonance frequency f_0 and above resonance frequency f_1 within the band, in which the directional characteristics are appropriate (at $3.0 < ka < 5.5$, as follows from plots in Figure 6.32 and Figure 8.20). For this frequency range calculation of transducer characteristics may be performed using the simplified equivalent circuit presented in Figure 8.22, in which contribution of the bending modes is neglected.

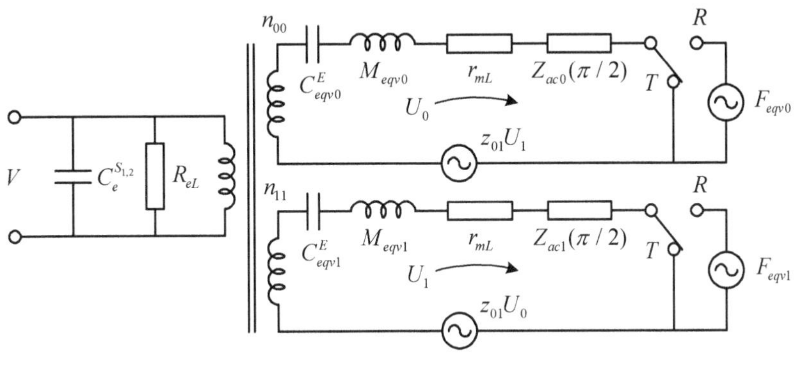

Figure 8.22: Equivalent circuit of the dual resonance frequency spherical baffled transducer. zero-mode operation $n_{00} = n_0$, $n_{11} = 0$; first-mode operation $n_{11} = n_1$, $n_{00} = 0$; 0+1 mode operation $n_{00} = n_0/2$, $n_{11} = -n_1/2$.

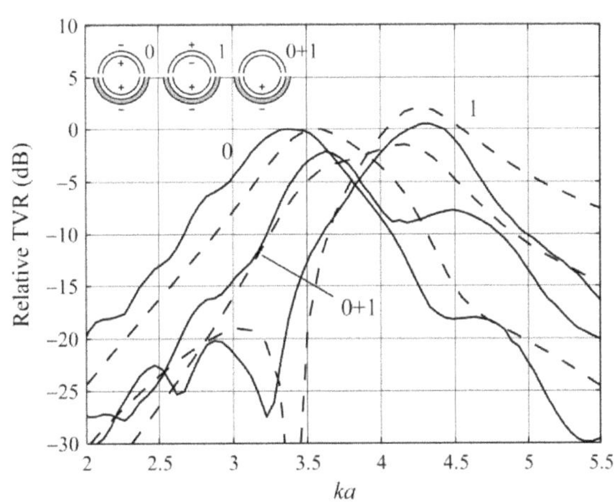

Figure 8.23: The transmit responses of baffled spherical transducer at different combinations of the halves of electrodes: for unipolar connection (0), for bipolar connection (1), for one half of electrodes located under the baffle (0+1). Measured characteristics are shown by the solid lines, calculated by the dashed lines.

Calculated transmit frequency responses of the transducer with the full-size electrodes in unipolar and bipolar connections and with one half of electrodes active are presented in Figure 8.23. Calculations are made for transducer that is assumed to be made of PZT-4 ceramics with dimensions $t/a = 0.1$ and $t = 3$ mm.

A remarkable property of the transducer operating in the 0+1 mode is that in a broad frequency range its directional factors almost do not change. This is illustrated in Figure 8.24 with results of measurements performed with transducer having the frequency responses shown in Figure 8.23.

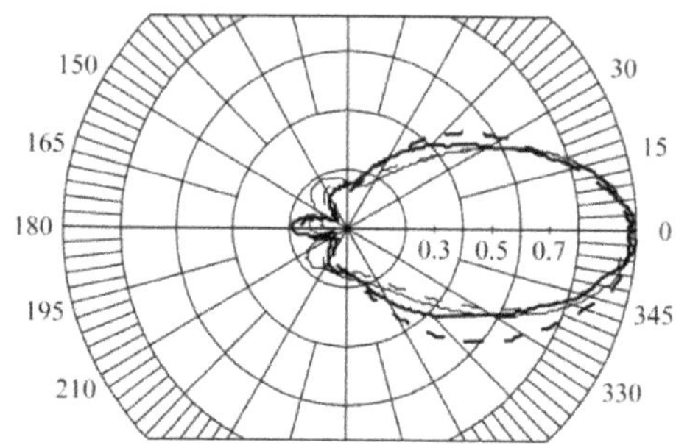

Figure 8.24: Measured directional factors of baffled spherical transducer in 0+1 mode of operation at different frequencies: $ka = 3.0$ (thin solid line), $ka = 3.5$ (thin dashed line), $ka = 4.0$ (thick solid line), $ka = 4.5$ (thick dashed line).

8.3.3 Hemispherical Transducer

Application of the transducers employing vibration of incomplete spherical shells is not common, although they may have some advantages as single unidirectional transducers, in which unidirectional operation is achieved without using acoustic baffles. This can be helpful given that effectiveness of the baffles may depend on operating depth. Using the hemispherical transducers was suggested in Refs. 7 and 8. We consider calculation of transducers in the shape of the hemisphere ($\varphi_0 = \pi / 2$) and results of experimental verification of their characteristics as this follows from Ref. 9.

As shown in Section 4.5.5.3, the spectrum of resonance frequencies of the hemisphere is the same as of the complete sphere vibrating in the odd modes. Thus, the closest to the first membrane resonance frequency, f_{1m}, is $f_{3m} \approx 1.9 f_{1m}$. Frequency f_{1m} is preceded by the frequencies f_{3b} and f_{5b} that belong to bending modes of vibration. An effective use of hemispherical transducers may be realized in the frequency range around their membrane resonance frequencies. In the case that PZT-4 ceramics is used the first resonance frequency for a hemisphere corresponds to $ka \approx 4.6$ ($D/\lambda \approx 1.5$). Resonance frequencies of the bending modes are much below the operating range ($(ka)_{3b} \approx 2.0$ and $(ka)_{5b} \approx 2.7$), therefore contribution of these modes can be neglected. Distribution of displacement for the first membrane mode is given by expressions (8.54). For calculating electromechanical characteristics of the transducer equivalent circuit can be used without the bending modes related contours. The equivalent parameters of the equivalent circuit are those presented by formulas (8.71)-(8.75) reduced by factor of two. Thus, calculating electromechanical characteristics of the hemispherical transducer can be performed using equivalent circuit shown in Figure 8.18 with following equivalent parameters:

$$M_{eqv1} = \pi a^2 t\rho ,, \; K^E_{eqv1} = \frac{3\pi t Y_1^E}{1-\sigma_1^E}, \; n_1 = \frac{3\pi a d_{31}}{s_{11}^E(1-\sigma_1^E)}; \tag{8.104}$$

$$M_{eqv3} = 1.2\pi a^2 t\rho , \; K^E_{eqv3} = \frac{8.9\pi t Y_1^E}{1-\sigma_1^E}, \; n_3 = -\frac{2.08\pi a d_{31}}{s_{11}^E(1-\sigma_1^E)}. \tag{8.105}$$

The wave size of the hemispherical transducer over its operating range is large enough to assume that to the first approximation the acoustic radiation related parameters of the transducer: the radiation impedance, diffraction coefficient and directional factors can be considered the same, as for the full size sphere correspondingly baffled, and having distribution of radial velocities according to expressions (8.54) on their open parts. This assumption can be supported by the following considerations.

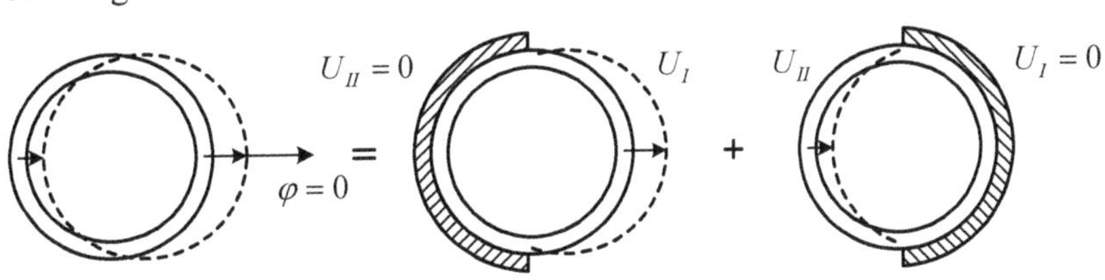

Figure 8.25: To the acoustic radiation of oscillating sphere as superposition of fields radiated by the coaxial baffled spherical transducers.

Acoustic field of the oscillating sphere can be imagined, as superposition of the fields generated by the two spherical transducers having the same acoustic center with rigid baffles on their half surfaces that vibrate with the same distribution of velocities on their open parts as the oscillating sphere has. This situation is shown in Figure 8.25.

Radiation impedance of the oscillating sphere can be represented as

$$Z_{ac11} = Z_{ac\,I,I} + Z_{acII,II} - 2Z_{ac\,I,II}, \tag{8.106}$$

where $Z_{ac\,I,I} = Z_{ac\,II,II}$ are the radiation impedances of the baffled first order transducers, and $Z_{ac\,I,II}$ is the mutual impedance between halves of the oscillating sphere. In terms of the non-dimensional coefficients of the radiation impedance relation (8.106) is equivalent to equations

$$\alpha_{11} = \alpha_{I,I} - \alpha_{I,II}, \tag{8.107}$$

$$\beta_{11} = \beta_{I,I} - \beta_{I,II} \tag{8.108}$$

(remember that $S_{eff11} = 4\pi a^2/3$, whereas $S_{eff\,I,I} = S_{eff\,II,II} = 2\pi a^2/3$).

Comparing the dependences of the nondimensional coefficients of the oscillating sphere in Figure 8.7 and of the baffled first order transducer in Figure 8.15 at $ka > 4$ leads to the conclusion that at these wave sizes $\alpha_{11} \approx \alpha_{I,I}$ and $\beta_{11} \approx \beta_{I,I}$. Thus, $\alpha_{I,II} \approx 0$ and $\beta_{I,II} \approx 0$, i.e., the mutual impedance between halves of the oscillating sphere is $Z_{ac\,I,II} \approx 0$, and they can be considered practically as acoustically independent. Thus, the conclusion can be made that radiation impedance of the hemispherical transducer at the wave sizes around and above its resonance frequency practically does not depend on whether it is backed by the rigid or compliant baffle. This is illustrated with Figure 8.26 since in these cases hemisphere does not interact with its images regardless of the phase of their vibration.

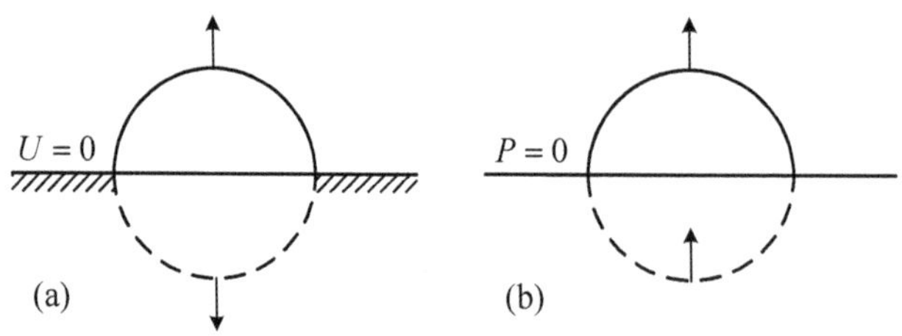

Figure 8.26: Hemispheres and their images that vibrate (a) in phase and (b) in antiphase.

By the analogous considerations, the dependence of the modal diffraction coefficient on *ka* for the first mode baffled transducer can be considered valid for the hemispherical transducers in the range of $ka > 4$. And all the considerations regarding the acoustic interaction between movement of sphere as a rigid body and active first mode of vibration are also applicable to hemispheres with free edges.

The calculated and measured transmit frequency responses of the hemispherical transducer are presented in Figure 8.27. The unintended active modes of vibration (the third membrane and bending modes) can be suppressed by reducing size of the electrodes to the coverage angle $\varphi_1 = 63.5°$, as it is illustrated for the case of the complete spherical transducer in Figure 8.9 and Figure 8.11. This will result in optimizing the effective coupling coefficient for the first mode and in broadening frequency range of non-corrupted directional characteristics. In this case $n_3 = 0$, and the circuit of Figure 8.18 is reduced to the single mechanical contour. The frequency response of the hemispherical transducer with reduced size of electrodes is shown in Figure 8.27 by the dash-dotted line.

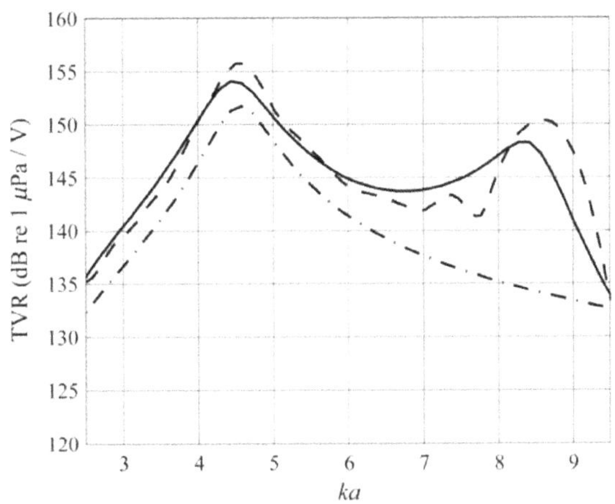

Figure 8.27: The transmit frequency response of the hemispherical transducer: calculated (solid line) and measured (dashed line). The resonance frequencies correspond to the first and third modes of vibration of the complete sphere. By the dash-dotted line the frequency response is shown of the transducer with size of electrodes reduced to $\varphi_1 = 63.5°$

Despite the large wave size of the hemispherical transducer some small level of back radiation exists, and it is approximately the same as in the case of the baffled spherical transducer of the same diameter. This assertion is supported by results of comparing the measured directional characteristics of a hemispherical transducer with modal directional factors calculated for the first mode baffled transducer in the range $3 < ka < 6$ that are shown in Figure 8.20 by the solid thin line.

8.4 References

1. S-H. Ko, G. A. Brigham, and J. L. Butler, "Multimode spherical hydrophone," J. Acoust. Soc. Am. **56**(6), 1890-1898 (1974).
2. S-H. Ko and H. L. Pond, "Improved design of spherical multimode hydrophone," J. Acoust. Soc. Am. **64**(5), 1270-1277 (1978).
3. C. H. Sherman and J. L. Butler, *Transducers and Arrays for Underwater Sound* (Springer, New York, 2007).
4. S. L. Ehrlich, "Spherical acoustic transducer," U.S. Patent No. 3,732,535 (1973); and J. L. Butler and S. L. Ehrlich, "Superdirective spherical radiator," J. Acoust. Soc. Am. **61**(6), 1427-1431, (1977).
5. B. Aronov, D. A. Brown, X. Yan, and C. L. Bachand, "Modal analysis of the electromechanical conversion in piezoelectric ceramic spherical shells," J. Acoust. Soc. Am. **130**(2), 753-763 (2011).
6. B. S. Aronov, "On the optimization of the effective electromechanical coupling coefficients of a piezoelectric body," J. Acoust. Soc. Am. **114**(2), 792-800 (2003).
7. J. J. Horan, J. R. Brown, Jr., and M. F. Pressler, "Open hemispherical transducers", U. S. Patent No. 3,230,504 (1966).
8. T. A. Henriquez, A. C. Tims, "Pressure release hemispherical piezoelectric type transducer," U.S. Patent No. 3,891,871 (1975).
9. B. Aronov, C. L. Bachand, X. Yan, and D. A. Brown, "Analysis of unidirectional broadband piezoelectric spherical shell transducers for underwater acoustics," J. Acoust. Soc. Am. **131**(3), 2079-2090 (2012).

CHAPTER 9

FLEXURAL PLATE TRANSDUCERS

9.1 Introduction

Piezoelectric flexural plate transducers have multiple applications as electroacoustic and electromechanical devices in many fields of engineering including underwater and air acoustics. Our treatment is intended predominantly for underwater applications that imply certain restrictions on configuration of the transducers mechanical systems. For this reason, only the lowest modes of vibration of beams and plates are considered as the most usable for underwater applications. Though the method of theoretical analysis used can be applied to a broader range of configurations and higher modes of vibration, which may be of interest for designing different electromechanical devices in a straightforward way. As to specifics that arise from requirements for the underwater projectors and receivers of the flexural type, they will be considered in Chapters 13 and 14. Besides, a comprehensive treatment of the underwater circular disk and rectangular beam transducers was made by R. Woollett in Refs. 1 and 2.

The most widely used variants of mechanical system configurations of the transducer that can meet different requirements for their operational parameters and environmental conditions are schematically shown in Figure 9.1 with examples of the circular plates for the case that their boundary is simply supported. The cross sections of the rectangular beams look in the same way, if to replace notations for coordinate r by x and for radius a by $l/2$, where l is the length of a beam. Therefore, references to the Figure 9.1 will be made for both the circular plates and rectangular beams.

The "bimorph" configuration (Figure 9.1(a)) can be considered as the basic design, and the "trilaminar" (Figure 9.1(b)) and "bilaminar" (Figure 9.1(c)) configurations as its modifications. Less typical for underwater applications are flexural type transducers shown in Figure 9.2.

Transducers of these types may be used as accelerometers in designs of the pressure gradient sensors, but mainly they are employed as electromechanical drivers and sensors for general applications.

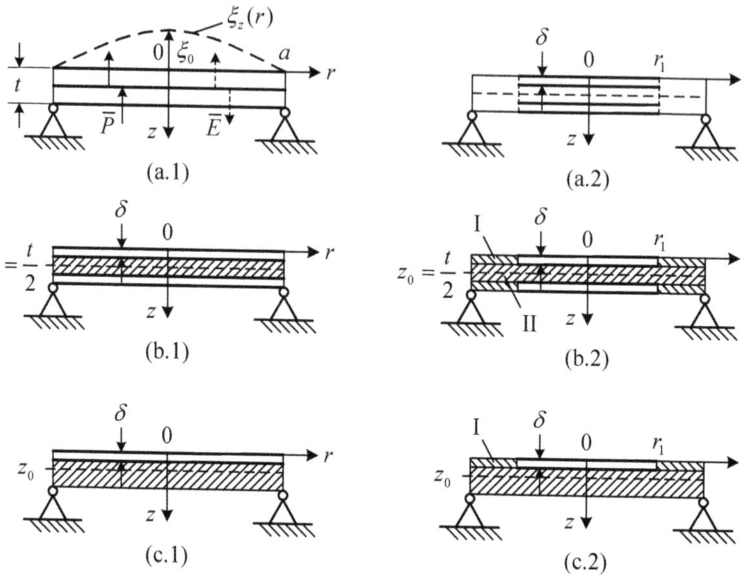

Figure 9.1: Illustration of the flexural plate transducers. (a) "Bimorph" mechanically uniform transducer considered as the basic design (1- fully active, 2- partially active). (b) Trilaminar transducers, (c) bilaminar transducers. Transducers (b.1) and (c.1) are radially uniform. Transducers (b.2) and (c.2) are radially nonuniform. Electrodes are shown by the solid bold lines, δ is the thickness of piezoceramic lamination, shaded parts indicate inactive material. Arrows indicate: $\rightarrow \overline{P}$ the vector of polarization, $- - \rightarrow \overline{E}$ the electric field vector.

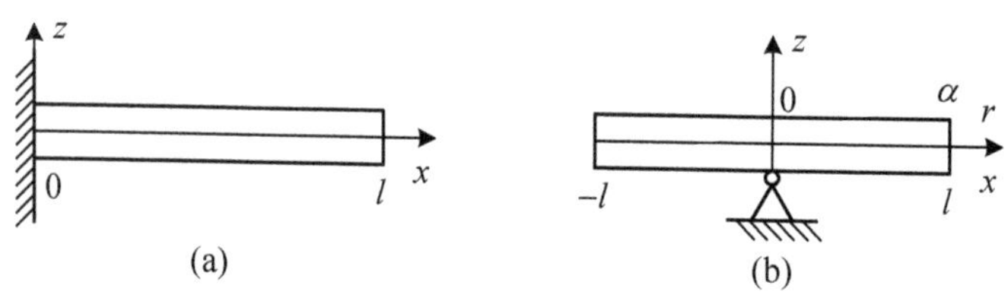

Figure 9.2: (a) Cantilever bimorph beam and (b) center supported (dual cantilever) beam, or circular plate. The dual cantilever design is a way of insuring clamping of the cantilever at one and (in the middle cross section) that otherwise is hard to achieve practically.

Transducers of the flexural type are used in the frequency range around and below the first resonance frequency. In this frequency range they mostly (with exclusion of highly nonuniform and virtually not usable for effective electromechanical transduction mechanical systems) can be considered as systems with one mechanical degree of freedom, and properties of the transducers as electromechanical devices can be represented in the transmit mode (for the projectors and drivers in general) and receive mode (for hydrophones and sensors in general) using the common equivalent electromechanical circuit with single mechanical contour that is repeatedly shown for convenience in Figure 9.3

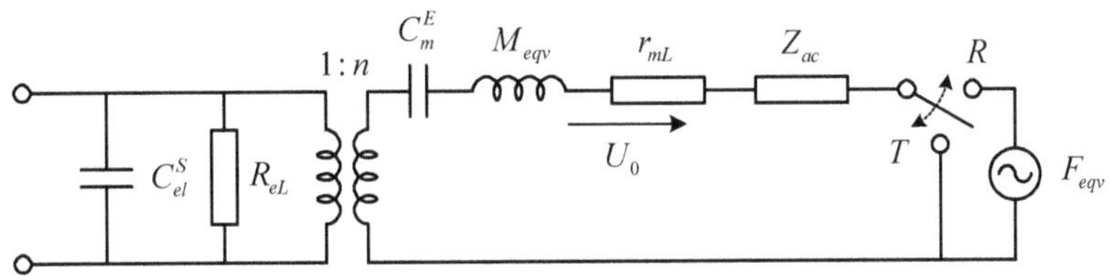

Figure 9.3: The equivalent electromechanical circuit of a flexural type transducer having one mechanical degree of freedom. Positions of the switch: *T* for the transmit, and *R* for the receive modes of operation.

Theory of the uniform bimorph and trilaminar modifications of the beam and circular plate flexural transducers for underwater applications was well developed by Woollett[2] though using a different methodical approach. Analysis of the uniform bimorph circular plate transducers using the energy method was made in Ref. 3.

As the parts of mechanical systems of the transducers located near their boundaries and neutral planes are less effective in terms of electromechanical energy conversion, an active material at those parts can be replaced by a passive material (e.g., by a metal) practically without loss or even with an increase of the effective coupling coefficient of the transducer. Thus, the length and radially nonuniform trilaminar and especially bilaminar transducers (shown schematically in Figure 9.1(b.2) and (c.2)) may become advantageous in terms of their reliability, cost effectiveness, and possibilities of manufacturing thin low frequency transducers. Assumption of applicability of the single degree of freedom approximation may become questionable regarding such mechanical systems, in general. But, as it is shown in Sections 4.5.6 and 4.5.7 with examples of the beams and circular plates nonuniform over their length and radius, this assumption can be adopted to the first approximation at reasonable for transducers designing degree of nonuniformity. A number of publications were devoted to considering nonuniform circular plate transducers predominantly for in air applications. A detailed list of references to these publications can be found, for example, in Refs. 4-6. A brief analysis of the approximate analytical methods that are used in these works is made in Ref. 8, where it is concluded that they are hardly applicable for treating transducers of more general types including underwater transducers. In the further treatment the approach used in Ref. 8 is accepted.

9.2 Electromechanical Conversion in the Flexural Transducers

Mechanical systems of the flexural transducers are in the shape of the circular plates or the rectangular beams and plates that are completely or partially made of piezoceramics, as shown in Figure 9.1. When considering rectangular plates, we will assume that they are made up of several beams mechanically separated. Examples of the flexural plate transducers of the basic design were used in Section 5.6 for illustrating the electromechanical conversion in nonuniformly deformed bodies. Here all the variants of transducer designs are analyzed in detail.

The electromechanical energy conversion in the plate transducers will be mainly considered under the transverse piezoelectric effect, which is typically used in transducers of this type. The exceptions may be rectangular beams (plates composed of the beams) that are tangentially polarized or made segmented in the manner as the rings were shown to be built, in which cases the longitudinal piezoelectric effect is used. In the analysis of electromechanical conversion, we assume that the plates are thin (so that elementary theory of bending is applicable) and that the boundary conditions are ideal. Effects due to deviation from these assumptions will be considered separately and resulting corrections will be introduced.

For calculating energies involved in the electromechanical conversion expressions (5.67) and (5.70) for the energy densities can be used, since in the case of the transverse piezoeffect for the beams it holds that $T_2 = T_3 = 0$ and for the circular plates $T_3 = 0$. But we will repeat their derivation here for the sake of convenience, moreover, this is sufficiently straightforward. At first, we consider rectangular beams.

9.2.1 Electromechanical Conversion in the Piezoceramic Beams

Conventional assumption for the beams in the elementary theory of bending is that $w, t \ll l$. Thus, it can be considered that $T_2 = T_3 = 0$, $T_1 \neq 0$ and the one-dimensional piezoelectric equations in the form

$$T_1 = \frac{1}{s_{11}^E} S_1 - \frac{d_{31}}{s_{11}^E} E_3, \tag{9.1}$$

$$D_3 = \frac{d_{31}}{s_{11}^E} S_1 + \varepsilon_{33}^T (1 - k_{31}^2) E_3 \,. \tag{9.2}$$

are valid. In Eq. (9.2) $\varepsilon_{33}^T (1 - k_{31}^2) = \varepsilon_{33}^{S_1}$ is the dielectric constant of piezoelectric material in the beam clamped in the direction of axis (at $S_1 = 0$).

Following Eq. (4.28) in the coordinate system shown in Figure 9.1 expression for strain is

$$S_1 = S_x = -(z - z_0) \frac{\partial^2 \xi_z}{\partial x^2} = -(z - z_o) \xi_o \frac{d^2 \theta}{dx^2} \,, \tag{9.3}$$

where z_0 is coordinate of the neutral plane, $\xi_z = \xi_o \theta(x)$, ξ_o is displacement of the reference point on the surface of the beam, and $\theta(x)$ is the mode shape of vibration. For the beams uniform over length the mode shapes of vibration for different boundary conditions on the ends are considered in Section 4.3.4. Vibration of the nonuniform over length beams is considered in Section 4.5.6. As a result, the conclusion is made that for extend of nonuniformity that is reasonably acceptable for the transducer designs the mode shapes of vibration to the first approximation are the same as for the uniform beams. Thus, the mode shapes $\theta(x)$ will be considered known for all the modifications of beams shown in Figure 9.1.

The stress at $E_3 = 0$ in Eq. (9.1) is

$$T_1^E = T_x = -(z - z_0) Y_1^E \xi_o \frac{d^2 \theta}{dx^2} \,. \tag{9.4}$$

In this form the expression for stress is valid for parts of a beam that are made of piezoceramics. For the nonuniform beams that are composed of active and passive materials formula (9.4) can be generalized as

$$T_1 = -(z - z_0) Y(z, x) \xi_o \frac{d^2 \theta}{dx^2} \,, \tag{9.5}$$

where $Y(z, x) = Y_p$ on the parts of the beam made of passive material with Young's modulus Y_p, and $Y(z, x) = Y_a^E$ on the active parts of the beam (at the transverse piezoeffect $Y_a^E = Y_1^E = 1/s_{11}^E$, at the longitudinal piezoeffect $Y_a^E = Y_3^E = 1/s_{33}^E$).

The general expression for the potential energy of deformation of a beam under flexure is

$$W_{pot} = \frac{1}{2} \int_{\tilde{V}} S_1 T_1 d\tilde{V} = \frac{\xi_o^2}{2} w \int_{-l/2}^{l/2} \int_{-t/2}^{t/2} (z - z_0)^2 Y(x, z) \left(\frac{d^2 \theta}{dx^2} \right)^2 dz dx = \frac{\xi_o^2}{2} K_{eqv}^E \,. \tag{9.6}$$

Here K_{eqv}^E is the equivalent rigidity of the beam. Superscript E indicates that elastic constant of piezoceramic part of a beam that in general may be nonuniform should be taken at $E_3 = 0$. For the beam with simply supported ends that is in full made of active ceramic material

$$K_{eqv}^E = \frac{\pi^4}{24}\frac{wt^3}{l^3}Y_1^E . \tag{9.7}$$

9.2.1.1 Electromechanical transformation coefficients

The electromechanical energy associated with the flexural deformation is

$$W_{em} = \frac{1}{2}\int_{\tilde{V}_a} D_3^E E_3(z,x)d\tilde{V}_a = \frac{1}{2}\int_{\tilde{V}_a}\frac{d_{31}}{s_{11}^E}S_1 E_3(z,x)d\tilde{V}_a = \frac{1}{2}\xi_o Vn . \tag{9.8}$$

Here $\tilde{V}_a$ is the volume of active material, and $E_3(z,x)$ is the electric field within the volume of active material, in other words, the volume confined between the electrodes; $D_3^E = (d_{31}/s_{11}^E)S_1$ is the charge density determined from Eq. (9.2) at $E_3 = 0$; V is the voltage applied, n is the coefficient of electromechanical transformation. The electric field can be represented as

$$E_3(z,x) = (V/\delta)\cdot\Omega_t(z)\cdot\Omega_l(x), \tag{9.9}$$

where δ is separation between electrodes, and $\Omega_t(z)$, $\Omega_l(x)$ are the nondimensional functions that depend on configuration of the electrodes in the volume of a beam and on their polarities. At passive parts of the beam, they vanish. At active parts their moduli are $|\Omega_t(z)| = |\Omega_l(x)| = 1$, and their signs depend on the mutual direction of vectors of polarization and electric field in the volume of ceramics between the electrodes. Now expression (9.8) for the electromechanical energy after substituting strain S_1 by formula (9.3) can be transformed to

$$W_{em} = \frac{1}{2}\xi_o V(-)\frac{wd_{31}}{\delta s_{11}^E}\left(\int_0^t \Omega_t(z)(z - z_0)dz\right)\cdot\left(\int_{-l_e/2}^{l_e/2}\Omega_l(x)\frac{d^2\theta}{dx^2}dx\right). \tag{9.10}$$

By comparing expressions (9.8) and (9.10) the electromechanical transformation coefficient of the beam can be represented as

$$n = -\frac{wd_{31}}{\delta s_{11}^E}\Omega_t\cdot\Omega_l , \tag{9.11}$$

where it is denoted

$$\Omega_t = \int_0^t \Omega_t(z)(z - z_0)dz\ , \qquad (9.12)$$

$$\Omega_l = \int_{-l_e/2}^{l_e/2} \Omega_l(x)\frac{d^2\theta}{dx^2}dx\ . \qquad (9.13)$$

These factors depend on the status of electrodes over the surface and over the thickness of a beam. For illustration consider the bimorph beams shown in Figure 9.1(a), that are composed of two identical laminates of piezoceramics. Coordinate of the neutral plane is $z_0 = t/2$. In terms of polarities of the electrodes and mutual direction of the vectors of polarization $\bar{\boldsymbol{P}}$ and electric field $\bar{\boldsymbol{E}}_3$ the variants possible that are shown in Figure 9.4 and Figure 9.5. The convention is that $\Omega_t(z) > 0$ and $\Omega_l(x) > 0$ in the parts of piezoceramic volume, in which directions of the vectors of polarization and electric field coincide ($\bar{E}_3 \uparrow\uparrow \bar{P}$), and $\Omega_t(z) < 0$, $\Omega_l(x) < 0$ in the parts, where they are in opposition ($\bar{E}_3 \uparrow\downarrow \bar{P}$).

At first, consider Figure 9.4, where different mutual directions of vectors of polarization and electric field are shown over the thickness.

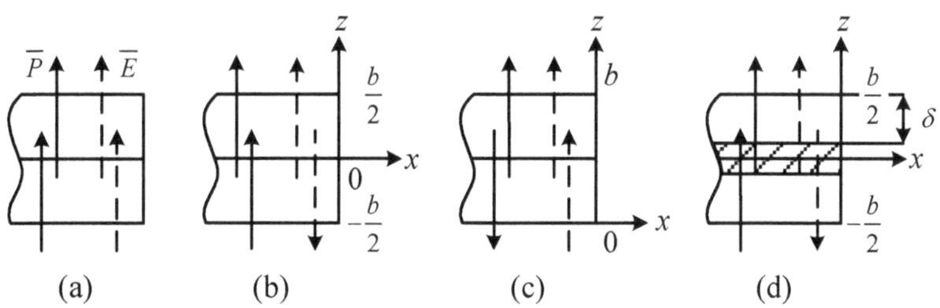

Figure 9.4: Variants of electrical connection of the piezoceramic laminates forming bimorph piezoelement: (a) distractive connection that does not result in generating the bending moment; (b) parallel connection of the halves of the laminates, $\delta = t/2$; (c) their series connection, $\delta = t$; (d) parallel connection of active laminates typical for trilaminar design, dashed area is passive material. The vector of polarization, $\bar{P}$, solid line; the electric field vector, $\bar{E}$, dashed line.

In the variant (a) $\Omega_t(z)$ does not change sign over the thickness, and as result of integrating in expression (9.12), in which $z_0 = t/2$, will be obtained that factor $\Omega_t = 0$, i.e., electromechanical conversion does not take place. In variants (b) and (c) $\Omega_t(z) = -1$ at $0 \le z < t/2$, and $\Omega_t(z) = 1$ at $t/2 < z \le t$ Result of integrating in (9.12) will be

$$\Omega_t = t^2/4\ . \qquad (9.14)$$

Difference between the variants is that in variant (b), which corresponds to the parallel connection of halves of the beam, separation between the electrodes is $\delta = t/2$; in variant (c), which corresponds to the series connection of the halves, $\delta = t$. As the result, the capacitances of the beams will be different. In variant (d) the electrodes are imbedded into volume of the beam, $\Omega_t(z) = -1$ at $0 \le z < \delta$, and $\Omega_t(z) = 1$ at $t - \delta < z \le t$. At $\delta < z < t - \delta$ function $\Omega_t(z) = 0$. Result of integrating in (9.11) is

$$\Omega_t = \delta(t - \delta). \tag{9.15}$$

Consider now variants of the electrode's configuration over the length of a beam that are shown in Figure 9.5.

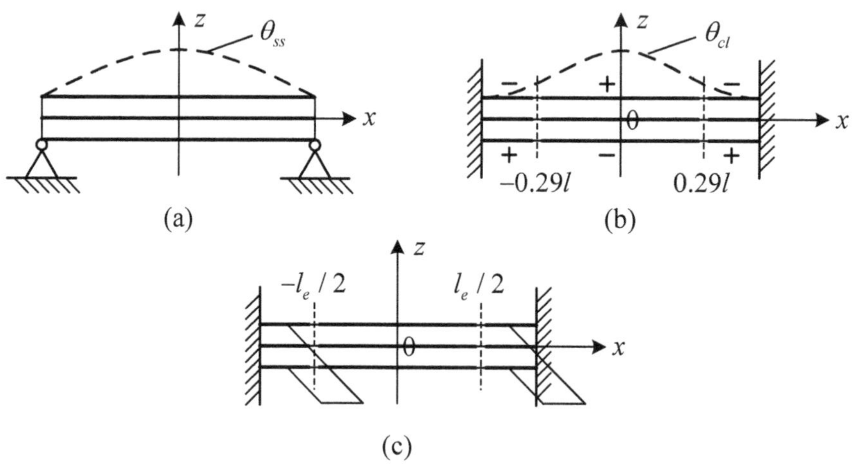

Figure 9.5: Variants of the electrodes configuration over the length of a beam: (a) full size unipolar electrodes; (b) the electrodes are split into parts, and their periphery parts are connected in anti-phase to the central part; (c) the external parts of the electrodes are short circuited, thus $E_3 = 0$ in the volume between them. $\theta_{ss}(x)$ and $\theta_{cl}(x)$ are the mode shapes of the beams with simply supported and clamped ends.

Note that the modes of vibration are presented in Figure 9.5 (a) and (b) in the coordinate system with origin on the left end of the beam unlike in Figure 9.1(a.1). In the variant of the full size unipolar electrode $\Omega_l(x)$ does not change sign along the length, and integrating in (9.13) performed in the interval [0, l) results in

$$\Omega_l = \left.\frac{d\theta}{dx}\right|_{x=l} - \left.\frac{d\theta}{dx}\right|_{x=0}. \tag{9.16}$$

Thus, this factor completely depends on the turning angles of cross sections of a beam on the edges of the electrodes (tangents to the mode shape on the edges of the electrodes), which in this case coincide with the ends of the beam. For a beam with simply supported ends $\theta_{ss}(x) = \sin(\pi x / l)$ and

$$\Omega_l = \frac{2\pi}{l}. \tag{9.17}$$

It is instructive to consider the electromechanical conversion performed by the vibrating beam with clamped ends. The boundary conditions on the clamped ends are $\theta(x)|_{x=\pm l/2} = 0$ and $(d\theta / dx)_{x=\pm l/2} = 0$. According to formula (9.16) $\Omega_l = 0$, and the overall effect of electromechanical conversion under the flexural deformation of the beam does not exist. This is due to the fact that the electrical charge density that is proportional to curvature of the deflection curve ($D_3(x) \sim S_1(x) \sim (d^2\theta / dx^2)$) according to Eqs. (9.2) and (9.3)) changes sign at the coordinates, at which the curvature changes sign (at point of inflection), and the total charge over the electrodes appears to be zero. The deflection curve for a beam with clamped ends (4.493) is

$$\theta(x) = \frac{16x^2}{l^2}\left(1 - \frac{2x}{l} + \frac{x^2}{l^2}\right). \tag{9.18}$$

Coordinates of the inflection points found from equation $(d^2\theta / dx^2) = 0$ are $x_1 = 0.21l,\ x_2 = 0.79l$. The electromechanical conversion can be "revived" by splitting the electrodes on the lines corresponding to these coordinates, and then by connecting the periphery parts to the central part in opposite phase, as it is shown in Figure 9.5(b). In this case the factor Ω_l may be calculated from expression (9.13) taking into consideration that $\Omega_l(x) = 1$ at $0.21l < x < 0.79l$, and $\Omega_l(x) = -1$ at $0.21l < x < 0.79l$ and $0.79l < x < l$. Simple manipulations result in

$$\Omega_l = \left(\left.\frac{d\theta}{dx}\right|_{x=0.79l} - \left.\frac{d\theta}{dx}\right|_{x=0.21l}\right) - 2\left(\left.\frac{d\theta}{dx}\right|_{x=0.21l} - \left.\frac{d\theta}{dx}\right|_{x=0}\right) = \frac{12.4}{l}. \tag{9.19}$$

The ideal clamped boundary conditions are hardly achievable in practical transducer designs, and the theoretical gain that is obtained by antiphase connection of parts of electrodes close to the ends may be not so impressive. More reliable way of avoiding a possible distractive contribution of end parts of a beam to electromechanical conversion in situation, when boundary conditions are not exactly known, is to use the central part of electrodes only, as is shown in Figure 9.5(c). In this case it is more convenient to use formula (9.13) in its original form that corresponds to position of $x = 0$ in the middle of a beam. For the simply supported beam this will result in

$$\Omega_l = \left.\frac{d\theta_{ss}}{dx}\right|_{x=l_e/2} - \left.\frac{d\theta_{ss}}{dx}\right|_{x=-l_e/2} = \frac{2\pi}{l}\sin\frac{\pi l_e}{2l}. \tag{9.20}$$

Combining general expressions (9.11) for electromechanical transformation coefficient and expressions (9.15) for Ω_t and (9.20) for Ω_l we obtain for the variant of simply supported beam shown in Figure 9.1(a.2)

$$n = \frac{2\pi w d_{31}}{l s_{11}^E}(t-\delta)\sin\frac{\pi l_e}{2l}. \tag{9.21}$$

The electromechanical transformation coefficient characterizes force of electromechanical origin that generates vibration of mechanical system of a transducer,

$$F_{em} = nV. \tag{9.22}$$

As it was noted in Section 2.6.1, sign (-) in the expressions (9.10) and therefore in formula (9.11) is in accord with the sign convention formulated in Section 1.5.2. It indicates that at the conditionally positive direction of the electric field) that coincides with direction of polarization and conditionally positive curvature (with convex down and bending moment positive in anti-clockwise direction on the right end) the electromechanical force is directed in positive direction at the transverse piezoeffect (given that d_{31}), or in negative direction at the longitudinal piezoeffect. This results in compression or tension in the upper layer of ceramics, respectively. The same explanation is valid for the circular plates under flexure. For calculating parameters of the single mode transducers these details do not matter, and the sign (-) in expressions for the transformation coefficients can be further omitted. Keeping the signs is important, when considering multimode transducers, in which case the modal transformation coefficients may have different signs. Such a situation takes place, for example, in calculating spherical transducer (see note under formulas (8.75)).

9.2.1.2 The Effective Coupling Coefficients

An important figure of merit that characterizes quality of electromechanical conversion of energy performed by the transducer is the effective coupling coefficient that can be calculated by formula

$$k_{eff}^2 = \frac{\alpha_c}{1+\alpha_c}, \text{ where } \alpha_c = \frac{n^2}{C_{el}^{S_1} K_{eqv}^E}. \tag{9.23}$$

It can be readily determined so far as the equivalent parameters of a transducer are known. The equivalent rigidity may be determined from Eq. (9.6). For determining the capacitance $C_{el}^{S_1}$ of the transducer at $S_1 = 0$ consider the electric energy of a beam clamped in direction *x*,

$$W_{el}^{S_1} = \frac{1}{2}\int_{\tilde{V}_a} \varepsilon_{33}^{S_1} E_3^2(z,x) d\tilde{V}_a = \frac{1}{2} C_{el}^{S_1} V^2, \tag{9.24}$$

Upon substituting expression (9.9) for the electric field and integrating over the active part of the beam we arrive at

$$C_{el}^{S_1} = \varepsilon_{33}^T (1-k_{31}^2) \frac{w}{\delta^2} \int_{-l_e/2}^{l_e/2} \Omega_l^2(x) dx \int_0^t \Omega_t^2(z)(z-z_0) dz. \tag{9.25}$$

For the case with embedded electrodes shown in Figure 9.1 (a2)

$$C_{el}^{S_1} = \varepsilon_{33}^T (1-k_{31}^2) \frac{2w}{\delta} l_e. \tag{9.26}$$

The absolute value of the effective coupling coefficient of the fully active simply supported beam following formulas (9.23), (9.21) for *n*, (9.26) for $C_{el}^{S_1}$, and (9.7) for K_{eqv}^E will be obtained as

$$k_{eff}^2 = \frac{6k_{31}^2}{\pi^2 - (\pi^2 - 6)k_{31}^2}. \tag{9.27}$$

For the beam made of PZT-4 $k_{eff} \approx 0.8 k_{31}$.

9.2.1.3 Employing the Longitudinal Piezoeffect

Consider now the peculiarity of the electromechanical conversion that arises from utilizing the longitudinal piezoelectric effect. This is typical of the rectangular beams intended for underwater projectors, operation of which is often electric field limited and therefore requires the electromechanical transformation coefficients to be as large as possible. Practically, the longitudinal piezoeffect can be realized in the designs shown in Figure 9.6 (a) and (b). In this case the beams must be cemented from the piezoelements that are polarized in the length direction and connected electrically as it is shown schematically in the Figure 9.6. Note that the piezoceramic bars that are tangentially polarized can also be used in the designs of Figure 9.6 in a straightforward way, but this variant would require a special approach to calculating electromechanical parameters, which is described in Section 7.2.2.1.2, and therefore will not be considered here.

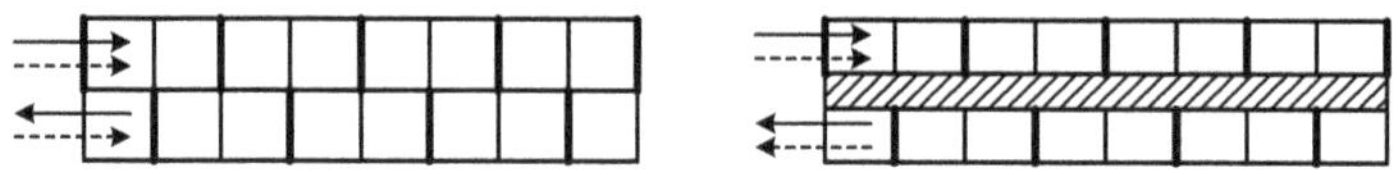

(a) (b)

Figure 9.6: The beam designs that utilize the longitudinal piezoeffect. (a) The beam is cemented out of two segmented bars, or of the prisms with electrodes separated analogous to those shown in Figure 7.1 (c). (b) The beam is cemented out of segmented bar and of the metal lamination with a dielectric substrate between them (trilaminar design). The electrodes having (+) polarity in process of polarization are shown by the solid lines. The solid arrow-line $\rightarrow$ indicates the direction of the polarization; the dashed arrow-line $--\rightarrow$ indicates the direction of the operating electric field.

In the variant of the longitudinal piezoeffect $T_1 \approx T_2 \approx 0$, $T_3 \neq 0$, and the appropriate piezoelectric equations are,

$$T_3 = \frac{1}{s_{33}^E} S_3 - \frac{d_{33}}{s_{33}^E} E_3, \tag{9.28}$$

$$D_3 = \frac{d_{33}}{s_{33}^E} S_3 + \varepsilon_{33}^T (1 - k_{33}^2) E_3 . \tag{9.29}$$

Here $\varepsilon_{33}^T (1 - k_{33}^2) = \varepsilon_{33}^{S_3}$ is the dielectric constant of piezoelectric material in the beam clamped in direction of axis x (at $S_3 = 0$).

Expressions for the equivalent parameters of a transducer utilizing the longitudinal piezoeffect may be readily obtained from those that were determined for the variant of the transverse effect. Thus, when calculating the equivalent rigidity from expression (9.6) it should be considered that $Y_a^E = 1 / s_{33}^E$. After denoting

$$E_3(z,x) = (V/\Delta)\cdot\Omega_t(z)\cdot\Omega_l(x), \tag{9.30}$$

where $\Delta = l_{el}/N$ (l_{el} is the total length of the segmented bar, N is the number of segments) and functions $\Omega_t(z)$ and $\Omega_l(x)$ have the same meaning as in the case of the transverse piezoeffect, the general expression for the electromechanical transformation coefficient will become

$$n = \frac{wd_{33}}{\Delta s_{33}^E}\Omega_t\cdot\Omega_l, \tag{9.31}$$

instead of the formula (9.11) for the transverse piezoeffect. In the variant of the simply supported beam formula for the transformation coefficient becomes

$$n = 2\pi\frac{wd_{33}}{s_{33}^E}\frac{(t-\delta)}{l}\frac{\delta}{\Delta}\sin\frac{\pi l_{el}}{2l}. \tag{9.32}$$

The capacitance of the beam at parallel connection of the segments is

$$C_{el}^{S_3} = \varepsilon_{33}^T(1-k_{33}^2)\frac{2w\delta}{\Delta}N. \tag{9.33}$$

As it was discussed in Section 5.5.2, an additional to K_{eqv}^E rigidity ΔK must be included that arises, if a nonuniform distribution of strain in direction of the electric field of polarization takes place. In the case under consideration the distribution of strain along the field of polarization (over the length of the beam) is ($S_3(x) \sim \cos(2\pi x/l)$), and the additional rigidity theoretically must be taken into account. But, as it follows from results presented in Section 5.5.2.1 in the case that piezoelement is segmented along the electric field lines, and number of segments is $N > 6$, the ΔK term is small and can be neglected compared with K_{eqv}^E. It is noteworthy that in the case of the transverse effect the distribution of strain through the thickness ($S_1(z) \sim z$), i. e. in direction of polarization, also results in an additional rigidity. According to formula (5.162) this rigidity has value

$$\Delta K = 0.25\frac{k_{31}^2}{1-k_{31}^2}K_{eqv}^E. \tag{9.34}$$

Given that for PZT piezoceramics k_{31}^2 is on the order of 0.1 this term can be also neglected.

Expression for the effective coupling coefficients under the longitudinal piezoeffect differs from expression (9.27) for the transverse piezoeffect by the absolute value of the coupling coefficient of piezoceramics. Thus, for the fully active simply supported beam

$$\alpha_c = \frac{6}{\pi^2}\frac{k_{33}^2}{1-k_{33}^2}, \quad k_{eff}^2 = \frac{6k_{33}^2}{\pi^2-(\pi^2-6)k_{33}^2}. \tag{9.35}$$

9.2.2 Electromechanical Conversion in the Piezoceramic Circular Plates

The plates are assumed to be thin compared with the radius, $t \ll a$, and free of stress on the major surfaces. Thus, the stress T_3 in the axial direction throughout the thickness can be neglected. The appropriate piezoelectric equations are

$$S_1 = s_{11}^E T_1 + s_{12}^E T_2 + d_{31}E_3, \tag{9.36}$$

$$S_2 = s_{12}^E T_1 + s_{11}^E T_2 + d_{31}E_3, \tag{9.37}$$

$$D_3 = d_{31}(T_1+T_2) + \varepsilon_{33}^T E_3. \tag{9.38}$$

The electrodes and the plate vibrations are axially symmetric. Therefore, the strains in the polar coordinates are according to formulas (4.182) and (4.183)

$$S_1 = S_{rr} = -(z-z_0)\xi_o \frac{\partial^2\theta}{\partial r^2}, \tag{9.39}$$

$$S_2 = S_{\varphi\varphi} = -(z-z_0)\xi_o \frac{1}{r}\frac{\partial\theta}{\partial r}, \tag{9.40}$$

where $\xi(z) = \xi_o\theta(r/a)$. Upon substituting $T_1 + T_2$ from Eqs. (9.36) and (9.37) into Eq. (9.38) we obtain

$$D_3 = \frac{d_{31}}{s_{11}^E + s_{12}^E}(S_1+S_2) + \varepsilon_{33}^{S_{1,2}}E_3, \tag{9.41}$$

where $\varepsilon_{33}^{S_{1,2}} = \varepsilon_{33}^T(1-k_p^2)$, and $k_p^2 = 2d_{31}^2/\varepsilon_{33}^T(s_{11}^E+s_{12}^E)$ is the square of the planar coupling coefficient of the piezoceramic material. From Eq. (9.41) the charge density at $E_3 = 0$ is

$$D_3^E = \frac{d_{31}}{s_{11}^E + s_{12}^E}(S_1+S_2). \tag{9.42}$$

It follows from Eqs. (9.36) and (9.37) that stress in piezoelectric material at $E_3 = 0$ are

$$T_1^E = \frac{Y_1^E}{1-(\sigma_1^E)^2}S_2 - \frac{\sigma_1^E}{1-(\sigma_1^E)^2}S_2, \tag{9.43}$$

$$T_2^E = \frac{\sigma_1^E}{1-(\sigma_1^E)^2} S_2 - \frac{Y_1^E}{1-(\sigma_1^E)^2} S_2 . \tag{9.44}$$

For the nonuniform plates that are composed of active and passive materials these formulas can be generalized in the same way, as it was done for the beams, namely,

$$T_1 = \frac{Y(z,r)}{1-\sigma(z,r)^2} S_1 - \frac{\sigma(z,r)}{1-\sigma(z,r)^2} S_2 , \tag{9.45}$$

$$T_2 = \frac{\sigma(z,r)}{1-\sigma(z,r)^2} S_1 - \frac{Y(z,r)}{1-\sigma(z,r)^2} S_2 . \tag{9.46}$$

Here $Y(z,r) = Y_p$ and $\sigma(z,r) = \sigma_p$ on the parts of the plate made of passive material with Young's modulus Y_p and Poison's ratio σ_p, and $Y(z,r) = Y_1^E$, $\sigma(z,r) = \sigma_1^E$ on the active parts of the beam ($Y_1^E = 1/s_{11}^E$, $\sigma_1^E = -s_{12}^E / s_{11}^E$). For brevity, the modified elastic modulus will be introduced

$$Y_\sigma = \frac{Y}{1-\sigma^2} , \tag{9.47}$$

as the combination of the Young's modulus and Poisson's ratio for the material. On the passive and active parts of a plate the modified elastic moduli are

$$Y_{\sigma p} = \frac{Y_p}{1-\sigma_p^2} \text{ and } Y_{\sigma a} = \frac{Y_1^E}{1-\sigma_1^{E2}} . \tag{9.48}$$

9.2.2.1 Electromechanical Transformation Coefficient

Expression for the electromechanical energy associated with deformation of a circular plate is

$$W_{em} = \frac{1}{2}\int_{\tilde{V}} D_3^E E_3 d\tilde{V} = \frac{1}{2}\frac{2\pi d_{31}}{s_{11}^E + s_{12}^E}\int_0^a\int_0^t (S_1 + S_2) E_3(r,z) r dr dz = \frac{1}{2}\xi_0 V n . \tag{9.49}$$

Here expression (9.42) for the charge density D_3^E is used, and strain S_1, S_2 are determined by formulas (9.39) and (9.40). Strength of the electric field $E_3(r,z)$ can be represented by expression analogous to (9.9) for the beams as

$$E_3(r,z) = (V/\delta)\Omega_t(z)\Omega_r(r/a) . \tag{9.50}$$

The function $\Omega_r(r/a)$ characterizes the geometry and polarity of the electrodes and their electrically isolated parts over the radius. It is similar to the function $\Omega_l(x)$ for the beams.

Suppose that the electrodes are divided into two parts, as shown in Figure 9.7. Within these parts $\Omega_r(r/a) = \pm 1$, where the sign depends on polarity of the electrodes. The variant of full size

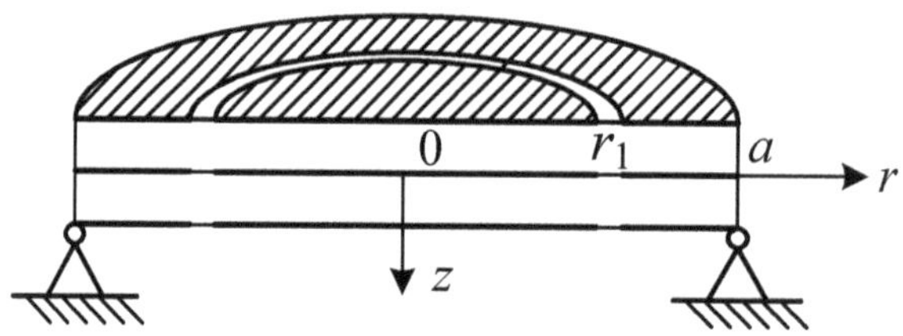

Figure 9.7: The piezoelectric ceramic plate with electrodes divided into two parts.

electrodes correspond to $r_1 = a$. If only the part of the electrode having radius r_1 is active, we assume that the remaining part is short circuited, in which case the rigidity of the transducer remains unchanged, as the condition of the constant electric field is fulfilled throughout the volume of the plate. If this part remains open circuited, its elastic constant must be determined at $D_3 = 0$ and, strictly speaking, the plate must be considered as nonuniform over the radius. After substituting expressions for S_1 and S_2 into formula (9.49) and integrating over the volume will be obtained that

$$W_{em} = \frac{1}{2}\xi_0 V \frac{2\pi d_{31}}{s_{11}^E + s_{12}^E}\Omega_t\Omega_r = \frac{1}{2}V\xi_0 n, \tag{9.51}$$

where

$$\Omega_t = \int_0^t \Omega_t(z)(z - z_0)dz \tag{9.52}$$

is the same quantity as by expression (9.12) for the rectangular beam; and

$$\Omega_r = \int_0^{r_1}\left(\frac{d^2\theta}{dr^2} + \frac{1}{r}\frac{d\theta}{dr}\right)rdr = \int_0^{r_1}\frac{d}{dr}\left(r\frac{d\theta}{dr}\right)dr = r_1\left.\frac{d\theta}{dr}\right|_{r=r_1}. \tag{9.53}$$

If both parts of electrodes in Figure 9.7 are used and connected in the opposite phase, then

$$\Omega_r = \int_0^{r_1}\frac{d}{dr}\left(r\frac{d\theta}{dr}\right)dr - \int_{r_1}^{a}\frac{d}{dr}\left(r\frac{d\theta}{dr}\right)dr = r_1\left.\frac{d\theta}{dr}\right|_{r=r_1} - \left(a\left.\frac{d\theta}{dr}\right|_{r=a} - r_1\left.\frac{d\theta}{dr}\right|_{r=r_1}\right). \tag{9.54}$$

Following Eq. (9.51) the electromechanical transformation coefficient is

$$n = \frac{2\pi d_{31}}{(s_{11}^E + s_{12}^E)\delta}\Omega_t\Omega_r \ . \tag{9.55}$$

In this section we will assume that the plate has bimorph design with active laminates connected in parallel. In this case $\delta = t/2$ and $\Omega_t = t^2/4$ according to formula (9.14) and

$$n = \frac{\pi d_{31} t}{s_{11}^E + s_{12}^E}\Omega_r \ . \tag{9.56}$$

The coefficient Ω_r and therefore the electromechanical transformation coefficient, is determined by the slope of the mode of vibration on the contours of electrodes. It may have a negative sign. Regarding nature of signs of the electromechanical transformation coefficient see the note under formula (9.22). Further the signs will be omitted unless noted otherwise. If the electrodes are divided into parts, it depends also on the polarity of the parts. The mode of vibration of the plate is defined by the boundary conditions, which may in practice vary significantly depending on the transducer design. In the case that the boundary is simply supported, the mode shape being approximated by the static deflection curve is given by expression (see (4.495))

$$\theta(r/a) = (1 - r^2/a^2)(1 - r^2/4a^2) \, . \tag{9.57}$$

and coefficient Ω_r is

$$\Omega_{r\,ss} = r_{el}\left.\frac{d\theta}{dr}\right|_{r=r_{el}} = -\frac{r_{el}^2}{a^2}\left(2.5 - \frac{r_{el}^2}{a^2}\right). \tag{9.58}$$

It is informative to consider the electromechanical parameters of circular plates for the boundary conditions other than simply supported. One of the reasons behind this is that in a practical transducer design it is hard to achieve the ideal boundary conditions, and it is good to have an estimate as how the change in boundary conditions can affect the transducer parameters.

Thus, for the clamped boundary conditions (Figure 9.8) the static deflection curve is (see (4.496))

$$\theta(r/a) = \left(1 - \frac{r^2}{a^2}\right)^2, \tag{9.59}$$

and

$$\Omega_{rcl} = -4\frac{r_{el}^2}{a^2}\left(1-\frac{r_{el}^2}{a^2}\right). \tag{9.60}$$

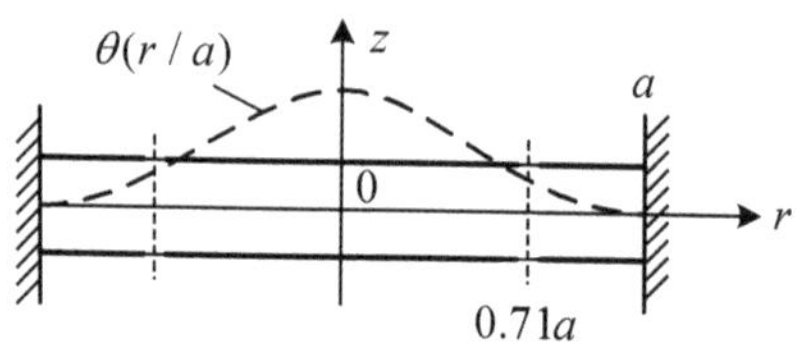

Figure 9.8: Illustration of the circular plate having clamped boundary with its mode shape of vibration and electrode configuration.

At $r_{el} = a$ the electromechanical transformation coefficient is zero by formula (9.49) because the slope is zero on the clamped boundary $[(d\theta / dr)_{r=a} = 0]$. The physical reason behind this is the same as was discussed regarding the rectangular beam with clamped ends. In the case of the circular plate the charge density on the surface of electrodes is proportional to $[r(d\theta / dr)]'$, and it changes sign at $r = a\sqrt{2}$. Theoretically, the maximum output from the clamped circular plate transducer can be obtained if the electrodes are split into two parts on the circle of radius $r = 0.71a$ and the parts are connected in antiphase. In this case the electromechanical transformation coefficient must be determined by formula (9.54), namely,

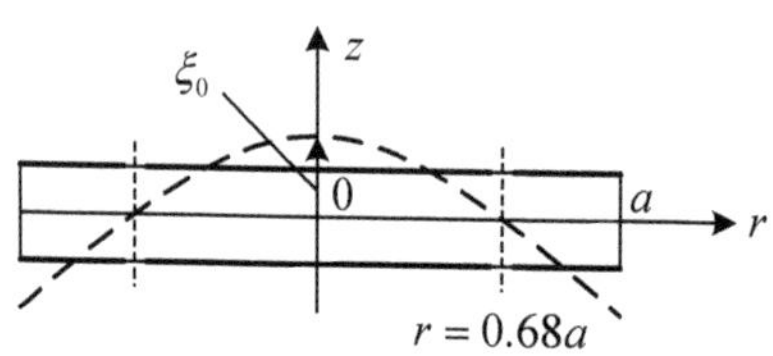

Figure 9.9: The mode shape of the circular plate with free boundary. The nodal line is at $r = 0.68a$.

$$n = \frac{\pi d_{31} t}{s_{11}^E + s_{12}^E} 2r_1 \left.\frac{d\theta}{dr}\right|_{r=r_1} = -\frac{2\pi d_{31}}{s_{11}^E + s_{12}^E}. \tag{9.61}$$

In the variant of free boundary (Figure 9.9) the mode shape of vibration at the first resonance frequency according to (4.200) is

$$\theta(r / a) = \frac{I_1(k_1 a) J_0(k_1 r) - J_1(k_1 a) I_0(k_1 r)}{I_1(k_1 a) - J_1(k_1 a)}, \tag{9.62}$$

where $J_i(k_1r)$ and $I_i(k_1r)$ are the first order Bessel and modified Bessel functions, respectively, and $k_1a = 3.01$. The coefficient Ω_r is

$$\Omega_r = r_{el}\left.\frac{d\theta}{dr}\right|_{r=r_{el}} = -\frac{r_{el}}{a}[3.28J_1(3r_{el}/a) + 0.27I_1(3r_{el}/a)]\,. \tag{9.63}$$

For the full size electrode (at $r_{el} = a$)

$$n = 2.18\frac{\pi d_{31}t}{s_{11}^E + s_{12}^E}\,. \tag{9.64}$$

9.2.2.2 The Effective Coupling Coefficients of the Circular Plates

In accordance with formulas (9.23) the coefficients

$$\alpha_c = \frac{n^2}{C_{el}^{S_{1,2}} K_{eqv}^E} \tag{9.65}$$

must be considered first. In this formula the equivalent rigidity, K_{eqv}^E, is determined by expression (4.391)

$$K_{eqv}^E = \frac{t^3 Y_{\sigma\,eqv}}{12}\cdot L(0.3)_{bc}\,. \tag{9.66}$$

Here $Y_{\sigma\,eqv}$ is the equivalent Young's modulus that depends on the composition of the plate through the thickness and is determined by expression (4.390). Factor $L(0.3)_{bc}$ depends on the boundary conditions (on the mode shape of vibration). It is determined by expression (4.382) at $\sigma = 0.3$. For the simply supported boundary $L(0.3)_{ss} = 7.2\pi/a^2$. The value of the equivalent rigidity does not depend on the electrodes size under the condition that remaining part of the plate surface is covered with electrodes, which are shot circuited. Capacitance of the clamped plate at parallel connection of the ceramic laminates is

$$C_{el}^{S_{1,2}} = \varepsilon_{33}^T(1-k_p^2)\frac{4\pi r_{el}^2}{t}\,. \tag{9.67}$$

The procedure of calculating the effective coupling coefficients and their optimizing will be considered for uniform over the radius plates. In principle, it remains the same as for the beams. After substituting the general expressions for the transformation coefficient (9.55), for the capacitance (9.67) and rigidity (9.66) into formula for coefficient α_c and normalizing to its value for the fully active simply supported plate we obtain

$$\frac{\alpha_{cbc}[(t/2), r_{el}]}{\alpha_{css}(t/2, a)} = \frac{a^2}{(1.5)^2} \frac{L(0.3)_{ss}}{L(0.3)_{bc}} \left.\frac{d\theta_{bc}}{dr}\right|^2_{r=r_{el}} \quad . \tag{9.68}$$

Here the subscript *bc* stands for the specific boundary conditions, the factor $\alpha_{css}(t/2, a)$ is the value of the coefficient for the simply supported fully active plate. The function depends on the diameter of the electrodes and the boundary conditions. For the simply supported plates using the mode shape described by function (9.57) we obtain

$$\frac{\alpha_{css}(t/2, r_{el})}{\alpha_{css}(t/2, a)} = \frac{r_{el}^2}{(1.5)^2 a^2}\left(2.5 - \frac{r_{el}^2}{a^2}\right)^2 . \tag{9.69}$$

For the clamped and free boundaries expressions (9.60) and (9.63) for the function Ω_r must be used, and values of coefficients $L(0.3)_{bc}$ that correspond to these boundary conditions according to expression (4.382) with all the other parameters of the plates equal. They are $L(0.3)_{cl} = 21.4\pi / a^2$ and $L(0.3)_{free} = 22.0\pi / a^2$. Thus, the results will be

for the clamped plate,

$$\frac{\alpha_{ccl}(t/2, r_{el})}{\alpha_{css}(t/2, a)} = 7.1\frac{r_{el}^2}{a^2}\left(1 - \frac{r_{el}^2}{a^2}\right)^2 \cdot \frac{3.6}{a^2 I_{cl}} , \tag{9.70}$$

and for the free plate,

$$\frac{\alpha_{c\,free}(t/2, r_{el})}{\alpha_{css}(t/2, a)} = 0.14 \cdot [3.28 J_1(3r_{el}/a) + 0.27 I_1(3r_{el}/a)]^2 . \tag{9.71}$$

These functions are presented in Figure 9.10. Note that the plots of these functions are valid for all the radially uniform plate designs. The maximum values of the effective coupling coefficients at different boundary conditions are achieved at:

$$r_{el} \approx 0.90a \text{ for the simply supported boundary,} \tag{9.72}$$

$$r_{el} \approx 0.58a \text{ for the clamped boundary,} \tag{9.73}$$

$$r_{el} \approx 0.72a \text{ for the free boundary.} \tag{9.74}$$

The coefficient α_{max} for the plate with clamped boundary theoretically can be doubled if the electrodes are split at radius $r_{el} \approx 0.58a$ and the parts are connected in antiphase. In this case

$$\frac{\alpha^*_{c\,cl}[t/2, 0.58a]}{\alpha_{c\,ss}(t/2, a)} = 0.6\,. \tag{9.75}$$

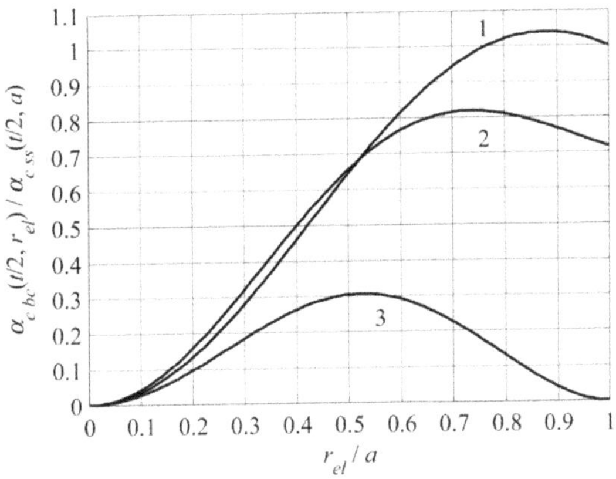

Figure 9.10: Plots of the functions that characterize relative dependences of the effective coupling coefficients on the radius of electrodes at different boundary conditions: (1) simply supported, (2) free, (3) clamped. $\alpha_{css} = 0.62k_p^2/(1-k_p^2)$.

Absolute maximum values of the coupling coefficients for different boundary conditions can be calculated using expressions for the coefficient $\alpha_{css}(t/2a)$ and $k_{eff\,bc}$ that are given by formulas (9.23), namely,

$$\alpha_{css} = 0.62\frac{k_p^2}{1-k_p^2} \text{ and } k^2_{eff\,bc} = \frac{\alpha_{c\,bc}}{1+\alpha_{c\,bc}} \tag{9.76}$$

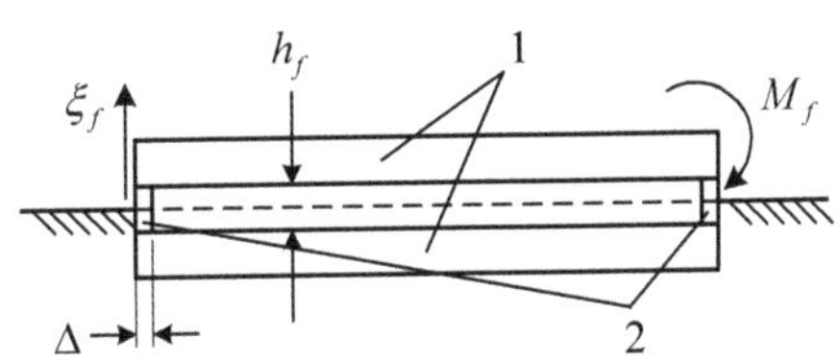

Figure 9.11: Double-plate balanced transducer design. (1) The identical bimorph plates, (2) common boundary foundation for the plates. Dashed line shows the plane of symmetry.

As the comparison shows, the plates with simply supported boundaries have an advantage in terms of the effective coupling coefficient. They are the most suitable for underwater applications. The simply supported conditions can be closely achieved in the symmetrical double sided design schematically shown in Figure 9.11 in case that dimensions of the supporting ring ensure small moment of resistance on the contour of the plate. The displacements of foundation $\xi_f = 0$ due to symmetry, and the moments $M_f \approx 0$ (is shown on one side), if the thickness of the foundation, Δ, is small enough compared with its height, h_f.

It is noteworthy that the clamped boundary conditions can provide greater mechanical strength ($K^E_{eqv\,cl} > K^E_{eqv\,ss}$), but not only are the clamped piezoelements less electromechanically effective, the clamped conditions hardly can be achieved in the real transducer design.

After the general expressions for parameters that characterize electromechanical energy conversion are obtained, the values of equivalent electromechanical parameters for particular transducer designs will be determined in the next two Sections for the transducers that are made of the rectangular beams and circular plates.

9.3 Rectangular Beam Transducers

In terms of peculiarities of calculating the rectangular beam transducers of configurations shown in Figure 9.1 fall into two groups: with mechanical systems uniform (a.1, b.1, c.1) and nonuniform (a.2, b.2, c.2) over the length. For all the transducers of the first group the mode shapes of vibration are the same at the same boundary conditions, and differences may arise only due to differences in composition of the beams through their thickness. Calculating transducers of the second group requires also considering differences between their modes of vibration.

9.3.1 Beams Uniform over the Length

The bimorph beam design (a.1) may be considered as the basic design within this group, with which all the other modifications of beams (trilaminar (a.2) and bilaminar (a.3)) is convenient to compare in terms of quality of electromechanical conversion. Examples of the bimorph beams and their parameters were considered in Section 2.5 and throughout Section 9.2.1 for illustrating the results that follow from the general expressions.

9.3.1.1 Simply Supported Fully Active Bimorph Beam

In this variant all the volume of the beam is active. Therefore, in the general formulas for the energies involved $\delta = t/2$, $l_e = l$, and throughout the volume $Y(x,z) = Y_a^E$ and $\rho(x,z) = \rho_a$

The normal modes of vibration of a simply supported beam are $\theta(x) = \sin(\pi xi / l)$. Upon substituting these values into expressions (9.6) and (9.10) for the potential and electromechanical energies and for the kinetic energy we arrive at the following summary for the equivalent parameters of a beam utilizing the transverse piezoeffect for the first and for the next closest active third mode of vibration

$$K_{eqv1}^{E} = \frac{1}{C_{eqv1}^{E}} = \frac{\pi^4 wt^3}{24l^3 s_{11}^{E}},\ K_{eqv3}^{E} = 9\frac{\pi^4 wt^3}{24l^3 s_{11}^{E}},\ M_{eqvi} = \frac{1}{2}wtl\rho_a = \frac{M}{2}, \tag{9.77}$$

$$n_1 = \pi w \frac{d_{31}}{s_{11}^{E}}\frac{t}{l},\ n_3 = 3\pi w \frac{d_{31}}{s_{11}^{E}}\frac{t}{l}. \tag{9.78}$$

Capacitance of the bimorph beam is

$$C_{el}^{S_1} = \varepsilon_{33}^{T}(1-k_{31}^2)\frac{4w}{t}l. \tag{9.79}$$

The expressions (9.78) and (9.79) are given for the case that halves of a beam are connected in parallel. For the case of series connection of the halves the value of transformation coefficient must be reduced by factor of 2, and value of capacitance by factor of 4.

For the beams that utilize longitudinal piezoeffect and have geometry shown in Figure 9.7 (a) the expressions for equivalent parameters that take into consideration comments made in Section 9.2.1.3 regarding contribution of the term ΔK are as follows (for the first mode of vibration).

$$K_{eqv}^{E} = \frac{1}{C_{eqv1}^{E}} = \frac{\pi^4 wt^3}{24l^3 s_{33}^{E}},\ M_{eqv} = \frac{1}{2}wtl\rho_a = \frac{M}{2}. \tag{9.80}$$

$$n = \pi w \frac{d_{33}}{s_{33}^{E}}\frac{t^2}{2l\Delta}, \tag{9.81}$$

$$C_{el}^{S_3} = \varepsilon_{33}^{T}(1-k_{33}^2)\frac{tw}{\Delta^2}l\ .\ C_{el}^{S_3} = \varepsilon_{33}^{T}(1-k_{31}^2)\frac{4lw}{t} \tag{9.82}$$

Useful equivalent parameters that depend solely on the geometry of the mechanical system and mode of vibration are the effective, S_{eff}, and average, S_{av}, surface areas that are introduced for the beams by formulas (2.120) and (2.126). For the simply supported beams vibrating in the first mode they are

$$S_{av} = 2lw/\pi\,,\ S_{eff} = lw/2\,. \tag{9.83}$$

The resonance frequencies of the bimorph beam calculated by formula $f_r = 1/2\pi\sqrt{M_{eqv}C^E_{eqv}}$ are

$$f_1 = 0.45\frac{t}{l^2}\frac{1}{\sqrt{s^E_{11}\rho_a}} = 0.45\frac{t}{l^2}c_{c1}\,,\ f_3 = 3\cdot 0.45\frac{t}{l^2}c_{c1}\,, \tag{9.84}$$

where c_{c1} is the sound speed of longitudinal vibration in the piezoceramic bar utilizing transverse piezoeffect. In the frequency range below and around the first resonance frequency contribution of the third and higher modes with sufficiently remote resonance frequencies to operational characteristics of the flexural beam transducer is negligible, and calculating properties of the transducer can be produced with help of the equivalent circuit shown in Figure 9.3. The acoustic field related parameters of the circuit will be considered later in the Chapter for all the flexural type transducers together. Following expressions (9.23) with reference to the above formulas for the equivalent parameters the values of the effective coupling coefficients for the bimorph benders utilizing the transverse and longitudinal piezoeffect will be determined as

$$\alpha_c = \frac{n^2}{C^{S_i}_{el}K^E_{eqvi}} = \frac{6}{\pi^2}\frac{k^2_{3i}}{1-k^2_{3i}}\,,\ \text{and}\ k^2_{eff\,i} = \frac{6}{\pi^2}k^2_{3i}\frac{1}{1-0.39k^2_{3i}}\,. \tag{9.85}$$

Thus, for the simply supported bimorph beams made of PZT-4 ceramics having $k_{31} = 0.33$ and $k_{33} = 0.7$, the respective values of the effective coupling coefficients are

$$k_{eff1} \approx 0.8k_{31}\,,\ k_{eff3} \approx 0.87k_{33}\,. \tag{9.86}$$

9.3.1.2 Cantilever Bimorph Beam

The basic design of the cantilever bimorph beam transducer as shown in Figure 9.20 (a) was considered in Section 2.5. Here the expressions for the equivalent parameters of the transducer are summarized for comparing with results for their trilaminar and bilaminar modifications. They are obtained from the general expressions by specifying the mode of static deflection of uniformly loaded beam with one end clamped, namely,

$$\theta(x) = 2(x/l)^2\left(1 - 2x/3l + x^2/6l^2\right). \tag{9.87}$$

The equivalent parameters of the cantilever beam employing transverse piezoeffect at parallel connection of the piezoelements are

$$K_{eqv}^{E}=3.2\frac{wt^{3}}{12s_{11}^{E}l^{3}},\ M_{eqv}=0.26M,\ n=\frac{2wd_{31}t}{3s_{11}^{E}l},\ C_{el}^{S_1}=\varepsilon_{33}^{T}(1-k_{31}^{2})\frac{4w}{t}l, \tag{9.88}$$

$$S_{av}=0.42lw,\ S_{eff}=0.26lw. \tag{9.89}$$

The resonance frequencies of the cantilever beam are

$$f=0.16\frac{t}{l^{2}}\frac{1}{\sqrt{s_{11}^{E}\rho_{a}}}. \tag{9.90}$$

Remember that the reference point is at the end of a beam. Therefore, the generalized velocity is U_l.

The effective coupling coefficients for the cantilever beam utilizing the transverse and longitudinal piezoeffect will be determined as

$$\alpha_{c}=0.42\frac{k_{3i}^{2}}{1-k_{3i}^{2}},\ \text{and}\ k_{eff\,i}^{2}=0.42k_{3i}^{2}\frac{1}{1-0.58k_{3i}^{2}},\ i=1,3. \tag{9.91}$$

Thus, for the cantilever beams made of PZT-4 ceramics values of the effective coupling coefficients are

$$k_{eff1}\approx 0.67k_{31},\ k_{eff3}\approx 0.77k_{33}, \tag{9.92}$$

respectively.

It is noteworthy that real clamping of a beam at the end is hard to accomplish. One of the practical ways of achieving this is the dual center supported beam design, as shown in Figure 9.2 (b). All the parameters of transducers are the same per each half of the design, as presented above. Remember that the reference point is at the end of a beam. Therefore, the generalized velocity is U_l.

9.3.1.3 Elastically Uniform Partially Active Beam

The basic variant of bimorph beam (Figure 9.1 (a.1)), in which case all the volume of the ceramic is active, was considered in Section 194. Consider now the variant of design with electrodes having variable length l_e and separation δ between them, which are embedded into the volume of ceramic, as shown in Figure 9.1(a.2). We assume that the Young's modulus and density of the ceramic are the same Y_a^E and ρ_a as in the bimorph design in the case that all the ceramic volume is confined between the electrodes. Such variant of design cannot be practically realized unless the electrodes in the passive parts exist and are short circuited. Bu it is useful in the methodical sense, as ideal base for comparison properties of the truly nonuniform over length transducers and for illustrating peculiarities of electromechanical conversion in the flexural transducers. In the last capacity it was used as an example in Section 9.2.1. Some of related results of Section 9.2.1 are repeated here for convenience of referencing. The bimorph design is the particular case of this transducer at $\delta = t/2$ and $l_e = l$. The equivalent mechanical parameters of the transducer are the same as bimorph transducer of the same geometry and at the same boundary conditions has. Namely, for simply supported beam

$$K_{eqv}^E = \frac{1}{C_{eqv1}^E} = \frac{\pi^4 wt^3}{24l^3 s_{11}^E}, \quad M_{eqv} = \frac{1}{2} wtl\rho_a = \frac{M}{2}. \tag{9.93}$$

The transformation coefficient and capacitance at parallel connection of the active laminates are

$$n = \frac{2\pi w d_{31}}{l s_{11}^E}(t-\delta)\sin\frac{\pi l_e}{2l} l_e = l, \tag{9.94}$$

$$C_{el}^{S_1} = \varepsilon_{33}^T(1-k_{31}^2)\frac{2w}{\delta} l_e \tag{9.95}$$

according to expressions (9.21) and (9.26). Coefficient α_c used for determining the effective coupling coefficient by formula (9.23) is

$$\alpha_c(\delta, l_e) = \frac{n^2}{C_{el}^{S_1} K_{eqv}^E} = \frac{6}{\pi^2}\frac{k_{31}^2}{1-k_{31}^2}\cdot\left[8\left(1-\frac{\delta}{t}\right)^2\frac{\delta}{t}\right]\cdot\frac{l}{l_e}\sin^2\left(\frac{\pi l_e}{2l}\right). \tag{9.96}$$

For the bimorph beam (at $\delta = t/2$,)

$$\alpha_c[(t/2), l] = \frac{6}{\pi^2}\frac{k_{31}^2}{1-k_{31}^2}. \tag{9.97}$$

The ratio $\alpha_c(\delta, l_e)/\alpha_c[(t/2), l]$ falls into product of two functions

$$\frac{\alpha_c(\delta, l_e)}{\alpha_c[(t/2), l]} = \frac{\alpha_c(\delta, l)}{\alpha_c[(t/2), l]}\cdot\frac{\alpha_c(t/2, l_e)}{\alpha_c[(t/2), l]}, \tag{9.98}$$

one of which,

$$\frac{\alpha_c(\delta,l_e)}{\alpha_c[(t/2),l]}=8\left(1-\frac{\delta}{t}\right)^2\frac{\delta}{t}, \tag{9.99}$$

depends only on position of the electrodes over the thickness of the beam, and another,

$$\frac{\alpha_c(\delta,l_e)}{\alpha_c[(t/2),l]}=\frac{l}{l_e}\sin^2\left(\frac{\pi l_e}{2l}\right), \tag{9.100}$$

depends only on the length of the electrodes and on the boundary conditions. Plots of these functions are presented in Figure 9.12 and Figure 9.13.

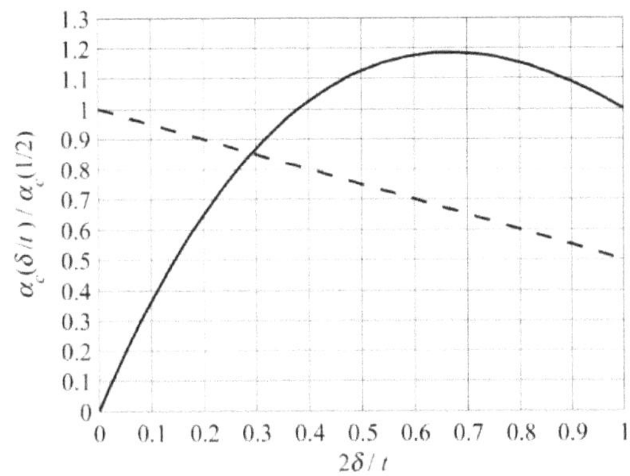

Figure 9.12: Plot of the function that characterizes dependence of the effective coupling coefficient on the relative separation between the electrodes (solid line). Function $0.5\cdot[n(\delta/t)/n(1/2)]$ $0.5\cdot[n(\delta/t)/n(1/2)]$ at l_e/l constant (dashed line).

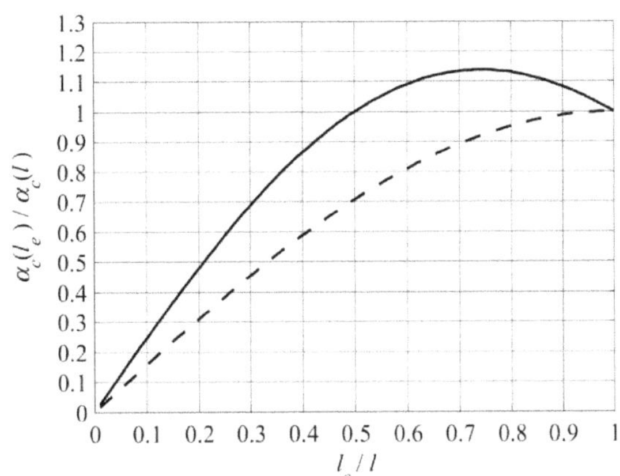

Figure 9.13: Plot of the function that characterizes dependence of the effective coupling coefficient on the relative length of the electrodes for a beam with simply supported ends (solid line). Function $n(l_e/l)/n(1)$ at δ/t constant (dashed line).

The dependence according to formula (9.99) is the same for all the transducers of the flexural type due to the same linear dependence of strain over their cross sections. Dependences of (9.100) type for different boundary conditions are presented in Figure 9.10.

As it follows from the plots, the maximum value of the effective coupling coefficient can be achieved at $\delta / t = 1/3$ and $l_{el} / l = 0.7$, i.e., with volume of active material reduced almost in factor of two, and it equals to

$$\alpha_{c\max} = 1.3 \cdot \alpha_c [(t/2), l] = \frac{7.8}{\pi^2} \cdot \frac{k_{31}^2}{1 - k_{31}^2} \cdot \qquad (9.101).$$

Thus, for the maximum value of the effective coupling coefficient we obtain

$$k_{eff\max}^2 = \frac{7.8}{\pi^2} k_{31}^2 \frac{1}{1 - 0.79 k_{31}^2} \cdot \qquad (9.102)$$

For the beam made of PZT-4 $k_{eff\max} \approx 0.93 k_{31}$, whereas for the bimorph (fully active) beam $k_{eff} \approx 0.80 k_{31}$.

It is also shown in Figure 9.12 and Figure 9.13, how the relative values of the electromechanical transformation coefficient, $n(l_e / l) / n(1)$ at δ / t constant and $n(\delta / t) / n(1/2)$ at l_e / l constant, change simultaneously with relative values of coefficients α_c. It is remarkable that the increase of effective coupling coefficient at reduction of the active volume of a beam is achieved with even some increase of the electromechanical transformation coefficient and thus of the driving electromechanical force (9.22). As it follows from formula (9.94), $n[(t/3), (0.7l)] \approx 1.3 \cdot n[(t/2), l]$. This effect is typical of flexural transducers due to highly non-uniform distribution of the dynamic stress in their mechanical systems.

9.3.1.4 Trilaminar Beams

The symmetric trilaminar beam transducer shown in Figure 9.1(a.2) is composed of two piezoelectric ceramic layers having equal thickness that are cemented to the central laminate made of a passive material. The elastic moduli and densities of the active and passive laminates will be denoted Y_a^E, Y_p, and ρ_a, ρ_p. The following notations will be used for brevity to characterize properties of the trilaminar beams

$$\frac{Y_p}{Y_a^E} = \gamma_Y, \quad \frac{\rho_p}{\rho_a} = \gamma_\rho, \quad \frac{\delta}{t} = y. \qquad (9.103)$$

Vibration of the passive trilaminar beams is considered in Section 4.5.6.1, where it is shown that in terms of the equivalent mechanical parameters trilaminar beam can be represented as uniform through the thickness beam having the same boundary conditions and equivalent Young's modulus, Y_{eqvt}, and density, ρ_{eqvt}, that are presented by formulas (4.319) and (4.321), namely,

$$Y_{eqvt} = Y_a^E + \left(Y_p - Y_a^E\right)\left(1 - \frac{2\delta}{t}\right)^3 = Y_a^E[1 + (\gamma_Y - 1)(1 - 2y)^3]. \tag{9.104}$$

$$\rho_{eqvt} = \frac{\rho_a 2\delta + \rho_p t_p}{t} = \rho_a[\gamma_\rho - 2(\gamma_\rho - 1)y]. \tag{9.105}$$

Thus, for the simply supported beam

$$K_{eqvt}^E = \frac{1}{C_{eqvt}^E} = \frac{\pi^4 w t^3 Y_{eqvt}}{24 l^3}, \quad M_{eqv\,t} = \frac{1}{2} w t l \rho_{eqvt}. \tag{9.106}$$

The resonance frequency of the trilaminar beam is

$$f_t = 0.45 \frac{t}{l^2}\sqrt{\frac{Y_{eqvt}}{\rho_{eqvt}}}. \tag{9.107}$$

Further we will mainly consider beams with simply supported ends keeping in mind that values of the equivalent electromechanical parameters determined for different boundary conditions differ by the corresponding factors that can be introduced separately.

Consider now the electromechanical conversion in the trilaminar beam. In the general expression (9.8) for the electromechanical energy and in the corresponding expression (9.11) for the electromechanical transformation coefficient only the value of factor Ω_t is different for the trilaminar design from those for the bimorph variant. The function $\Omega_t(z)$ that characterizes the distribution of electric field through the thickness at parallel connection of active layers in this case is shown in Figure 9.4 (d). Therefore, we arrive at value (9.15) of the factor $\Omega_t(z)$. Thus, expression (9.11) for the electromechanical transformation coefficient becomes

$$n_t = 2\pi \frac{w d_{31}}{s_{11}^E} \frac{(t - \delta)}{l}. \tag{9.108}$$

Related to the quantity of the effective coupling coefficient function α_{ct} that is defined by formula (9.23) after substituting expressions (9.108) for n_t, (9.106) for K_{eqvt}^E, and

$$C_{el}^{S_1} = \varepsilon_{33}^{T}(1-k_{31}^2)2wl/\delta \tag{9.109}$$

will be determined as

$$\alpha_{ct} = \frac{48}{\pi^2}\frac{k_{31}^2}{1-k_{31}^2}\frac{Y_a^E}{Y_{eqvt}}\cdot\left(1-\frac{\delta}{t}\right)^2\frac{\delta}{t} = \frac{48}{\pi^2}\frac{k_{31}^2}{1-k_{31}^2}\beta_t(\delta/t). \tag{9.110}$$

The factor $\beta_t(\delta/t)$, which after using notations (9.103) and expression (9.104) for Y_{eqvt} may be transformed to

$$\beta_t(y) = \frac{(1-y)^2 y}{1+(\gamma_Y-1)(1-2y)^3}, \tag{9.111}$$

depends on the structure of the beam cross section only. This function and, hence, the effective coupling coefficient of the transducer have maximum at value of y that can be determined from condition $\beta_t(y)'=0$. This results in equation

$$\frac{1-4y+3y^2}{4y^3(y-1)} = 1-\frac{1}{\gamma_Y}. \tag{9.112}$$

This value depends on ratio of elastic moduli of the passive and active materials, γ_Y. Plots of the function $\beta_t(y)$ are presented in Figure 9.14 for different combinations of active (PZT-4) and passive (aluminum, steel, alumina, G-10) materials. They are normalized to value of the function at $y=0.5$, i.e., to its value for the bimorph beam of the same size.

Parameters of materials used for calculations are presented in Table 9.1 (see Appendix A.)

Table 9.1. Parameters of materials used for calculations.

	PZT-4	Alumina	Aluminum	Steel	G10
Y $(1/s_{11}^E)$, 10^9 N/m^2	81	300	70	210	12.0
ρ, 10^3 kg/m^3	7.6	3.7	2.7	7.8	2.0

It is seen from the plots in Figure 9.14 that position of maximum of function $\beta_t(y)$ and its maximum value depend on parameter γ_Y. The smaller this parameter, the greater the maximum value and it is achieved at a smaller ratio of thicknesses of the active to passive material. This means that maximum of the coupling coefficient is achieved at relatively smaller amounts of the active material. But the effective coupling coefficient is not the only important parameter of a transducer that depends on the ratio $(\delta / t) = y$. The resonance frequency and the electromechanical transformation coefficient n, to which the electromechanical force generating vibration of the mechanical system of the transducer is proportional, also depending on this ratio. Both these quantities must be considered as requirements for the transducer design in accord with the coupling coefficient.

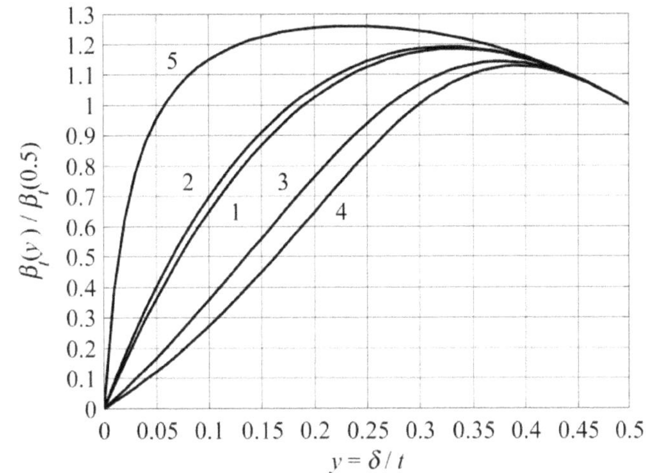

Figure 9.14: Functions $\beta_t(y) / \beta_t(0.5)$ for different combinations of the active (PZT-4) and passive materials: (1) material with properties of PZT-4, (2) aluminum, (3) steel, (4) alumina, (5) G 10.

Expressions for the resonance frequency and transformation coefficient vs. y obtained from relations (9.107) and (9.108), respectively, are as follows

$$\frac{f_t(y)}{f_t(0.5)} = \sqrt{\frac{1+(\gamma_Y - 1)(1-2y)^3}{\gamma_\rho - 2(\gamma_\rho - 1)y}}, \tag{9.113}$$

$$\frac{n_t(y)}{n_t(0.5)} = 2(1-2y). \tag{9.114}$$

Here $f_t(0.5)$ and $n_t(0.5)$ are the values of these quantities for the fully active bimorph beam.

The plots for $f_r(y) / f_r(0.5)$ vs. combinations of the active and passive materials are presented in Figure 9.15. Relation (9.114) shows that the transformation coefficient increases with reducing the relative thickness of the active layer. The following comment is appropriate in this regard. As the force that generates vibration of the mechanical system of the transducer is

$F = Vn_t$, the sound pressure radiated per unit voltage applied increases proportionally with an

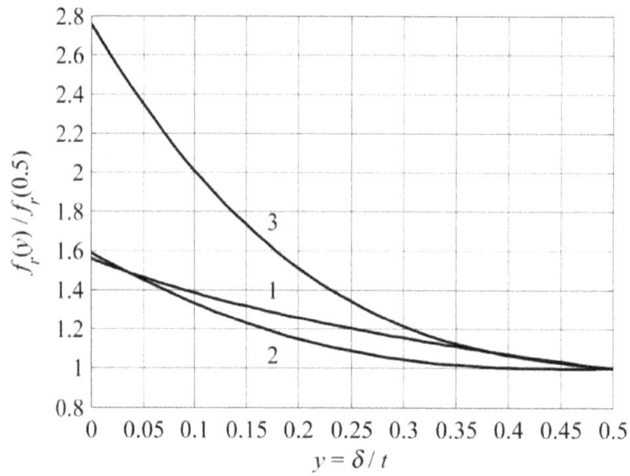

Figure 9.15: The resonance frequency of the trilaminar beam normalized to the resonance frequency of the fully active beam for different combinations of active and passive materials: PZT-4 with: (1) aluminum, (2) steel, and (3) alumina.

increase of the transformation coefficient (that is the TVR of the transducer increases). However, the magnitude of the operating electric field, $E_{op} = V / \delta$, that is required for generating of a certain force is

$$E_{op}(y) = \frac{F}{\delta \cdot n_t(y)}, \tag{9.115}$$

and

$$\frac{E_{op}(y)}{E_{op}(0.5)} = \frac{t}{2\delta} \cdot \frac{n_t(0.5)}{n_t(y)} = \frac{1}{4y(1-y)}. \tag{9.116}$$

This dependence shows that a reduction of the amount of active material, which both optimizes the effective coupling coefficient and increases TVR, is not possible, if radiating the greatest acoustic power under condition of limited electric field is required. It is a tradeoff. Increasing the electromechanical coupling coefficient (and hence useable bandwidth) can be achieved at the expense of reduction of the maximum operating acoustic power of the transducer. The lesser the requirement of high acoustic power, the greater is the opportunity for optimizing the effective coupling coefficient by replacing parts of the active material with passive material. Such optimizing the effective coupling coefficient is the most beneficial for implementing in the hydrophone designs.

9.3.1.5 Bilaminar Beams

The bilaminar beam is composed of active and passive parts having different mechanical properties, as shown Figure 9.1 (c.1). The main reason for using bilaminar design of a bender transducer intended for underwater application is in increasing its mechanical strength under action of the hydrostatic pressure by replacing piezoceramics in the parts of mechanical system that experience tension with a passive material (usually metal) having a higher yield stress. Yet another reason may be in reducing cost and gaining technological benefits in manufacturing the transducers in the case that they are used as hydrophones (receivers in general), or drivers that are not electric field limited (usually for air applications).

Essential peculiarity of the bilaminar design is that the neutral plane under bending does not coincide with the middle plane, as it was in previously considered bimorph and trilaminar beams. Therefore, in order to be able to use the general expressions for the equivalent electromechanical parameters, location of the neutral plane (coordinate z_0, as shown in Figure 9.16) must be determined at first. Analysis of deformation of the passive bilaminar beams for determining position of the neutral plane is performed in Section 4.5.6.3. Results of this analysis are summarized here.

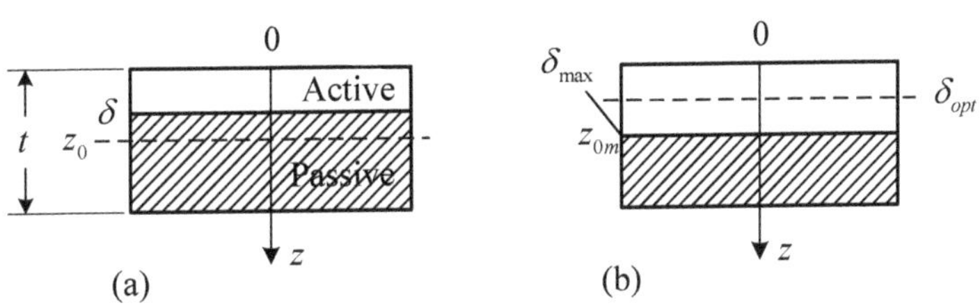

Figure 9.16: Bilaminar beam configuration and position of the neutral plane: (a) the general case, (b) the case that corresponds to maximum reasonable thickness of the active layer $\delta_{\max}$.

The relative coordinate of the neutral plane is determined as

$$\frac{z_0}{t} = \frac{1}{2}\frac{\gamma_Y + (1-\gamma_Y)y^2}{\gamma_Y + (1-\gamma_Y)y}, \tag{9.117}$$

where notations (9.103) are used. For a rational transducer design it should be $z_0 \geq \delta$

(otherwise the electromechanical effects in the piezoelectric element above and below the neutral plane would be in opposite phase). The value of z_0 that is equal to δ is denoted as z_{0m}. Obviously, $z_{0m} = \delta_{\max}$ is the maximum reasonable thickness of piezoceramic layer for a given combination of active and passive materials. It follows from equation (9.117) that

$$\frac{z_{0m}}{t} = \frac{\sqrt{\gamma_Y}}{1+\sqrt{\gamma_Y}}. \tag{9.118}$$

The mechanical equivalent parameters of a transducer with the piezoelectric and passive laminates having different relative thickness are as follows.

The equivalent Young's modulus is

$$Y_{eqvb}(z_0) = 4\left\{\left[\left(\frac{\delta}{t}-\frac{z_0}{t}\right)^3+\left(\frac{z_0}{t}\right)^3\right]Y_a^E+\left[\left(1-\frac{z_0}{t}\right)^3+\left(\frac{z_0}{t}-\frac{\delta}{t}\right)^3\right]Y_p\right\}. \tag{9.119}$$

In the case that $z_0 = z_{0m} = \delta_{\max}$,

$$Y_{eqvb}(z_{0m}) = 4\left[Y_a^E\left(\frac{z_{0m}}{t}\right)^3+Y_p\left(1-\frac{z_{0m}}{t}\right)^3\right] = 4Y_a^E[y^3+(1-y)^3\gamma_Y]. \tag{9.120}$$

The equivalent density of the beam is

$$\rho_{eqvb} = \frac{\rho_a\delta+\rho_p(t-\delta)}{t} = \rho_a[\gamma_\rho-(\gamma_\rho-1)y]. \tag{9.121}$$

Expressions (9.88) and (9.90) for the equivalent parameters and resonance frequency of the bimorph beam (we rename them as K_{eqva}^E, M_{eqva}, f_{ra}) remain valid for those of the bilaminar beam, if to replace $Y_a^E = 1/s_{11}^E$ and ρ_a therein by Y_{eqvb} and ρ_{eqvb}. Thus,

$$M_{eqvb} = M_{eqva}\cdot[\gamma_\rho-(\gamma_\rho-1)y]; \tag{9.122}$$

$$K_{eqvb}^E = K_{eqva}^E\cdot Y_{eqvb}(z_0)/Y_a^E, \tag{9.123}$$

where $Y_{eqvb}(z_0)$ is given by formula (9.119) in the general case, and

$$K_{eqvb}^E = 4K_{eqva}^E[y^3+(1-y)^3\gamma_Y] \tag{9.124}$$

in the variant that $z_0 = z_{0m} = \delta_{\max}$. For the same variant

$$f_{rb} = f_{ra} \cdot 2\sqrt{\frac{y^3 + (1-y)^3 \gamma_Y}{\gamma_\rho - (\gamma_\rho - 1)y}}. \tag{9.125}$$

Consider now the electromechanical energy associated with deformation of the bilaminar beam

$$\begin{aligned} W_{em} &= \frac{\xi_0}{2}\frac{V}{\delta} w \frac{d_{31}}{s_{11}^E} \int_0^\delta (z_0 - z)\, dz \int_0^l \left[\frac{d^2\theta}{dx^2}\right] dx \\ &= \frac{1}{2}\xi_0 V \frac{w d_{31}}{2 s_{11}^E}(2z_0 - \delta)\left[\frac{d\theta}{dx}\right]\Bigg|_{x=0}^{x=l} = \frac{1}{2}\xi_0 V n_b, \end{aligned} \tag{9.126}$$

where from expression for the electromechanical transformation coefficient is

$$n_b(\delta) = \frac{w d_{31}}{2 s_{11}^E}(2z_0 - \delta)\left[\frac{d\theta}{dx}\right]\Bigg|_{x=0}^{x=l}. \tag{9.127}$$

In the case that $z_0 = z_{0m} = \delta_{\max}$

$$n_b(\delta_{\max}) = \frac{w d_{31}}{2 s_{11}^E}\delta_{\max}\left[\frac{d\theta}{dx}\right]\Bigg|_{x=0}^{x=l}. \tag{9.128}$$

The same question arises regarding maximizing the effective coupling coefficient k_{eff} of the bilaminar beam for a particular combination of the active and passive materials, as in the previous case of symmetric trilaminar beam. In order to determine the conditions for optimizing the effective coupling coefficient by changing the relative thickness of the active layer, δ / t, consider parameter

$$\alpha_{cb}(\delta) = \frac{n_b^2(\delta)}{C_{el}^{S_1}(\delta) K_{eqvb}^E(\delta)} \tag{9.129}$$

in comparison with its value at $z_0 = z_{0m} = \delta_{\max}$, namely, the function

$$\frac{\alpha_{cb}(\delta)}{\alpha_{cb}(\delta_{\max})} = \frac{n_b^2(\delta)}{n_b^2(\delta_{\max})} \frac{K_{eqvb}^E(\delta_{\max})}{K_{eqv\,b}^E(\delta)} \frac{C_{el}^{S_1}(\delta_{\max})}{C_{el}^{S_1}(\delta)}. \tag{9.130}$$

Considering expressions for n_b, capacitance

$$C_{el}^{S_1} = \varepsilon_{33}^T (1 - k_{31}^2) wl / \delta, \tag{9.131}$$

and relation

$$K^E_{eqvb}(\delta)/K^E_{eqvb}(\delta_{\max}) = Y_{eqvb}(z_{0m})/Y_{eqvb}(z_0), \tag{9.132}$$

we arrive at the expression for the normalized coefficient α_c α_c

$$\frac{\alpha_{cb}(\delta)}{\alpha_{cb}(\delta_{\max})} = \frac{[(2z_0/t)-y]^2 y}{(z_{0m}/t)^3}\frac{Y_{eqvb}(z_{0m})}{Y_{eqvb}(z_0)}. \tag{9.133}$$

Here $Y_{eqvb}(z_0)$ and $Y_{eqvb}(z_{0m})$ are defined by formulas (9.119) and (9.120).

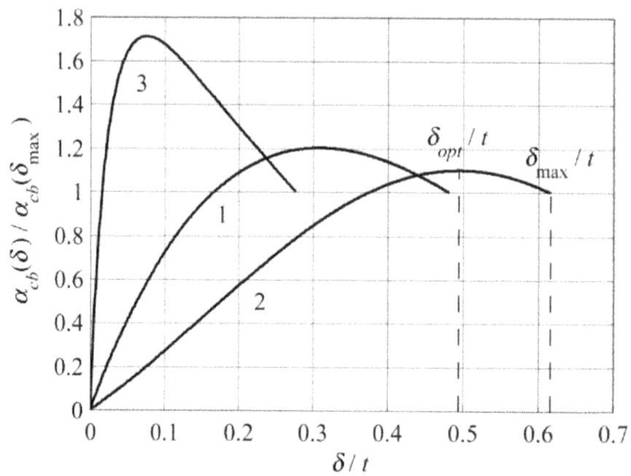

Figure 9.17: To optimizing the effective coupling coefficient of the bilaminar beam: function (9.133) vs. ratio of active to total thickness $y = \delta/t$ for different combinations of materials: (1) aluminum, (2) steel, (3) G-10. Positions of $\delta_{\max}/t$ and δ_{opt}/t are shown regarding curve 2.

Dependencies of functions (9.133) from $\delta/t = y$ are depicted in Figure 9.17 for combinations of different materials with PZT-4 ceramics. Note that G-10 as a passive layer is presented for illustrating results of combining PZT-4 with material having significantly smaller Young's modulus. This should not be considered as a recommendation for its practical use.

All the expressions for the equivalent parameters of the trilaminar beams and for the bilaminar beams in case that $z_0 = z_{0m} = \delta_{\max}$ can be represented through parameters of the bimorph (fully active) beam in the following way:

$$K^E_{eqvt} = K^E_{eqva}\cdot Y_{eqvt}/Y^E_a = K^E_{eqva}[1+(\gamma_Y-1)(1-2y)^3], \tag{9.134}$$

$$K^E_{eqvb} = K^E_{eqva}\cdot Y_{eqvb}/Y^E_a = 4K^E_{eqva}[y^3+(1-y)^3\gamma_Y], \tag{9.135}$$

$$M_{eqvb} = M_{eqva}\cdot \rho_{eqvb}/\rho_a = M_{eqva}\cdot[\gamma_\rho-(\gamma_\rho-1)y]; \tag{9.136}$$

$$n_t = n_a\cdot(1-y), \tag{9.137}$$

$$n_b = n_a \cdot [(2z_0 / t) - y]; \tag{9.138}$$

$$f_{rt} = f_{ra} \cdot \sqrt{\frac{1+(\gamma_Y - 1)(1-2y)^3}{\gamma_\rho - 2(\gamma_\rho - 1)y}}, \tag{9.139}$$

$$f_{rb} = f_{ra} \cdot 2\sqrt{\frac{y^3 + (1-y)^3 \gamma_Y}{\gamma_\rho - (\gamma_\rho - 1)y}}. \tag{9.140}$$

9.3.2 Beams of Finite Thickness

So far all the treatments in this chapter were performed within the elementary theory of bending, i.e., under the assumption that the thickness to length aspect ratio of the beams, t/l, is small enough for neglecting effects of the shear stresses T_5 and rotary inertia of the cross sections under flexure on the energies of beam deformations. These effects were discussed in Section 4.2.2. In some cases, an increase of the aspect ratio is required from operational and/or environmental considerations. With increase of the relative thickness of a beam, contribution of the above listed effects to the energy balances may become significant, so that corrections must be made to the values of equivalent parameters of the beams previously determined.

9.3.2.1 Corrections for the Bender Parameters Due to finite Thickness of the Beams

Corrections for the equivalent parameters of uniform passive beams were introduced in Section 4.3.5. They are completely applicable to bimorph beams with corresponding replacement of the elastic constants. Peculiarity in this regard is that with electric field E_3 applied in the direction of polarization the share deformation is not electromechanically active, and therefore the values of shear moduli must be used for non-polarized ceramics, which are not specified. Assumingly a good approximation to these values gives substituting $\mu = 1/s_{66}$, where $s_{66} = 2(s_{11}^E - s_{12}^E)$ (see Table 5.1). Thus, according to relations (4.140) for the simply supported beams

$$M'_{eqva} = M_{eqva}\left(1 + \frac{\pi^2}{12}\frac{t^2}{l^2}\right), \tag{9.141}$$

$$K'^E_{eqva} = K^E_{eqva}\left(1 - \frac{\pi^2}{20}\frac{t^2}{l^2} Y_a^E s_{66}\right), \tag{9.142}$$

where M'_{eqva} and K'^E_{eqva} are the equivalent parameters with corrections due to the finite thickness. Noteworthy is to remind that the resulting rigidity K'^{E}_{eqva} was presented in Section 4.3.5 as

$$K'^{E}_{eqva} = K_{eqvb} - \Delta K_{eqvs}, \tag{9.143}$$

where $K_{eqvb} = K^{E}_{eqva}$ is the rigidity proportional to the energy of pure bending, and ΔK_{eqvs} is the correction rigidity term that is accounted for reduction of this energy due to shear deformation. Thus, the contribution of energy of shear deformation to the total energy of pure bending can be estimated as

$$\frac{\Delta W_s}{W_b} = \frac{\pi^2}{20}\frac{t^2}{l^2} Y^E_a s_{66}. \tag{9.144}$$

It is shown (Eq. (4.34)) that the shear stress is

$$T_5 = \frac{\xi_o}{2} Y^E_a \left(\frac{t^2}{4} - z^2\right)\frac{d^3\theta}{dx^3}. \tag{9.145}$$

The shear strain is

$$S_5 = T_5 s_{66} = \psi_s(x,z), \tag{9.146}$$

where ψ_s is the additional tilt of elements of distorted cross section of the deformed beam relative to their in plane position according the elementary theory, as shown in Figure 9.18.

The distortions of a cross section result in the additional displacements in direction of axis x,

$$\xi_{xs}(z,x) = \psi_s(z,x)\,z. \tag{9.147}$$

The related linear strain in x direction is

$$S_{xs}(z,x) = d\xi_{xs}(z.x)/dx. \tag{9.148}$$

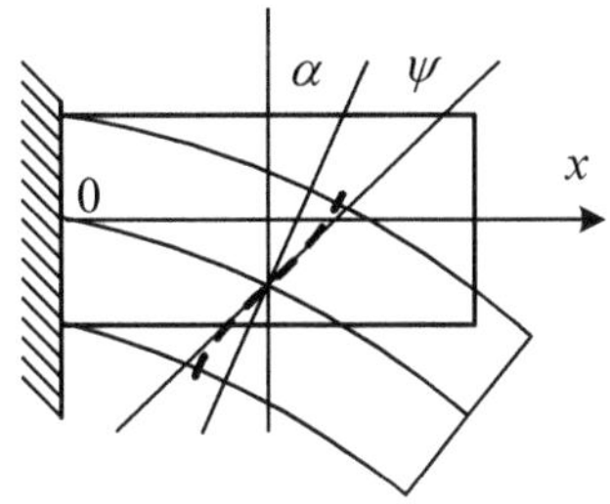

Figure 9.18: To distortion of the beam cross section due to shear deformation. Tilt of not distorted cross section according to elementary theory - α . The dashed line shows distortion of the cross section due to finite thickness, and ψ_s is the additional tilt in the point on the neutral plane.

Considering expression (9.145),

$$\xi_{xs}(z,x) = \psi_s(z,x)z = \frac{\xi_o}{2} Y_a^E s_{66} \left(\frac{t^2}{4} - z^2 \right) z \frac{d^3\theta}{dx^3}, \tag{9.149}$$

and, consequently, the additional strain due to the shear deformation is

$$S_{xs}(z,x) = \frac{d\xi_{xs}}{dx} = \frac{\xi_o}{2} Y_a^E s_{66} \left(\frac{t^2}{4} - z^2 \right) z \frac{d^4\theta}{dx^4}. \tag{9.150}$$

As follows from this expression, $S_{xs}(z = \pm t/2) = S_{xs}(0) = 0$, and it has the maximum value at $z \approx \pm 0.3t$. The total strain due to the bending and shear is

$$S_\Sigma = S_{xb} + S_{xs}, \tag{9.151}$$

where $S_{xb}(s_{66}/s_{11}^E) \approx 2.7$ (subscript *b* stays for *bending*) is given by Eq. (9.3).

The electromechanical energy of a beam having finite thickness must be calculated by the general formula (9.8), in which S_1 is replaced by the total strain S_Σ. Thus, we obtain

$$W_{em} = \frac{1}{2} w \frac{d_{31}}{s_{11}^E} \int_{-t/2}^{t/2} \int_{-l/2}^{l/2} [S_{xb}(x,z) + S_{xs}(x,z)] E_3(z)\, dz\, dx = \frac{1}{2} \xi_o V n', \tag{9.152}$$

where n' is the electromechanical transformation coefficient for the beam that includes correction due to the shear deformation. For the simply supported beam after substituting expressions for strain (9.151) and performing integration we obtain

$$n' = n_a \left(1 - \frac{\pi^2}{8} \frac{t^2}{l^2} Y_a^E s_{66} \right). \tag{9.153}$$

Here n_a is the electromechanical transformation coefficient determined for the thin bimorph beam. The correction factor in formula (9.153) is valid for either the transverse or longitudinal piezoeffect with $Y_a^E = 1/s_{11}^E$ or $1/s_{33}^E$. The factor s_{66}/s_{ii}^E for PZT-4 ceramics is and $(s_{66}/s_{33}^E) \approx 2.0$ for the transverse and longitudinal piezoeffect, respectively.

The effective coupling coefficient for a beam of the finite thickness, k'_{eff}, may be determined using expression for the factor α'_c that includes the equivalent parameters of the beam with corrections due to the finite thickness taken into account,

$$\alpha'_c = \frac{n'^2}{K'^E_{eqv} \cdot C^S_{el}}. \tag{9.154}$$

Upon substituting expressions (9.153) for n', (9.142) for K'^E_{eqv} and $C^S_{el} = C^{S_1}_{el\,a}$ this factor becomes

$$\alpha'_c = \frac{n_a^2}{K^E_{eqva} C^{S_1}_{el\,a}} \cdot \frac{[1-(\pi^2 t^2/8l^2) Y_a^E s_{66}]^2}{1-(\pi^2 t^2/20l^2) Y_a^E s_{66}} = \alpha_{ca} \cdot \gamma(t/l), \tag{9.155}$$

where α_{ca} is this factor for the bimorph beam at small t/l, and

$$\gamma(t/l) = \frac{[1-(\pi^2 t^2/8l^2) Y_a^E s_{66}]^2}{1-(\pi^2 t^2/20l^2) Y_a^E s_{66}}. \tag{9.156}$$

According to (9.85)

$$\alpha_{ca} = \frac{6}{\pi^2} \frac{k_{3i}^2}{1-k_{3i}^2}. \tag{9.157}$$

The corrected value of effective coupling coefficient of the simply supported beam will be obtained from equation

$$k'^2_{eff\,i} = \frac{\alpha_{ca}\gamma(t/l)}{1+\alpha_{ca}\gamma(t/l)}. \tag{9.158}$$

Dependences of the functions $\gamma(t/l)$ and corresponding effective coupling coefficients on the thickness to length ratio for the beams made of PZT-4 ceramics at the transverse and longitudinal piezoeffects are presented in Figure 9.19.

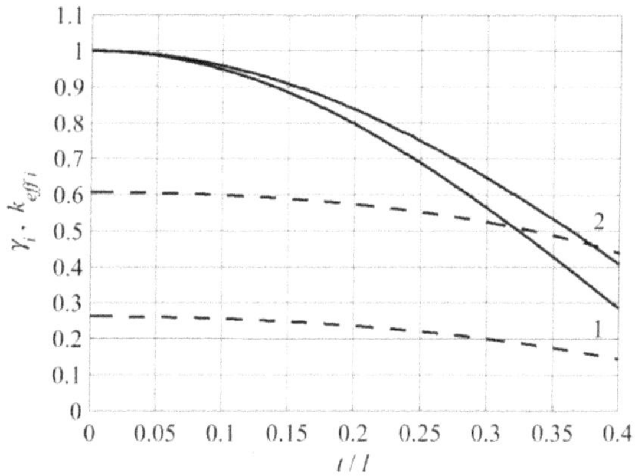

Figure 9.19: Dependences of the functions $\gamma(t/l)$ (solid lines) and effective coupling coefficients (dashed lines) on the thickness to length ratio: (1) transverse piezoeffect, (2) longitudinal piezoeffect. PZT-4 ceramics.

9.3.2.2 Employing Shear Deformation

With increase of the relative thickness of a beam the portion of total energy of deformation that is due to shear gradually increases, as it follows from relation (9.144), and at some point it may become profitable to use electromechanical conversion of the shear deformations instead of or in combination with deformations due to pure bending. In addition to relation (9.144) it must be taken into consideration that $k_{15}^2 \approx 0.5$, whereas $k_{31}^2 \approx 0.1$, and results of the electromechanical conversion of energies of shear and bending may become comparable. Example of the bender transducer that employs electromechanical conversion of the shear deformations is described in Ref. 7. It is informative to consider the design principle and way of calculating parameters of such transducer. Moreover, qualitatively they may be applied to different mechanical systems vibrating in flexure, such as the relatively thick circular plates and rings.

For employing electromechanical conversion of the shear deformation, the direction of operating electric field in the piezoelement must be perpendicular to the direction of polarization, as it is shown in Figure 9.20 with example of the partially active simply supported beam.

This can be achieved by removing the electrodes used for polarization and then applying electrodes on the perpendicular surfaces of the piezoelement. It is noteworthy that after this procedure all the elastic constants related to the extensional deformations due to bending must be changed to quantities with superscript "*D*" instead of "*E*".

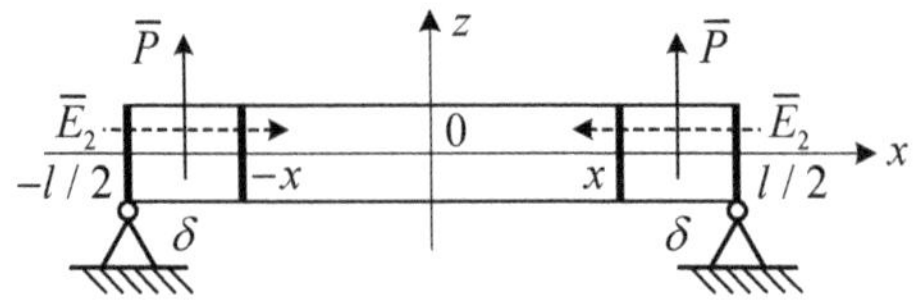

Figure 9.20: Configuration of piezoceramic beam vibrating in flexure that employs electromechanical conversion of the shear deformations. Operating electrodes are shown by the solid bold lines. $\overline{P}$ is the vector of polarization, $\overline{E}_2$ is the vector of operating electric field.

Putting aside details of the transducer design considers the electromechanical energy associated with the shear deformation of the piezoelements within beam vibrating in flexure. The piezoelectric equations that describe electromechanical conversion in the piezoelement are

$$S_5 = s_{55}^E T_5 + d_{15} E_2 , \tag{9.159}$$

$$D_2 = d_{15} T_5 + \varepsilon_{11}^T E_2 . \tag{9.160}$$

Substituting the expression for T_5 that follows from Eq. (9.159) into Eq. (9.160) results in

$$D_2 = \frac{d_{15}}{s_{55}^E} S_5 + \varepsilon_{11}^T (1 - k_{15}^2) E_2 , \tag{9.161}$$

where $k_{15}^2 = d_{15}^2 / s_{55}^E \varepsilon_{11}^T$. The electromechanical energy is

$$W_{em} = \frac{1}{2} w \frac{d_{15}}{s_{55}^E} \int_{-t/2}^{t/2} \int_{-l/2}^{l/2} S_5(z,x) E_2(x) dz dx . \tag{9.162}$$

After substituting the strain $S_5 = T_5 s_{55}^E$ with T_5 from expression (9.145), where Y_1^E must be replaced by Y_1^D, the expression for electromechanical energy will be

$$W_{em} = \frac{1}{2} \xi_o \frac{w Y_1^D d_{15}}{2} \int_{-t/2}^{t/2} \int_{-l/2}^{l/2} \left(\frac{t^2}{4} - z^2 \right) E_2(x) \frac{d^3 \theta}{dx^3} dz dx = \frac{1}{2} \xi_o V n_s . \tag{9.163}$$

As it is follows from Figure 9.20, $E_2(x) = -E_2(-x) = V / \delta$ in the active part of the beam, and $E_2(x) = 0$ in the interval $[x, -x]$. After integrating the transformation coefficient for the simply supported beam will be obtained as

$$n_s = \frac{\pi^2}{6} \frac{w t Y_1^D d_{15}}{\delta} \frac{t^2}{l^2} \cos \frac{\pi x}{l} . \tag{9.164}$$

When determining the equivalent rigidity, we will assume that the entire beam is made from ceramics that was polarized through the thickness, and afterwards the electrodes used for polarization were removed and working electrodes were applied to active parts of the beam. Under this assumption the expression (9.143) for the equivalent rigidity of the beam remains the same as in the case that the bending deformations were active, but the elastic moduli related to bending deformation must be taken with superscripts *D*,. Thus, the corresponding rigidity will be K_{eqvb}^{D}, whereas the portion of rigidity related to shear deformation must be calculated with shear module $1/s_{55}^{E}$, i.e., can be denoted ΔK_{eqvs}^{E}. Therefore, the total equivalent rigidity will be

$$K'_{eqv} = K_{eqvb}^{D} - \Delta K_{eqvs}^{E} = K_{eqvb}^{D}\left(1 - \frac{\pi^2 t^2}{20l^2} Y_1^D s_{55}^E\right). \tag{9.165}$$

This equivalent rigidity, the equivalent mass by formula (9.141), electromechanical transformation coefficient n_s by formula (9.164) and capacitance, which in this case is

$$C_{el}^{S_5} = \varepsilon_{11}^{T}(1-k_{15}^2)\frac{2wt}{\delta} = \varepsilon_{11}^{T}(1-k_{15}^2)\frac{4wt}{l[1-(2x/l)]}, \tag{9.166}$$

form the full set of equivalent parameters that are sufficient for calculating transducer electromechanical parameters using regular one-dimensional equivalent circuit.

The effective coupling coefficient of the shear mode driven flexural transducer can be estimated using expression

$$\frac{k_{eff\,s}^2}{1-k_{eff\,s}^2} = \alpha_{cs} = \frac{n_s^2}{C_{el}^{S_5} K_{eqv}^{D}} \cdot \frac{1}{1-(\Delta K_{eqv\,s}^{E}/K_{eqv}^{D})} C_{el}^{S_5}. \tag{9.167}$$

At first consider dependence of the coefficient α_{cs} from coordinate x. After substituting n_s, K'_{eqv} and from expressions (9.164), (9.165) and (9.166), respectively, we arrive at

$$\alpha_{cs}(x/l) = \frac{2}{3}\frac{k_{15}^2}{1-k_{15}^2} \cdot \frac{t^2}{l^2} \cdot \frac{Y_1^D s_{55}^E}{1-(\pi^2 t^2/20l^2)Y_1^D s_{55}^E} \cdot \frac{\cos^2(\pi x/l)}{1-(2x/l)}. \tag{9.168}$$

Dependence of the ratio

$$\frac{\alpha_{cs}(x/l)}{\alpha_{cs}(0)} = \frac{\cos^2(\pi x/l)}{1-(2x/l)} \tag{9.169}$$

is depicted in Figure 9.21 (solid line). This function has maximum at $x = 0.125l$, and at

$x = 0.25l$ has the same value, as for the fully active beam.

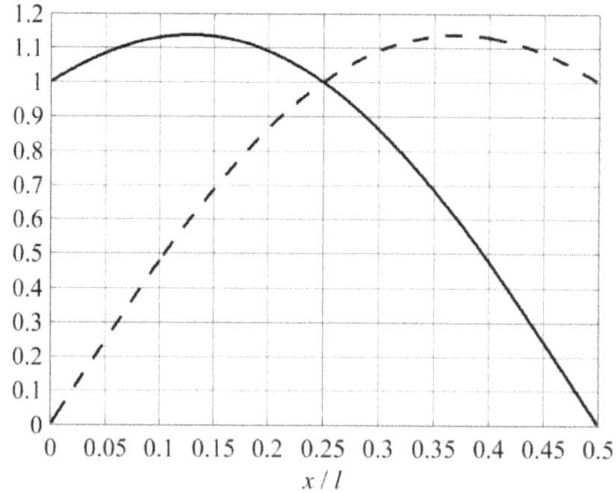

Figure 9.21: The normalized functions $\alpha_{cs}(x/l)/\alpha_{cs}(0)$ that characterize dependence of the effective coupling coefficient due to shear from the length of active part of the piezoelement (solid line), and $\alpha_{cb}(x/l)/\alpha_{cb}(0.5)$ due to bending (dashed line).

Compare function (9.169) with analogous function for the variant of the "bending activated" beam that can be obtained by replacing l_{el} by $2x$ in formula (9.100), namely,

$$\frac{\alpha_{cb}(x/l)}{\alpha_{cb}(0.5)} = \frac{\sin^2(\pi x/l)}{(2x/l)}. \tag{9.170}$$

This function is also depicted in Figure 9.21. The plots in the figure show that the parts of the beam, which are the most effective for utilizing the bending deformations, are the least effective for using the shear deformations. This means that the overall effective coupling coefficient of a bender transducer can be increased by combining active elements that utilize the bending and shear deformations. This can be achieved in the design schematically shown in Figure 9.22

For estimating the resulting benefit of such a transducer design, the maximum values of coefficients α_{cs},

$$\alpha_{cs}(x/l) = \frac{2}{3}\frac{k_{15}^2}{1-k_{15}^2}\cdot\frac{t^2}{l^2}\cdot\frac{Y_1^D s_{55}^E}{1-(\pi^2 t^2/20l^2)Y_1^D s_{55}^E}, \tag{9.171}$$

and α'_{cb} (see (9.155)),

$$\alpha'_{cb} = \frac{6}{\pi^2}\frac{k_{3i}^2}{1-k_{3i}^2}\cdot\frac{[1-(\pi^2 t^2/8l^2)Y_i^E s_{66}]^2}{1-(\pi^2 t^2/20l^2)Y_i^E s_{66}}, \tag{9.172}$$

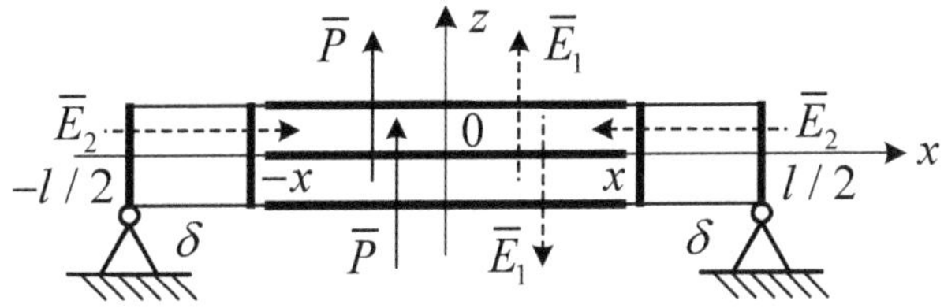

Figure 9.22: Configuration of a beam, in which electromechanical conversion due to the bending and shear deformations takes place simultaneously.

must be compared depending on the thickness to length ratio. As the result, the corresponding effective coupling coefficients are presented in Figure 9.23 for the case that PZT-4 is used. The design variant with the transverse piezoeffect used for conversion of the bending deformations is shown in Figure 9.22 as an example. More practical is the design with the longitudinal effect in segmented piezoelement used in combination with elements employing shear. The coupling coefficients are presented in Figure 9.23 for both variants for comparison.

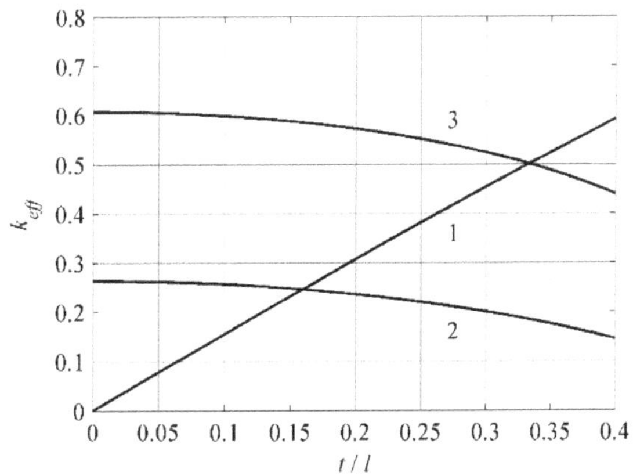

Figure 9.23: Comparison of the effective coupling coefficients of the beam transducers employing the shear (curve 1) and bending deformations at the transverse (curve 2) and longitudinal (curve 3) piezoeffect. The comparison is made for the case that PZT-4 is used.

Although the results obtained are quantitatively specified for the rectangular beams, they are qualitatively applicable to similar mechanical systems vibrating in flexure, such as circular plates and thick rings.

9.3.3 Nonuniform Over the Length Beams

As it was shown above, the effective coupling coefficient of a bender transducer can be maximized by placing a reduced amount of the active material in the parts of a beam, where it experiences greater stress. Therefore, in the cases that maximizing the effective coupling coefficient is desirable even at expense of a reduction of the electromechanical force (for example, for receivers or projectors operating in air with small mechanical load), the parts of active material that contribute the least to the electromechanical conversion can be replaced by passive material. This can make the transducer more reliable and cost effective. Two variants of transducer designs pursuing such goals are illustrated with Figure 9.1 (b.2), which will be called variant *A*, and (c.2), which will be called variant *B*. In both cases the beams become nonuniform over the length. Vibration of the passive nonuniform over the length beams is considered in Sections 4.5.6.2 and 4.5.6.4. The main results obtained therein are that for both transducer designs and variants of nonuniformity the mode shapes of vibration remain practically the same, as for the uniform over length beams of the same type, at least up to the size of active laminates $l_e / l \approx 0.6$ and for their relative thicknesses δ_{opt} / t that are optimal in terms of the effective coupling coefficients. Based on these results the expressions for the equivalent rigidities and masses of the nonuniform beams were obtained. Knowing the mode shapes of vibration for nonuniform beams allows calculating electromechanical transformation coefficients and effective coupling coefficients of the transducers, and thus completing all the set of parameters required for transducers designing. This is done in the following sections.

9.3.3.1 Trilaminar Beams

The equivalent masses and rigidities of the nonuniform over the length trilaminar beams in variant *A* according to expressions (4.332) and (4.337) are

$$M_{1tA} = M_{1t}\left[1+\frac{2(\gamma_\rho - 1)y}{\gamma_\rho - 2(\gamma_\rho - 1)y}F_1(l_e/l)\right], \tag{9.173}$$

$$K_{1tA} = K_{1t}\left[1+\frac{(\gamma_Y - 1)[1-(1-2y)^3]}{1+(\gamma_Y - 1)(1-2y)^3}F_1(l_e/l)\right], \tag{9.174}$$

where

$$M_{1t} = \frac{1}{2} wtl\rho_{eqvt} \, . \; K_{1t} = \frac{\pi^4 wt^3 Y_{eqvt}}{24l^3}, \tag{9.175}$$

and function $F_1(l_e/l)$ is

$$F_1(l_e/l) = 1 - (l_e/l) - (1/\pi)\sin(\pi l_e/l) \, . \tag{9.176}$$

The same expressions (9.173) and (9.174) are valid for the masses and rigidities in the variant *B*, if to change subscripts *A* for *B*, and to set to zero γ_ρ and γ_Y in numerators in the brackets.

The electromechanical transformation coefficient can be determined by formula (9.21) as

$$n_1 = \frac{2\pi w d_{31}}{l s_{11}^E}(t-\delta)\sin\frac{\pi l_e}{2l}, \tag{9.177}$$

because the mode shape of vibration is the same to the first approximation, as for the uniform beam with simply supported ends. Capacitance of the active laminates connected in parallel is

$$C_{el}^{S_1} = \varepsilon_{33}^T(1-k_{31}^2)2wl_e/\delta \, . \tag{9.178}$$

With all the equivalent parameters known, the effective coupling coefficient can be determined by formula (9.23) using the coefficient α_c that for variant *A* is

$$\alpha_{c1tA}(y, l_e/l) = \frac{n_{1t}^2}{K_{1tA} C_{el}^{S_1}(l_e)} = \frac{n_t^2}{K_{1t} C_{el}^{S_1}(l)} \cdot \psi[y, (l_e/l)] \, . \tag{9.179}$$

Here

$$\frac{n_t^2}{K_{1t} C_{el}^{S_1}(l)} = \alpha_{ct}(y) \tag{9.180}$$

is the coefficient for trilaminar beam uniform over the length according to formula (9.110). Its dependences on the relative thickness of the active laminate, $y = \delta/t$, are presented in normalized form of the function $\beta(\delta/t)$ in Figure 9.14 for different combinations of PZT-4 and passive materials. The function $\psi_t[y, (l_e/l)]$ in the expression (9.179) is,

$$\psi_t[y, (l_e/l)] = \left[1 + \frac{(\gamma_Y - 1)[1-(1-2y)^3]}{1+(\gamma_Y - 1)(1-2y)^3} F_1(l_e/l)\right]^{-1} \cdot \frac{l}{l_e}\sin^2\left(\frac{\pi l_e}{2l}\right). \tag{9.181}$$

This function depends on the relative length of the active laminate l_e/l for each value of its relative thickness *y*. Expression (9.181) for the function ψ_t is valid in variant *B* of the beam nonuniformity if to set $\gamma_Y = 0$ in the nominator of the term in brackets.

Dependences of the equivalent parameters and of the resonance frequencies of nonuniform over length trilaminar beams from l_e / l are shown in Figure 9.24–Figure 9.26 for values of $y_{opt} = (\delta / t)_{opt}$, which correspond to the optimal value of the effective coupling coefficient for the combinations of PZT-4 with different passive materials. They are normalized to parameters of uniform over length trilaminar beams having the same length, thickness and $y_{opt} = (\delta / t)_{opt}$ ratio. The plots can be used for practical calculating transducers utilizing the trilaminar nonuniform beams. They allow making an informed choice of the geometry of the beams. Plots of function $\psi_t[y_{opt}, (l_e / l)]$ are presented in Figure 9.27. They show that an additional gain or loss in the effective coupling coefficient of the nonuniform beam can be obtained compared with the case that the beam is uniform, and the length of electrodes is reduced to the same extent. This depends on what passive material is used. Qualitatively this result could be foreseen, because the value of rigidity of nonuniform beam K_{1tA} reduces or increases (see Figure 9.24) in comparison with those of uniform beam at all other conditions equal in expression for α_{c1tA}.

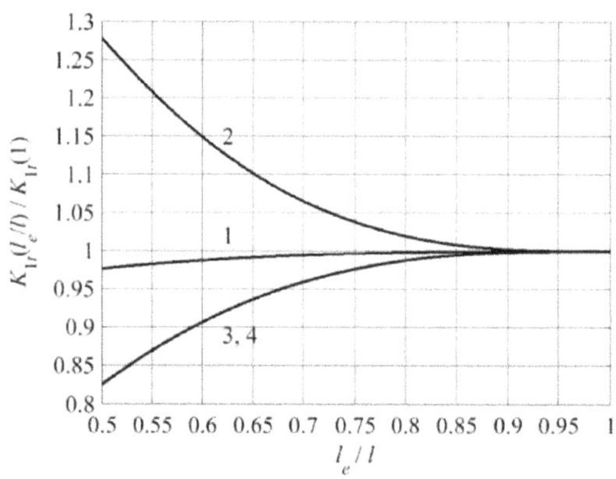

Figure 9.24: The equivalent rigidities of the nonuniform over length trilaminar beams normalized to the equivalent rigidities of the uniform trilaminar beams that correspond to the maximum effective coupling coefficient, vs. l_e / l for different passive materials used for replacement of piezoceramics. Variant *A*: (1) aluminum, (2) steel. Variant *B*: (3) aluminum, (4) steel.

Within range of values of l_e / l, at which the first approximation to the mode of vibration is applicable ($l_e / l > (0.5 - 0.6)$ at y_{opt}), calculating the nonuniform trilaminar beam transducer can be produced using the single mechanical degree of freedom equivalent circuit of Figure 9.3. All the parameters of the circuit are determined except for those related to acoustic field. They will be considered in Section 9.5.

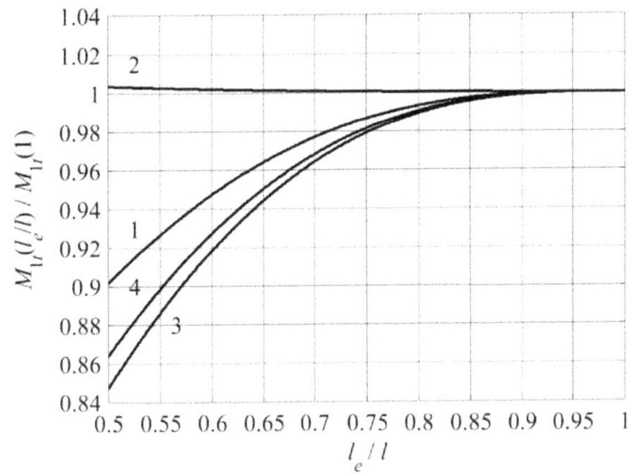

Figure 9.25: The equivalent masses of the nonuniform over length trilaminar beams normalized to the equivalent masses of the uniform trilaminar beams that correspond to the maximum effective coupling coefficient, vs. l_e/l for different passive materials used for replacement of piezoceramics. Variant *A*: (1) aluminum, (2) steel. Variant *B*: (3) aluminum, (4) steel.

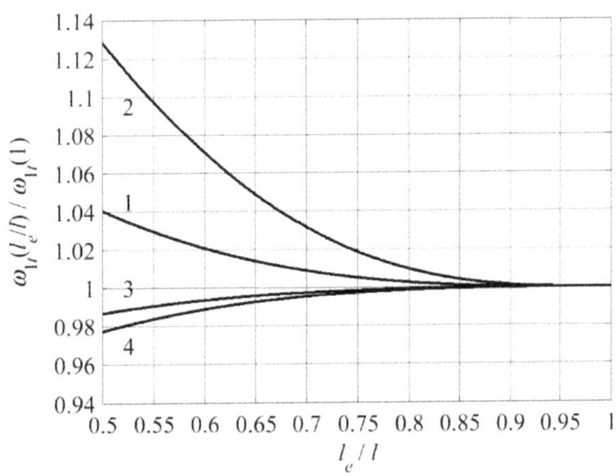

Figure 9.26: The resonance frequencies of the nonuniform over length trilaminar beams normalized to the resonance frequencies of uniform trilaminar beams that correspond to the maximum effective coupling coefficient, vs. l_e/l for different passive materials used for replacement of ceramics. Variant *A*: (1) aluminum, (2) steel. Variant *B*: (3) aluminum, (4) steel.

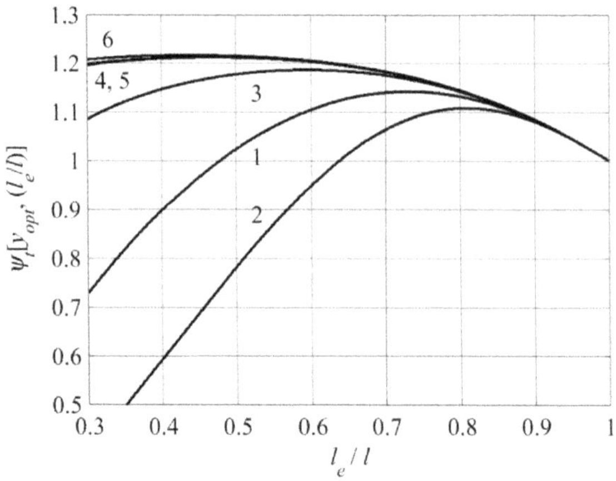

Figure 9.27: Dependences of the function $\psi_t[y_{opt}, (l_e/l)]$ on the relative length of the active laminate for different combinations of PZT-4 and passive materials. Variant *A*: (1) aluminum, (2) steel, (3) G-10. Variant *B*: (4) aluminum, (5) steel, (6) G-10.

If it is needed by some reasons to use beams with even smaller relative length of the active laminates, then the second approximation to the mode of vibration must be taken into consideration, as it is shown in Section 4.5.6.2, and the transducer must be considered as having two mechanical degrees of freedom in order to increase the accuracy of calculations.

9.3.3.2 Bilaminar Beams

The procedure of considering the bilaminar nonuniform over the length beam transducers after the mechanical part of the problem is solved in Section 4.5.6.4 is the same, as for the trilaminar beams. For simplicity we will consider the case, in which the thickness of active laminate is maximum, i.e., $z_0 = z_{0m} = \delta_{\max}$. Thus, it is assumed that in formulas below $y = z_{0m}/t$. The general case, in which the thickness of active laminate can be reduced ($\delta < \delta_{\max}$) for optimizing effective coupling coefficient, can be considered in analogous way just with more cumbersome calculations, as it was illustrated for the uniform bilaminar beams in Section 9.3.1.5.

The equivalent masses and rigidities of the nonuniform over the length bilaminar beams in variant *A* according to expressions (4.364) and (4.368) are

$$M_{1bA} = M_{1b}\left[1 + 2y\frac{\gamma_\rho - 1}{\gamma_\rho - (\gamma_\rho - 1)y}F_1(l_e/l)\right], \tag{9.182}$$

$$K_{1bA} = K_{1b}\left[1+\frac{(\gamma_Y-1)y^3}{y^3+(1-y)^3\gamma_Y}F_1(l_e/l)\right], \tag{9.183}$$

where

$$M_{1b} = \frac{1}{2}wtl\rho_{eqvb},\ K_{1b} = \frac{\pi^4wt^3Y_{eqvb}}{24l^3}, \tag{9.184}$$

and function $F_1(l_e/l)$ is determined by Eq. (9.176). The same Eqs. (9.182) and (9.183) are valid for the masses and rigidities in the variant B, if to change subscripts A for B, and to set to zero γ_ρ and γ_Y in numerators in the brackets.

The electromechanical transformation coefficient to the first approximation remains the same as for the uniform bilaminar beam. Thus, in the general case (see Eq. (9.129))

$$n_b(\delta) = \frac{wd_{31}}{2s_{11}^E}(2z_0-\delta)\left[\frac{d\theta}{dx}\right]_{x=-l_e/2}^{x=l_e/2}. \tag{9.185}$$

In the case that $z_0 = z_{0m} = \delta_{\max}$ and thus $y = z_{0m}/t = y_{\max}$

$$n_b(z_{0m}) = \frac{wtd_{31}}{2s_{11}^E}y\frac{d\theta}{dx}\bigg|_{x=-l_e/2}^{x=l_e/2}. \tag{9.186}$$

If to assume that the beam is simply supported and the mode shape of displacement remains the same as for a uniform beam, which is the case to the first approximation, then

$$n_{1b} = \pi\frac{wtd_{31}}{s_{11}^E}\frac{y}{l}\sin\frac{\pi l_e}{2l}, \tag{9.187}$$

Capacitance of the active laminate is

$$C_{el}^{S_1} = \varepsilon_{33}^T(1-k_{31}^2)wl_e/\delta. \tag{9.188}$$

Coefficient α_{c1bA} for determining effective coupling coefficient in the variant A is

$$\alpha_{c1bA} = \frac{n_{1b}^2}{K_{1bA}C_{el}^{S_1}(l_e)} = \frac{n_b^2}{K_{1b}C_{el}^{S_1}(l)}\cdot\psi_b(l_e/l), \tag{9.189}$$

where the first factor is coefficient $\alpha_{cb}(\delta_{\max})$ for the uniform bilaminar beam (see formula (9.129)), and function $\psi_b(l_e/l)$ is

$$\psi_b(l_e/l) = \left[1+\frac{(\gamma_Y-1)y^3}{y^3+(1-y)^3\gamma_Y}F_1(l_e/l)\right]^{-1}\cdot\frac{l}{l_e}\sin^2\left(\frac{\pi l_e}{2l}\right). \tag{9.190}$$

Expression for the function ψ_b in variant B of the beam nonuniformity differs from expression (9.190) by setting $\gamma_Y = 0$ in the nominator of the term in brackets.

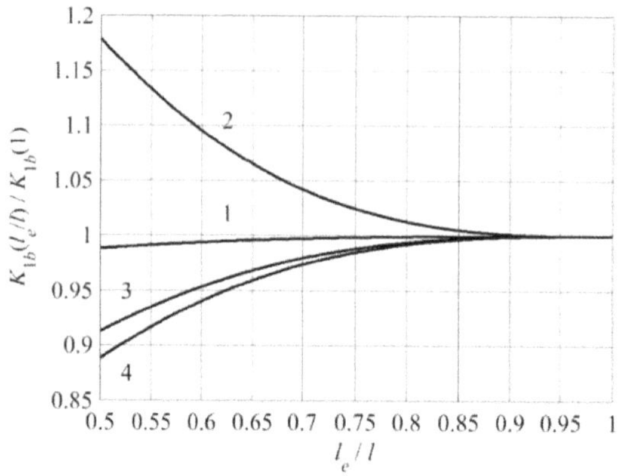

Figure 9.28: The equivalent rigidities of the nonuniform over length bilaminar beams normalized to the equivalent rigidities of the analogous uniform bilaminar beams vs. l_e / l for different passive materials used for replacement of piezoceramics. Variant A: (1) aluminum, (2) steel. Variant B: (3) aluminum, (4) steel.

Dependences of the equivalent parameters, resonance frequencies and function $\psi_b[y_{max}, (l_e / l)] / \psi_b[y_{max}, 1]$ for nonuniform over length bilaminar beams from are shown in Figure 9.28-Figure 9.31. Remember that it is assumed that in all these cases $y = z_{0m} / t = y_{max}$. The functions are normalized to parameters of the analogous uniform over length bilaminar beams having the same length, thickness and $y = z_{0m} / t$ ratio vs. l_e / l.

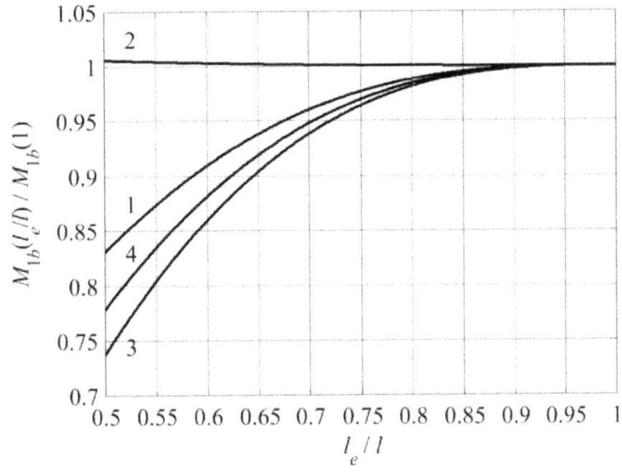

Figure 9.29: The equivalent masses of the nonuniform over length bilaminar beams normalized to the equivalent masses of the uniform bilaminar beams vs. l_e / l for different passive materials used for replacement of piezoceramics. Variant *A*: (1) aluminum, (2) steel. Variant *B*: (3) aluminum, (4) steel.

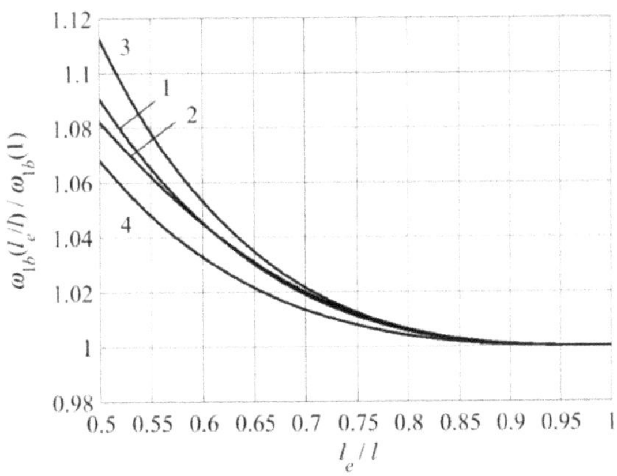

Figure 9.30: The resonance frequencies of the nonuniform over length bilaminar beams normalized to the resonance frequencies of analogous uniform bilaminar beams vs. l_e / l for different passive materials used for replacement of piezoceramics. Variant *A*: (1) aluminum, (2) steel. Variant *B*: (3) aluminum, (4) steel.

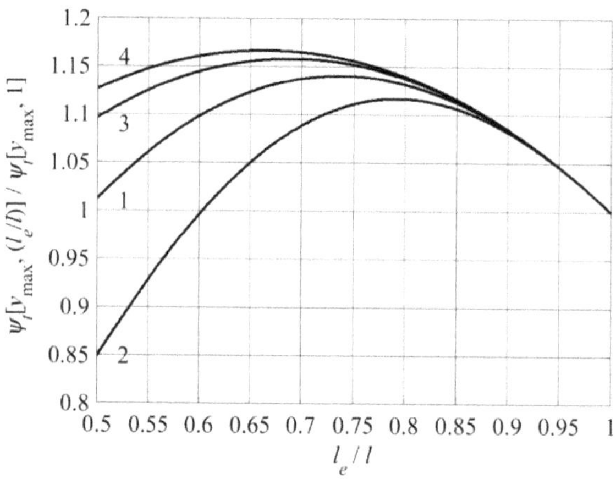

Figure 9.31: Dependences of the function $\psi_b[y_{\max}, (l_e / l)] / \psi_b[y_{\max}, 1]$ on the relative length of the active laminate for different combinations of PZT-4 and passive materials. Variant *A*: (1) aluminum, (2) steel. Variant *B*: (3) aluminum, (4) steel.

The same considerations regarding applicability of the equivalent circuit shown in Figure 9.3 for calculating parameters of trilaminar beams are valid for the bilaminar beams. The peculiarity of the bilaminar beams is that the range of values of relative lengths of active laminates, at which the first approximation to the mode of vibration is applicable, is wider than for trilaminar beams, as it is shown in Section 4.5.6.4.

9.4 Circular Plate Transducers

The modifications of the circular flexural type transducers to be considered are presented in Figure 9.1. They are analogous to those of the rectangular beam transducers and will be treated in the same sequence: uniform over radius transducers with bimorph plate as the basic design, and transducers nonuniform over the radius. The main results of this Section were presented in Ref. 8.

9.4.1 Radially Uniform Transducers

The basic bimorph design of uniform through the thickness circular plate transducer was considered in Section 2.6.3 under the simply supported boundary conditions. The procedure of determining the mechanical equivalent parameters of nonuniform through the thickness plates (trilaminar and bilaminar) was illustrated with example of the rectangular beam in Section 4.5.6. The general expressions for calculating electromechanical conversion related parameters (electromechanical transformation and effective coupling coefficients) are presented in Section 9.2.2. The goal of this section is to summarize all the set of circular transducers equivalent parameters under different boundary conditions.

9.4.1.1 Equivalent Electromechanical Parameters of the Circular Plate Transducers

General expressions for the equivalent mechanical parameters of the radially uniform and non-uniform by the thickness circular plates were obtained in Section 4.5.7.1 in the form of

$$K^E_{eqv} = \frac{t^3 Y_{eqv}}{12} \cdot L(0.3)_{bc}, \tag{9.191}$$

$$M_{eqv} = \rho_{eqv} t S_{eff\,bc}. \tag{9.192}$$

Here $Y_{\sigma eqv}$ and ρ_{eqv} are the equivalent Young's modulus and density that depend on composition of the plate through the thickness. They are defined by formulas (4.390) and (4.393), respectively. In terms of nonuniformity of properties through the thickness the variants of bimorph, trilaminar and bilaminar plates must be considered that are shown in Figure 9.1 (a.1), (b.1), and (c.1). Quantities $L(0.3)_{bc}$ and $S_{eff\,bc}$ depend on the boundary conditions for the plates. They are defined by formulas (4.382) at $\sigma = 0.3$ and (4.394) and reproduced here for the sake of convenience.

$$L(\sigma)_{bc} = 2\pi \int_0^a F(\theta) r dr, \tag{9.193}$$

where

$$F(\theta) = \left(\frac{d^2\theta}{dr^2}\right)^2 + 2\sigma \frac{1}{r}\frac{d\theta}{dr}\frac{d^2\theta}{dr^2} + \left(\frac{1}{r}\frac{d\theta}{dr}\right)^2, \tag{9.194}$$

$$S_{eff\,bc} = 2\pi \int_0^a \theta^2 (r/a)\, r dr. \tag{9.195}$$

The general expression for the electromechanical transformation coefficient is (9.55)

$$n = \frac{2\pi d_{31}}{(s^E_{11} + s^E_{12})\delta} \Omega_t \Omega_{r\,bc}. \tag{9.196}$$

Here Ω_t depends on configuration of electric field through the thickness, and Ω_{rbc} depends on the boundary conditions and on the size of the electrodes. We will assume that in general the radially uniform plates may have reduced electrodes. (As it was mentioned before, strictly speaking, the remaining parts of the plate surface are supposed to be covered with electrodes that are short circuited in order to keep value of the elastic constants of ceramic unchanged along the radius and equal to s_{ik}^E.) These coefficients are defined by formulas (9.53), and will be denoted $\Omega_{rbc}(r_1)$.

9.4.1.2 Bimorph Plates

The bimorph plates are fully active uniform in thickness, therefore, $Y_{eqv} = Y_{\sigma a}$ and $\rho_{eqv} = \rho_a$. Due to symmetry the neutral plane coincides with the middle plane of a plate. It is convenient to place the origin of the coordinate system in the neutral plane. As the result, $z_0 = 0$ and integration in all the general expressions must be performed over the interval $(t/2) \geq z \geq -(t/2)$.

Table 9.2. Factors accounting for different boundary conditions.

	$L(0.3)_{bc}$, π / a^2	$S_{eff\,bc}$, πa^2	S_{avbc}, πa^2	$\Omega_{rbc}(a)$	$\Omega_{rbc}(r_1)$
ss	7.2	0.28	0.45	1.5	$-\frac{r_1^2}{a^2}\left(2.5-\frac{r_1^2}{a^2}\right)$
cl	21.4	0.18	0.33	0	$-4\frac{r_1^2}{a^2}\left(1-\frac{r_1^2}{a^2}\right)^*$
Free	22	0.26	0	2.18	(9.63)
Center supported**	6.6	0.46	0.63	0.97	

* Maximum value at $r_1 = a/\sqrt{2}$. **Reference point is on the edge.

In the expression (9.196) for the electromechanical transformation coefficient $\delta = t/2$, and $\Omega_t = t^2/4$ for parallel connection of the half plates according to formula (9.14). Thus, the equivalent parameters of the bimorph plate become

$$K_{eqva}^E = \frac{t^3 Y_{\sigma a}^E}{12} \cdot L(0.3)_{bc}\,, M_{eqva} = \rho_a t S_{ef\,bc}\,,\; n_a = \frac{\pi d_{31} t}{(s_{11}^E + s_{12}^E)} \Omega_{rbc}\,, \tag{9.197}$$

where the factors $L(0.3)_{bc}$, $S_{eff\,bc}$ and Ω_{rbc} depend on the boundary conditions. Values of these factors are presented in Table 9.2. The subscripts "a" indicates that the parameters belong to the basic bimorph plate design for distinguishing them in case of comparing with analogous parameters of different plate modifications.

Together with expression for the capacitance,

$$C_{el}^{S_{1,2}} = \varepsilon_{33}^{T}(1-k_p^2)\frac{4\pi a^2}{t}, \tag{9.198}$$

we obtain the full set of electromechanical parameters of the bimorph circular plate transducer.

In the variant of the simply supported boundary, which will be used for reference as the basic design,

$$K_{eqv}^{E} = \frac{0.6\pi t^3 Y_{\sigma a}^{E}}{a^2}, M_{eqv} = 0.28\pi a^2 t\rho_a = 0.28M, \ n = 1.5\frac{\pi d_{31} t Y_a^E}{1-\sigma_a^E}. \tag{9.199}$$

The resonance frequency of the plate is

$$f_{res\,a} = \frac{1}{2\pi}\sqrt{\frac{K_{eqv\,a}^{E}}{M_{eqv\,a}}} = 0.23\frac{t}{a^2}\sqrt{\frac{Y_{\sigma a}^{E}}{\rho_a}}. \tag{9.200}$$

The effective coupling coefficients of the bimorph plates under different boundary conditions were considered as an example in Section 9.2.2.2. For the simply supported bimorph plate with the full-size electrodes ($r_{el} = a$) the coefficient $\alpha_{c\,ss}(a)$ and effective coupling coefficients are

$$\alpha_{c\,ss}(a) = 0.62\frac{k_p^2}{1-k_p^2}, \text{ and, } k_{eff\,ss}^2(a) = \frac{\alpha_{c\,ss}(a)}{1+\alpha_{c\,ss}(a)} = 0.62\frac{k_p^2}{1-0.38k_p^2}, \tag{9.201}$$

It the plate is made of PZT-4 ceramics ($\sigma_a^E = 0.33$, $k_p = 0.58$), $k_{eff\,ss}(a) \approx 0.49$.

In Figure 9.10 are presented dependences from radius of electrodes of the relative values of coefficients, which characterize effective coupling coefficients, $\alpha_{c\,bc}(r_{el})/\alpha_{c\,ss}(a)$, for different boundary conditions. For the simply supported plate the maximum of this ratio is achieved at $r_{el} = 0.91a$. The maximum value of the effective coupling coefficient is

$$k_{eff}^2 = 0.64\frac{k_p^2}{1-0.36k_p^2}, \tag{9.202}$$

and it remains about the same up to $r_{el} \approx 0.8$.

9.4.1.3 Bimorph Center Supported (Clamped) Plate

Vibration of the center supported plate made of a passive material was considered in Section 4.5.8.1. The first resonance mode of vibration was determined by expression (4.438)

$$\xi(r/a)=\frac{r^2}{a^2}-0.91\frac{r^3}{a^3}+0.27\frac{r^4}{a^4}=\xi(a)\theta(r/a), \tag{9.203}$$

where $\xi(a)=0.36$ is displacement of the reference point on the edge. Thus, the mode shape of vibration is

$$\theta(r/a)=\frac{1}{0.36}\left(\frac{r^2}{a^2}-0.91\frac{r^3}{a^3}+0.27\frac{r^4}{a^4}\right). \tag{9.204}$$

Following general expressions (9.193) and (9.195) the coefficients L_{bc} and $S_{eff\,bc}$ calculated for this mode of vibration will be $L_{bc}=17\pi/a^2$ and $S_{eff\,bc}=0.46\pi a^2$. Expressions for the equivalent rigidity, mass and resonance frequency of the transducer are, accordingly,

$$K^E_{eqvcs}=\frac{6.6\pi t^3 Y^E_{\sigma a}}{12a^2}=\frac{6.6\pi t^3}{12a^2 s^E_{11}(1-\sigma_1^{E2})}, M_{eqvcs}=\rho t S_{eff}=0.46M,$$
$$f_{rcs}=0.17\frac{t}{a^2}\sqrt{\frac{1}{s^E_{11}\rho(1-\sigma_1^{E2})}}, \tag{9.205}$$

i.e., they are the same as for the passive plate, (see (4.440) and (4.441), with elastic modulus replaced by its analog for the active material. Besides $S_{av}=0.63\pi a^2$.

The electromechanical transformation coefficient must be determined by formula (9.196). In the case of parallel connection of the piezoelements and under general assumption that a part of the electrodes having radius r_1 is active the expression for the coefficient will be

$$n_{cs}=\frac{\pi d_{31}t}{(s^E_{11}+s^E_{12})}\Omega_{r\,bc}, \tag{9.206}$$

where function $\Omega_{r\,bc}$ is determined by formula (9.53). Taking into account expression (9.205) for the mode shape,

$$\Omega_r=r_1\frac{d\theta}{dr}\bigg|_{r=r_1}=\frac{r_1^2}{0.36a^2}\left(2-2.73\frac{r_1}{a}+1.08\frac{r_1^2}{a^2}\right). \tag{9.207}$$

Thus,

$$n_{cs} = \frac{\pi d_{31} t}{0.36 s_{11}^E (1-\sigma_1^E)} \frac{r_1^2}{a^2} \left(2 - 2.73 \frac{r_1}{a} + 1.08 \frac{r_1^2}{a^2} \right). \tag{9.208}$$

For the fully active electrodes, at $r_1 = a$, $\Omega_r = 0.97$, and

$$n_{cs} = 0.97 \frac{\pi d_{31} t}{s_{11}^E (1-\sigma_1^E)}. \tag{9.209}$$

Effective coupling coefficient of the transducer can be commonly calculated through the coefficient

$$\alpha_{ccs} = n_{cs}^2 / C_{el}^{S_{1,2}} K_{eqvcs}^E. \tag{9.210}$$

After substituting parameters by formulas (9.209), (9.205) and capacitance $C_{el}^{S_{1,2}} = \varepsilon_{33}^T (1-k_p^2) 4\pi r_1^2 / t$, we arrive at

$$\alpha_{ccs} = 2.3 \frac{k_p^2}{1-k_p^2} \cdot \frac{r_1^2}{a^2} \left(2 - 2.73 \frac{r_1}{a} + 1.08 \frac{r_1^2}{a^2} \right)^2. \tag{9.211}$$

At $r_1 = a$

$$\alpha_{ccs}(a) = 0.28 \frac{k_p^2}{1-k_p^2}. \tag{9.212}$$

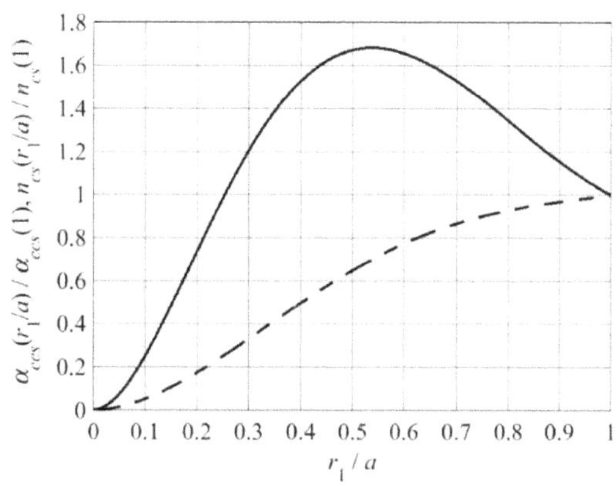

Figure 9.32: Dependencies of the ratio $\alpha_{ccs}(r_1/a)/\alpha_{ccs}(1)$ (solid line) and electromechanical transformation coefficient n (dashed line) from the relative radius of the active part of the electrodes.

Function (9.211) has maximum at value of (r_1/a) that may be found from equation $\alpha'_{ccs}(r_1/a) = 0$. This value appears to be $r_1 = 0.54a$, and the maximum values of the function

α_{ccs} and corresponding effective coupling coefficient are, respectively,

$$\alpha_{ccs}(0.54a) = 0.47\frac{k_p^2}{1-k_p^2} \text{ and } k_{eff} = \sqrt{\frac{\alpha_{ccs}}{1+\alpha_{ccs}}} = \frac{0.53k_p}{\sqrt{1-0.72k_p^2}}. \tag{9.213}$$

In the case that piezoceramic PZT-4 is used ($k_p = 0.58$) the coupling coefficients with fully active (at $r_1 = a$) and optimized (at $r_1 = 0.54a$) electrodes are $k_{eff} = 0.6k_p$ and $0.76k_p$. Plot of the ratio $\alpha_{ccs}(r_1/a)/\alpha_{ccs}(1)$ vs. r_1/a is shown in Figure 9.32. At the same Figure, the dependence of electromechanical transformation coefficient $n_{cs}(r_1/a)/n_{cs}(1)$ vs. relative size of the electrodes is depicted that shows, how the electromechanical force generated by the transducer changes simultaneously with the effective coupling coefficient.

9.4.1.4 Bimorph Plate Center Supported by the Post of a Finite Radius

The center supported (clamped) plate is the ideal model. In practical applications, such as mechanical system of a hydrophone (accelerometer), real supporting element has a finite diameter. Vibration of a passive plate supported by the post of a finite diameter was considered in Section 4.5.8.2, where the first resonance mode of displacement was determined as

$$\theta(r/b) = \left[\left(1-\frac{r^2}{b^2}\right)\ln\frac{r}{b} + A\left(1-\frac{r^2}{b^2}\right)^2\right]\cdot\left[\left(1-\frac{a^2}{b^2}\right)\ln\frac{a}{b} + A\left(1-\frac{a^2}{b^2}\right)^2\right]^{-1}. \tag{9.214}$$

Coefficient A in this expression depends on the ratio b/a. A reasonable estimation for value of b/a in application to electromechanical transducers may be about $b/a = 0.1$. For this case $A = 0.01$. We will use this example for comparing results with those obtained by using analytically more straightforward ideal case of the center supported plate. Comparison of the mode shapes of vibration and equivalent parameters of the center supported and clamped on radius $b/a = 0.1$ plates are made in Figure 4.33 and in Table 4.5. Thus, at $b = 0.1a$.

$$K_{eqv01}^E = \frac{9.6}{6.6}\cdot K_{eqv0}^E(1-b^2/a^2) \approx 1.45K_{eqvcs}^E, \tag{9.215}$$

$$M_{eqv01} = \frac{0.37}{0.46}M_{eqv0}(1-b^2/a^2) \approx 0.8M_{eqvcs}, \tag{9.216}$$

$$S_{av0.1} = \frac{0.53}{0.63}(1-b^2/a^2)S_{av0} \approx 0.84S_{avcs}, \tag{9.217}$$

$$f_{r0.1} = \left(\frac{2.25}{1.94}\right)^2 f_{rcs} = 1.34 f_{rcs}. \tag{9.218}$$

The coefficient of electromechanical transformation at radius of electrodes $r_{el} = r_1$ must be determined by formula (9.206), where

$$\Omega_{rb/a}(r_1) = r_1 \left. \frac{d\theta(r/b)|_{b/a}}{dr} \right|_{r=r_1}. \tag{9.219}$$

The coefficient $\alpha_{cb/a}$, which is linked to the effective coupling coefficient, is

$$\alpha_{cb/a}(r_1) = n_{b/a}^2(r_1) / C_{el\,b/a}^{S_{1,2}}(r_1) K_{eqv\,b/a}^E. \tag{9.220}$$

Here $C_{el\,b/a}^{S_{1,2}} = \varepsilon_{33}^T (1-k_p^2) 4\pi a^2 (1-b^2/a^2)/t$ is the capacitance of bimorph plate at parallel connection of laminates. Using expressions (9.214) and (9.219) for the fully active electrodes at $b/a = 0.1$, will be obtained that $\Omega_r = 1.25$. Thus, we arrive at the following expressions for the coefficients of electromechanical transformation, $n_{0.1}$, and $\alpha_{c0.1}$:

$$n_{0.1} = 1.25 \frac{\pi d_{31} t}{s_{11}^E (1-\sigma_1^E)} \approx 1.3 n_{cs}, \tag{9.221}$$

$$\alpha_{c0.1}(a) = n_{0.1}^2(a) / C_{el\,0.1}^{S_{1,2}}(a) K_{eqv\,0.1}^E \approx 1.2 \alpha_{ccs}(a). \tag{9.222}$$

The values of $K_{eqv\,0.1}^E$ and $\alpha_{ccs}(a)$ are given by formulas (9.205) and (9.212), respectively. Thus,

the effective coupling coefficient with electrodes of the full size is

$$k_{eff\,01} = \sqrt{\frac{\alpha_{c01}(a)}{1+\alpha_{c01}(a)}} = \sqrt{\frac{1.2\alpha_{ccs}(a)}{1+1.2\alpha_{ccs}(a)}} = \frac{0.58 k_p}{\sqrt{1-0.66 k_p^2}}. \tag{9.223}$$

Comparison with the effective coupling coefficient of the center supported plate (see (9.213)) shows that for the plates with full size electrodes made of PZ-4 ceramics $k_{eff\,0.1} = 1.09 k_{eff\,cs}$.

Summarizing results of the last two sections we may conclude that increase of relative radius of the supporting central post up to $b/a = 0.1$ produces increase of the resonance frequency of the transducer (according to (9.218)) and slight improvement of its electromechanical characteristics.

9.4.1.5 Trilaminar Plates

Structure of the trilaminar circular plate through the thickness is the same as of the trilaminar beam considered in Section 9.3.1.4. Therefore, all the relations that characterize properties of the equivalent parameters that depend on this structure are the same. Thus, the equivalent Young's modulus and density of the trilaminar plate according to (9.104) and (9.105) are

$$, \rho_{eqvt} = \rho_a \cdot [\gamma_\rho - 2(\gamma_\rho - 1)y] \,. \tag{9.224}$$

(Subscript t stands for trilaminar and all the notations γ_Y , γ_ρ and y are defined by relations (9.103)). The mechanical equivalent parameters of the trilaminar plates must be determined by formulas (9.191) and (9.193) using these expressions, namely,

$$K_{eqvt}^E = \frac{t^3 Y_{\sigma a}}{12} L(0.3)_{bc} \cdot [1 + (\gamma_Y - 1)(1 - 2y)^3] = K_{eqva}^E \cdot [1 + (\gamma_Y - 1)(1 - 2y)^3] \,, \tag{9.225}$$

$$M_{eqvt} = \rho_a t S_{eff\,bc} \cdot [\gamma_\rho - 2(\gamma_\rho - 1)y] = M_{eqv\,a} \cdot [\gamma_\rho - 2(\gamma_\rho - 1)y] \,. \tag{9.226}$$

And the resonance frequency of the trilaminar plate is

$$f_{res\,t} = \frac{1}{2\pi}\sqrt{\frac{K_{eqvt}^E}{M_{eqvt}}} = f_{res\,a}\sqrt{\frac{1 + (\gamma_Y - 1)(1 - 2y)^3}{\gamma_\rho - 2(\gamma_\rho - 1)y}} \,. \tag{9.227}$$

Dependence of ratio $f_{res\,t} / f_{res\,a}$ for combinations of PZT-4 ceramics with different passive materials is the same as for the trilaminar beam and it is shown in Figure 9.16.

$Y_{eqvt} = Y_{\sigma a} \cdot [1 + (\gamma_Y - 1)(1 - 2y)^3]$ Electromechanical transformation coefficient of the trilaminar plate is determined by expression (9.196), where coefficient Ω_t for the trilaminar plate it is $\Omega_t = \delta(t - \delta)$ according to (9.15). Thus, the coefficient of electromechanical transformation is

$$n_{t\,bc} = \frac{2\pi d_{31} t}{s_{11}^E + s_{12}^E}(1 - y)\Omega_{r\,bc}(r_1) \tag{9.228}.$$

Given that the capacitance is

$$C_{el}^{S_{1,2}} = \varepsilon_{33}^T (1 - k_p^2)\frac{2\pi a^2}{ty} \,, \tag{9.229}$$

the coefficient $\alpha_{ct\,bc}$ will be found to be

$$\alpha_{ctbc} = \frac{n_{tbc}^2}{C_{el}^{S_{1,2}} K_{eqvt}^E} = \left[1.5\pi(1+\sigma_a)\frac{k_p^2}{1-k_p^2}\right] \cdot \frac{\Omega_{rbc}^2(r_1)}{a^2 L(0.3)_{bc}} \cdot \frac{8y(1-y)^2}{1+(\gamma_Y-1)(1-2y)^3}. \quad (9.230)$$

The value of this expression at $y = 0.5$,

$$\alpha_{cabc} = \left[1.5\pi(1+\sigma_a)\frac{k_p^2}{1-k_p^2}\right] \frac{\Omega_{rbc}^2}{a^2 L(0.3)_{bc}}, \quad (9.231)$$

corresponds to the case of the bimorph (fully active) plate design (denoted with subscript "*c a bc*"). The last factor in expression (9.230) is the same function $\beta_t(y)$ that is introduced by Eq. (9.111) and characterizes dependence of ratio of the coefficients α_c for the trilaminar and bimorph plate designs from composition of the trilaminar design through the thickness at the same boundary conditions. Plots of this function for combinations of PZT-4 ceramics with different passive materials are presented in Figure 9.14. Numerical values of the last factor in expression (9.231) can be calculated for different boundary conditions using data from Table 9.2.

As it follows from the data presented in Figure 9.10 and Figure 9.14, the thickness and radius of the active laminates of the plates vibrating in flexure can be significantly reduced without reduction, and even with some increase of the effective coupling coefficient. Thus, in the case of combination PZT-4 – aluminum under the simply supported boundary conditions it remains the same at $\delta = 0.2t$, $r_1 = 0.8a$, and achieves its maximum value, which reaches $k_{eff} = 0.52$ vs. 0.48 for the full size piezoelement, at $\delta = 0.2t$, $r_1 = 0.9a$.

It is noteworthy that minimizing the volume of active material with the goal of improving (or at least of keeping the same) the coupling coefficient is possible only in the case that required acoustic (mechanical) power generated is not electric field limited. This can be the case of receivers (sensors), projectors operating in air and drivers at small mechanical loads, while fulfilling of this condition can be questionable for underwater projectors. Comments that are made in this regard in Section 9.3.1.4 about tradeoff between acoustic power radiated and minimizing the volume of active material without loss of the effective coupling coefficient are completely applicable to the case of the circular plates.

9.4.1.6 Bilaminar plates

Structure of the bilaminar circular plate through the thickness is the same as of the bilaminar beam considered in Section 9.3.1.5 and shown in Figure 9.16. Therefore, all the characteristics of the bilaminar plates that depend on this structure are the same. In particular the position of the neutral plane (coordinate z_0) is determined by relation (9.117),

$$\frac{z_0}{t} = \frac{1}{2}\frac{\gamma_Y + (1-\gamma_Y)y^2}{\gamma_Y + (1-\gamma_Y)y}, \tag{9.232}$$

in the general case, and by relation (9.118),

$$\frac{z_{0m}}{t} = \frac{\sqrt{\gamma_Y}}{1+\sqrt{\gamma_Y}}, \tag{9.233}$$

in the case that the thickness of the active laminate $\delta = z_0$ (these values are denoted $\delta_{\max}$ and z_{0m}). Values of the ratio

$$\frac{z_{0m}}{t} = \frac{\delta_{\max}}{t} = y_{\max} \tag{9.234}$$

for different combinations of the active and passive materials are presented in Figure 9.17.

Expressions for the equivalent Young's modulus and density of the bilaminar plates following Eqs. (9.119)-(9.121) are

$$Y_{eqvb}(z_0) = 4Y_{\sigma a}^E\left\{\left[\left(y-\frac{z_0}{t}\right)^3+\left(\frac{z_0}{t}\right)^3\right]+\left[\left(1-\frac{z_0}{t}\right)^3+\left(\frac{z_0}{t}-y\right)^3\right]\gamma_Y\right\}, \tag{9.235}$$

in the general case that $\delta < z_0$, and

$$Y_{eqvb}(z_{0m}) = 4Y_{\sigma a}^E[y_{\max}^3 + (1-y_{\max}^3)\gamma_Y] \tag{9.236}$$

in the case that $\delta_{\max} = z_{0m}$, i. e., $z_{0m}/t = y_{\max}$;

$$\rho_{eqvb} = \rho_a[\gamma_\rho - (\gamma_\rho - 1)y], \tag{9.237}$$

(subscript b stands for bilaminar).

Accordingly, the expressions (9.191) and (9.192) for the equivalent rigidity and equivalent mass become:

$$K_{eqvb}^E = \frac{t^3 Y_{eqvb}(z_0)}{12}\cdot L(0.3)_{bc} \tag{9.238}$$

for the general case with $Y_{eqvb}(z_0)$ determined by formula (9.235), and

$$K_{eqvb}^{E} = \frac{t^3 Y_{\sigma a}^{E}}{12} \cdot 4[y_{\max}^3 + (1 - y_{\max}^3)\gamma_Y] \cdot L(0.3)_{bc} = K_{eqva}^{E} \cdot 4[y_{\max}^3 + (1 - y_{\max}^3)\gamma_Y], \qquad (9.239)$$

for the case that $z_{0m} / t = y_{\max}$;

$$M_{eqvb} = t\rho_a[\gamma_\rho - (\gamma_\rho - 1)y]S_{eff\,bc} = M_{eqva} \cdot [\gamma_\rho - (\gamma_\rho - 1)y]. \qquad (9.240)$$

Resonance frequency of the bilaminar plate for the case that $z_{0m} / t = y_{\max}$ is

$$f_{res\,b} = f_{res\,a}\sqrt{\frac{4[y_{\max}^E + (1 - y_{\max}^3)\gamma_Y]}{\gamma_\rho - (\gamma_\rho - 1)y_{\max}}}. \qquad (9.241)$$

The transformation coefficient n_b must be determined from formula (9.196), where coefficient

$$\Omega_t = \frac{\delta}{2}(\delta - 2z_0) \qquad (9.242)$$

is calculated by formula (9.12) and coefficient $\Omega_{r\,bc}(r_1)$ is determined by formula (9.53). Thus,

$$n_{b\,bc}(z_0) = \frac{2\pi d_{31} t}{s_{11}^E + s_{12}^E}\left(2\frac{z_0}{t} - y\right)\Omega_{r\,bc}(r_1) \qquad (9.243)$$

in the general case, and

$$n_{b\,bc}(z_{0m}) = \frac{2\pi d_{31} t}{s_{11}^E + s_{12}^E}\frac{z_{0m}}{t}\Omega_{r\,bc}(r_1) \qquad (9.244)$$

in the case that $\delta_{\max} = z_{0m}$.

The effective coupling coefficient of the bilaminar plate related coefficient α_{cb} can be commonly presented by formula (9.65), using expressions for the transformation coefficient n_b, rigidity K_{eqvb}^E and capacitance

$$C_{el}^{S_{1,2}} = \varepsilon_{33}^T (1 - k_p^2)\pi r_1^2 / \delta. \qquad (9.245)$$

This results in

$$\alpha_{cb\,bc}(y, z_0) = \left[24\pi(1 + \sigma_1^E)\frac{k_p^2}{1 - k_p^2}\right] \cdot \left[[(2z_0 / t) - y]^2 y \frac{Y_{\sigma a}^E}{Y_{eqvb}(z_0)}\right] \cdot \frac{\Omega_{r\,bc}^2(r_1)}{r_1^2 L(0.3)_{bc}} \qquad (9.246)$$

in the general case, and

$$\alpha_{cbbc}(y = z_{0m}/t) = \left[24\pi(1+\sigma_1^E)\frac{k_p^2}{1-k_p^2}\right] \cdot \frac{1}{4\left\{1+[(t/z_{0m})-1]^3 \gamma_Y\right\}} \cdot \frac{\Omega_{rbc}^2(r_1)}{r_1^2 L(0.3)_{bc}}. \tag{9.247}$$

in the case that $\delta_{max} = z_{0m}$. The second factor in both expressions depends on the correlation between the thicknesses of the layers of active and passive materials for a given combination of materials. It can be optimized by the corresponding choice of thicknesses. Dependence of the coefficient $\alpha_{cbbc}(y, z_0)$ from the thickness of the active layer at constant total thickness of a plate normalized to $\alpha_{cbbc}(y = z_{0m}/t)$ is the same as those for the bilaminar beam that is presented in Figure 9.17. Dependences of the third factor from the relative radius of electrodes for different boundary conditions are the same, as presented in Figure 9.10.

9.4.2 Radially Nonuniform Plates

Vibration of the nonuniform over the radius circular plates made of passive materials is considered in Section 4.5.7.2. As the result it was concluded that parameters of the nonuniform active-passive plates for a wide range of their dimensions can be calculated to the first approximation using the mode shapes of vibration of uniform plates. This conclusion is valid at least up to ratios $r_1/a \geq 0.6$ even for the maximum reasonable thicknesses of the active laminates and for the most critical case *B* (with part of active material removed).

Expressions for the mechanical equivalent parameters (rigidity and mass) of the bilaminar plates were determined in Section 4.5.7.3. They are expressed through the equivalent rigidity and mass of the uniform bilaminar plates as follows:

$$K_{eqv\,b1}(z_0) = K_{eqvb}^E\left[1-\left(1-\frac{(1-y)^3 Y_{\sigma p}}{Y_{eqvb}(z_0)}\right)\frac{F_{r_1 bc}}{L(0.3)_{bc}}\right] \tag{9.248}$$

in the general case, and

$$K_{eqvb1}(z_{0m}) = K_{eqvb}^E\left[1-\left(1-\frac{\gamma_Y[1-(z_{0m}/t)]^3}{4(z_{0m}/t)^2}\right)\frac{F_{r_1 bc}}{L(0.3)_{bc}}\right] \tag{9.249}$$

in the case that $\delta_{max} = z_{0m}$;

$$M_{eqvb1} = M_{eqvb}\left[1-\frac{y}{\gamma_\rho-(\gamma_\rho-1)y}\frac{S_{eff\,r_1 bc}}{S_{eff\,bc}}\right]. \tag{9.250}$$

Here the subscripts "*eqv b*1" denote equivalent parameters of the nonuniform plates to the first approximation, and notations are introduced in addition to (9.193) and (9.194):

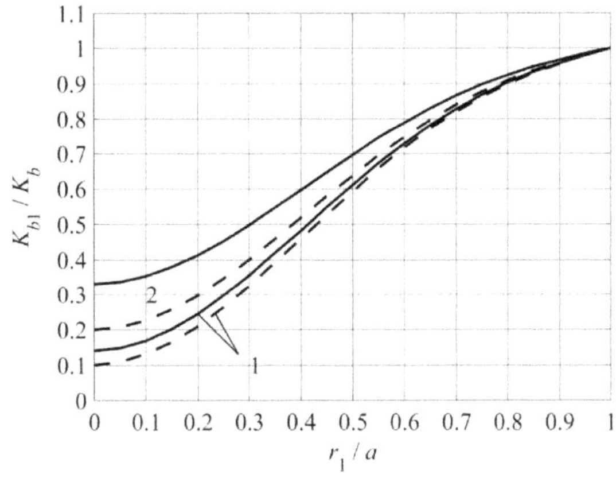

Figure 9.33: The equivalent rigidities of the nonuniform over length bilaminar circular plates normalized to the equivalent rigidity of the uniform bilaminar plates having the same radius, thickness and $y = z_{0m} / t$ ratios vs. r_1 / a for different passive materials used for replacement of piezoceramic. Aluminum (solid lines), steel (dashed lines). Curves 1 for $y = z_{0m} / t$, curves 2 for $y = (\delta / t)_m$.

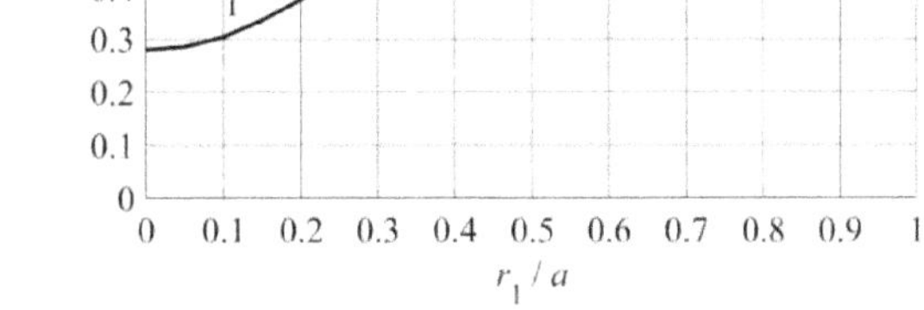

Figure 9.34: The equivalent masses of the nonuniform over length bilaminar circular plates normalized to the equivalent masses of the uniform bilaminar plates having the same radius, thickness and $y = z_{0m} / t$ ratios vs. r_1 / a for different passive materials used for replacement of piezoceramic. Aluminum (solid lines), steel (dashed lines). Curves 1 for $y = z_{0m} / t$, curves 2 for $y = (\delta / t)_m$.

$$F_{r_1 bc} = 2\pi \int_{r_1}^{a} F(\theta) r dr \,, \tag{9.251}$$

$$S_{eff\, r_1 bc} = 2\pi \int_{r_1}^{a} \theta^2 (r/a) r dr \,, \tag{9.252}$$

For the particular values of $y = z_{0m}/t$ and $y = (\delta/t)_m$ (the relative thickness of active laminate, at which the effective coupling coefficient reaches maximum (see Figure 9.17)) dependences of the equivalent parameters and the resonance frequencies of nonuniform bilaminar plates on r_1/a normalized to parameters of uniform bilaminar plates having the same overall dimensions are shown in Figure 9.33 - Figure 9.35.

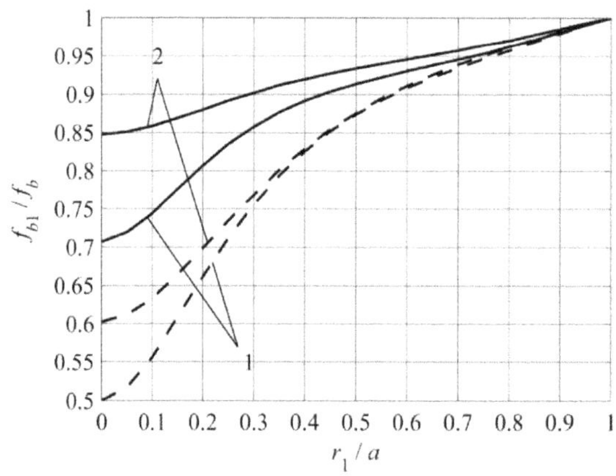

Figure 9.35: The resonance frequencies of the nonuniform over radius bilaminar plates normalized to the resonance frequencies of uniform bilaminar plates having the same radius, thickness and $y = z_{0m}/t$ ratio vs. r_1/a for different passive materials used for replacement of piezoceramic. Aluminum (solid lines), steel (dashed lines). Curves 1 for $y = z_{0m}/t$, curves 2 for $y = (\delta/t)_m$.

They are calculated using formulas (9.248) and (9.249) that are valid to the first approximation. Note, that values of the equivalent parameters at $(r_1/a) = 0$ belong to the uniform plates made of the corresponding passive material having thickness $(t - \delta)$.

Expressions for the electromechanical transformation coefficient of radially nonuniform plate to the first approximation are the same as (9.243) and (9.244) for a radially uniform plate, because the mode of vibration is assumed to be the same, i.e., $n_{b\,1bc}(z_0) = n_{b\,bc}(z_0)$ and $n_{b\,1bc}(z_{0m}) = n_{b\,bc}(z_{0m})$. The capacitance of the active laminate also remains the same.

The coefficient α_c that determines the effective coupling coefficient will be denoted as

α_{cb1bc} for the nonuniform bilaminar plate to the first approximation, and it is

$$\alpha_{cb1bc} = n_{b1bc}^2 / C_{el}^{S_{1,2}} K_{eqvb1} . \tag{9.253}$$

Substituting the parameters involved in this expression results in

$$\alpha_{cb1bc} = \alpha_{cbbc} (K_{eqvb1} / K_{eqvb}^E)^{-1} , \tag{9.254}$$

where α_{cbbc} are determined by expressions (9.246), (9.237), and the factor in parenthesis is determined by expressions (9.248), (9.249) for the cases that $\delta < z_0$ and $\delta = z_{0m}$, respectively. This factor is presented as reversed in Figure 9.36 for several combinations of materials and values of the relative thicknesses of active laminate. Plots of function (9.254) vs. ratio r_1 / a are presented in Figure 9.36 for different combinations of materials under assumption that the boundary is simply supported.

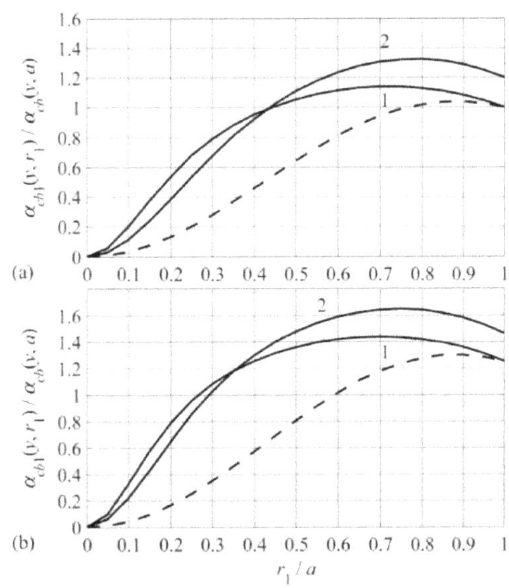

Figure 9.36: Plots of the function $\alpha_{cb1} / \alpha_{cb}$ by formula (9.254) for different combinations of materials ((a) aluminum-PZT-4, (b) steel- PZT-4) and for different relative thicknesses of the active layers: 1 - $y = (z_{0m} / t)$, 2 - $y = (\delta / t)_m$. The plots are normalized to value of α_{cb} for the uniform bilaminar plate made from combination of aluminum and PZT-4 ceramic with relative thickness of ceramic $y = (z_{0m} / t) = 0.48$ that have the full size electrodes. Dependence of this coefficient from relative radius of the electrodes, r_1 / a, (see Figure 9.10 for the simply supported boundary) is shown in the figure (a) and (b) by the dashed lines.

Several conclusions regarding the effective coupling coefficients of the bilaminar circular flexural plates can be made based on the data presented in Figure 9.36. First, they are greater for the combination steel – PZT-4 compared with aluminum – PZT-4 for the same categories of the relative thicknesses of the active laminates. This reflects the general fact that the higher the Young's modulus of the passive material is the greater the effective coupling coefficient of the bilaminar plate. Second, the effective coupling coefficients are greater for nonuniform by radius plates compared with the uniform plates having the same radiuses of electrodes. The maximum of coupling coefficients for the nonuniform plates shifts towards the smaller relative radiuses of the active elements. (Remember that we are considering the rational designs of the bilaminar plates that are restricted by the condition that $\delta \le z_{0m}$).

As to other variants of the transducers (bilaminar and trilaminar of A modification), they were considered in Section 4.5.7.3 using the same approximation and analogous calculating procedures. Thus, for the bilaminar plates of A modification (with removed parts of active laminates replaced by passive materials) expressions for the equivalent rigidities and masses were obtained in the following forms (see expressions (4.421) - (4.423)):

for the rigidity

$$\frac{K_{eqvb1}}{K^E_{eqvb}} = 1 - \left(1 - \frac{Y_{\sigma p}}{Y_{eqvb}(z_0)}\right)\frac{F_{r_1 bc}}{L(0.3)_{bc}} \tag{9.255}$$

in the general case that $\delta \le z_{0m}$, and

$$\frac{K_{eqvb1}}{K^E_{eqv\,b}} = 1 - \left(1 - \frac{\gamma_Y}{4(z_{0m}/t)^2}\right)\frac{F_{r_1 bc}}{L(0.3)_{bc}} \tag{9.256}$$

in the case that $\delta = z_{0m}$; and for the mass

$$\frac{M_{eqv\,b1}}{M_{eqvb}} = 1 + \frac{y(\gamma_\rho - 1)}{\gamma_\rho - (\gamma_\rho - 1)y}\frac{S_{eff\,r1}}{S_{eff\,1}}. \tag{9.257}$$

The electromechanical transformation coefficients remain the same as for modification B because the mode of vibration does not change.

It is noteworthy that designs of A modification have the advantage of greater static and dynamic strength in comparison with B modification. This is due to avoiding concentration of stress that occurs in the region, where the cross section of the mechanical system changes abruptly.

9.4.3 Corrections for Parameters Due to Finite Thickness of the Plates

With increase of the relative thickness t/a of the circular plates corrections to the energies associated with their bending deformations must be made that account for the rotary inertia and transverse shear deformation of their cross sections. These effects for the plates made of passive materials were considered in Section 4.4.3.1. In application to the electromechanical transducers that employ flexural vibrations of the circular plates the expressions for the additional energies presented therein must be specified for active materials used. For definiteness we will consider fully active bimorph plates. Therefore, elastic parameters of ceramics only must be used in the expressions for the energies, namely, the Young's modulus $Y_{\sigma a} = s_{11}^E / (s_{11}^{E2} - s_{12}^{E2})$ and the shear modulus $\mu = 1/s_{44}^D$.

After substituting expressions for displacement in the form of

$$\xi_z(r,t) = \xi_o(t)\theta(r) \tag{9.258}$$

into expressions for the energy densities (4.203) and (4.208) and after integrating over volume of the plate we obtain the corrections for the kinetic and potential energies in the form

$$\Delta W_{kin} = \frac{1}{2}\dot{\xi}_o^2 \frac{\pi t^3 \rho}{6} \int_0^a \left(\frac{\partial \theta}{\partial r}\right)^2 r dr = \frac{1}{2}\dot{\xi}_o^2 \cdot \Delta M , \tag{9.259}$$

$$\Delta W_{pot} = -\frac{1}{2}\xi_o^2 \frac{2\pi Y_{\sigma a}^2 s_{44}^D t^5}{120} \int_0^a \left(\frac{d^3\theta}{dr^3}\right)^2 r dr = \frac{1}{2}\xi_o^2 \cdot \Delta K . \tag{9.260}$$

Correction for the electromechanical energy of the plate due to its finite thickness must be calculated by the general formula (9.49) with bending strain S_{rb} replaced by the additional strain S_{rs} by formula (4.212). After changing elastic moduli by those of the piezoceramics and considering that for the fully active bimorph plate at parallel connection of ceramic laminates $E_3(z,r) = 2V/t$, we arrive at

$$\Delta W_{em} = \frac{1}{2}\xi_o V \frac{2\pi d_{31}}{t(s_{11}^E + s_{12}^E)} Y_{\sigma a} s_{44}^D \int_{-t/2}^{t/2} \left(\frac{t^2}{4} - z^2\right) z dz \int_0^a \frac{d^4\theta}{dr^4} r dr = \frac{1}{2}\xi_o V \cdot \Delta n_s . \tag{9.261}$$

Here Δn_s is correction for the electromechanical transformation coefficient of a thick plate.

After performing integration, we will obtain that

$$\Delta n_s = \frac{\pi}{16} \cdot \frac{d_{31} t^3}{s_{11}^E + s_{12}^E} Y_{\sigma a} s_{44}^D \int_0^a \frac{d^4\theta}{dr^4} r dr \ . \tag{9.262}$$

As examples of quantitative estimations, we consider corrections due to the finite thickness for equivalent parameters of the bimorph plates with simply supported and clamped boundaries, and of the center supported plates. For this purpose we will use expressions for the mode shapes of displacements for these plates by formulas (9.57), (9.59). and (9.204), respectively, that are reproduced here

$$\theta(r/a) = (1 - r^2/a^2)(1 - r^2/4a^2) \ . \tag{9.263}$$

$$\theta(r/a) = \left(1 - \frac{r^2}{a^2}\right)^2 , \tag{9.264}$$

$$\theta(r/a) = \frac{1}{0.36}\left(\frac{r^2}{a^2} - 0.91\frac{r^3}{a^3} + 0.27\frac{r^4}{a^4}\right) . \tag{9.265}$$

Table 9.3. Values of integrals $In1$, $In2$ and $In3$ for the plates with different boundary conditions.

Boundary conditions	S.s.	Cl.	C.s.
$In1$	0.85	2/3	0.62
$In2,\ 1/a^4$	9.0	144	13.8
$In3,\ 1/a^2$	3.0	12	9.0

Expressions (9.197) together with data from Table 9.2 will be used for the original equivalent parameters of the plates considered as thin. Values of integrals in expressions (9.259), (9.260) and (9.262), which we denote for brevity as $In1$, $In2$ and $In3$, are presented in Table 9.3 for the corresponding boundary conditions.

Using expressions (9.197) for equivalent parameters of the thin plates and expressions (9.259), (9.260), (9.262) for the corrections, we will finally obtain the following results for the parameters of plates having finite thickness to radius ratio.

For the plates with simply supported boundaries

$$M'_{eqv\,ss} = M_{eqv\,ss}[1 + 0.5 \cdot (t^2/a^2)] \, , \ K'_{eqv\,ss} = K^E_{eqv\,ss}[1 - 0.25 \cdot Y_{\sigma a} s^D_{44}(t^2/a^2)] \, . \tag{9.266}$$

$$n'_{ss} = n_{ss}[1 - 0.39 Y_{\sigma a} s^D_{44}(t/a)^2], \tag{9.267}$$

$$f'_r = \frac{1}{2\pi}\sqrt{\frac{K'_{eqv}}{M'_{eqv}}} = f_r\sqrt{\frac{1 - 0.25 Y_{\sigma a} s^D_{44}(t/a)^2}{1 + 0.5(t/a)^2}}. \tag{9.268}$$

For the plates with clamped boundaries

$$M'_{eqv\,cl} = M_{eqv\,cl}[1 + 0.6\cdot(t^2/a^2)],\; K'_{eqv\,cl} = K^E_{eqv\,cl}[1 - 1.3\cdot Y_{\sigma a} s^D_{44}(t^2/a^2)]. \tag{9.269}$$

The coefficient of electromechanical transformation for the clamped thin bimorph plate having full size electrode is zero. Distortions of the cross section of the plate due to finite thickness result in some electromechanical activity of the clamped plate that is characterized by the electromechanical transformation coefficient

$$\Delta n_{cl} = 1.8\cdot\frac{d_{31}t}{s^E_{11} + s^E_{12}} Y_{\sigma a} s^D_{44}(t^2/a^2). \tag{9.270}$$

For the center supported plates

$$M'_{eqv\,cs} = M_{eqv\,cs}[1 + 0.22\cdot(t^2/a^2)],\; K'_{eqv\,cs} = K^E_{eqv\,cs}[1 - 0.41\cdot Y_{\sigma a} s^D_{44}(t^2/a^2)],$$

$$n'_{cs} = n_{cs}[1 - 0.58 Y_{\sigma a} s^D_{44}(t/a)^2]. \tag{9.271}$$

For PZT-4 and PZT-5 ceramic compositions

$$Y_{\sigma a} s^D_{44} = \frac{s^D_{44}}{s^E_{11}}\frac{1}{1 - (s^E_{12}/s^E_{11})^2} \approx 1.75, \tag{9.272}$$

and reduction of resonance frequency of simply supported plate is determined by ratio

$$f'_r/f_r = \sqrt{\frac{1 - 0.44(t/a)^2}{1 + 0.5(t/a)^2}}. \tag{9.273}$$

For relatively thin for practical underwater applications simply supported piezoceramic plate with $(t/a) = 0.2$ this ratio already is $f'_r/f_r \approx 0.98$ and it drops with increase of the relative thickness of the plate.

It is noteworthy that presenting data on the resonance frequencies for the plates by formulas obtained from thin plates theory to a greater accuracy does not make sense without information about their relative thickness. Moreover, this is true regarding the resonance frequencies of higher modes of vibration, for which the corrections for finite thickness significantly increase, as it is shown in Ref. 9 with example of beams.

Consider dependence of the effective coupling coefficient of a simply supported plate of finite thickness from the thickness to radius ratio. The related coefficient $\alpha'_{c\,ss}$ is

$$\alpha'_{c\,ss} = \frac{n'^2_{ss}}{C_{el}^{S_{1,2}} K'^E_{eqv\,ss}} = \frac{0.62k_p^2}{1-k_p^2} \cdot \frac{[1-0.39Y_{\sigma a}s^D_{44}(t^2/a^2)]^2}{1-0.25 \cdot Y_{\sigma a}s^D_{44}(t^2/a^2)}. \tag{9.274}$$

Here the first factor represents coefficient $\alpha_{c\,ss}$ for the thin bimorph plate according to formula (9.201). Thus, for the plate made of PZT-4 ceramics

$$\frac{\alpha'_{c\,ss}}{\alpha_{c\,ss}} = \frac{[1-0.68(t^2/a^2)]^2}{1-0.44(t^2/a^2)}. \tag{9.275}$$

At $t/a = 0.2$ this ratio is 0.96, and at $t/a = 0.3$ it is already 0.9.

9.5 Acoustic Field Related Issues

The radiation impedances of the bender transducers depend significantly on the size of a single transducer unit and on the configuration of the system of transducers operating collectively. The radiation impedances of the typical systems of the bender transducers are depicted schematically in Figure 9.37.

In terms of the peculiarity of determining the radiation impedances they can be divided into three categories: (1) single transducers of small wave size and small groups of such transducers closely spaced, (2) column like transducers with length of the column comparable with wavelength, (3) arrays of a large wave size populated with the bender transducers.

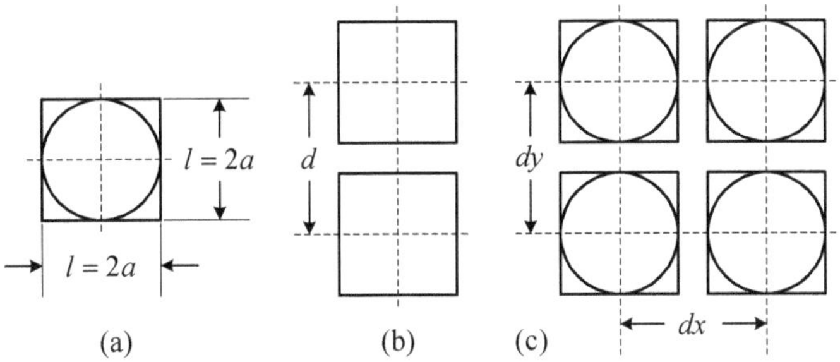

Figure 9.37: Categories of the bender transducers that differ in terms of determining their radiation impedances: (a) single transducer units and their combinations; (b) the column like array densely populated by the single transducer units, (c) the flat array of a large size populated by the single transducer units.

9.5.1 Radiation Impedances of the Single Transducers

Dimensions of the single bender transducers are usually much less than the wavelength λ. Using formula (9.200) for the resonance frequency of the simply supported circular plate it can be concluded that at resonance $D/\lambda \approx t/a$. For estimating the maximum wave size of a transducer, we assume that $t/a < 1/5$, then $D/\lambda < 0.2$. The overall thickness of a single double plate bender transducer, t_{tr}, is usually about two thicknesses of the comprising plates, i.e., $t_{tr}/\lambda < 0.04$. The rectangular plate benders predominantly consist of beams. We will assume that a single transducer unit has approximately the same wave size as a circular plate having the same resonance frequency ($l = w = D$). Though this assumption is not quite rigorous, it allows a sufficiently fair comparison between the benders.

By symmetry considerations, one side of the double sided circular plate pulsating projector has the same radiation impedance as a circular piston vibrating in an infinite rigid plane baffle that is (see Section 6.3.3.2)

$$Z_{ac1} = r_{ac1} + jx_{ac1}, \tag{9.276}$$

where

$$r_{ac1} = \rho c \pi a^2 \left[1 - \frac{J_1(2ka)}{ka}\right], \tag{9.277}$$

and

$$x_{ac1} = \frac{S_1(2ka)}{ka}. \tag{9.278}$$

Here $J_1(2ka)$ and $S_1(2ka)$ are the Bessel and Struve functions of the first order (subscripts 1 for the radiation impedance and its components indicate that they are related to one side of the double-sided transducer). The approximate estimations that follow from these relations for one side of the projector at $ka < 0.6$ that corresponds with the maximum size of a bender plate transducer are

$$r_{ac1} \approx \rho c\, \pi a^2 \frac{(ka)^2}{2} = 2\pi\rho c \frac{(\pi a^2)^2}{\lambda^2}, \tag{9.279}$$

within 5% accuracy, and

$$x_{ac1} \approx \rho c\, \pi a^2 \frac{8}{3\pi} ka = \omega \rho \frac{8}{3} a^3 = \omega M_{ac1}, \tag{9.280}$$

within 10% accuracy.

These formulas can be generalized for transducers of small wave size that have distribution of velocity on their surface by replacing the surface area $S = \pi a^2$ with the average surface area S_{av}, because the volume velocity $U_{\tilde{V}} = S_{av}\xi_o$ is what counts in terms of the radiation for small sources. Thus, further the expressions for the radiation resistance per one plate of the double-sided bender transducer (as of a simple source radiating in half space) will be used

$$r_{ac1} \approx 2\pi\rho c \frac{(S_{av})^2}{\lambda^2}, \quad x_{ac1} \approx \rho c\, S_{av} \frac{8}{3\pi} ka. \tag{9.281}$$

It is noteworthy that these values of radiation impedances must be used when calculating the magnitude of vibration of one side of the double-sided bender transducer from the equivalent electromechanical circuit. The acoustic radiation impedance for the entire transducer must be doubled. Then for the pulsating piston

$$r_{ac} = \pi\rho c (2\pi a^2)^2 / \lambda^2, \tag{9.282}$$

and taking into account that now the full radiating surface $S = 2\pi a^2$ must be in the general case replaced by S_{av}, we obtain the known formula for the radiation resistance of a simple source vibrating in free space in the form

$$r_{ac} = \pi \rho c \frac{(S_{av})^2}{\lambda^2}. \qquad (9.283)$$

This value will be used for the benders of finite size having $(D/\lambda) < 0.2$ with the stated estimated accuracy. Formulas (9.262) and (9.283) are introduced under the assumption that the acoustic mass is proportional to the volume of water forced out by a pulsating body, i.e., it is proportional to its volume velocity, or to the average radiating surface.

The above estimations are made for a single bender transducer that can be approximated as a simple source. In practice, the low frequency transducer designs often are composed of a number of elementary single sources, in our case from a number of single benders, to form radiators of larger size. The radiation impedance of the combination of several single bender transducers can be determined considering the acoustic interaction between simple sources by the formula (see Section 6.4.3)

$$Z_{12} = r_{11}\left(\sin kd + j\cos kd\right)/ kd , \qquad (9.284)$$

where Z_{12} the mutual radiation impedance between two sources separated by distance *d,* and r_{11} is the radiation resistance of a single source given by expression (9.281). For example, in the case of a transducer comprised of three simultaneously operating identical equally spaced single benders their radiation impedances can be calculated by formulas

$$Z_1 = Z_2 = Z_3 = Z_{11} + 2Z_{12}. \qquad (9.285)$$

Thus, the radiation impedance of a single unit can be significantly increased.

9.5.2 Radiation Impedances of the Column Like Transducers

In practice a low frequency transducer may have a comparable with wavelength linear size being composed of a number of elementary single sources in the column like fashion, as shown in Figure 9.37(b). The column like design of such kind is typical for the rectangular beam benders. If the total height of the transducer exceeds wavelength, the components of radiation impedance per unit height can be calculated approximately as for infinitely long pulsating strip.

This problem is considered in Section 6.3.4.4. Based on the results obtained therein, the radiation impedance per one side of the double-sided bender transducer can be presented as

$$Z_{ac}(kl) = r_{ac}(kl) + jx_{ac}(kl) = \rho c S_{eff1}[\alpha(kl) + j\beta(kl)], \qquad (9.286)$$

where S_{eff1} is the effective radiating surface area per unit height. For the transducer comprised of simply supported beams that vibrate with velocity distribution $U(x) = U\cos(\pi x / l)$, $S_{eff} = l/2$ per unit height. Numerical values of the nondimensional coefficients of the radiation

Table 9.4. The nondimensional coefficients of the radiation impedance.

kl	0.1	0.2	0.3	0.4	0.5	0.6	0.7	0.8
$\alpha(kl)$	0.04	0.08	0.12	0.16	0.20	0.24	0.28	0.32
$\beta(kl)$	0.10	0.18	0.24	0.29	0.33	0.37	0.40	0.43

kl	0.9	1.0	1.1	1.2	1.3	1.4	1.5	
$\alpha(kl)$	0.36	0.39	0.43	0.47	0.51	0.54	0.57	
$\beta(kl)$	0.45	0.47	0.49	0.51	0.52	0.53	0.54	

resistance and reactance, $\alpha(kl)$ and $\beta(kl)$, are presented in Table 9.4 up to values $kl = 2$ that is sufficient for a practical range of the bender transducer sizes.

It follows from the table that the nondimensional coefficient of the radiation resistance can be presented as $\alpha \approx 0.4kl$ up to $kl < 1.5$ within 4% accuracy. Considering that per unit height of the entire double-sided transducer $S_{eff1} = (l/2)\cdot 2 = l$ and $S_{av1} = (2l/\pi)\cdot 2$, in this range of kl the radiation resistance may be represented by formula

$$r_{ac} = \rho c S_{ef1f} \cdot \alpha(kl) \approx \frac{\pi}{2}\rho c \frac{(S_{av1})^2}{\lambda}, \tag{9.287}$$

which is the expression (6.358) for radiation resistance of the cylindrical simple source.

For determining the radiation impedances of the finite size transducers, having height comparable with wavelength, the expressions (9.286) and (9.287) must be multiplied by its height H. (Note that S_{eff1} and S_{av1} per unit height have dimensions of length).

The radiation impedance of several column like bender transducers of a large height operating simultaneously may be determined considering the acoustic interaction between the cylindrical simple sources. The mutual radiation impedance between two transducers according to formula (6.359) is

$$(z_{12}/r_{11}) = J_0(kd) - jN_0(kd), \tag{9.288}$$

where d is the separation between their axes.

9.5.3 Radiation Impedance in Flat Arrays of a Large Size

One of the possible applications of the bender transducers is their employment in flat low frequency arrays of a large size, as shown in Figure 9.37(c). The radiating surface of the array can be considered as a plane uniformly vibrating with average volume velocity, because dimensions of the single bender transducers are small compared with wavelength, and the volume velocity generated by the benders is what counts in terms of radiation. The volume velocity of a single transducer unit is $U_{\tilde{V}1} = S_{av}\dot{\xi}_0$ (here $S_{av} = (2l/\pi)\cdot l$ according to the dimensions of a single block shown in Fig. 9.37 (a)). The volume velocity of a part of the radiating flat array that is occupied by a single transducer unit is $U_{\tilde{V}array} = d_x d_y \dot{\xi}_{array\ av}$, where d_x and d_y are the separations between centers of the benders, and $\dot{\xi}_{array\ av}$ is the average velocity of array vibration. It follows from equality of these volume velocities that $\dot{\xi}_{array\ av} = (S_{av}/d_x d_y)\dot{\xi}_0$.

The radiation impedance of a large uniformly vibrating plane array is active, and acoustic radiation resistance per unit area is $r_{ac} = (\rho c)_w$. Thus, the acoustic energy radiated per part of the array surface that corresponds to a single transducer unit is $\dot{W}_{ac} = (\rho c)_w d_x d_y \dot{\xi}^2_{array\ av} =$ $= (\rho c)_w (S^2_{av}/d_x d_y)\dot{\xi}^2_0$. This amount of energy is equal to the energy radiated by the single bender, $\dot{W}_{ac1} = r_{ac1}\dot{\xi}^2_o$. From equality of these energies the radiation resistance per one side of the double-sided benders that populate a large size array will be determined as

$$r_{ac1} = (\rho c)_w \frac{S^2_{av}}{d_x d_y}. \tag{9.289}$$

It must be remembered that so far the double sided bender design was considered symmetrical in respect to its middle plane with both sides equally loaded. In case of application in a flat array the symmetry becomes broken because one side of the transducer is loaded acoustically whereas the loading status of another side depends on the array design. Typically, it can be loaded by input impedance of acoustic baffle that is used for eliminating back radiation of the array, and the load may be influenced by structures, to which the transducers are fixed. The transducer design must be considered, strictly speaking, as asymmetrical under these conditions.

9.6 References

1. R. S. Woollett, *Theory of the Piezoelectric Flexural Disk Transducer with Applications to Underwater Sound*, USL Research Report No. 490, Naval Undersea Warfare Center, Newport (1960).
2. R. S. Woollett, *The Flexural Bar Transducers*, Naval Underwater Systems Center, Newport (1986).
3. B. S. Aronov, "The energy method for analyzing the piezoelectric electroacoustic transducers II. (With the Examples of the flexural plate transducer)," J. Acoust. Soc. Am. **118**(2), 627-637 (2005).
4. S. A. N. Prasad et al, "Analytical Electroacoustic Model of a Piezoelectric Composite Circular Plate," AIAA J. **44(**10), 2311-2318 (2006).
5. Yu. T. Antonyak and M. E. Vassergisser, "Calculation of the characteristics of a membrane type flexural-mode piezoelectric transducer," Sov. Phys. Acoust., **28**(3), 176-180 (1982).
6. Shuo Hung Chang and B. C. Du, "Optimization of asymmetric bymorphic disk transducers," J. Acoust. Soc. Am. **109**(1), 194-202 (2001).
7. B. S. Aronov and L. B. Nikitin, "Electromechanical wafer transducer using shear strain," Sov. Phys. Acoust. **35**(2), 207-210 (1989).
8. B. S. Aronov, "Nonuniform piezoelectric circular plate flexural transducers with underwater applications", J. Acoust. Soc. Am. **138**(3), 1570-1584 (2015).
9. S. P. Timoshenko, *Vibration Problems in Engineering*, 2nd Ed. (Van Nostrand, New York, 1937).

CHAPTER 10

LENGTH EXPANDER TRANSDUCERS AND THEIR MODIFICATIONS

10.1 Introduction

Under the category of the length expander transducers, we will assume the bar transducer (as the basic design) and its modifications that perform one-dimensional vibrations along one of coordinates with other coordinates not working (no energy flow in their directions). The basic transducer designs are shown in Figure 10.1(a) as the bars made of piezoceramics that have the lateral dimensions much smaller than their length ($t, w \ll l$) to the extent that the stresses being zeros on the free side surfaces can be considered the same throughout the lateral dimensions of the bar. These transducers are used in several modifications: with electrodes on the side surfaces (transverse piezoeffect) (a.1), with electrodes on the ends (a.2), or made of segments (longitudinal piezoeffect) (a.3). The mechanical boundary conditions in the first case are $T_2 = T_3 = T_4 = T_5 = T_6 = 0$, and in the second $T_1 = T_2 = T_4 = T_5 = T_6 = 0$. (The share stresses don't exist in all modifications of the length expander transducers, and further will be omitted.)

Another transducer of the category of one-dimensional length expander can be imagined as the thin strip shown in Figure 10.1(b) that has the length large enough for considering strain being negligible along the strip and stress negligible throughout the thickness. The transducer may employ the transverse (b.1) and longitudinal (b.2) piezoeffect with mechanical boundary conditions $T_3 = 0$, $S_2 = 0$ and $T_2 = 0$, $S_1 = 0$, respectively. One more length expander transducer is the thickness vibrating plate (Figure 10.1(c)) that has the lateral dimensions large enough for considering side surfaces of the transducer clamped, i.e., vibrating under boundary conditions $S_1 = S_2 = 0$.

Vibration of the thin prolonged strip and of the plate having large lateral dimensions can be treated in the same way as one-dimensional vibrations of the thin bars. Difference in the mechanical boundary conditions results only in changing the equivalent parameters of the transducers. These changes will be considered in Section 10.2. The assumption regarding the one-dimensional vibration of strips and plates is, strictly speaking, valid in the case that their length and lateral dimensions, respectively, are infinitely large. And what is "large enough" for using these approximations needs to be determined. This is done in Section 10.3, where vibration of these systems is considered as coupled two-dimensional, and where it is shown at what aspect ratios results of the calculations are approaching the limiting cases of infinitely large dimensions with sufficient accuracy. Transition of vibrating mechanical system between the extreme limiting one-dimensional cases is schematically illustrated with Figure 10.2.

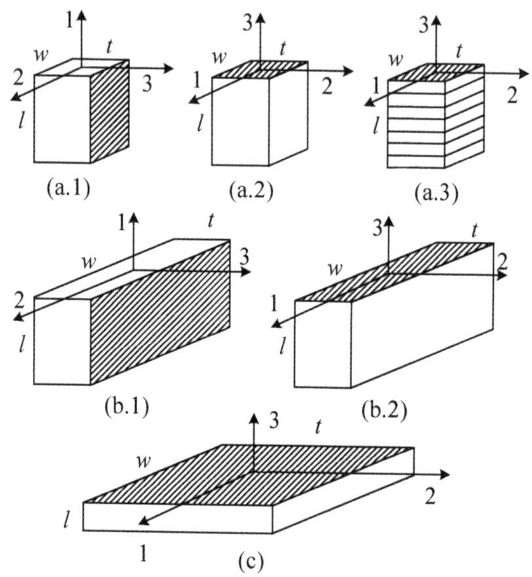

Figure 10.1: The length expander transducers as one-dimensional vibrating systems having different mechanical boundary conditions on their side surfaces. (a) Bars: (a.1) $T_2 = T_3 = 0$ (transverse piezoeffect), (a.2) and (a.3) $T_1 = T_2 = 0$ (longitudinal piezoeffect). (b) Strips: (b.1) $T_3 = 0$, $S_2 = 0$ (transverse piezoeffect; (b.2) $T_2 = 0$, $S_1 = 0$ (longitudinal piezoeffect). (c) Thickness vibrating plate: $S_1 = S_2 = 0$.

It is noteworthy that dimensions of the mechanical systems in Figure 10.1 and Figure 10.2 are denoted differently intentionally. In the first case they are chosen for the ideal extreme mechanical systems in order to comply with previously used notations for the bar as basic transducer design (l – in direction of wave propagation, w and t – as dimensions of the radiating surface). In the second case the dimensions are denoted in accordance with directions of the crystallographic axis (L_1, L_2, L_3) for convenience of considering the coupled vibrations in course of transition between the extreme mechanical systems.

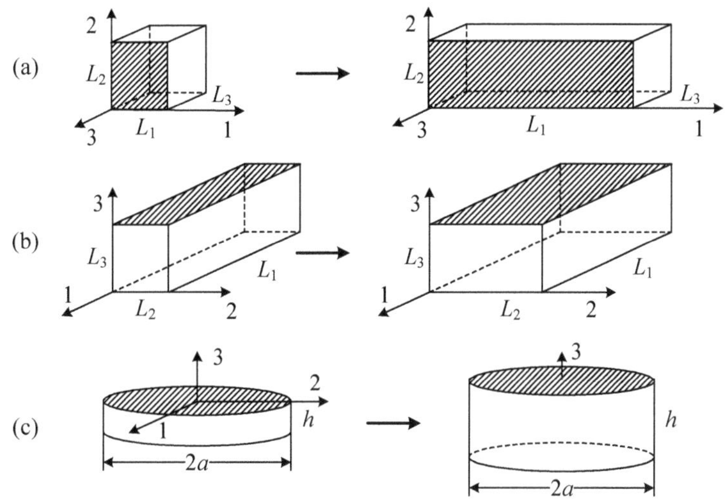

Figure 10.2: Transition between the one-dimensional vibrating mechanical systems having ideal extreme dimensions through the two-dimensional coupled mechanical systems: (a) between the thin bar ($L_1 / L_3 \to 0$) and strip ($L_1 / L_3 \to \infty$); (b) between strip ($L_2 / L_3 \to 0$) and plate ($L_2 / L_3 \to \infty$); (c) between thickness vibrating thin circular disk (aspect ratio $h/2a \to 0$) and axial vibrating long bar ($h/2a \to \infty$).

Uniform over the length bar transducers intended for underwater applications appear to be greatly under loaded by radiation impedances, and therefore their acoustic power generated is the dynamic stress limited, while significant reserves of the electrical strength exist (see Chapters 3 and 13). Therefore, the internal mechanical impedances of the transducers must be matched with acoustic field for effective operation. The matching elements may be imagined as parts of transducer mechanical system having different cross section areas ("Tonpilz" design, mass loaded transducer considered in Section 2.5), or as the matching layers of the same cross section attached to the transducer surface. In both the cases the mechanical system of the bar transducer becomes nonuniform over the length, as is shown in Figure 10.3. The bar transducers nonuniform over the length are considered in Section 10.4. The most widely used nonuniform over the length Tonpilz transducer design is described only in principle, because such transducer are considered to detail in vast literature. In this respect only an alternative approach to optimizing matching the transducer with acoustic field is considered in Section 10.4.

Equations of one-dimensional vibration of length expander bars and equivalent circuits for calculating their operating characteristics were considered as examples of employing the general approaches in Sections 5.7.3.1 and 5.8.3. Some of the results obtained there will be repeated in this Chapter for convenience of reading.

Figure 10.3: Mechanical systems of bar transducers with matching elements: (a) with variable cross section, (b) with cross section uniform over the length. 1 – piezoelement, 2 and 3 – matching elements.

10.2 Transducers Uniform over the Length

10.2.1 Energy Densities in the Deformed Piezoelements

For calculating energy status of the length expander piezoelements deformed under different boundary conditions expressions for the corresponding energy densities that were derived in

Section 5.2 can be used (see Eqs. (5.53) - (5.67). However, we will use the constituting piezoelectric equations of state for this purpose, moreover that the derivations are straightforward, and the piezoelectric equations will be needed for reference, when considering coupled vibrations in mechanical systems of the transducers in case that they have finite aspect ratios.

All the length expander piezoelements under the ideal boundary conditions experience one-dimensional deformation with a single working strain S_1 or S_3. Thus, the governing equations for them can be represented in the generalized form

$$T = K_{\Delta}^{E} S - n_{\Delta} E_3 , \tag{10.1}$$

$$D_3 = n_{\Delta} S + C_{e\Delta}^{S} E_3 , \tag{10.2}$$

where K_{Δ}^{E} is analog of elastic modulus of isotropic passive body in one-dimensional deformation at particular boundary conditions,

$$D_3^E = n_{\Delta} S \tag{10.3}$$

is the charge density and $C_{e\Delta}^{S}$ is the dielectric constant for the ceramics at a particular combination of boundary conditions. According to expression (5.99)

$$C_{e\Delta}^{S_i} = C_{e\Delta}^{T}(1-k_m^2), \tag{10.4}$$

where k_m is the coupling coefficients of the ceramic composition at particular boundary conditions. Notations for the coefficients k_m will be used according to the classification of boundary conditions accepted in Ref. 2.

In the case that the boundary conditions are $T_2 = T_3 = 0$ or $T_1 = T_2 = 0$ it is convenient to use the piezoelectric equations with independent stress in the form

$$S_i = s_{ii}^E T_i + d_{3i} E_3, \tag{10.5}$$

$$D_3 = d_{3i} T_i + \varepsilon_{33}^T E_3, \tag{10.6}$$

where $i = 1, 3$ for the case of the transverse and longitudinal piezoeffect, respectively. From Eqs. (10.1) and (10.5) at $E_3 = 0$

$$K_\Delta^E = 1/s_{ii}^E = Y_i^E . \tag{10.7}$$

After substituting expression for stress T_i from Eq. (10.5) into Eq. (10.6) we arrive at

$$D_3 = \frac{d_{3i}}{s_{ii}^E} S_i + \varepsilon_{33}^T \left(1 - \frac{d_{3i}^2}{\varepsilon_{33}^T s_{ii}^E}\right) E_3 = D_3^E(S_i) + \varepsilon_{33}^{S_i} E_3 . \tag{10.8}$$

Thus,

$$D_3^E(S_i) = (d_{3i}/s_{ii}^E) S_i, \; n_\Delta = d_{3i}/s_{ii}^E, \tag{10.9}$$

and

$$C_{e\Delta}^{S_i} = \varepsilon_{33}^{S_i} = \varepsilon_{33}^T (1 - k_{3i}^2) \tag{10.10}$$

is the dielectric constant of ceramics in the bar clamped in direction of deformation ($S_i = 0$).

In case that the boundary conditions are $S_1 = S_2 = 0$ (infinite plate vibrating over the thickness at longitudinal piezoeffect), appropriate are piezoelectric equations with independent strain that at these boundary conditions reduce to the form

$$T_3 = c_{33}^E S_3 - e_{33} E_3, \tag{10.11}$$

$$D_3 = e_{33} S_3 + \varepsilon_{33}^S E_3, \tag{10.12}$$

where from follows that

$$K_{\Delta}^{E} = c_{33}^{E},\ n_{\Delta} = e_{33},\ C_{e\Delta}^{S_3} = \varepsilon_{33}^{S} = \varepsilon_{33}^{T}(1-k_t^2) \tag{10.13}$$

The analogous elastic modulus for isotropic plate deformed through the thickness according to (4.18) is

$$Y_{pl} = \frac{Y(1-\sigma)}{(1+\sigma)(1-2\sigma)}. \tag{10.14}$$

The following respective expressions may be obtained from Eqs. (5.57)–(5.64) for the case of a bar element in the form of a thin strip vibrating at the following boundary conditions: (1) $T_1 = 0$, $S_2 = 0$ that correspond to the longitudinal piezoeffect; (2) $T_3 = 0$, $S_2 = 0$ that correspond to the transverse piezoeffect; (3) variants of the transverse piezoeffect with axis 3 parallel to the length and vibration taking place in the direction of axis 2 (variant at $T_1 = 0$, $S_3 = 0$ that corresponds to the width mode of vibration) and of axis 1 (variant at $S_2 = 0$, $S_3 = 0$ that corresponds to the thickness mode).

$$K_{\Delta 3}^{E} = \frac{s_{11}^{E}}{s_{11}^{E}s_{33}^{E} - s_{13}^{E2}} = \frac{Y_3^E}{1-\sigma_3\sigma_{13}},\ n_{\Delta 3} = \frac{d_{33}s_{11}^{E} - d_{31}s_{13}^{E}}{s_{11}^{E}s_{33}^{E} - s_{13}^{E2}},\ C_{e\Delta}^{S_{3,1}} = \varepsilon_{33}^{S_{3,1}}; \tag{10.15}$$

$$K_{\Delta}^{E} = \frac{s_{11}^{E}}{s_{11}^{E2} - s_{12}^{E2}} = \frac{Y_1^E}{1-\sigma_1^2},\ n_{\Delta} = \frac{d_{31}}{s_{11}^{E} + s_{12}^{E}},\ C_{e\Delta}^{S_{1,2}} = \varepsilon_{33}^{S_{1,2}}; \tag{10.16}$$

$$K_{\Delta 1}^{E} = \frac{s_{33}^{E}}{s_{11}^{E}s_{33}^{E} - s_{13}^{E2}} = \frac{Y_1^E}{1-\sigma_3\sigma_{13}},\ n_{\Delta 1} = \frac{d_{31}s_{33}^{E} - d_{33}s_{13}^{E}}{s_{11}^{E}s_{33}^{E} - s_{13}^{E2}},\ C_{e\Delta}^{S_{1,3}} = \varepsilon_{33}^{S_{1,3}}; \tag{10.17}$$

Note that parameters for the thickness mode of variant 3 are presented in Table 5.3 in the column numbered II.2 and they are: $K_{\Delta}^{E} = c_{11}^{E}$, $n_{\Delta} = e_{31}$, $C_{e\Delta}^{S_1} = \varepsilon_{33}^{S}$;. $k_{31t}'' = e_{31}/\sqrt{\varepsilon_{33}^{S}c_{11}^{D}}$.

The analogous elastic modulus for the isotropic strip according to Eq. (4.15) is

$$Y_{st} = \frac{Y}{1-\sigma^2}. \tag{10.18}$$

In accordance with relation (10.4) parameters $C_{e\Delta}^{S_i}$ can be represented as

$$C_{e\Delta}^{S_{3,1}} = \varepsilon_{33}^{S} + \frac{e_{31}^2}{c_{11}^{E}} = \varepsilon_{33}^{S_{3,1}} = \varepsilon_{33}^{T}(1-k_{33}'^2), \tag{10.19}$$

$$C_{e\Delta}^{S_{1,2}} = \varepsilon_{33}^{S} + \frac{e_{33}^2}{c_{33}^{E}} = \varepsilon_{33}^{S_{1,2}} = \varepsilon_{33}^{T}(1-k_{31}'^2), \tag{10.20}$$

$$C_{e\Delta}^{S_{1,3}} = \varepsilon_{33}^{S_{1,3}} = \varepsilon_{33}^{T}(1-k_{31w}''^2). \tag{10.21}$$

(notations for the coupling coefficients of piezoceramics at different boundary conditions are the same as in Ref. 1).

With all the parameters K_{Δ}^{E}, $C_{e\Delta}^{S_i}$ and n_{Δ} known, expressions for the energy densities in a deformed bar may be presented as

$$w_{mch}^{E} = \frac{1}{2} K_{\Delta}^{E} S^2, \; w_{el}^{S} = \frac{1}{2} C_{e\Delta}^{S_i} E_3^{\,2}, \; w_{em} = \frac{1}{2} n_{\Delta} E_3 S . \tag{10.22}$$

The coupling coefficients of piezoceramic material that characterize electromechanical energy conversion in a unit volume of a bar deformed under particular boundary conditions may be calculated according to expression (5.127) as

$$\frac{k_m^2}{1-k_m^2} = \frac{w_{em}^2}{w_{el}^{S_i} w_m^E} = \frac{n_{\Delta}^2}{C_{e\Delta}^{S_i} K_{\Delta}^{E}} . \tag{10.23}$$

Taking into consideration relation (10.4) and notations (10.19)-(10.21) for elementary capacitances at different boundary conditions we obtain

$$k_m^2 = \frac{n_{\Delta}^2}{\varepsilon_{33}^{T} K_{\Delta}^{E}} . \tag{10.24}$$

After substituting parameters by formulas (10.15)-(10.17) into this expression we will arrive at the following expression for the coupling coefficients of piezoceramics:

$$k_{33}'^2 = \frac{n_{\Delta 3}^2}{\varepsilon_{33}^{T} K_{\Delta 3}} = \frac{1}{1-\sigma_3 \sigma_{13}} \cdot (k_{33} - \sqrt{\sigma_3 \sigma_{13}} \, |k_{31}|)^2 , \tag{10.25}$$

$$k_{31}'^2 = \frac{n_{\Delta 1}^2}{\varepsilon_{33}^{T} K_{\Delta 1}} = k_{31}^2 (1+\sigma_1)/(1-\sigma_1) , \tag{10.26}$$

$$k_{31w}''^2 = \frac{1}{1-\sigma_3 \sigma_{13}} \cdot (-|k_{31}| + \sqrt{\sigma_3 \sigma_{13}} \, k_{33})^2 , \tag{10.27}$$

$$k_{31t}''^2 = \frac{e_{31}^2}{\varepsilon_{33}^{S} c_{11}^{D}} , \tag{10.27a}$$

(the subscripts w and t indicate the with ant thickness modes).

Although the coupling coefficients are essentially positive quantities, sometimes k_{31} is considered having sign (-) of the corresponding piezoelectric modulus d_{31}. In expressions (10.25) and (10.27) k_{31} is taken by modulus in order to avoid this confusion. For PZT-4

$k_{33}' = 0.62$, $k_{31}' = 0.46$, $k_{31w}'' = 0.06$, $k_{31t}'' = 0.18$.

It is noteworthy that the expressions (10.25) and (10.27) for k_{33}' and k_{31w}'' differ by the factor $1/\sqrt{1-k_{3i}^2}$ from those presented for this coefficients in Ref.1 This results from the erroneous using expression (5.126) instead of (5.127) that is appropriate for determining the coupling coefficients for the boundary conditions under consideration. For a detailed explanation of these issues see Section 5.4.3.

10.2.2 Equations of Vibrations in the Generalized Coordinates

Equations of vibration of the length expander bar transducers of the basic design shown in Figure 10.1 (at extreme case of large length to width aspect ratio) were considered in detail in Section 5.7.3.1. The results presented therein are completely applicable to all the one-dimensional length expander transducers with mechanical systems vibrating under different boundary conditions with corresponding changes of values of their equivalent parameters. Considering the expressions (10.22) for the energy densities and definitions for the equivalent parameters it can be obtained, as results of integrating these independent of coordinates quantities over volume of a mechanical system, that per unit of radiating surface of the piezoelement

$$K_{eqv}^E = K_{eqvb}^E (K_\Delta^E / K_{\Delta b}^E)\ ,\ n = n_b (n_\Delta / n_{\Delta b})\ ,\ C_{el}^S = C_{elb}^S (C_{e\Delta}^S / C_{e\Delta b}^S)\ , \qquad (10.28)$$

where parameters with subscript "b" are related to the basic design. The equivalent mass per unit surface does not depend on the boundary conditions and remains the same, as for the basic design. The resulting expressions for the modal equivalent parameters and corresponding resonance frequencies are presented in the following sections.

10.2.2.1 The Basic Bar Transducer Designs

Summary of equivalent parameters for the designs depicted in Figure 10.1(a) is as follows. (Note that in all the cases subscripts $i = 1,3,...$ denote number of the mode of vibration assuming that only odd modes are active.)

For the transverse piezoeffect (Figure 10.1 (a.1)):

$$K_{eqvi}^E = \frac{i^2\pi^2 wt}{2l} \cdot Y_1^E\ ,\ n_i = 2d_{31}w_2Y_1^E\ ,\ C_{el}^{S_1} = \varepsilon_{33}^T(1-k_{31}^2)wl/t\ , \qquad (10.29)$$

$$M_{eqvi} = M/2, \; f_{ri} = \frac{i}{2l}\sqrt{\frac{Y_1^E}{\rho}}. \tag{10.30}$$

For the longitudinal piezoeffect (Figure 10.1(a.2), (a3)):

$$K_{eqvi}^E = \frac{i^2\pi^2 S_{cs}}{2l} \cdot Y_3^E, \; n_i = \frac{2S_{cs}Nd_{33}}{l} \cdot Y_3^E, \tag{10.31}$$

$$C_{el}^{S_3} = \varepsilon_{33}^T(1-k_{33}^2)S_{cs}N^2/l, \; M_{eqvi} = M/2. \tag{10.32}$$

Here $S_{cs} = wt$ for the rectangular bars, and $S_{cs} = \pi a^2$ for round bars; N is number of segments. Additional rigidity term ΔK must be used that depends on number of the segments. This dependence is given in Section 5.5.2 and is replicated here in Figure 10.4.

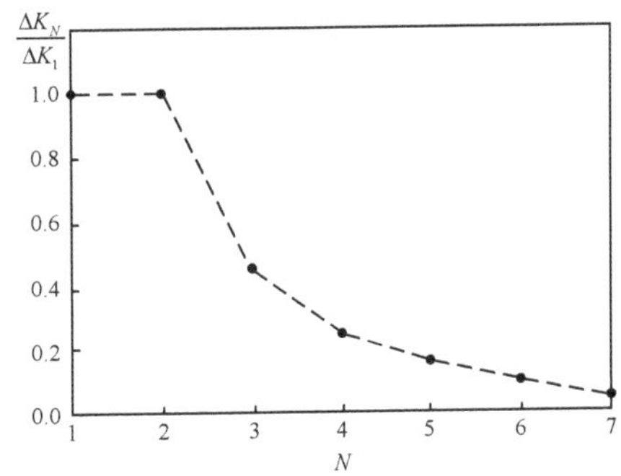

Figure 10.4: Ratio $\Delta K_N / \Delta K_1$ as a function of number of segments N.

As it follows from the plot, at $N \geq 6$ on the half wave of deformation the ΔK term can be neglected. At $N = 1$

$$\Delta K = K_{eqvi}^E \frac{k_{33}^2}{1-k_{33}^2}\left(1 - \frac{8}{i^2\pi^2}\right). \tag{10.33}$$

Thus, the resonance frequency is

$$f_{ri} = \frac{i}{2l}\sqrt{\frac{Y_3^E}{\rho}} \cdot \sqrt{1 + \left(1 - \frac{8}{i^2\pi^2}\right)\frac{k_{33}^2}{1-k_{33}^2}}. \tag{10.34}$$

Otherwise, the dependence of ΔK from a number of segments must be taken into account.

10.2.2.2 Width Vibrating Long Strip of Small Thickness

For the transverse piezoeffect (Figure 10.1 (b.1) the energy densities are determined by formulas (10.16)) and the equivalent parameters are:

$$K^E_{eqvi} = \frac{i^2\pi^2 wt}{2l}\cdot\frac{Y_1^E}{1-\sigma_1^2},\ n_i = 2d_{31}w\cdot\frac{Y_1^E}{1-\sigma_1},\ C_{el}^{S_{1,2}} = \varepsilon_{33}^T(1-k_{31}'^2)wl/t, \tag{10.35}$$

$$M_{eqvi} = M/2,\ f_{ri} = \frac{i}{2l}\sqrt{\frac{Y_1^E}{\rho(1-\sigma_1^2)}}. \tag{10.36}$$

For the longitudinal piezoeffect (Figure 10.1 (b.2) the energy densities are given by formulas (10.15)) and the equivalent parameters are:

$$K^E_{eqvi} = \frac{i^2\pi^2 wt}{2l}\cdot\frac{Y_3^E}{1-\sigma_3\sigma_{13}}, \tag{10.37}$$

$$n_i = \frac{2wtNd_{33}[1-(d_{31}/d_{33})\sigma_3]}{l}\cdot\frac{Y_3^E}{1-\sigma_3\sigma_{13}}, \tag{10.38}$$

$$C_{el}^{S_{3,1}} = \varepsilon_{33}^T(1-k_{33}'^2)wtN^2/l. \tag{10.39}$$

The additional rigidity term at $N = 1$ has to be calculated by formula (10.33) with replacing the coupling coefficient, namely,

$$\Delta K = K^E_{eqvi}\frac{k_{33}'^2}{1-k_{33}'^2}\left(1-\frac{8}{i^2\pi^2}\right). \tag{10.40}$$

The resonance frequencies are

$$f_{ri} = \frac{i}{2l}\sqrt{\frac{Y_3^E}{\rho(1-\sigma_3\sigma_{13})}}\cdot\sqrt{1+\left(1-\frac{8}{i^2\pi^2}\right)\frac{k_{33}'^2}{1-k_{33}'^2}}. \tag{10.41}$$

10.2.2.3 Thickness Vibrating Plate

For the longitudinal piezoeffect (Figure 10.1 (c) the energy densities are by formulas (10.13)) and the equivalent parameters are:

$$K^E_{eqvi} = \frac{i^2\pi^2 wt}{2l}c_{33}^E,\ M_{eqvi} = M/2,\ n_i = \frac{2wte_{33}}{l},\ C_{el}^{S_{1,2}} = \varepsilon_{33}^T(1-k_t^2)wt/l. \tag{10.42}$$

The additional rigidity is

$$\Delta K = K^E_{eqvi}\frac{k_t^2}{1-k_t^2}\left(1-\frac{8}{i^2\pi^2}\right), \tag{10.43}$$

and the resonance frequencies are

$$f_{ri} = \frac{i}{2l}\sqrt{\frac{c_{33}^E}{\rho}} \cdot \sqrt{1+\left(1-\frac{8}{\pi^2}\right)\frac{k_t^2}{1-k_t^2}} \,. \tag{10.44}$$

It must be remembered that all the expressions for the equivalent parameters are accurate within limits of the aspect ratios, at which the vibrations can be considered one-dimensional. The conclusions about these limits can be made based on the results presented in Section 10.3, where the estimations are obtained regarding the first modes of vibration. For these estimations to be valid for higher modes, they must be related to the half wave of deformation. Thus, while for a bar vibrating in the first mode, it is sufficient for the length to be approximately twice as large as the lateral dimensions, for the mode with number *i*, this condition must be valid for *l/i.*

10.2.2.4 Equivalent Electromechanical Circuits

Equations of vibration and the corresponding multicontour equivalent electromechanical circuit of the length expander transducers are the same, as where considered in Section5.7 in general, and in Section 5.7.3.1 with example of the bar transducers (see Figure 5.25). The general equivalent circuit of Figure 5.25 is specified here to the case that the transducers are intended for underwater applications. In this case the external forces and mechanical loads are the acoustic equivalent force, F_{eqvi}, and radiation impedance, Z_{aci}. We will assume that they are acting on one end of the transducer (for certainty on the end at $x = 0$) with another end remaining free of external actions, which is usually the case. The modified circuit is shown in Figure 10.5. The equivalent parameters in the circuit correspond to piezoelements at the particular boundary conditions considered above. The equivalent compliance, C_{eqvi}^E, in general is $C_{eqvi}^E = 1/(K_{eqvi}^E + \Delta K_i)$. As the radiating surfaces vibrate uniformly at all the modes of vibration, the equivalent forces and self- radiation impedances are the same for all the modes. Thus, according to formulas (5.253)-(5.255) F_i and Z_{ii} must be replaced by F_{eqv} and Z_{aci},

$$Z_{aci} = Z_{ac}[1+\sum_{l\neq i}(U_l/U_i)], \tag{10.45}$$

where Z_{ac} is the self-impedance of the mode of vibration and the second term in brackets represents contribution of mutual impedances between the modes, introduced impedance $Z_{inli} = Z_{ac}(U_l/U_i)$. Information regarding the radiation impedances of piston like vibrating surfaces of transducers in the form of circular and rectangular pistons, and strip of large length is presented in Section 6.3.3.2-6.3.3.5 for the single transducers, and in Section 6.4.5-6.4.7 for the acoustic interacting transducers.

The length expander transducers are greatly underloaded by radiation impedances and have large mechanical quality factor Q_m. (As it was noted in Ch. 3 (Table 3.1) even under the greatest acoustic impedance $Z_{ac} = (\rho c)_w S_{cs}$ this quantity is about 25.) Therefore, in vicinity of resonance frequency of ith mode in expression (10.45) $U_i \gg U_l$, and contribution of other modes can be neglected. The contours in Figure 10.5 can be considered as independent, and for calculating operational characteristics of transducer in the range around its resonance frequency this circuit may be reduced to a single contour, as shown in Figure 10.6(a). The equivalent parameters in the circuit depend on the above considered boundary conditions.

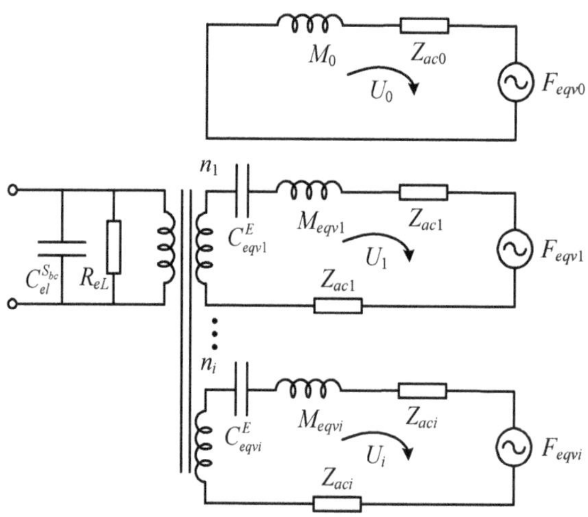

Figure 10.5: Multicontour equivalent circuit of a length expander transducer.

When calculating the output voltage of a transducer in the receive mode at low frequencies much below the first resonance, contribution of all the modes must be taken into consideration in the way as was shown in Section 5.7.3.1 for a bar under the transverse piezoeffect and resulted in formula (5.256). For the length expander transducers at different boundary conditions the analogous result is

$$V_{out} = \frac{1}{\omega C_{el}^{S_{bc}}} \cdot n_{bc} \cdot \left(\sum_{i=1}^{\infty} \frac{\omega}{K_{eqv\,i\,bc}^{E}} \right) \cdot (F_{eqv} + F_{in}) = \frac{1}{\omega C_{el}^{S_{bc}}} \cdot n_{bc} \cdot \frac{\omega}{K_{\Sigma bc}^{E}} \cdot (F_{eqv} + F_{in}), \tag{10.46}$$

where

$$K_\Sigma^E = \frac{l}{4S_{cs}Y_{bc}} = \frac{1}{C_\Sigma^E} \tag{10.47}$$

is the static rigidity of the mechanical system clamped at the middle section, and subscript *bc* indicates correspondence of the parameter to certain boundary conditions in accordance with formulas (10.30), (10.32), (10.36), (10.39) and (10.40). As to Y_{bc}, it is defined as $Y_{bc} = K^E_{eqvi\,bc} \cdot (2L / i^2\pi^2 wt)$. Equivalent circuit that corresponds to the result by formula (10.46)

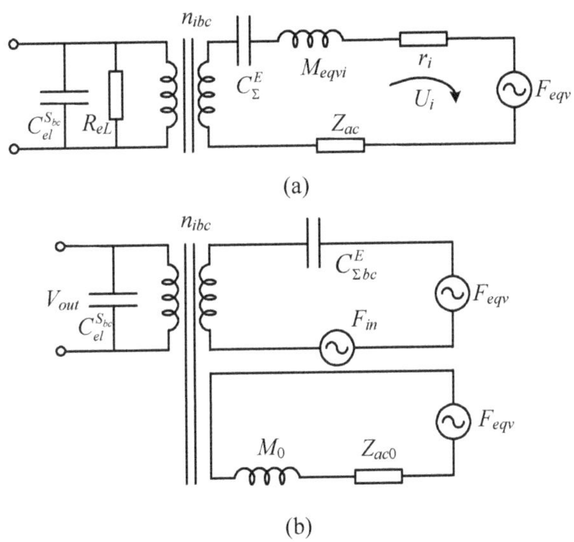

Figure 10.6: Equivalent circuits of the length expander transducer: (a) for the frequency range around resonance frequency, (b) at low frequencies much below the first resonance.

is shown in Figure 10.6 (b). The "zero mode" contour represents vibration of mechanical system of the transducer without deformation unless it is not fixed in the middle section. This movement does not produce a direct effect on the electrical side of the transducer, though introduces acoustically induced force $F_{in} = Z_{ac}U_0$ in the active contour.

10.2.3 Equations of Vibrations in the Geometry Coordinates

10.2.3.1 Three-Port T-Network Equivalent Circuits

Equations of vibration of piezoceramic bodies in the rectangular coordinates are considered in Section 5.8. It is shown that for one-dimensional vibrating piezoelements, which is the case for the length expander transducers, the equations of vibration are the same as for the passive mechanical systems of the same configuration except for difference in the values of sound speed. As is shown with example of longitudinally vibrating bar (Eq. (5.348)), the sound speed is

$$c^2 = K_\Delta^E / m_\Delta = (c^E)^2 \tag{10.48}$$

under the transverse piezoeffect and under the longitudinal piezoeffect in segmented bars with sufficient number of segments. For the solid bars under the longitudinal piezoeffect

$$c^2 = K_\Delta^D / m_\Delta = (c^D)^2 . \tag{10.49}$$

(K_Δ^E and K_Δ^D are analogs of elastic modulus of isotropic passive body at $E = 0$ and $D = 0$, respectively, and $m_\Delta = \rho$).

The equivalent electromechanical circuits that can be put in correspondence with equations of motion and conditions of mechanical loading on the ends formulated in the rectangular coordinates were introduced in Section 5.8 in the form of three-port networks shown in Figure 5.38. The circuits are valid for all the length expander transducers, if to replace parameters of the circuits by those with subscripts "*bc*." The circuits with modified notations are presented in Figure 10.7 together with their schematic representations that can be used for brevity.

It must be remembered that the general requirement for the aspect ratios of the piezoelements employed that ensure their one-dimensional mode of vibration must be fulfilled.

In the circuit (a)

$$Z_{1bc}^E = -j\rho c_{bc}^E S_{cs} / \sin k_{bc}^E l \, , \; Z_{2bc}^E = j\rho c_{bc}^E S_{cs} tan(k_{bc}^E l / 2) ; \tag{10.50}$$

$$N_{1bc} = n_{1bc} / 2 \, , \; C_{el1}^{S_{bc}} = \varepsilon_{33}^T (1 - k_{31bc}^2) wl / t . \tag{10.51}$$

In the circuit (b) for not segmented transducer ($N = 1$)

$$Z_{1bc}^D = -j\rho c_{bc}^D S_{cs} / \sin k_{bc}^D l \, , \; Z_{2bc}^D = -j\rho c_{bc}^D S_{cs} / \sin k_{bc}^D l ; \tag{10.52}$$

$$N_{3bc} = n_{3bc} / 2 \, , \; C_{el3}^{S_{bc}} = \varepsilon_{33}^T (1 - k_{33bc}^2) S_{cs} / l . \tag{10.53}$$

For the segmented transducer

$$N_{3bc} = N \cdot n_{3bc} / 2 \, , \; C_{el3}^{S_{bc}} = \varepsilon_{33}^T (1 - k_{33bc}^2) S_{cs} N^2 / l . \tag{10.54}$$

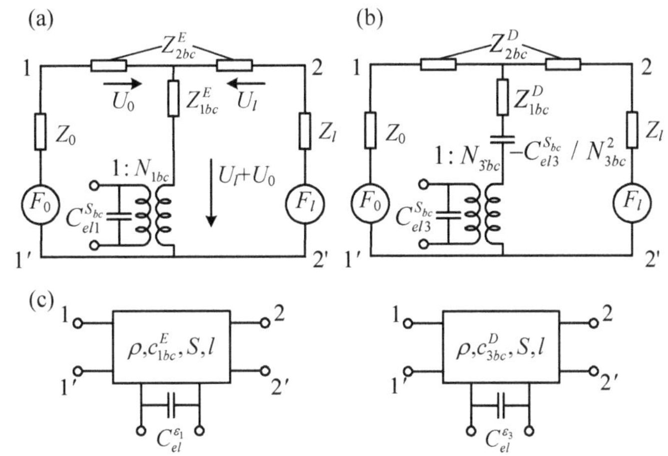

Figure 10.7: Equivalent electromechanical three-port networks of the length expander transducers: (a) for the transverse piezoelectric effect, (b) for the longitudinal piezoelectric effect, (c) schematic representation of the networks.

Expressions for the parameters in the circuits can be rewritten by generalizing expressions (5.385), (5.387) and (5.392), (5.393) for particular boundary conditions considered above.

As was noted in Section 5.8.3, for a segmented transducer under the longitudinal piezoelectric effect at number of segments $N \geq 6$, the circuit shown in Figure 10.7 (a) is valid if to replace N_{1bc} by N_{3bc} and $C^{S_{bc}}_{el1}$ by $C^{S_{bc}}_{el3}$. In the circuits F_0, F_l and Z_0, Z_l are the forces and impedances acting on the ends of the transducer. In particular, F_l and Z_l are F_{eqv} and Z_{ac}, assuming that radiating surface is at $x = l$.

Advantage of application of the three-port equivalent circuits, as well as analogous two-port T-networks for passive elements, which can be used in concert, is that they allow determining operating characteristics of a transducer under mechanical loads on the ends that may vary in a broad range. In particular, the loads may constitute input impedances of the bars made of passive materials, which may also be represented by their two-port T-networks. In this case the cascade connection of the equivalent circuits can be used instead of solving new equations of motion and considering mating conditions on the boundaries of the two systems that may have different cross section areas and/or different elastic properties. The mating conditions in this case are met automatically. This is illustrated with examples in Section 10.4, where a possible approach to calculating nonuniform over length transducers is considered.

The disadvantage of using these equivalent circuits is in a quite formal way of determining the operating characteristics of the transducers, a lack of possibility of predicting and physical interpreting results unless multiple calculations are made. In this sense the advantages are on the side of the generalized coordinates approach. In practice the positive features of both methods may be combined.

10.2.3.2 Modifications of the T-Network Equivalent Circuits

10.2.3.2.1 Bar Transducers with Different Configuration of Electrodes

For investigating possibility of optimizing the effective coupling coefficient and governing the frequency response of a bar transducer by changing configuration of electrodes (these issues where considered in Section 5.6.2 in the generalized coordinates), the T-network circuits may be modified as shown in Figure 10.8.

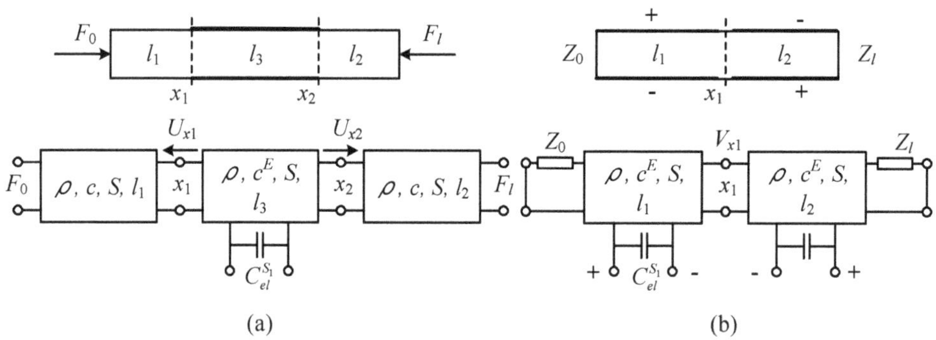

Figure 10.8: Modified equivalent T-network circuits for different variants of electrodes configuration: (a) transducer with partial electrodes; (b) transducer with electrodes connected for generating the second mode of vibration.

The vibration velocities of the cross sections at $x = x_1$ and $x = x_2$ correspond to the "currents" $U(x_1)$ and $U(x_2)$ in the respective branches of the combined equivalent circuit. Proper values of elastic constants for the deprived of electrodes end sections must be used in the passive networks. The variant of electrical connection of parts of electrodes in the opposite polarities, which is shown in the Figure 10.8 (b), illustrates application of the T-equivalent circuits to considering effects of different electrodes arrangement on the frequency response and effectiveness of the electromechanical conversion. In general, various voltages can be applied to the separated parts of the electrodes.

10.2.3.2.2Operation of Transducer in the Frequency Range around Its Resonance Frequency

The equivalent circuits in Figure 10.7 and their analysis may be simplified under certain loading conditions and in the case that transducer operates in the frequency range around its resonance frequency. Thus, in the typical variant of the side electroded bar with one end under action of load Z_L and force F_L and another end free, which is considered in detail in Ref. 1, the equivalent circuit in Figure 10.7 (a) (in this case terminals 1, $1'$ must be short circuited) may be transformed into the circuit shown in Figure 10.9 (a). The series impedance in this circuit is $Z_s^E = -j2\rho c_1^E S / \tan(k_1^E l/2)$. The impedance Z_2^E is the same as in the original circuit, namely, $Z_{2bc}^E = j\rho c_{1bc}^E S_{cs} tan(k_{1bc}^E l/2)$. Note that at low frequencies, such that $\tan kl/2 \approx kl/2$, the parallel impedance in the circuit transforms into inertia shunting element $2Z_2^E \approx j\omega M$, which is analogous to the passive parallel contour in Figure 10.5.

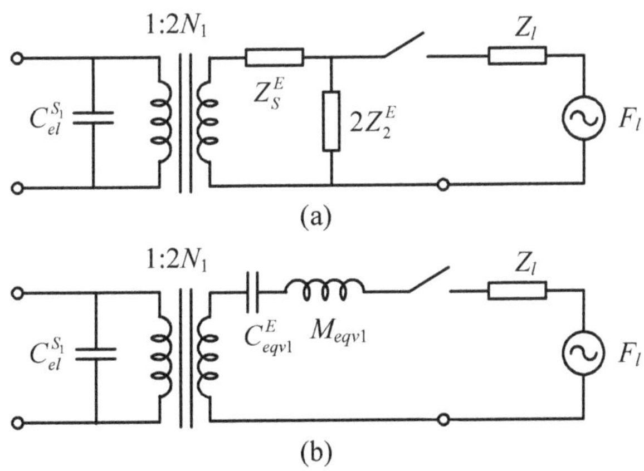

Figure 10.9: Equivalent circuits of the transducer under transverse piezoeffect with one end unloaded: (a) for a broad frequency range, (b) for the frequency range around its first resonance.

At frequencies around the first resonance further simplification of this equivalent circuit may be produced by expansion of trigonometric functions Z_s^E and Z_2^E into a series in terms of small deviations from the resonance frequency. This results in the circuit shown in Figure 10.9 (b), where for the basic transducer design (bar at transverse piezoeffect)

$$K_{eqv1}^E = \frac{\pi^2 wt}{2ls_{11}^E}, \quad M_{eqv} = M/2,$$

$$2N_1 = 2wd_{31} / s_{11}^E = n_1 \,,\; C_{el}^{S_1} = \varepsilon_{33}^T (1-k_{31}^2) wl/t \,. \tag{10.55}$$

This circuit coincides with the circuit in Figure 10.6 (a), which was obtained in generalized coordinates for the first mode of vibration in more explicit way. For different boundary conditions the equivalent parameters must be changed accordingly.

If to apply the analogous derivation to the equivalent circuit in Figure 10.7 (b) for the longitudinal piezoelectric effect, we obtain the circuits presented in Figure 10.10 (a) and (b), where

$$M_1 \to M_{eqv1} \,,\; 2N_3 = 2S_{cs} d_{33} / s_{33}^E = n_3 \,,\; C_{el}^{S_3} = \varepsilon_{33}^T (1-k_{33}^2) S_{cs} / l \;. \tag{10.56}$$

With terminals 2, 2' in Figure 10.10 (b) short-circuited, we obtain circuits of a resonator with free ends shown in Figure 10.10 (c) and (d). Physical meaning of the "negative capacitance" $(-C_{el}^{S_3})$ in the equivalent circuit of Figure 10.10 (a) is not clear, but it may be clarified as result of the following manipulations.

At first, we transform the "negative capacitance" into the mechanical contour, as shown in Figure 10.10 (c), where $C' = -C_{el}^{S_3} / 4N_3^2$. After performing the calculations in course of which the relation $s_{33}^D = s_{33}^E (1-k_{33}^2)$ is taken into account, we arrive at the circuit representation shown in Figure 10.10 (d), where

$$\begin{aligned} \frac{1}{j\omega C_{eqv3}^D} - \frac{4N_3^2}{j\omega C_{el}^{S_3}} &= \frac{\pi^2 S_{cs}}{j\omega 2l}\left[\frac{1}{s_{33}^D} - \frac{8}{\pi^2}\frac{k_{33}^2}{s_{33}^E(1-k_{33}^2)}\right] = \\ &= \frac{1}{j\omega C_{eqv3}^E} + \frac{1}{j\omega C_{eqv3}^E}\left(1-\frac{8}{\pi^2}\right)\frac{k_{33}^2}{s_{33}^E(1-k_{33}^2)} \end{aligned} , \tag{10.57}$$

$$\frac{1}{C_{eqv3}^E} = K_{eqv3}^E = \frac{\pi^2 S_{cs}}{2ls_{33}^E} \tag{10.58}$$

and

$$\frac{1}{\Delta C} = \Delta K = K_{eqv3}^E \left(1-\frac{8}{\pi^2}\right)\frac{k_{33}^2}{(1-k_{33}^2)} \,. \tag{10.59}$$

This is the same quantity ΔK as by formulas (5.156) and (10.40). And the circuits in Figure 10.10 (b) and (c) are the same, as those in Figure 5.7 that were obtained directly by using the first normal mode of vibration of the end-electroded bar. Thus, the negative capacitance on the electrical side has to do with the electrical interaction in the solid (not segmented) end-electroded length expander piezoelement.

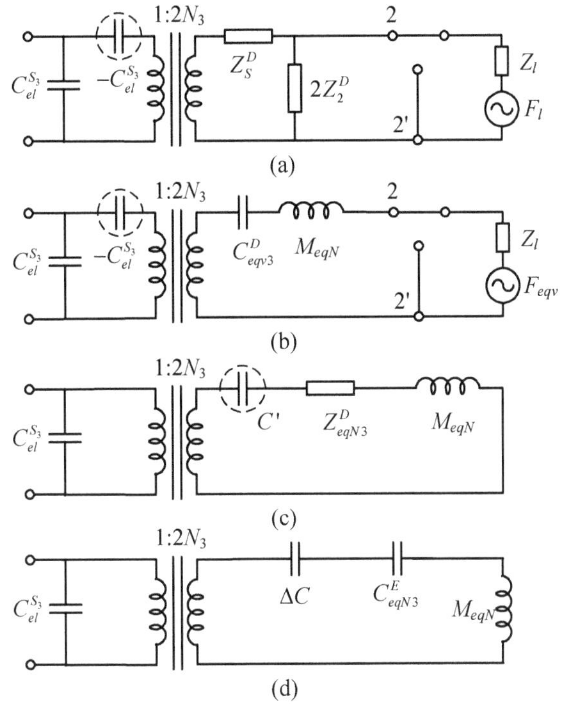

Figure 10.10. Equivalent circuits of the transducer under the longitudinal piezoelectric effect: (a) with one end unloaded, and (b) of the sane transducer at frequencies around the first resonance; (c) and (d) around the first resonance with both ends unloaded (terminals 2, 2' are short-circuited). $C' = -C_{el}^{S_3} / 4N_3^2$.

10.2.3.2.3 The Small Wavelength of the Piezoelement Approximation.

If the size of a length expander piezoelement (let it be a bar for simplicity) loaded on the ends appears to be small as compared to the wavelength, the equivalent circuits in Figure 10.7 can be simplified. At this condition $\tan(kl/2) \approx kl/2$, and independently of the mode of polarization

$$Z_2 = j\rho c S_{cs} \tan(kl/2) \approx j\omega\rho S_{cs} l/2 = j\omega M/2, \tag{10.60}$$

where M is the mass of the piezoelement. When transforming impedance Z_1 at small kl, for more accurate result we represent $\sin kl \approx kl - (kl)^3/6$. Thus, Z_1^E for the side-electroded bar becomes

$$Z_1^E = \frac{\rho c_1^E S_{cs}}{j \sin k_1^E l} \approx \frac{\rho c_1^E S_{cs}}{j k_1^E l[1-(k_1^E l)^2/6]} = \frac{1}{j\omega C_1^E} - j\omega\frac{M}{6}, \tag{10.61}$$

where M is the mass of the bar, and (remembering that $c_1^E = \sqrt{1/\rho s_{11}^E}$)

$$C_1^E = \frac{l s_{11}^E}{S_{cs}} \tag{10.62}$$

This is the static compliance of the bar at $E = 0$.

For the end-electroded bar in the same way we obtain

$$Z_1^D = \frac{\rho c_3^D S_{cs}}{j \sin k_3^D l} \approx \frac{1}{j\omega C_3^D} - j\omega \frac{M}{6}, \tag{10.63}$$

where

$$C_3^D = \frac{l s_{33}^D}{S_{cs}}. \tag{10.64}$$

Taking into account the compliance term ($-C_{el}^{S_3} / N_3^2$) in the equivalent circuit in Figure 10.7 (b) transformed from electrical side, we will find that

$$\frac{1}{j\omega C_3^D} - \frac{N_3^2}{j\omega C_{el}^{S_3}} = \frac{S_{cs}}{j\omega l}\left[\frac{1}{s_{33}^D} - \frac{k_{33}^2}{s_{33}^E(1-k_{33}^2)}\right] = \frac{1}{j\omega C_3^E}, \tag{10.65}$$

since $s_{33}^D = s_{33}^E(1-k_{33}^2)$. Thus, under the assumptions of small wavelength of the piezoelement the equivalent circuits of the length expander transducer for both the transverse and longitudinal piezoelectric effect simplify to the form shown in Figure 10.11.

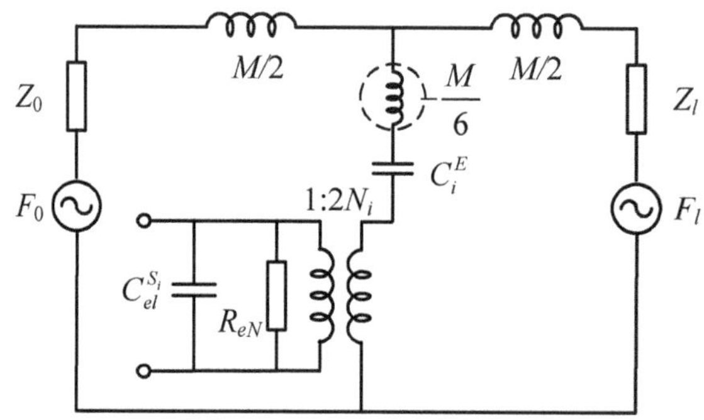

Figure 10.11. Equivalent circuit for the length expander transducer loaded on the ends under the condition that length of the piezoelement is much smaller than wavelength: $i = 1, 3$ for the transverse and longitudinal piezoeffect, respectively.

The difference between the circuits for the longitudinal and transverse piezoelectric effect due to existence of the negative capacitance (of the elastic impedance $-N_3^2 / j\omega C_{el}^{S_3}$)

disappeared. As it is shown in Section 5.5, there should not be such differences. if the strains do not change along the lines of force of the electric field. In the case under consideration the strains along the length of the piezoelement practically do not change.

10.3 Coupled Vibrations in the Transducers

10.3.1 Statement of the Problem

So far, the electromechanical conversion and vibration in the mechanical systems of the length expander transducers were considered in one-dimensional approximation, i.e., under the assumption that their deformations occur along a single geometry coordinate. In many cases this assumption does not hold, or at least is questionable, and deformations must be treated as coupled two-dimensional. The main goal of this treatment is in determining the aspect ratios of the mechanical systems of the transducers, at which the one-dimensional approximation remains applicable to a certain extent, and thus in avoiding harmful effects of the coupled vibrations on the operational characteristics of the transducers. In case that effects of the coupled vibration are tolerable or even useful, considering the transducers calculation in the two-dimensional approximation may be also of interest.

General technique of treating the coupled vibration in isotropic passive mechanical systems was considered in Section 4.6, where outline of the theory of coupled vibrations is given (Section 4.6.1), and effects of the coupled vibrations are illustrated with examples of several mechanical system configurations in Section 4.6.2.2 (vibration of rectangular plates), and in Section 4.6.2.3 (vibration of cylindrical discs and solid rods). The results obtained therein will be specialized in this Section for the anisotropic mechanical systems made of piezoceramics and will be used for determining electromechanical parameters of the corresponding transducers and for developing equivalent circuits that allow calculating their operational characteristics. The variants of mechanical systems that are of special interest for practical applications and must be considered in general as two-dimensional are presented in Figure 10.1 and Figure 10.2. Transitions between the one-dimensional vibrating piezoelements having ideal extreme dimensions through the two-dimensional piezoelements performing coupled vibration: (a) between the thin bar ($L_1 / L_2 \to 0$) and strip ($L_1 / L_2 \to \infty$), (b) between strip ($L_2 / L_3 \to 0$) and plate (

$L_2 / L_3 \to 0$), (c) between thickness vibrating thin circular disk (aspect ratio $h/2a \to 0$) and axial vibrating long bar ($h/2a \to \infty$) were considered in Refs. 2-4. The analytical results obtained therein, and results of their experimental verification are presented in this section.

As a prerequisite for treating effects of coupled vibration in the active piezoelements, it is useful to remind the procedure for considering the coupled vibrations in passive mechanical systems that was used in Section 4.6.2.2. General two-dimensional vibrations of an isotropic plate were presented in Section 4.6.2.2 as superposition of vibration of two partial mechanical systems with corresponding distribution of displacements and strain in the form (with example of the plate in Ref. 2, and considering only the first fundamental modes of vibration)

$$\xi(x,y) = \xi_{1p} \sin(\pi x / L_1) + \xi_{2p} \sin(\pi y / L_2), \qquad (10.66)$$

$$S_1 = \xi_{1p}(\pi / L_1)\cos(\pi x / L_1),\ S_2 = \xi_{2p}(\pi / L_2)\cos(\pi y / L_2). \qquad (10.67)$$

Here ξ_{1p} and ξ_{2p} are the generalized coordinates - maximum displacements in the partial systems. The boundary condition for a partial system is determined by the condition that another partial system is clamped. Therefore, for the partial system #1 it should be $S_2 = 0$. Together with condition $T_3 = 0$ due to small thickness of the plate this means that the partial system is the infinitely long in y direction thin strip. Likewise, for the partial system #2 the boundary conditions are $S_1 = 0$, $T_3 = 0$, and the partial system is the strip infinitely long in x direction. The partial systems are shown in Figure 10.12 by the dashed lines.

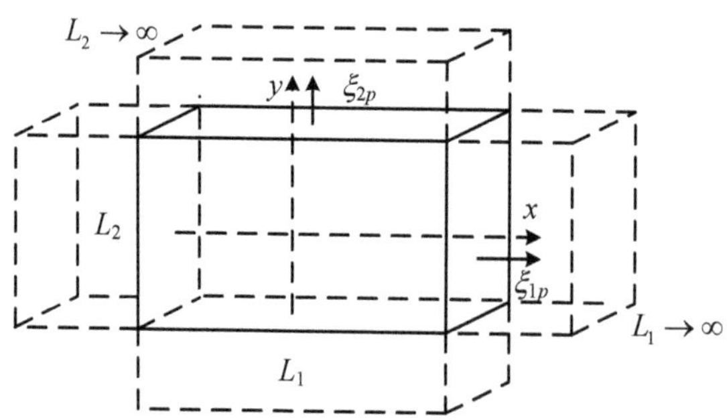

Figure 10.12: To the coupled vibration of a thin plate, as superposition of vibration of two partial systems– infinitely long thin strips vibrating through their widths. The partial systems are shown by the dashed lines.

It is noteworthy that in transition between two extreme one-dimensional systems only one of them coincides with a partial system. For example, in the transition of one-dimensional bar and infinite strip shown in Figure 10.12 (a) this is the strip at $L_1 / L_2 \to \infty$. As it is shown in Section 4.6.1, for determining all the characteristics of the coupled vibrations it is necessary (and sufficient) to know resonance frequencies of the partial systems, f_{ip}, and coefficient of coupling between the partial systems, γ, which can be readily obtained from considering expressions for the potential and kinetic energies of the coupled system (4.520) and (4.521) that we rewrite in the form

$$W_{pot}^E = \frac{1}{2}\int_{\tilde{V}}\left(T_1^E S_1 + T_2^E S_2\right)d\tilde{V} = \frac{1}{2}K_1^E \xi_{1p}^2 + K_{12}\xi_{1p}\xi_{2p} + \frac{1}{2}K_2^E \xi_{2p}^2, \tag{10.68}$$

$$W_{kin} = \frac{1}{2}M_1\dot{\xi}_{1p}^2 + M_{12}\dot{\xi}_{1p}\dot{\xi}_{2p} + \frac{1}{2}M_2\dot{\xi}_{2p}^2. \tag{10.69}$$

Here K_1^E, K_2^E and M_1, M_2 are the equivalent rigidities and masses of the partial systems; K_{12} and M_{12} are the parameters that characterize the elastic and inertial coupling mechanism between the partial systems. In our cases $M_{12} \approx 0$ and the coupling may be considered as pure elastic. The partial resonance frequencies are determined by formulas $f_{ip} = (1/2\pi)\sqrt{K_i / M_i}$, and the coefficient of elastic coupling between the partial systems by formula $\gamma = K_{12} / \sqrt{K_1 K_2}$. The factual values of partial resonance frequencies and coupling factors depend on configuration of the partial systems and on the boundary conditions imposed. Thus, for the partial systems in form of isotropic passive strips expressions for the partial resonance frequencies, rigidities and coupling factors were determined in Section 4.6.2.2 as

$$f_{ip} = \frac{1}{2L_i}\sqrt{\frac{Y}{\rho(1-\sigma^2)}}, \quad i = 1, 2; \tag{10.70}$$

$$K_1 = \frac{\pi^2 Y t}{2(1-\sigma^2)}\frac{L_2}{L_1}, \; K_2 = \frac{\pi^2 Y t}{2(1-\sigma^2)}\frac{L_1}{L_2}, \; K_{12} = \frac{4Y\sigma t}{1-\sigma^2}; \tag{10.71}$$

$$\gamma = \frac{K_{12}}{\sqrt{K_1 K_2}} = \frac{8\sigma}{\pi^2}. \tag{10.72}$$

Using these quantities two branches of the resonance frequencies, $\Omega(\beta)$, and corresponding mode shape coefficients, $ms(\beta) = \xi_{2p} / \xi_{1p}$, were found for the coupled systems, as functions of the aspect ratio factor, $\beta = f_{2p} / f_{1p}$. All these functions of the aspect ratio factor have to be determined for the partial systems made of piezoceramic taking into consideration anisotropy of elastic and piezoelectric properties of the material that leads to different results for variants, in which the transverse and longitudinal piezoeffect is used. After this is done, the mode shapes of vibration of the coupled system that correspond to the resonance frequencies may be obtained from expression (10.66) as

$$\begin{aligned} \xi(x,y) &= \xi_{1p}\sin(\pi x_1/L_1)+\xi_{2p}\sin(\pi x_2/L_2) = \\ &= \xi_{1p}[\sin(\pi x_1/L_1)+ms(\beta)\cdot\sin(\pi x_2/L_2)] = \xi_{1p}\theta(x_1,x_2;\beta) \end{aligned}, \tag{10.73}$$

and all the electromechanical parameters of transducer including the effective coupling coefficient can be calculated using the previously described technique. These procedures are considered in Section 10.3.2 and 10.3.3 for the situations shown in Figure 10.2 (a) and (b). Coupled vibrations in the isotropic passive disks and rods were considered in Section 4.6.2.3 with essentially the same general approach. Specifics of coupled vibrations in the piezoceramic cylinders in course of their transition from thickness vibrating thin disks to one-dimensional long bar is considered in Section 10.3.4.

10.3.2 Transition Between the Thin Bar and Infinite Strip Vibrating over Its Width

10.3.2.1 Transverse Piezoeffect

In the variant of transverse piezoeffect that is shown in Figure 10.2 (a), in which case the strip is polarized in perpendicular to its radiating surface direction, the elastic properties of piezoceramic in the plane are isotropic and characterized by the constants Y_1^E and σ_1^E that are analogous to the Young's modulus and Poisson's ratio for isotropic passive material. Therefore, all the results presented in Section 4.6.2.2 regarding coupled vibrations in isotropic plates are valid for the piezoceramic plates of the same geometry, if to replace Y and σ by Y_1^E and σ_1^E in all the related expressions. (Note that names strip and plate can be used interchangeably in the context of this section. The difference is in their extreme configurations.) Thus, the expressions (10.70)-(10.72) become

$$f_{ip} = \frac{1}{2L_i}\sqrt{\frac{Y_1^E}{\rho(1-\sigma_1^{E2})}}, \quad i=1,2; \tag{10.74}$$

$$K_1 = \frac{\pi^2 Y_1^E t}{2(1-\sigma_1^{E2})}\frac{L_2}{L_1}, \; K_2 = \frac{\pi^2 Y_1^E t}{2(1-\sigma_1^{E2})}\frac{L_1}{L_2}, \; K_{12} = \frac{4Y_1^E\sigma_1^E t}{1-\sigma_1^{E2}}; \tag{10.75}$$

$$\gamma = \frac{K_{12}}{\sqrt{K_1 K_2}} = \frac{8\sigma_1^E}{\pi^2}. \tag{10.76}$$

Dependencies of the resonance frequencies and mode shape coefficients from the plate aspect ratio (in this case the aspect ratio factor is $\beta = f_{2p} / f_{1p} = L_1 / L_2$), which are shown in Figure 4.46 and Figure 4.47 by solid lines, are also valid for the piezoceramic plates because they were calculated at the value of $\sigma_1^E = 0.33$ for PZT-4 ceramics.

It is noteworthy that after the assumption is made that distribution of displacements in the plate having finite dimensions can be represented in the form of expression (10.66) all the same results regarding parameters of the coupled vibration, which are predicted by the general coupling theory, can be obtained by using calculating procedure typical of application of the energy method to the transducer with two mechanical degrees of freedom. The advantage of such approach is that it allows presenting results in the common form of equivalent electromechanical circuit and calculating operating characteristics of the transducer under an external loading. We will illustrate this approach with example under consideration though the procedure is general. At first, all the energies associated with transducer performance must be determined. The potential and kinetic energies are presented in the form of expressions (10.68) and (10.69).

The electrical energy of a transducer is

$$W_{el}^{S_{1,2}} = \frac{1}{2}\int_{\tilde{V}} \varepsilon_{33}^{S_{1,2}} E_{33}^2 \, d\tilde{V} = \frac{1}{2} C_{el}^{S_{1,2}} V^2, \tag{10.77}$$

where $C_{el}^{S_{1,2}} = \varepsilon_{33}^{S_{1,2}} L_1 L_2 / t$. The dielectric constant of ceramic $\varepsilon_{33}^{S_{1,2}}$ depends on boundary conditions for the mechanical system. It must be noted that uncertainty exists in presenting the dielectric constant of the coupled system. Hypothesis, on which the approach to treating coupled vibrations in mechanical systems is based, does not apply to calculating the electrical energy $W_{el}^{S_{1,2}}$ of the partially (depending on mechanical boundary conditions) clamped piezoelectric body. Thus, in the extreme case of infinitely long strip (one of the partial systems for variant of transducer under consideration) according to (10.15) and (10.16) $\varepsilon_{33}^{S_{1,2}} = \varepsilon_{33}^S + e_{33}^2 / c_{33}^E$ and may be represented as $\varepsilon_{33}^{S_{1,2}} = \varepsilon_{33}^T (1 - k_{31}'^2)$. But such value of the dielectric constant is questionable for the plate of finite size in course of transition between the extreme cases of thin bar and infinite strip. It is more logical to assume that for the plate of finite size the dielectric constant may be represented as

$$\varepsilon_{33}^{S_{1,2}} = \varepsilon_{33}^{T}[1 - k_l^2(\beta)], \tag{10.78}$$

where $k_l(\beta)$ is the effective coupling coefficient that corresponds to the mode shape associated with coefficient ms_l for each value of aspect ratio factor β. (The subscript $l = I, II$ denotes number of one of two branches of resonance frequencies for a coupled system.) At least this expression holds for the extreme cases. Determining the coupling coefficients $k_l(\beta)$ is one of the goals of solving the coupled vibration problem for mechanical system made of piezoceramics. Thus, expression for the capacitance will be

$$C_{el}^{S_{1,2}} = \varepsilon_{33}^{T}[1 - k_l^2(\beta)]L_1 L_2 / t. \tag{10.79}$$

The electromechanical energy of a transducer is

$$W_{em} = \frac{1}{2}\int_{\tilde{V}} D_3^E(S_1, S_2)E_3 d\tilde{V} = \frac{1}{2}V(n_1\xi_{1p} + n_2\xi_{2p}). \tag{10.80}$$

Following expressions (10.14) and (10.16), for this case

$$D_3^E(S_1, S_2) = \frac{d_{31}}{s_{11}^E + s_{12}^E}(S_1 + S_2), \tag{10.81}$$

and after integrating over the volume, it will be obtained that

$$n_1 = \frac{2Y_1^E d_{31} L_2}{1 - \sigma_1^E}, \quad n_2 = \frac{2Y_1^E d_{31} L_1}{1 - \sigma_1^E}. \tag{10.82}$$

After all the energies associated with functioning of the transducer as an electromechanical device are determined, the Lagrange's equations that describe its vibration, can be represented in the following form (without considering the energy losses and external loads)

$$(j\omega M_1 + K_1 / j\omega)U_{1p} + (K_{12} / j\omega)U_{2p} = n_1 V, \tag{10.83}$$

$$(K_{12} / j\omega)U_{1p} + (j\omega M_2 + K_2 / j\omega)U_{2p} = n_2 V, \tag{10.84}$$

$$I = j\omega C_{el}^{S_{1,2}} V + n_1\dot{\xi}_{1p} + n_2\dot{\xi}_{2p}. \tag{10.85}$$

Here I is the input current of the transducer, $U_i = j\omega\xi_{ip}$. Considering the losses and external loads is not essential for illustrating the coupled vibration analysis. They can be included in the Lagrange's equations in the same general way as it was done before, when necessary.

In order to determine the resonance frequencies and the resonance mode shapes of the transducer, we first consider free vibration of its mechanical system. The Eqs. (10.83) and (10.84) at $V = 0$ can be presented in the form of

$$(f_{1p}^2 - f^2)U_{1p} + (K_{12} / K_1) f_{1p}^2 U_{2p} = 0, \tag{10.86}$$

$$(K_{12} / K_2) f_{2p}^2 U_{1p} + (f_{2p}^2 - f^2) U_{2p} = 0, \tag{10.87}$$

where from the frequency equation follows

$$(f_{1p}^2 - f^2)(f_{2p}^2 - f^2) - \gamma^2 f_{1p}^2 f_{2p}^2 = 0. \tag{10.88}$$

The elastic coupling factor γ is defined by formula (10.76). Thus, the dependences of the resonance frequencies and mode shape coefficients of a transducer on the aspect ratio of the plate may be determined, as it is done in Section 4.6.2.2. Assuming for definiteness that the dimension L_2 (i. e., f_{2p}) is kept constant while L_1 changes, after denoting $f^2 / f_{2p}^2 = \Omega$ as the normalized nondimensional resonance frequency factor and introducing aspect ratio coefficient $\beta = (f_{2p} / f_{1p})$, Eq. (10.88) may be transformed to

$$\Omega^2 - (1 + 1/\beta^2)\Omega + (1 - \gamma^2)/\beta^2 = 0. \tag{10.89}$$

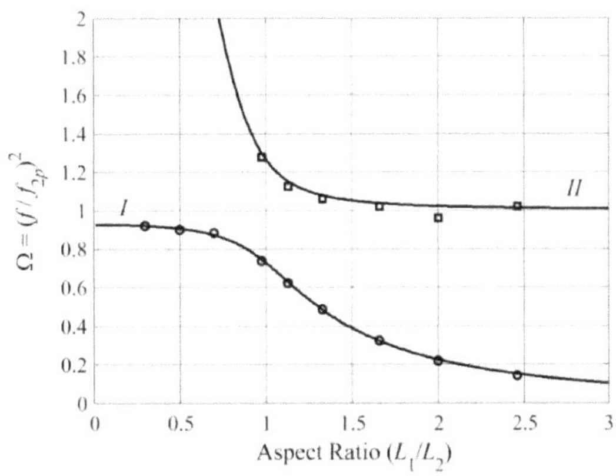

Figure 10.13: Dependence of the normalized resonance frequencies on aspect ratio for lower (*I*) and upper (*II*) frequency branches.

From this equation two branches of resonance frequencies corresponding to the solutions Ω_I and Ω_{II} may be found as functions of the aspect ratio for the strip. We accept the convention that $\Omega_I(\beta)$ forms the lower and $\Omega_{II}(\beta)$ the upper frequency branches. The frequency dependencies calculated at value $\gamma = (8/\pi^2)\sigma_1 = 0.27$ of the coupling factor for PZT-4 are shown in Figure 10.13.

In particular, it follows from Eq. (10.89) that at $\beta = 1$

$$\Omega_I = 1 - \gamma\,,\ \Omega_{II} = 1 + \gamma\,. \qquad (10.90)$$

The experimental data that are shown by markers in the figure were obtained in Ref. 2 with samples made of PZT-4 and PZT-8. The plots are normalized to values of frequencies at the limiting case of $\beta \ll 1$ to avoid differences between parameters of ceramic compositions and absolute values of dimensions. The results show good agreement between the predicted and experimental data regardless of some difference (about 5%) in coefficients σ_1^E for PZT-4 and PZT-8 ceramics.

After the resonance frequencies are determined, the corresponding mode shape factors, which will be defined as

$$ms_l = (U_{1p} / U_{2p})_{at\ \Omega=\Omega_l} = (\xi_{1p} / \xi_{2p})_{at\ \Omega=\Omega_l}\ \ (l = I, II), \qquad (10.91)$$

may be found from one of Eqs. (10.86), (10.87). Namely,

$$ms_l = -\frac{K_{12}}{K_1}\frac{1}{1-\Omega_l\beta^2}\,. \qquad (10.92)$$

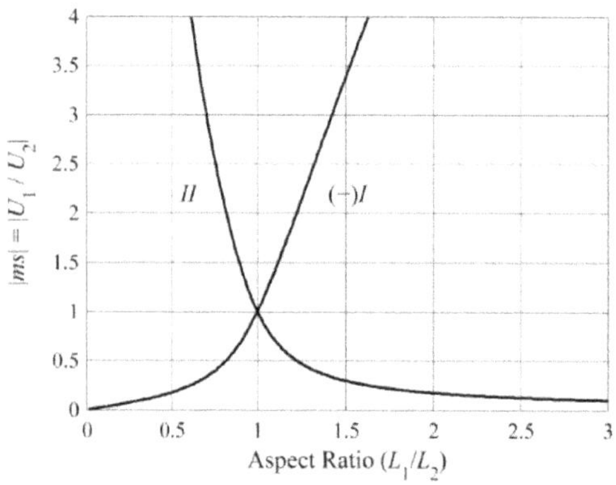

Figure 10.14: Dependence of the mode shape factors on aspect ratio for lower (*I*) and upper (*II*) frequency branches. At branch *I* $ms_I < 0$. At branch *II* $ms_{II} > 0$.

Using expressions (10.75) and (10.76) it may be concluded that $(K_{12} / K_{11}) = \gamma\beta$, and hence

$$ms_l = -\frac{\gamma\beta}{1-\Omega_l\,\beta^2}\,. \tag{10.93}$$

The dependence of the mode shape factors on the aspect ratio β is shown in Figure 10.14. It follows from the figure that at frequencies pertaining to the lower branch the velocities U_{1p} and U_{2p} are in anti-phase. In particular, it follows from expressions (10.93) and (10.90) that at $\beta = 1$

$$ms_I = -1 \text{ and } ms_{II} = 1\,. \tag{10.94}$$

This means that at higher and lower resonance frequencies the velocities have the same magnitude but are in-phase at the higher frequency and in anti-phase at the lower frequency.

The effective electromechanical coupling coefficient of a transducer can be found from the generalized relation (10.23)

$$\frac{W_{em}^2}{W_{el}^{S_{1,2}}\,W_m^E} = \frac{k_l^2(\beta)}{1-k_l^2(\beta)} = \alpha_{cl}(\beta)\,, \tag{10.95}$$

where expressions for energies W_{pot}^E, $W_{el}^{S_{1,2}}$, W_{em} are given by formulas (10.68), (10.77) and (10.80), respectively. The subscripts $l = I$ or II correspond to the lower and upper frequency branches. The expressions for energies W_{pot}^E and W_{em} can be modified by using the mode shape coefficients ms_l and aspect ratio factor β as follows

$$W_{pot}^E = \frac{1}{2}K_2\xi_{2p}^2\left(1+\frac{2K_{12}}{K_2}ms_l+\frac{K_1}{K_2}ms_l^2\right) = \frac{1}{2}\frac{K_2\xi_{2p}^2}{\beta^2}(\beta^2+2\gamma\beta ms_l+ms_l^2)\,, \tag{10.96}$$

$$W_{em} = \frac{1}{2}V n_2\xi_{2p}\left(\frac{n_1}{n_2}ms_l+1\right) = \frac{1}{2}V\frac{n_2\xi_{2p}}{\beta}(ms_l+\beta)\,. \tag{10.97}$$

In the expression for electric energy the capacitance must to be presented by formula (10.79).

Substituting all the expressions for energies into relation (10.95) results in

$$\frac{k_l^2(\beta)}{1-k_l^2(\beta)} = \frac{n_2^2}{C_{el}^{S_{1,2}}K_2}\,\frac{(ms_l+\beta)^2}{ms_l^2+2\gamma\beta\, ms_l+\beta^2} = A_{em}\,A_l(\beta)\,. \tag{10.98}$$

The right side of this relation appears to be represented as a product of two factors, one of which, $A_{em} = n_2^2 / C_{el}^{S_{1,2}} K_{22}$, depends on the electromechanical properties of the piezoelement and another, $A_l(\beta)$, is a function of the aspect ratio factor,

$$A_l(\beta) = \frac{(ms_l + \beta)^2}{ms_l^2 + 2\gamma\beta ms_l + \beta^2}. \tag{10.99}$$

After substituting expressions for the parameters K_2, $C_{el}^{S_{1,2}}$ and n_2 given by formulas (10.75), (10.79) and (10.82), the factor A_{em} for the transverse piezoeffect becomes

$$A_{em} = \frac{8}{\pi^2}\frac{k_{31}^2}{1-k_l^2(\beta)}\cdot\frac{1+\sigma_1^E}{1-\sigma_1^E}. \tag{10.100}$$

(Note that the dielectric constant of the two-dimensional piezoelement depends on the effective coupling coefficient to be determined according to expression (10.79).)

Thus, the resulting expression for the effective coupling coefficient as function of the aspect ratio factor at given coefficient of coupling between the partial systems is

$$k_l^2(\beta) = \frac{8}{\pi^2}\,k_{31}^2\cdot\frac{1+\sigma_1^E}{1-\sigma_1^E}\cdot\frac{(ms_l + \beta)^2}{ms_l^2 + 2\gamma\beta\, ms_l + \beta^2}. \tag{10.101}$$

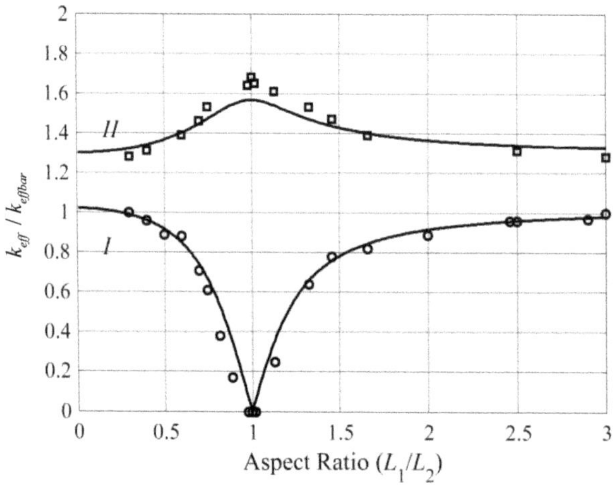

Figure 10.15: Dependences of the effective coupling coefficient on aspect ratio for lower (*I*) and upper (*II*) frequency branches.

Dependences of the effective coupling coefficients on the aspect ratio for the upper and lower resonance frequency branches are shown in Figure 10.15 The coupling coefficients are normalized to the effective coupling coefficient of a thin side electroded bar, $k_{eff\,bar}$, that represents the limiting case of a strip at $\beta \to 0$ and L_2 constant, which is known as

$$k_{eff\,bar}^2 = \frac{1}{1+(\pi^2/8)(1-k_{31}^2)/k_{31}^2}. \tag{10.102}$$

This formula is also valid for the effective coupling coefficient of the strip, $k_{eff\,strip}$, that represents another limiting case at $\beta \to \infty$ and L_2 constant, if to replace k_{31} by k'_{31}, i.e.,

$$k_{eff\,strip}^2 = \frac{1}{1+(\pi^2/8)(1-k'^2_{31})/k'^2_{31}}. \tag{10.103}$$

It is interesting to estimate the correlation between the obtained solutions for the resonance frequencies and the coupling coefficients and known results for the limiting one-dimensional configurations of a strip. At $\beta = L_1/L_2 \to 0$ the extreme configurations are: thin bar (L_2 constant, $L_1 \to 0$) for the lower frequency branch *I*, and infinite strip (L_1 constant, $L_2 \to \infty$) for the upper branch *II* (the strip is the partial system #1). At $\beta = L_1/L_2 \to \infty$ the extreme configurations are: infinite strip (L_2 constant, $L_1 \to \infty$) for the upper frequency branch *II* (the strip is the partial system #2), and thin bar (L_1 constant, $L_2 \to 0$) for the lower branch *I*.

It is shown in Section 4.6.2.2 that at $\beta \to 0$

$$\Omega_I \to 1-\gamma^2, \ \Omega_{II} \to 1/\beta^2. \tag{10.104}$$

Therefore,

$$f_I \to f_{2p}\sqrt{1-\gamma^2} \ \text{and} \ f_{II} \to (f_{2p}/\beta) = f_{1p}. \tag{10.105}$$

Considering expressions (10.70) for the partial frequencies f_{ip}, we arrive at the expression for the resonance frequency of the infinitely thin bar of the length L_2

$$f_I = \frac{1}{2L_2}\sqrt{\frac{Y_1^E(1-\gamma^2)}{\rho(1-\sigma_1^{E2})}}, \tag{10.106}$$

and for the infinitely long strip of the width L_1

$$f_{II} = \frac{1}{2L_1}\sqrt{\frac{Y_1^E}{\rho(1-\sigma_1^{E2})}}. \tag{10.107}$$

Given that the exact value for the resonance frequency of the bar must be $f_{bar} = (1/2L_2)\sqrt{Y_1^E/\rho}$ and that $\gamma = 0.81\sigma_1$, for PZT-4 ceramics ($\sigma_1^E = 0.33$, $\gamma = 0.27$) we obtain $f_1 = 1.02 f_{bar}$. Thus, the error of the current approach can be considered as negligible for the limiting case at $\beta \to 0$. For another extreme case (long strip of width L_1) the value of resonance frequency obtained by formula (10.107) is exact.

In terms of the effective coupling coefficients determined from the two-dimensional and one-dimensional approximations, comparison of the values for the limiting cases of the transducers in the shape of a bar and of a strip gives the following results. Using expressions (10.91) for the mode shape coefficients and (10.104) for the normalized frequencies Ω_I, at $\beta \to 0$ we arrive at

$$ms_I \to -\gamma\beta \text{ and } ms_{II} \to 1/\gamma\beta \,. \tag{10.108}$$

From formula (10.99) for $A_I(\beta)$ follows that

$$A_I(\beta)_{\beta\to 0} \to (1-\gamma)/(1+\gamma) \text{ and } A_{II}(\beta)_{\beta\to 0} \to 1, \tag{10.109}$$

respectively. Thus, from Eq. (10.101) we obtain

$$k_I^2(\beta)_{\beta\to 0} = \frac{8}{\pi^2}\, k_{31}^2 \cdot \frac{1+\sigma_1^E}{1-\sigma_1^E} \cdot \frac{1-\gamma}{1+\gamma}, \tag{10.110}$$

$$k_{II}^2(\beta)_{\beta\to 0} = \frac{8}{\pi^2}\, k_{31}^2 \cdot \frac{1+\sigma_1^E}{1-\sigma_1^E}\,. \tag{10.111}$$

At another extreme case of $\beta \to \infty$ from Eq. (4.587) follows (using relation $\sqrt{1-\alpha} \approx 1-\alpha/2$ at $\alpha \ll 1$) that

$$\Omega_I \to \frac{1-\gamma^2}{\beta^2},\ \Omega_{II} \to 1\,. \tag{10.112}$$

Therefore, the lower and upper resonance frequencies for this extreme case are

$$f_I = \frac{1}{2L_1}\sqrt{\frac{Y_1^E(1-\gamma^2)}{\rho(1-\sigma_1^{E2})}} = f_{bar}\sqrt{\frac{(1-\gamma^2)}{(1-\sigma_1^{E2})}} \text{ and } f_{II} = f_{2p}\,. \tag{10.113}$$

In order to estimate values of the effective coupling coefficients, the mode shape coefficients must be determined by formula (10.93) using the normalized frequencies (10.112). This results in the expressions

$$ms_I = -\beta/\gamma \text{ and } ms_{II} = \gamma/\beta\,. \tag{10.114}$$

Now from formula (10.99) for $A_I(\beta)$ follows

$$A_I(\beta)_{\beta\to\infty} \to (1-\gamma)/(1+\gamma) \text{ and } A_{II}(\beta)_{\beta\to\infty} \to 1, \tag{10.115}$$

i.e., the same results as were obtained for the case of $\beta \to 0$ by formulas (10.109). Therefore, values of the effective coupling coefficients in the extreme cases are equal for both frequency branches, i.e.,

$$k_l(\beta)_{\beta\to\infty} = k_l(\beta)_{\beta\to o}, \quad (10.116)$$

as it can be seen in Figure 10.15. This could be expected because the partial systems for both extreme cases have the same configuration. In the case that PZT-4 ceramics is used ($k_{31} = 0.33$, $\gamma = 0.27$) it follows from expressions (10.110) and (10.111) that $k_I(0) \approx 0.31$, whereas for the bar in one-dimensional approximation it should be $k_{eff\,bar}$ =0.30 (by formula (10.102)); and $k_{II}(0) = 0.42$, whereas in one-dimensional approximation it should be for the strip $k'_{eff\,strip}$ ≈0.43 (by formula (10.103) and given that $k'_{31} = 0.46$). The obtained values of the effective coupling coefficients for these extreme cases are somewhat different from those obtained by one-dimensional approximation. But in the dependencies on the aspect ratio that are normalized to value $k_{eff\,bar}$ this systematic error is avoided, and they appear to be accurate enough, as results of experimenting show. Besides, the accurate absolute values of the effective coupling coefficients for the extreme case configurations must be calculated from the one-dimensional approximations anyway.

10.3.2.2 Longitudinal Piezoeffect

The variant of longitudinal piezoeffect is shown in Figure 10.16.In this case the strip is polarized in direction of its width, the elastic properties of piezoceramics in the plane are not isotropic, and results of calculating parameters of the coupled isotropic system cannot be directly implemented for the piezoceramic strip.

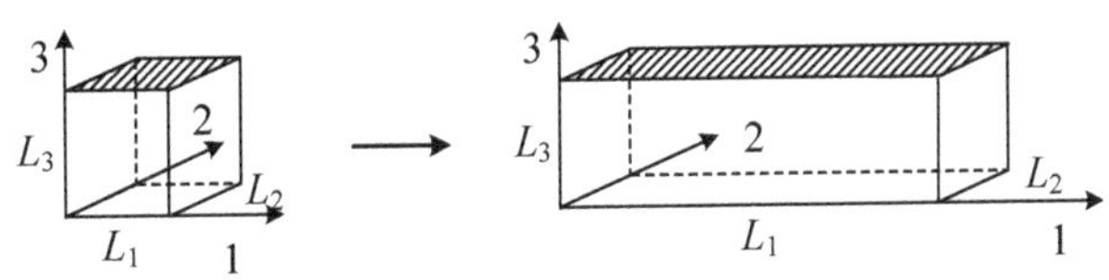

Figure 10.16. To the transition between the thin end-electroded bar ($L_1 / L_3 \to 0$) and strip polarized through the width ($L_1 / L_3 \to \infty$).

In this variant one of the partial systems is determined by the condition $S_1 = 0$ and constitutes the strip that is infinite in the x direction and vibrates along its width L_3. Similarly, another partial system at $S_3 = 0$ is the strip that is infinite in the z direction and vibrates along dimension L_1. After substituting displacement analogous to those by Eq. (10.66) into formula for the potential energy we will obtain the following expressions for the rigidities of the partial systems at these boundary conditions,

$$K_1^E = \frac{\pi^2 Y_1^E t}{2(1-\sigma_3^E \sigma_{13}^E)} \frac{L_3}{L_1}, \quad K_3^E = \frac{\pi^2 Y_3 t}{2(1-\sigma_3 \sigma_{13})} \frac{L_1}{L_3}, \quad K_{13} = \frac{4 Y_1^E \sigma_3^E t}{1-\sigma_3^E \sigma_{13}^E}. \tag{10.117}$$

As in the partial system at $S_1 = 0$ deformation changes along the direction of polarization, the additional rigidity $\Delta K_3 \approx 0.2 \cdot K_3^E k_{33}'^2 / (1-k_{33}'^2)$ must be taken into account according to relation (10.40), and the effective rigidity K_3 will be $K_3 = K_3^E + \Delta K_3$. For PZT-4 ceramics $K_3 \approx 1.12 K_3^E$. Thus, the partial resonance frequencies are

$$f_{3p} = \frac{1.06}{2L_3}\sqrt{\frac{Y_3^E}{\rho(1-\sigma_3^E \sigma_{13}^E)}}, \quad f_{1p} = \frac{1}{2L_1}\sqrt{\frac{Y_1^E}{\rho(1-\sigma_3^E \sigma_{13}^E)}}, \tag{10.118}$$

respectively. (Remember that it is denoted $Y_1^E = 1/s_{11}^E$, $Y_3^E = 1/s_{33}^E$, $\sigma_3^E = -s_{13}^E / s_{33}^E$, $\sigma_{13}^E = -s_{13}^E / s_{11}^E$). The ratio of the partial resonance frequencies is

$$\beta = \frac{f_{3p}}{f_{1p}} = 1.06 \frac{L_1}{L_3}\sqrt{\frac{Y_3^E}{Y_1^E}} = 1.06 \frac{w}{h}\sqrt{\frac{Y_3^E}{Y_1^E}}. \tag{10.119}$$

For the strip made of PZT-4 ceramics $\beta = 0.94 \cdot w/h$. The elastic coupling factor that is determined from expression for the potential energy of the coupled system (10.68) is

$$\gamma = 0.76 \sigma_3^E \sqrt{Y_1^E / Y_3^E}. \tag{10.120}$$

After the coefficients of aspect ratio and of coupling between the partial systems are determined, all the calculations of the resonance frequencies, mode shape coefficients, and effective coupling coefficients can be performed as it is done above. The resulting dependencies of these parameters are as follows.

Dependences of the resonance frequencies from the aspect ratio coefficient β in the case that L_3 is kept constant normalized to the partial frequency f_{3p} are presented in Figure 10.17. Note that at the extreme value of the aspect ratio, $\beta \to \infty$, the upper branch resonance frequency coincides with the partial resonance frequency f_{3p}, and at $\beta \to 0$ the lower branch resonance frequency corresponds to the resonance frequency of the thin longitudinally vibrating bar.

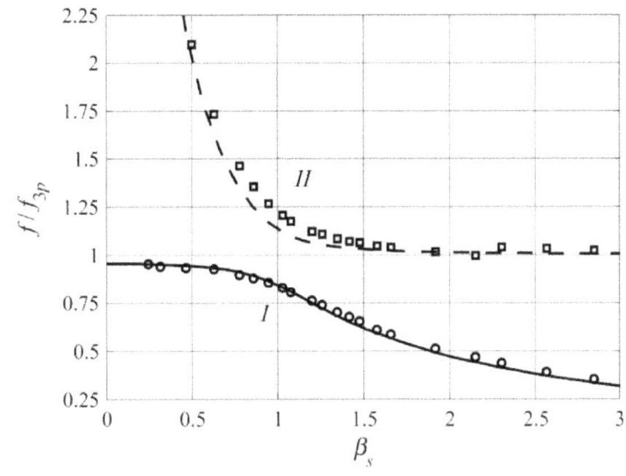

Figure 10.17: Dependence of the nondimensional resonance frequencies on the aspect ratio coefficient β for the width poled strip: (*I*) for the lower frequency branch and (*II*) for the upper branch. The lines - results of calculation, markers - experimental results from Ref. 3. Note that $\beta = 0.94 \cdot w/h$

To determine the electromechanical energy and electromechanical transformation coefficients from Eq. (10.80), expression for the charge density $D_3^E(S_1, S_3) = n_{\Delta 1}S_1 + n_{\Delta 3}S_3$ must be used, where $n_{\Delta 1}$ and $n_{\Delta 3}$ are (see the formulas (10.15) and (10.17))

$$n_{\Delta 1} = \frac{d_{31}s_{33}^E - d_{33}s_{13}^E}{s_{11}^E s_{33}^E - s_{13}^{E2}} = \frac{d_{31}Y_1^E[1 + (d_{33}/d_{31})\sigma_{13}^E]}{1 - \sigma_3^E \sigma_{13}^E}, \tag{10.121}$$

$$n_{\Delta 3} = \frac{d_{33}s_{11}^E - d_{31}s_{13}^E}{s_{11}^E s_{33}^E - s_{13}^{E2}} = \frac{d_{33}Y_3^E[1 + (d_{31}/d_{33})\sigma_3^E]}{1 - \sigma_3^E \sigma_{13}^E}. \tag{10.122}$$

Integration in (10.80) results in the following expressions for the electromechanical transformation coefficients

$$n_1 = 2n_{\Delta 1}t, \tag{10.123}$$

$$n_3 = 2n_{\Delta 3}t \cdot (L_1 / L_3) = 2n_{\Delta 3}t \cdot \beta. \tag{10.124}$$

Ratio of the coefficients is

$$\frac{n_1}{n_3} = \frac{n_{\Delta 1}}{n_{\Delta 3}} \cdot \beta. \tag{10.125}$$

The effective coupling coefficients that correspond to the lower (at $l = I$) and upper (at $l = II$) frequency branches may be found as result of manipulations analogous to those performed in the previous case in the form

$$k_l^2(\beta) = \frac{8}{\pi^2}\, k_{33}'^2 \cdot \frac{[(n_{\Delta 1}/n_{\Delta 3})\cdot ms_l + \beta]^2}{ms_l^2 + 2\gamma\beta\, ms_l + \beta^2}\,. \tag{10.126}$$

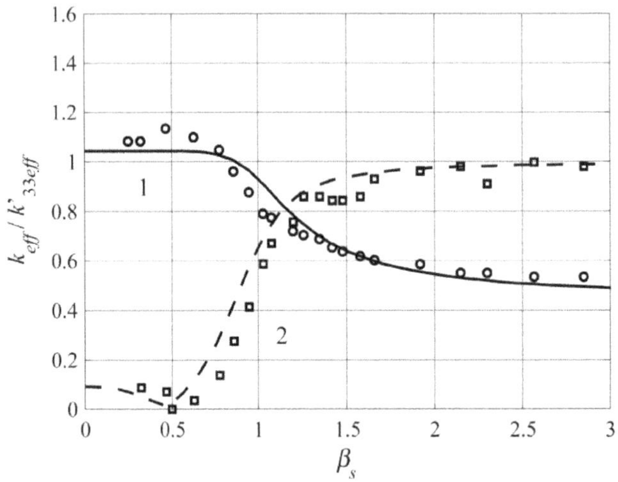

Figure 10.18: Dependences of the effective coupling coefficients on the aspect ratio coefficient β for the width poled strip: solid line (*I*) corresponds to the lower frequency branch, the dashed line (*II*) corresponds to the upper branch. The lines - results of calculation, the markers – experimental data from Ref. 3. Note that $\beta = 0.94 \cdot w/h$.

Dependencies of the effective coupling coefficients on the aspect ratio coefficient β are presented in Figure 10.18. The values of the coupling coefficients in the Figure are normalized to the effective coupling coefficient k'_{33eff} for the partial system that is the strip infinite in the *x* direction and vibrating in direction of polarization. The calculations were made for the strips made from PZT-4 ceramics with $k'_{33eff} = 0.58$, but in the normalized form the plots may be used for different ceramic compositions to a great accuracy. At the extreme case of $\beta \to \infty$ the effective coupling coefficient that corresponds to the upper frequency branch (branch II) coincides with those for the partial system, at $\beta \to 0$ the effective coupling coefficient that corresponds to lower frequency branch (branch I) is

$$k_{33eff} = \sqrt{8/\pi^2}\, k_{33}\,, \tag{10.127}$$

as for a thin bar at longitudinal piezoeffect.

It can be concluded based on the results presented in the figures of this section that the bars can be considered in one-dimensional approximation at values of aspect ratio about $\beta < 0.5$, i.e., at $L_2 / L_1 > 2.0$ under the transverse and $L_3 / L_1 > 2.1\sqrt{s_{11}^E / s_{33}^E}$ under the longitudinal piezoeffect.

10.3.3 Transition from the Thin Infinitely Long Strip to the Infinite Plate Vibrating over the Thickness

Geometry and the electrodes location of the problem are shown in Figure 10.19 for variant of the longitudinal piezoeffect (vibration over width of the strip and thickness of the plate). For the transverse piezoeffect the dimensions L_1 and L_3 must be reversed and location of electrodes must be changed correspondingly. The condition $S_2 = 0$ is held in course of the transition in both cases. Distribution of displacement in the piezoelements is assumed in the form of Eq. (10.66) with subscript 2 replaced by 3.

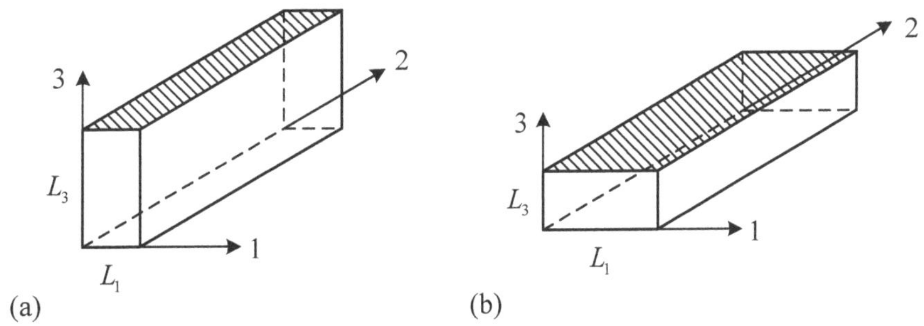

Figure 10.19. Transition from the one-dimensional thin infinitely long strip (extreme case at $L_1 / L_3 \to 0$) to the one-dimensional over the thickness infinite plate (extreme case at $L_1 / L_3 \to \infty$). Longitudinal piezoeffect.

The partial systems in this case are the infinite plates x, y ($S_1 = 0$, $S_2 = 0$) and z, y ($S_3 = 0$, $S_2 = 0$). Coupled vibrations in the isotropic mechanical system under these conditions are considered in Section 4.6.2.2, and expressions for the parameters that characterize the coupled vibrations are summarized by formulas (4.598) - (4.601). To present these parameters for the system made of piezoceramics, appropriate are piezoelectric equations (5.8), which for our case have the form

$$T_1 = c_{11}^E S_1 + c_{13}^E S_3 - e_{31} E_3 , \qquad (10.128)$$

$$T_2 = c_{12}^E S_1 + c_{13}^E S_3 - e_{31} E_3 , \qquad (10.129)$$

$$T_3 = c_{13}^E S_1 + c_{33}^E S_3 - e_{33} E_3 , \qquad (10.130)$$

$$D_3 = e_{31} S_1 + e_{33} S_3 + \varepsilon_{33}^S E_3 . \qquad (10.131)$$

Results of application of the equations are different for variants of the longitudinal and transverse piezoeffect.

10.3.3.1 Longitudinal Piezoeffect

For the partial system at $S_3 = 0$, $S_2 = 0$:

$$K_{\Delta 1}^E = c_{11}^E ,\ n_{\Delta 1} = e_{31} ,\ C_{e\Delta}^{S_1} = \varepsilon_{33}^{S_1} \text{ (or according to (10.4) } C_{e\Delta}^{S_1} = \varepsilon_{33}^T (1 - k_{31}''^2) \text{)}. \qquad (10.132)$$

For the partial system at $S_1 = 0$, $S_2 = 0$:

$$K_{\Delta 3}^E = c_{33}^E ,\ n_{\Delta 3} = e_{33} ,\ C_{e\Delta}^{S_3} = \varepsilon_{33}^{S_3} \text{ (or according to (10.4) } C_{e\Delta}^{S_3} = \varepsilon_{33}^T (1 - k_t^2) \text{)}. \qquad (10.133)$$

After substituting $T_1^E S_1 + T_3^E S_3$ into expression (10.68) for the potential energy we will obtain

$$K_1^E = \frac{\pi^2}{2} c_{11}^E \frac{L_2 L_3}{L_1} ,\ K_3^E = \frac{\pi^2}{2} c_{33}^E \frac{L_2 L_1}{L_3} ,\ K_{13}^E = 4 c_{13}^E L_2 . \qquad (10.134)$$

Strictly speaking, the additional rigidity $\Delta K_3 \approx 0.2 \cdot K_3^E k_t^2 / (1 - k_t^2)$ must be taken into account due to changing deformation along the direction of electric field. With PZT-4 ceramics used this would result in increase of rigidity up to $K_3 \approx 1.07 K_3^E$. In terms of numerical estimating effects of the coupled vibrations such correction is not significant and will be neglected.

Thus, expressions for the partial resonance frequencies, aspect ratio factor and for the coupling factor are:

$$f_{1p} = \frac{1}{2L_1}\sqrt{\frac{c_{11}^E}{\rho}} ,\ f_{3p} = \frac{1.03}{2L_3}\sqrt{\frac{c_{33}^E}{\rho}} \approx \frac{1}{2L_3}\sqrt{\frac{c_{33}^E}{\rho}} ,\ \Omega = \frac{f^2}{f_{3p}^2} ; \qquad (10.135)$$

$$\beta_l = \frac{L_1}{L_3}\sqrt{\frac{c_{33}^E}{c_{11}^E}} , \qquad (10.136)$$

$$\gamma = \frac{8}{\pi^2}\frac{c_{13}^E}{\sqrt{c_{33}^E c_{11}^E}} . \qquad (10.137)$$

Note that the real aspect ratio is $(L_1 / L_3) = (w / h)$. Thus, in this case the correlation between factor β_l (the subscript l stays for "longitudinal") and aspect ratio for piezoelements made of PZT-4 ceramics is $\beta_l = 0.9 \cdot (w / h)$.

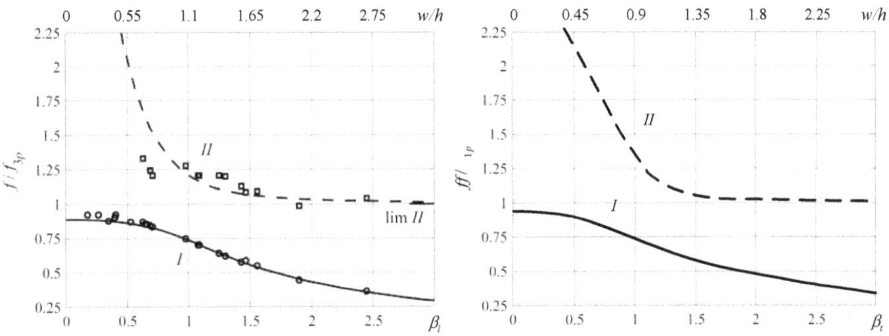

(a) (b)

Figure 10.20: Dependences of the nondimensional resonance frequencies on the aspect ratio factors β_l and β_t in course of transition from infinite strip to the thickness vibrating plate: (a) at the longitudinal piezoeffect, markers are the experimental data; (b) at the transverse piezoeffect. The lower frequency branches (I) – solid lines, the upper branches (II) - dash lines. The piezoelements are made of PZT-4 ceramics, $\beta_l = 0.9 \cdot (w / h)$ and $\beta_t = 1.1 \cdot (w / h)$.

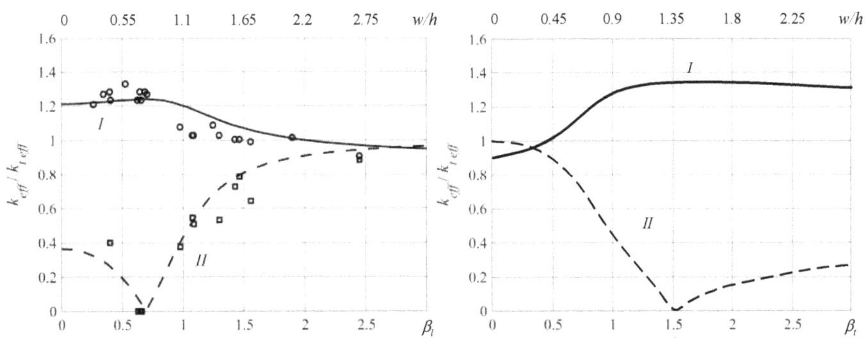

(a) (b)

Figure 10.21: Dependences of the effective coupling coefficients for the long strip on the aspect ratio coefficients β_l and β_t that correspond to lower (*I*) and upper (*II*) frequency branches: (a) at the longitudinal piezoeffect, markers are the experimental data; (b) at the transverse piezoeffect. The values are normalized to the effective coupling coefficient for the partial systems that are the plates infinite in the *x* and *y* directions polarized accordingly. Under the longitudinal polarization it is $k_{t\,eff}$. Under the transverse effect it is $k_{t\,eff}$. The piezoelements are made of PZT-4 ceramics, $\beta_l = 0.9 \cdot (w / h)$ and $\beta_t = 1.1 \cdot (w / h)$.

The modal resonance frequencies for the piezoelement, mode shape coefficients and effective coupling coefficients vs. aspect ratio factor β_l now can be calculated following the above-described procedures. The results of calculating the resonance frequencies for the piezo-elements made from PZT-4 ceramics are presented in Figure 10.20 (a).

Dependences of the modal effective coupling coefficients for the long strips made from PZT-4 ceramics from the aspect ratio coefficient β_l are depicted in Figure 10.21 (a). The values of the coupling coefficients in Figure 10.21 (a) are normalized to the effective coupling coefficient for the partial system that is the plate infinite in the *x* and *y* directions, which is $k_{t\,eff}$.

The results presented in Figure 10.20 (a) and Figure 10.21 (a) are restricted to the fundamental modes of vibration of the partial systems, and as such are applicable for frequency range near and below the resonance frequencies of the lowest modes of their vibration. The analysis is not intended for a much broader frequency range that could be of interest for a general treatment of the vibration of piezoelectric plates and bars per se.

The overtones of the partial systems are neglected because the corresponding modes of vibration are typically not suitable for practical or effective electromechanical transduction. While the response of vibrations at higher modes that correspond to the overtone frequencies of the partial systems were not considered here, their effects (although very weak) were observed during experimentation. It is understood that the closest and most influential overtones are due to the third and fifth harmonics of the partial system along the x axis, which are coupled with the lowest mode of vibration in z direction. The strongest coupling between these modes of vibration must be expected at the aspect ratios that correspond to $\beta \approx 1.5$ and $\beta \approx 2.5$, at which the partial resonance frequencies of the third and fifth harmonics of vibration in x direction are equal to the partial resonance frequency of the first mode of vibration in the z direction, respectively. Some peculiarities in the trend of coupling coefficient for the first mode can be seen in Figure 10.18 and Figure 10.21 (a) in the range of aspect ratios around these values.

The coupling coefficients are normalized to the coefficient $k_{t\,eff}$ because its value is specified. For PZT-t ceramics $k_{t\,eff} = 0.46$.

10.3.3.2 Transverse Piezoeffect

The case of the transverse piezoeffect can be considered in analogous way. As it follows from Figure 10.19 (b), the partial systems remain the same, as for the case of the longitudinal piezoeffect, if to reverse the dimensions L_1 and L_3. Therefore, the expressions for the rigidities, partial resonance frequencies, aspect ratio factor and for the coupling factor also remain the same, if to revers numbers 1 and 3 in their subscripts. Thus, the nondimensional frequency factor is now $\Omega = f^2 / f_{1p}^2$. Noteworthy is that though the correlation between expressions for the aspect ratio factors is $\beta_t = 1/\beta_l$ (the subscripts stay for the "transverse" and "longitudinal"), their expressions through the real aspect ratios are

$$\beta_l = \frac{w}{h} \cdot \sqrt{\frac{c_{33}^E}{c_{11}^E}} \text{ and } \beta_t = \frac{w}{h} \cdot \sqrt{\frac{c_{11}^E}{c_{33}^E}} . \tag{10.138}$$

In case that PZT-4 ceramics is used $\beta_l = 0.9 \cdot (w/h)$ and $\beta_t = 1.1 \cdot (w/h)$.

The results of calculations in the variant of transverse piezoeffect are presented in Figure 10.20 (b) for the resonance frequencies, and in Figure 10. 21 (b) for the effective coupling coefficients. For comparison they are placed next to analogous figures for the variant of the longitudinal piezoeffect.

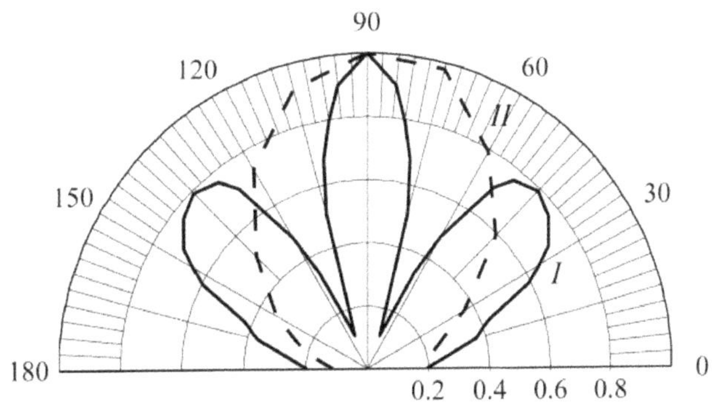

Figure 10.22: Directional factors of the piezoelements having different aspect ratios: $\beta = 1.3$ (solid line), $\beta = 0.65$ (dashed line).

Based on the results presented, it can be concluded that the best electromechanical characteristics can be obtained with piezoelements utilizing longitudinal piezoeffect (radiating from the electroded surface) at aspect ratios $\beta_l < 0.7$ ($w/h < 0.8$), approximately. At greater aspect ratios a contribution of the coupled vibrations in the transverse direction increases. This may result in substantial growth of the side lobes. This effect is illustrated by Figure 10.22, where the experimental directional factors from Ref. 3 are presented, that were obtained at the same operating frequency with piezoelements having different aspect ratios. In the case that the width of radiating surface of a projector needs to be increased for obtaining a required directivity pattern, the projector must be composed of mechanically isolated piezoelements having smaller aspect ratios. Approach to an analytical estimating of this effect of coupled vibrations on the modes of vibration of the radiating surfaces and thus on the directional factors is considered in Section 10.3.5.

In the variant of transducer employing the transverse piezoeffect in opposite a smaller contribution of motion in the transverse direction can be expected at larger aspect ratios (at $\beta \approx 1.3 \div 1.5$). For more accurate estimations see Section 10.3.5.

10.3.4 Transition from the One-Dimensional Vibrations of a Thin Circular Disk to the One-Dimensional Vibration of a Long Bar

Geometry of the problem is shown in Figure 10.23. Coupled vibrations of isotropic elastic finite size cylinders in this range of aspect ratios were considered in Section 4.6.2.3.

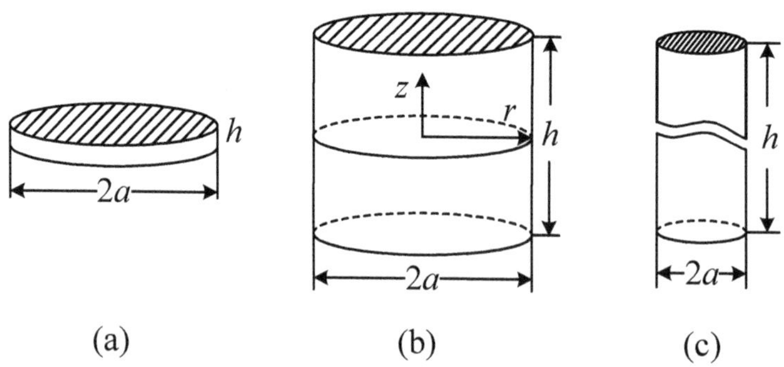

Figure 10.23: To the transition from the thickness vibrating thin disk at $h/2a \ll 1$ (a) to the one-dimensional axial vibrating bar at $h/2a \gg 1$ (c).

The assumed distribution of displacements in the finite size cylinders was therein accepted in the form

$$\xi(r,z) = \xi_a J_1(k_1 r)/J_1(k_1 a) + \xi_h \sin(\pi z/h), \tag{10.139}$$

where ξ_a and ξ_h are the generalized partial coordinates- maximum displacements in the radial and axial directions. The partial systems are: partial system a (at $h \to \infty$, $\xi_h = 0$) - radial vibrating infinite cylinder of radius a, and partial system h (at $a \to \infty$, $\xi_a = 0$) – axial vibrating disk of infinite radius, or thickness vibrating infinite plate. The coupled vibration technique can be applied to a system that can be considered as two-dimensional. The range of aspect ratios $h/2a$, in which piezoelements in the shape of finite size cylinders (as the piezoelements of the geometries under consideration will be called regardless of the factual value of their aspect ratio) can be treated as two-dimensional was analyzed in Ref. 4.

Expressions for strain in the body of cylinder are

$$S_1 = \frac{\partial \xi(r,z)}{\partial r} = \xi_a \frac{1}{J_1(k_1 a)} \frac{\partial J_1(k_1 r)}{\partial r} = \xi_a \frac{k_1}{J_1(k_1 a)} \left[J_0(k_1 r) - \frac{J_1(k_1 r)}{k_1 r} \right], \tag{10.140}$$

$$S_2 = \frac{\xi(r,z)}{r} = \xi_a \frac{k_1}{J_1(k_1 a)} \frac{J_1(k_1 r)}{k_1 r}, \tag{10.141}$$

$$S_3 = \frac{\partial \xi(r.z)}{\partial z} = \xi_h \frac{\pi}{h} \cos\frac{\pi z}{h}. \tag{10.142}$$

For determining eigenvalues of the radial modes for the partial system a (a tall cylinder at $S_3 \approx 0$) Eq. (4.166) is valid with value of the Poisson's ratio $\sigma_{rd}^E = c_{12}^E / c_{11}^E$ (the subscript "*rd*" stands for "rod" in distinction from $\sigma_1^E = -s_{12}^E / s_{11}^E$ for a thin disc). For the piezoceramic compositions PZT-4, PZT-5, and PZT-8 the values of σ_{rd}^E are 0.56, 0.62 and 0.54, respectively. The eigenvalues obtained from Eq. (4.166) at these values of σ_{rd}^E for these compositions are $k_i a = \lambda_i = 2.2; 5.4; 8,6; \ldots$ with deviation within less than 1%. Thus, $k_1 a = 2.2$ may be taken for calculating strains S_1 and S_2.

It is suitable for this case to use piezoelectric equations of state in the form

$$T_1 = c_{11}^E S_1 + c_{12}^E S_2 + c_{13}^E S_3 - e_{31} E_3, \tag{10.143}$$

$$T_2 = c_{12}^E S_1 + c_{11}^E S_2 + c_{13}^E S_3 - e_{31} E_3, \tag{10.144}$$

$$T_3 = c_{13}^E S_1 + c_{13}^E S_2 + c_{33}^E S_3 - e_{33} E_3, \tag{10.145}$$

$$D_3 = e_{13}(S_1 + S_2) + e_{33} S_3 + \varepsilon_{33}^S E_3, \tag{10.146}$$

Equivalent parameters of the cylindrical piezoelements will be obtained from expressions for the related energies. Omitting the straightforward calculating procedures, the following results will be obtained.

From the expression for potential energy,

$$W_{pot}^{E}=\frac{1}{2}2\pi\int_{-h/2}^{h/2}\int_{0}^{a}(S_1T_1^E+S_2T_2^E+S_3T_3^E)rdrdz=\frac{1}{2}(K_a^E\xi_a^2+2K_{ah}\xi_a\xi_h+K_h^E\xi_h^2)\,, \quad (10.147)$$

follows that the equivalent rigidities are

$$K_a^E=13.4\,hc_{11}^E,\quad K_{ah}=4.6\pi ac_{13}^E,\quad K_h^E=\frac{\pi^3a^2}{2h}c_{33}^E\,. \quad (10.148)$$

We will assume that cylindrical piezoelements under consideration are solid, which is usually the case for the most interesting range of aspect ratios around the point of strongest coupling ($h/2a$ is about unity). The value of the equivalent rigidity related to the axial vibration under this condition must be changed to

$$K_h'=K_h^E\left[1+\left(1-\frac{8}{\pi^2}\right)\frac{k_{33}^2}{(1-k_{33}^2)}\right], \quad (10.149)$$

as the deformation in this case changes along the electric field. For the piezoelements made of PZT-4 or PZT-8 ceramics $K_h'\approx 1.2K_h$.

The equivalent masses obtained from expression for the kinetic energy,

$$W_{kin}=\frac{1}{2}2\pi\int_{-h/2}^{h/2}\int_{0}^{a}\rho\dot{\xi}(r,z)^2rdrdz=\frac{1}{2}(M_a\dot{\xi}_a^2+2M_{ah}\dot{\xi}_a\dot{\xi}_h+M_h\dot{\xi}_h^2)\,, \quad (10.150)$$

are

$$M_a=0.86M,\quad M_h=0.5M,\quad M_{ah}=0\,. \quad (10.151)$$

The electromechanical transformation coefficients determined from the expression for electromechanical energy,

$$W_{em}=\frac{1}{2}\int_{\tilde{V}}D_3^E(S_1,S_2,S_3)E_3\,d\tilde{V}=\frac{1}{2}V(n_a\xi_a+n_h\xi_h)\,, \quad (10.152)$$

are

$$n_a=2\pi ae_{31},\quad n_h=(2\pi a^2/h)e_{33}\,. \quad (10.153)$$

With addition of expression for the capacitance of the clamped piezoelement,

$$C_{el}^S=(\pi a^2/h)\varepsilon_{33}^S\,, \quad (10.154)$$

the set of the equivalent parameters that are needed for calculating the resonance frequencies, effective coupling coefficients and operating characteristics of a cylindrical finite-size piezoelement is completed. Following general expressions for the partial resonance frequencies are,

$$f_a = \frac{1}{2\pi}\sqrt{\frac{K_a^E}{M_a}} = \frac{1.1}{\pi a}\sqrt{\frac{c_{11}^E}{\rho}}, \quad f_h = \frac{1}{2\pi}\sqrt{\frac{K_h'}{M_h}} = \frac{1.1}{2h}\sqrt{\frac{c_{33}^E}{\rho}}. \tag{10.155}$$

(Note that f_h coincides with the resonance frequency of a tall cylinder for the extreme case of the aspect ratio $(h/2a) \gg 1$.) The aspect ratio factor is

$$\beta = \frac{f_a}{f_h} \approx 1.4\frac{h}{2a}. \tag{10.156}$$

(for PZT-4 and PZT-8 ceramics the numerical factor differs by 2%). The coefficient of coupling between the partial systems is

$$\gamma = (K_{ah} / \sqrt{K_a^E K_h^E}) \approx \frac{c_{13}^E}{\sqrt{c_{11}^E c_{33}^E}}. \tag{10.157}$$

For piezoelements made of ceramics PZT-4 and PZT-8 it can be taken that $\gamma = 0.6$.

After all the parameters in expressions for energies associated with vibration of the cylindrical piezoelement are determined (the energies of the electrical and mechanical losses are not included for simplicity, because they are not essential at this phase of the treatment), the Lagrange's equations that govern operation of the piezoelement as electromechanical transducer will be obtained in the form of Eqs. (10.83)-(10.85), in which the subscripts 1 and 2 must be replaced by a and h. From the frequency equation (10.89) of free vibration of the piezoelement, where $\Omega = f^2 / f_a^2$, two branches of nondimensional resonance frequency factor, $\Omega_I(\beta)$ and $\Omega_{II}(\beta)$ follow. Plots of these frequency branches are presented in Figure 10.24.

The plots are calculated under the assumption that $2a$ remains constant and h is changing. The plots are the same for piezoelements made from PZT-4 and PZT-8 ceramics, as it is accepted that $\gamma = 0.6$ in both cases. With resonance frequencies known, the corresponding mode shape factors, which we define according to formulas (10.91) and (10.92) as

$$ms_I = (\xi_a / \xi_h)\big|_{at\ \Omega=\Omega_I} = -\frac{K_{ah}}{K_a^E}\frac{1}{1-\Omega_I}, \tag{10.158}$$

may be found. Dependences of the mode shape coefficients from the aspect ratio $h/2a$ are depicted in Figure 10.25.

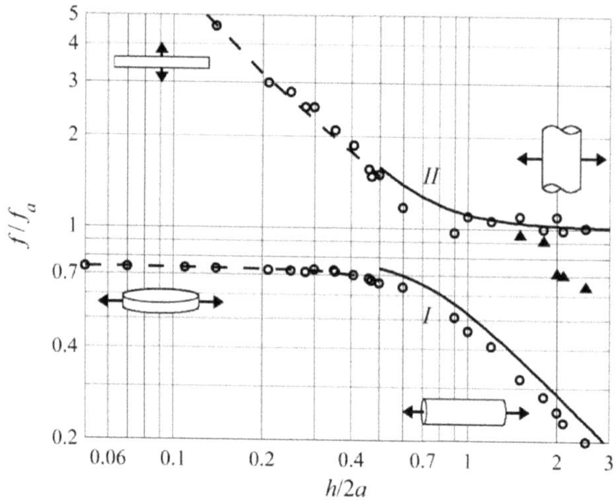

Figure 10.24: The resonance frequency branches for cylinders made from PZT-4 and PZT-8 ceramics normalized to frequencies f_a. The circles mark the experimental data[24]. The solid lines show theoretical results, the dashed lines show the trends revealed by experimenting. The triangles mark the experimental data related to the third mode of the longitudinal vibration. The icons illustrate directions of one-dimensional vibration for the extreme values of the aspect ratios

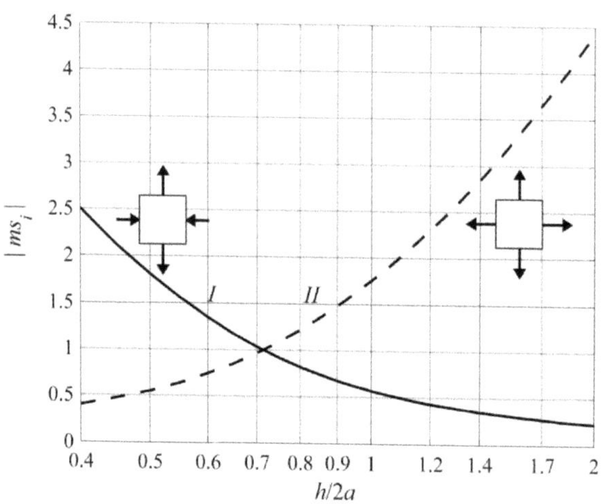

Figure 10.25: Dependences of modulus of the mode shape coefficients from the aspect ratio. At branch *I* $ms_I < 0$. At branch *II* $ms_{II} > 0$. The icons illustrate relative directions of the radial and axial vibration related to these branches.

As it follows from Eq. (10.158), the mode shape coefficients are negative for the lower frequency branch (branch *I*) and positive for the higher branch (branch *II*). This means that vibrations of the cylinder surfaces related to branch *I* occur in anti-phase, and in-phase related to branch *II*. This is shown with icons in Figure 10.25

The effective electromechanical coupling coefficients that correspond to the mode shapes of the coupled vibration can be determined from basic relation, which for this case will be modified in the way analogous to those employed in Section 10.3.3.1 to the expression

$$\frac{k_l^2(\beta)}{1-k_l^2(\beta)}=\frac{n_h^2}{K_h' C_{el}^S}\frac{[(n_a/n_h)ms_l+1]^2}{(K_a^E/K_h')ms_l^2+2(K_{ah}/K_h')ms_l+1}. \tag{10.159}$$

Given that capacitance of the clamped piezoelement can be represented as

$$C_{el}^S=C_{el}^T[1-k_l^2(\beta)], \tag{10.160}$$

$$k_l^2(\beta)=\frac{n_h^2}{K_h' C_{el}^T}\frac{[(n_a/n_h)ms_l+1]^2}{(K_a^E/K_h')ms_l^2+2(K_{ah}/K_h')ms_l+1}. \tag{10.161}$$

Note that in case the piezoelement is built of segments in number ≥ 6 connected in parallel, K_h' should be replaced by K_h^E here and in all the preceding formulas.

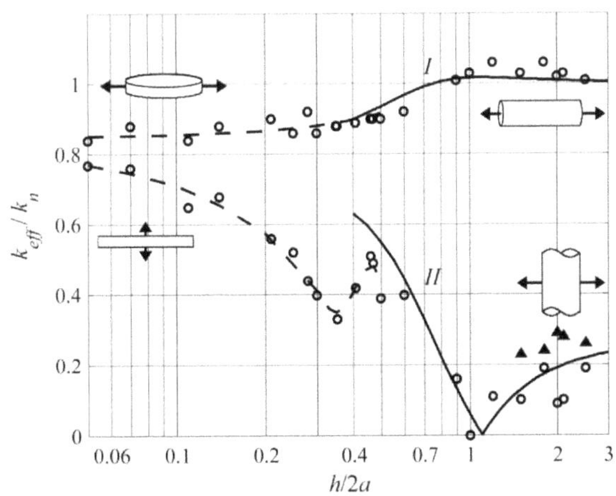

Figure 10.26: Dependences of the effective coupling coefficients that correspond to the frequency branches *I* and *II* for the piezoelements made of PZT-4 and PZT-8 that are normalized to coupling coefficients of a tall bar. The circles mark experimental data. The solid lines show theoretical results, the dashed lines show the trends revealed by experimenting. Triangles show experimental results related to the third mode of axial vibrations. The icons illustrate directions of one-dimensional vibration for the extreme values of the aspect ratio.

All the parameters in relation (10.161) are already determined as functions of aspect ratio. Results of calculating the effective coupling coefficients for the range of aspect ratios $h/2a >$ 0.5 that correspond to the frequency branches at $l = I, II$ and results of the experimental investigation made for a broader range of aspect ratios are presented in Figure 10.26.

The data are normalized to the value of effective coupling coefficient $k_{eff} = 0.59$ for a tall bar made of PZT- 4 These results may be sufficiently accurate for piezoelements made of PZT-8 ceramics considering that they are normalized to $k_{eff} = 0.57$. Peculiarities in the trend of the effective coupling coefficients for the axial mode of vibration in the range of aspect ratios from approximately 0.1 up to 0.5 that are revealed by the experimenting confirm the conclusion, which is generally recognized by the transducer industry, that this range of aspect ratios should be avoided for designing of axially vibrating transducers due to harmful effects of the coupled vibrations. Such a conclusion was made previously based on highly nonuniform distributions of displacement on the surface of a transducer (see Ref. 5) that take place by the same reason of coupling between the axial and radial modes.

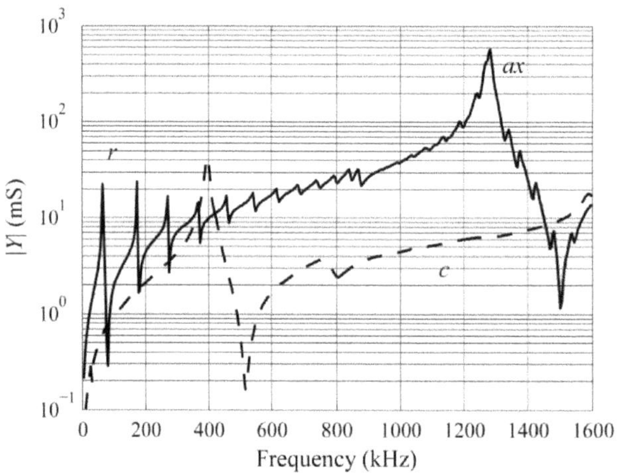

Figure 10.27: Plots of moduli of input admittances of the disc made of PZT-4 ceramics at aspect ratio $h/2a = 0.05$ (solid line), and for the composite disc (dashed line) made of 3-1 composite material having aspect ratio $h/2a = 0.13$

The coupled effects can be eliminated by slicing the circular discs into rods and thus achieving uniform distribution of displacements and getting high effective coupling coefficient.

The greatest positive effect in terms of effective coupling coefficient may be obtained, if the height to average diameter ratio of the rods is about unity or more, as the theoretical and experimental results presented in Figure 10.26 show. The same effect of decoupling the vibrations is achieved in the composite piezoelement designs. Plot of the modulus of admittance for such a circular disc having aspect ratio $h/2a = 0.13$ is depicted in Figure 10.27, where the first radial resonance of the solid disk is marked as "*r*" and the axial – as "*ax*". The plot shows that vibrations of the composite disc are one-dimensional in all the range of aspect ratios with effective coupling coefficient having maximum possible value for a particular piezoceramic composition.

10.3.5 Equivalent Electromechanical Circuit accounting for Effects of Coupled Vibrations

Content of this Section will be illustrated with example of the finite-size cylindrical piezoelement. But it will be seen that the developed approach to the problem can be generalized for other two-dimensional coupled systems in a straightforward way.

An important characterization of the electromechanical transducer that employs finite-size cylindrical piezoelement is distribution of displacement in axial direction over its end surfaces. The extent, to which vibration in the radial direction may contribute to nonuniformity of this distribution, can be estimated qualitatively by the values of ratio ξ_a/ξ_h of magnitudes of displacements in the radial and axial directions, i.e., by the mode shape coefficients. Their dependences on the aspect ratio that are presented in Figure 10.25 show that at $\beta < 1$ $((h/2a) < 0.8)$ the smaller the aspect ratio, the larger is ratio ξ_a/ξ_h, and more pronounced nonuniformity of the axial vibration can be expected. At $\beta > 1$ in opposite ξ_h gradually becomes much larger than ξ_a, and the axial vibration of the cylinder surface becomes more and more uniform. Experimental confirmation of these considerations can be found in Ref. 5.

Note, that as it follows from Eq. (10.145) for a thin disc (at $T_3 = 0$),

$$S_3(r) = -\frac{c_{13}^E(S_1 + S_2)}{c_{33}^E} \sim \xi_a J_0(k_1 r). \tag{10.162}$$

Given that for a finite-size cylinder $S_3(z) \sim \xi_h$, the distribution of axial displacement due to the radial deformation will be denoted $\xi_{ha}(r)$. It can be assumed that $\xi_{ha}(r) \sim J_0(k_1 r)\xi_a$. Likewise, it follows from Eqs. (10.143) and (10.144) for the tall rod (at $T_1 = T_2 = 0$) that

$$S_1 + S_2 = -\frac{2c_{13}^E}{c_{11}^E + c_{12}^E} S_3 \sim \xi_h \cos\frac{\pi z}{h}. \tag{10.163}$$

Given that $(S_1 + S_2) \sim \xi_a$, the distribution of radial displacement due to the axial deformation, which we denote $\xi_{ah}(z)$, may be assumed to be $\xi_{ah}(z) \sim \xi_h \cos(\pi z / h)$.

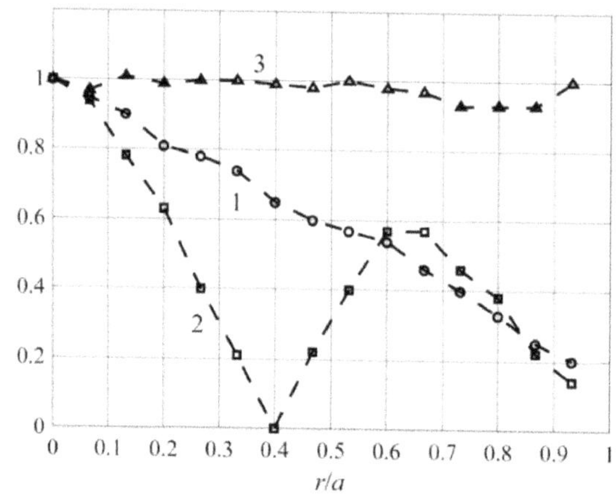

Figure 10.28: Plots of the mode shapes of vibration in the axial direction measured on surface of cylindrical piezoelement: (1) - low frequency branch, $h/2a = 0.5$; (2) – high frequency branch, $h/2a = 0.5$; (3) – low frequency branch, $h/2a = 1.5$.

Using these considerations for qualitative prediction of the mode shapes of vibration is illustrated with results of measuring distributions of axial displacements on the surfaces of disks with aspect ratios 0.5 and 1.5 presented in Figure 10.28. The results obtained are in a qualitative agreement with predictions that follow from data on the mode shape coefficients presented in Figure 10.25.

At $h/2a = 0.5$ at low frequency branch the radial mode dominates in accordance with results shown in the Figure for the mode shape coefficients, and the mode shape of vibration in axial direction reflects predominant contribution of this mode. At high frequency branch (at frequency that corresponds to axial resonance) the contribution of radial mode remains significant, and superposition of vibrations due to axial and radial modes results in changing the phase of vibration in axial direction.

Thus, radiation in the axial direction can be more effective at radial resonance mode than at the axial mode of vibration, as could be expected, if not to consider effects of coupled vibration. Moreover, that effective coupling coefficient for the radial mode is also much greater (see k_{eff} for branch *I* vs for branch *II* at $h/2a = 0.5$ Figure 10.15). At aspect ratio $h/2a = 1.5$ in the low frequency branch the axial mode of vibration dominates according to Figure 10.25. Contribution of the radial mode is negligible, and vibration in the axial direction is uniform within accuracy of the measurements.

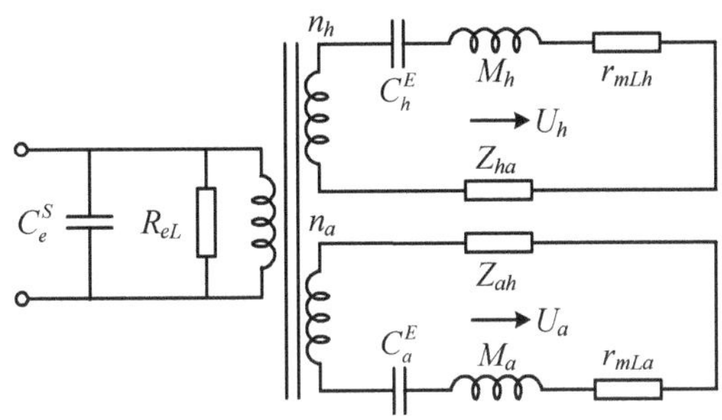

Figure 10.29: Equivalent electromechanical circuit of the finite size cylindrical transducer with effects of coupled vibration taken into consideration.

The relation between displacements in the partial systems (modal displacements ξ_a and ξ_h) in operational mode of a transducer, i.e., at forced vibration under action of applied voltage, differ from those determined for free vibration. Determining this relation and overall operational characteristics of finite cylinder based electromechanical transducers require considering their forced vibration with reaction of acoustic field included. The operational electromechanical properties of a transducer that employs the finite-size cylindrical piezoelement can be calculated using the set of Eqs. (10.83)-(10.85) (with subscripts that correspond to each particular case), or from the electromechanical circuit presented in Figure 10.29, which is equivalent to this set of equations.

We assume that the side surface of the cylinder is free of a load, and the mechanical (acoustic) load, Z_m, that is applied to one end surface (it is included in the equivalent circuit by observation as well as the resistances of the electrical and mechanical losses, R_{eL} and r_{mL}) has magnitude small enough as not to change distribution of displacement in the axial direction, which is usually the case for electroacoustic transducers. The impedances $Z_{ha} = z_{ha}(U_a / U_h)$ and $Z_{ah} = z_{ah}(U_h / U_a)$, where $z_{ah} = z_{ha} = K_{ah} / j\omega$, are the introduced impedances that characterize coupling between contours that correspond to the partial vibrating systems; z_{ah} is the mutual impedance between the partial systems. Velocities U_h and U_a result from the combined effect of the "self" velocities of the partial systems (i.e., without effect of coupling) and their mutual contributions due to coupling between the systems. Thus, U_h can be represented, as

$$U_h = \dot{\xi}_h + \dot{\xi}_{ah}, \tag{10.164}$$

where $\dot{\xi}_h$ is the self-partial axial velocity and $\dot{\xi}_{ha}$ is the additional axial velocity due to coupling with the radial vibrating partial system.

Likewise,

$$U_a = \dot{\xi}_a + \dot{\xi}_{ah}, \tag{10.165}$$

where $\dot{\xi}_a$ is the self-partial radial velocity and $\dot{\xi}_{ah}$ is the additional radial velocity due to coupling with the axial vibrating partial system. The velocities U_h and U_a must be calculated from the equivalent circuit in Figure 10.29. The self-partial velocities $\dot{\xi}_a$ and $\dot{\xi}_h$ must be determined from the equivalent circuit considering the contours as independent. After this is done, the velocities $\dot{\xi}_{ha}$ and $\dot{\xi}_{ah}$ can be found. We will assume that distributions of velocity on the surfaces of the finite-size cylinder may be presented following the notes regarding relations (10.162) and (10.163) as

$$U_h(r) = \dot{\xi}_h + \dot{\xi}_{ha} J_0(k_1 r), \tag{10.166}$$

$$U_a(z) = \dot{\xi}_a + \dot{\xi}_{ah} \cos(\pi z / h). \tag{10.167}$$

If to accept these assumptions, then calculating of all the parameters of the transducer as an electroacoustic device including the directional factor becomes straightforward in principle.

Employing the equivalent circuit based calculating technique is appropriate for the range of aspect ratios $0.5 < h/2a < 1.5$. It can be concluded following the results presented in Figure 10.24 and Figure 10.26 that the axial vibrations of the cylinder can be considered as one-dimensional at aspect ratios larger than 1.5, because values of the resonance frequencies and effective coupling coefficients practically reach their values for the extreme case at these aspect ratios. This conclusion is supported by results of measuring the mode shape of vibration in the axial direction of a cylinder at $h/2a = 1.5$ that proved to be uniform, as shown in Figure 10.28.

10.4 Transducers Nonuniform over the Length

Inherent disadvantage of the uniform in direction of vibration length expander transducers for underwater applications is that they are greatly under loaded by radiation impedances, as it was noted in Ch. 3. As the result, the acoustic power radiated is limited by the dynamic mechanical strength of the transducers, while they have significant reserves of the electrical strength. Other operational characteristics of the transducers such as efficiency and ability to work in a broad frequency range are also far from optimal. Improving the operational characteristics of the transducers can be achieved by employing matching elements that may produce a required step up transformation of the acoustic load. For low frequency transducers (approximately in the range from 2 to 30 kHz) this function can be fulfilled by the passive bars (length expander vibrating systems in general) vibrating one-dimensionally (in the piston like mode). Being cemented to the ends of the length expander piezoelement they perform transformation of loads and thus allow optimizing the transducer operational properties by proper choice of their parameters. Resulting structure presents the length expander transducer nonuniform over the length ("Tonpilz" transducer) that is shown schematically in Figure 10.30 (a).

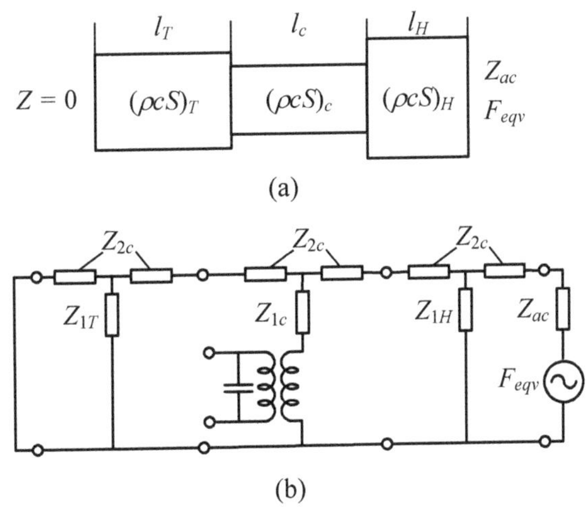

Figure 10.30: (a) Nonuniform length expander transducer composed of piezoelement with parameters marked by subscript "*c*" and two passive bars, with parameters marked by subscripts "*H*" (for Head) and "*T*" (for Tail), respectively; (b) cascade equivalent circuit of the transducer.

Calculating operating characteristics of the transducer can be performed using cascade connection of the T-network equivalent circuits of the length expander active and passive vibrating systems (see Section 10.2.3.1) that is presented in Figure 10.30 (b) for the basic design of the nonuniform transducer. Such transducer designs and calculating principle are considered to great detail in vast literature, for example, in Refs. 6-8 and in the works referenced therein. Using the cascade equivalent circuit allows calculating all the parameters of the transducer if dimensions and material properties of its parts are known. This problem can be called the direct problem of transducer design. The cascade circuit representation is very well suited for solving direct problems by application of matrix analysis. An outline of this approach to calculating transducer parameters can be found in Ref. 8. In the process of application of the matrix analysis it must be remembered that all the formulas for impedances in the particular T-networks are valid under the assumption that the corresponding parts vibrate in piston like mode (in one-dimensional fashion).

In the case that the assumption of one-dimensional mode of vibration of the transducer parts is not valid or is questionable, the alternative to analytical approach method of Finite Element Analysis (FEA) operating characteristics of the transducers is widely used. An introduction to employing the FEA for nonuniform bar transducer designing, and vast bibliography on the related issues including references to programs available for transducer analysis can be found in Ref. 8.

Being powerful means for solving the direct problem of transducer designing, the matrix and FEA methods do not provide means for physical clear and adequate approach to solving the inverse problem, which consists in optimizing transducer operating properties in accordance with particular requirements. Although a formal solution of this problem can be obtained by analyzing results of multiple calculations in the direct way, this procedure does not develop engineering intuition that can be based on understanding of underlying physics. In contrast is the situation with uniform transducers, in which case the conditions for their optimal matching with acoustic load may be clearly formulated from solutions in the generalized coordinates. But in this case, there is no reasonable technical means for realizing the necessary loading for relatively low frequency transducers. This contradiction can be partially resolved in the way of combining both approaches as it was suggested in Ref. 9 and is considered below.

10.4.1 Equivalent Transformation between Nonuniform and Uniform Length Expander Transducers

It can be shown that the problem of calculating nonuniform transducer depicted in Figure 10.30 (a) in the frequency range near to resonance can be reduced to already discussed calculation of a homogenous piezoceramic transducer, in which only a portion of piezoceramic volume is used as active. With this goal at first consider input impedance of one-dimensionally vibrating bar loaded on the opposite end (Figure 10.31 (a)). As it was shown in Section 4.3.3, the input impedance is

$$Z_{in} = \frac{j\rho c S_{cs} \tan kl + Z_L}{1 + j(Z_L / \rho c S_{cs}) \tan kl}, \tag{10.168}$$

where Z_L is impedance of the load; ρ, c, S_{cs}, l are the parameters of material and dimensions of the bar. The circuit representation that corresponds with this expression is shown in Figure 10.31(b). Correlation between the input and output velocities in this circuit is

$$\frac{U_l}{U_0} = \frac{1}{[1 + j(Z_L / \rho c S_{cs}) \tan kl] \cos kl}. \tag{10.169}$$

If to assume that

$$|Z_L| << |Z_1 + Z_2| = (\rho c S_{cs}) / |\tan(kl)|, \tag{10.170}$$

which usually holds for acoustic loading, then the equivalent circuit of the input impedance can be represented following the Thevenin's theorem in the form of circuit shown in Figure 10.31(c), where

$$Z \approx j(\rho c S_{cs}) \tan(kl),\ Z_L^{'} \approx Z_L / \cos^2(kl),\ F^{'} \approx F / \cos(kl), \tag{10.171}$$

$$U_0 = U_l / \cos(kl). \tag{10.172}$$

Note that formulas for Z_L' and F' follow directly from conservation of energy in course of manipulations The mechanical energies associated with load and force acting on the end are

$$\bar{W}_L = Z_L |U_l|^2 = Z_L' |U_0|^2, \tag{10.173}$$

$$\bar{W}_F = F U_l^* = F' U_0^*, \tag{10.174}$$

where from formulas (10.171) for Z_L' and F' follow due to relation (10.172). The input impedance of an unloaded bar obtained from formula (10.168) is

$$Z = j\left(\rho c S_{cs}\right)\tan\left(kl\right). \tag{10.175}$$

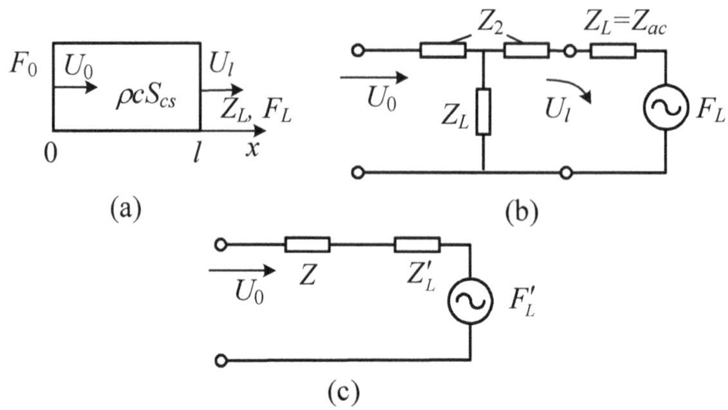

Figure 10.31: To the input impedance of a bar loaded on one end. (a) Geometry of the bar, (b) T-network representation of the input impedance, (c) the two-terminal equivalent circuit of input impedance after application of the Thevenin's theorem.

Now the method accepted in the theory of electric long lines can be applied, according to which a reactive load used to be replaced with a segment of the same line having the input impedance equal to impedance of the load. Let us imaginary replace the actual passive bars with imaginary bars having the cross section area, density and sound speed equal to those of the active bar. And let us assume that the input impedances of the imaginary bars are equal to the input impedances of the actual passive bars under action of real loads and forces on their ends at resonance frequency of the transducer. Usually only one of the ends of the combined transducer is operating, i.e., subjects to action of a load and/or an external force (let it be the end of bar H in Figure 10.30 (a), to which external actions $Z_L = Z_{ac}$ and $F_L = F_{eqv}$ are applied), while another end is free of loads ($Z_L = 0$ on the end of bar T). The input impedance of bar T from the side of the piezoelement is by formula (10.175) $Z_{inT} = j\left(\rho c S_{cs}\right)_T \tan\left(kl\right)_T$.

The input impedance of bar H in accordance with circuit in Figure 10.31 (c) and expressions (10.171) is

$$Z_{inH} = j\left(\rho c S_{cs}\right)_H \tan\left(kl\right)_H,\ Z_L' = Z_{ac} / \cos^2\left(kl\right)_H,\ F' = F_{eqv} / \cos\left(kl\right)_H. \tag{10.176}$$

The lengths l_1 and l_2 of ceramic bars that are equivalent to the passive bars in terms of their input impedances at the resonance frequency will be found from the relations

$$(\rho c S_{cs})_H \tan(kl)_H = (\rho c^E S_{cs})_c \tan k^E l_1, \tag{10.177}$$

$$(\rho c S_{cs})_T \tan(kl)_T = (\rho c^E S_{cs})_c \tan k^E l_2. \tag{10.178}$$

The subscript "c" stands for ceramics having the same parameters as the active piezoelement, from which the equivalent bars are supposed to be built. Impedance Z_L' and force F' defined by formulas (10.176) are applied to the boundary of the active segment and the passive bar H. Being transformed to the operating end of the imaginary homogenous bar 1, these quantities become

$$Z_L'' = Z_L' \cos^2 k^E l_1 = Z_{ac} \cos^2 k^E l_1 / \cos^2 (kl)_H, \tag{10.179}$$

$$F'' = F' \cos k^E l_1 = F_{eqv} \cos k^E l_1 / \cos (kl)_H. \tag{10.180}$$

Here the considerations expressed by relations (10.173) and (10.174) are applied. For definiteness we will further assume that the transducer operates in array of a big size, and therefore $Z_{ac} = (\rho c)_w S_H$ and $F_{eqv} = 2P_o S_H$, where P_o is the sound pressure in the free acoustic field.

Thus, the initial Tonpilz transducer appeared to be converted to the uniform piezoceramic bar with partial electrodes and with external actions transformed according to formulas (10.179) and (10.180), as shown in Figure 10.32.

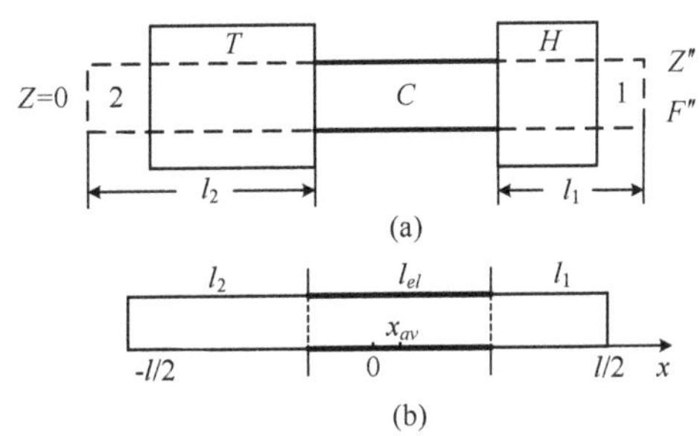

Figure 10.32: (a) The equivalent conversion of the composite transducer into uniform ceramic bar with partial electrodes. By the dashed lines are shown extensions of the ceramic section of the uniform bar that are equivalent to real passive parts of the composite transducer in terms of their input impedances. (b) The resulting uniform ceramic bar with partial electrodes.

The transducers are equivalent in terms of their calculated operating electromechanical characteristics, strictly speaking, at the resonance frequency, but it can be expected that the characteristics will not differ significantly in some frequency range around the resonance frequency.

Further, parameters of the uniform bar transducer can be found using a single contour equivalent circuit considered in Section 5.7.3.1, if calculations are restricted to the frequency region around the resonance frequency of the imaginary uniform bar with appropriately located active section and under transformed actions $Z_L^{''}$ and $F^{''}$ applied to the end of the bar.

Thus, replacement of the parts of the uniform piezoceramic bar by the passive parts made of a certain material and having a certain size can be regarded as a way of matching the transducer to a load. This is possible for the uniform bar transducers for two reasons. Firstly, because they have a significant reserve of electrical strength, as was noted in Ch. 3, and reducing this reserve with simultaneous increase of otherwise not sufficient mechanical strength of transducer due to increase of acoustic load is highly beneficial. Secondly, contribution to the electromechanical conversion of the parts of a length expander piezoelement located close to the ends is relatively small, and their replacement by passive material may even increase the effective coupling coefficient of the transducer, as it was shown in Section 5.6.2 and is illustrated by plot in Figure 10.33.

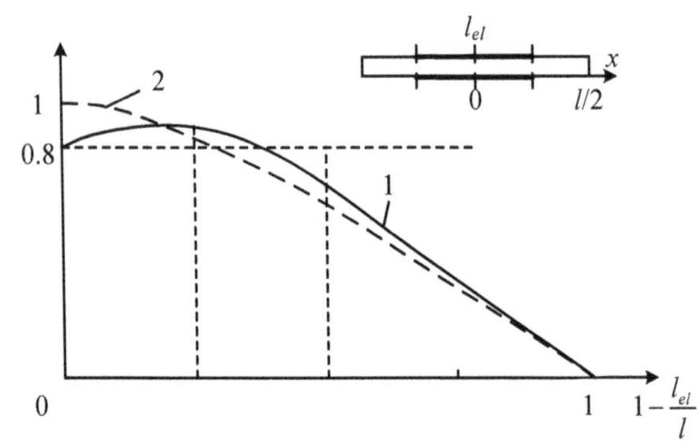

Figure 10.33. Dependence of the normalized effective coupling coefficient $k_{eff}^2(l_e)/k_{3i}^2$ (solid line) and electromechanical transformation coefficient $n(l_{el})/n(l)$ (dashed line) on the length of the symmetrically positioned active part of length expander piezoelement.

In case that active part of the uniform transducer is displaced from its center, as is shown in Figure 10.32, the electromechanical transformation coefficient will change according to formula

$$n = n(l_{el}) \cdot \cos(\pi x_{av} / l), \qquad (10.181)$$

where $n(l_{el})$ is the value for the symmetrically located active part and x_{av} is displacement of the center of the active part. The effective coupling coefficient will reduce proportionally to $\cos^2(\pi x_{av} / l)$.

Summarizing the above considerations, we will illustrate the procedure of optimal designing of the composite transducer with idealized example. Definitely, a real design is more complicated and, probably, may require several approximations for achieving desirable results. But our goal is to illustrate the idea of the approach.

Suppose that for the composite transducer requirements are formulated for the resonance frequency, f_r, radiating surface area, S_H, and operating characteristics (power radiated, frequency range of operation, overall length of the transducer and so on). We assume that from condition of operating in an array of a big wave size acoustic load of the composite transducer acts on one end and is equal to $Z_{ac} \approx (\rho c)_w S_H$. For simplicity we will also assume that the composite transducer will be symmetrical relative to its middle section, as it is shown in Figure 10.34. These assumptions are not essential in principle, though the symmetrical configuration of the electrodes can be recommended, as the most effective in terms of electromechanical conversion. If it is desirable to have not symmetrical design for meeting requirements for transducer, then the active part of the equivalent uniform bar may be displaced from the center, as it is shown in Figure 10.32 (b). This will result in reducing the electromechanical transformation coefficient in accordance with formula (10.181) and corresponding reduction of the effective coupling coefficient.

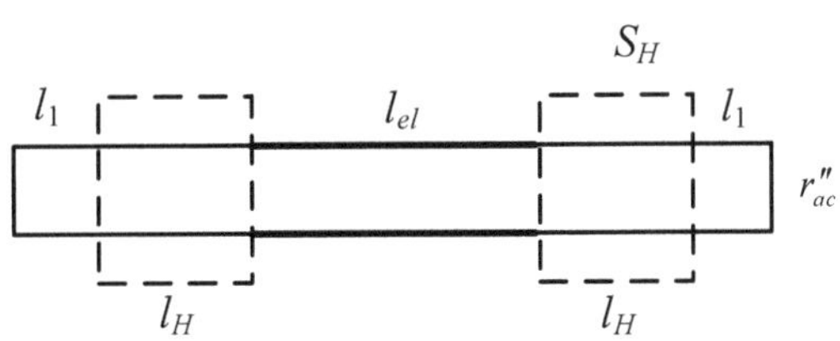

Figure 10.34: To the equivalent conversion of the uniform bar into composite transducer. By the solid line is shown imaginary uniform bar, by the dash lines are shown passive parts of composite transducer that are equivalent to their ceramic counterparts in terms of the input impedances.

At first, we consider uniform bar transducer (shown in Figure 10.34 by solid line) that has the required resonance frequency. Its overall length must be

$$l = l_{el} + 2l_1 = c_c^E / 2f_r \,. \tag{10.182}$$

Next, the required value of the acoustic load for the uniform bar, r''_{ac}, which ensures optimal matching of the transducer for achieving specified operating characteristics, may be determined in the way that was discussed in Ch. 3.

In doing this the length of the active part of the uniform transducer, l_{el}, of its cross section area, S_{cs}, and mode of polarization of piezoelement may be chosen from condition of reducing excessive reserve of the electrical strength to an acceptable minimum, in order to maximize the lengths l_1 of parts of the bar that are supposed to be replaced by the passive parts of combined transducer. This may result in a greater opportunity of tailoring the passive parts for achieving required transformation of acoustic load. It is noteworthy that the length of electrodes and cross section area of a piezoelement are related through the electromechanical transformation coefficient. It must be kept in mind in the process of determining their values that reducing the cross section area may be desirable, because increase of ratio S_H / S_c can make it easier optimizing parameters of the passive parts. Example that illustrates calculating the uniform transducer under an optimal load for achieving certain requirements will be given in Section 13.2.2.1.

Assuming that r''_{ac}, l_{el} and S_{cs} are known, the following equations can be formulated that have to be met by the material properties and length of the passive part of the combined transducer (its surface area is supposed to be known from requirements for the transducer):
from condition (10.179) of the proper impedance transformation

$$\cos^2(kl)_H = (\rho c)_w \frac{S_H}{r''_{ac}} \cos^2 k_c^E \frac{l - l_{el}}{2} \,; \tag{10.183}$$

from condition (10.177) of equality of the input impedances.

$$(\rho c S)_H \tan(kl)_H = \left(\rho c^E S_{cs}\right)_c \tan k_c^E \frac{l - l_{el}}{2} \,. \tag{10.184}$$

Thus, we have two equations for determining values of ρ_H, c_H and l_H. One more condition can be derived from practical considerations regarding material or geometry of the passive part. For example, this can be requirement for its piston like vibration, as at a given surface area, S_H, the length (thickness) of the passive part cannot be too small to avoid a harmful contribution of its flexural deformation. Several examples that illustrate dependence of modes of vibration of circular disks that may be used as passive parts of the combined transducers from their relative dimensions are considered in the next section.

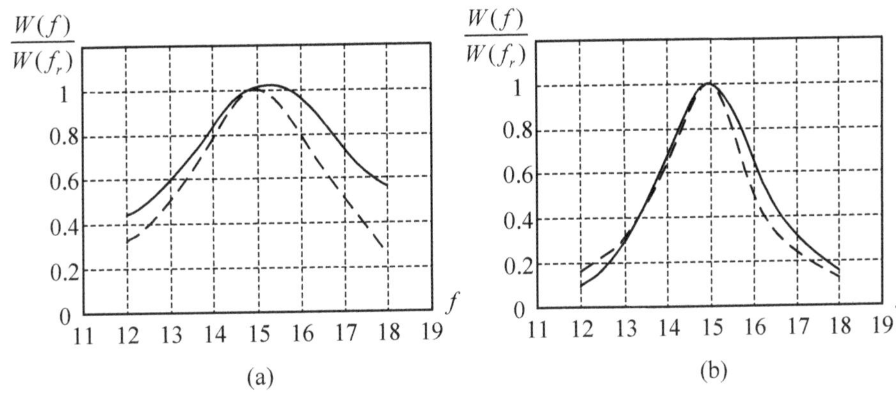

Figure 10.35: Results of calculating the frequency responses of the power radiated by two Tonpilz transducers having the same resonance frequencies and radiating surface areas: (a) at $\alpha_r = 4.0$, (b) at $\alpha_r = 2.5$. Solid lines-calculations are made using regular technique. Dashed lines-calculations are made for the equivalent uniform bar.

After the estimation of geometry of the combined transducer is done, results of calculating characteristics of the imaginary uniform transducer in the operating frequency range may be verified using the common cascade T-network technique for the resulting combined transducer. These characteristics can differ to some extent because the conditions of equivalency strictly speaking are held at resonance frequency only. Plots in Figure 10.35 illustrate the extent of difference of results obtained by employing both methods for calculating frequency responses of two transducer prototypes that have the same resonance frequencies and radiating surface areas, but different ratios of the load transformed to surface of the equivalent uniform bar and the real acoustic load, namely, that correspond to nondimensional coefficients of radiation resistance $\alpha_r = (r''_{ac} / \rho c S_c) = 2.5$ and 4.0. As can be seen from the plots, the agreement between the results is fairly good in a broad frequency range. The discrepancy increases with increase of the transformed acoustic load. On the positive side is that characteristics of the real operational band look more optimistic than predicted by the suggested approximate method.

10.4.2 Input Impedance and Mode of Vibration of the Circular Disk

The passive matching part of the bar transducer can be imagined as a circular disk cemented to the end of the active driver. Variants of typical configuration of the ends of the drivers are schematically shown in Figure 10.36.

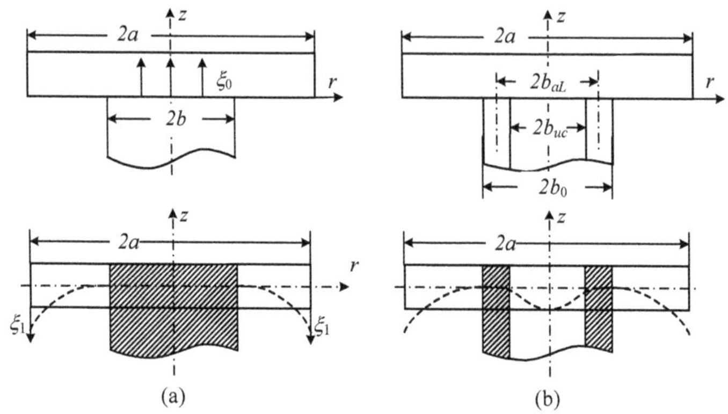

Figure 10.36: The schematic view of the disks vibrating under action of the drivers having different configuration: (a) solid piezoceramic bar, (b) cylindrical piezoelement having various mean diameter. The variant (b) is more common because it allows placing a strengthening bolt inside.

All the previous considerations were made under the assumption that the matching disk vibrates in the piston like manner, with uniform distribution of the displacements on its surface. Validity of this assumption may be questionable for real transducer designs, moreover, it is often desirable to have the disk as thin as can be acceptable, in which case a possibility of the flexural vibration of the disk that may violate this assumption must be considered. A trade off takes place between the thickness to diameter ratio of the matching disk and uniformity of vibration of its surface. The flexural vibrations of the disk under action of the drivers depend essentially on the boundary conditions on the area of contact between the driver and the disk. It is hardly possible to formulate these conditions analytically and to get an exact analytical solution to the problem. The adequate numerical solution of the problem can be obtained by FEA. But it seems beneficial to use a combined approach. Firstly, to formulate reasonably idealized boundary conditions and to obtain a physically clear approximate analytical solution, and secondly, to verify the obtained result and to make it more accurate by using FEA.

In this section we will use an approximate analytical method for estimating the acceptable dimensions of the matching disk and for obtaining the value of its input impedance by taking flexural deformations into consideration. The goal will be to determine the smallest possible thickness of the disk that insures a permissible deviation of displacement from uniform distribution over the surface of the disk. The following assumptions will be made. Let the diameter to wavelength ratio foe the disk be $D_d / \lambda = 1/2$ for definiteness. The actual size of radiating surface depends on application of the transducer and may vary. Desirable is to have smaller wave diameter for application in a big array with steerable directional pattern, but for the goal of estimating the minimal thickness the bigger diameter is more demanding. We will assume that displacement (denoted it ξ_0) is uniform over cross section of the driver on the area of contact with the disk. Overall displacement of the surface of the disk will be represented as

$$\xi_r(r) = \xi_0 + \xi_1\theta(r), \tag{10.185}$$

where $\theta(r)$ is the lowest mode of flexural vibration of the disk under the boundary conditions imposed by the driver, ξ_1 is the maximum deflection in the flexural vibration of the plate. As we are interested in close to piston like vibration of the matching disk, the condition $|\xi_1| \ll \xi_0$ must be fulfilled.

Under the accepted assumptions the disk can be considered as a system with two coupled degrees of freedom. The partial systems are: the rigid disk that vibrates as a piston with displacements ξ_0, and the plate that performs vibration $\xi(r) = \xi_1\theta(r)$. The boundary conditions that are imposed on the flexural vibrations of the disk by a driver is the matter of principle in this treatment. It is clear that real boundary conditions depend on the geometry of the driver. They are likely to be the clamped conditions on the contour of contact with the driver assuming that the moment of resistance of the driver is large enough for both variants of the drivers, as is shown in Figure 10.36 (a) and (b). Though what is large enough is not clear, we will consider that the clamped conditions are in place. Validity of this assumption, as well as of assumption that displacement is uniform over the cross section of a driver may be checked by experimental verifying results of calculations or/and by FEA.

At first, we consider variant (a), in which case the part of the disk at $r > b$ vibrates in flexure as the annular plate that is free on the outer surface and clamped on the inner surface. Determine energy status of the disk. As the piston like (without deformation) vibration of the disk does not contribute to the potential energy, the potential energy is

$$W_{pot} = \frac{1}{2}\xi_1^2 K_{eqv1}, \tag{10.186}$$

where K_{eqv1} is equivalent rigidity of the disk in flexural vibration under prescribed boundary conditions. The kinetic energy is

$$W_{kin} = \frac{1}{2}\cdot 2\pi\rho t\int_0^a [\dot{\xi}_0 + \dot{\xi}_1\theta(r)]^2 r dr = \frac{1}{2}[M_d\dot{\xi}_0^2 + 2\dot{\xi}_0\dot{\xi}_1 M_{01} + M_{eqv1}\dot{\xi}_1^2], \tag{10.187}$$

where $M_d = \rho t\cdot\pi a^2 = \rho t S_d$ is the mass of the disk, M_{eqv1} is the equivalent mass of the disk in flexure, and M_{01} is the mutual mass between the piston like and flexural vibration of the disk. The equivalent and mutual masses are commonly expressed as $M_{eqv} = \rho t S_{eff}$ and $M_{01} = \rho t S_{av}$, where

$$S_{eff} = 2\pi\int_{\Sigma}\theta^2(r) r dr, \tag{10.188}$$

$$S_{av} = 2\pi\int_{\Sigma}\theta(r) r dr. \tag{10.189}$$

Here Σ is the surface area of the part of the disk that vibrates in flexure (it depends on geometry of the driver). Usually, S_{eff} and S_{av} are presented among equivalent parameters of the plates at different boundary conditions. The energy of external action is

$$W_e = f_{dr}\xi_0, \tag{10.190}$$

where f_{dr} is the total force, with which the driver is acting on the area of contact.

The Lagrange equations after converting to the complex form ($\xi = U/j\omega$, $\ddot{\xi} = j\omega U$) are

$$j\omega M_d U_0 + j\omega M_{01} U_1 = F_{dr}, \tag{10.191}$$

$$j\omega M_{01} U_0 + j\omega M_{eqv1}[1-(\omega_{b/a}/\omega)^2]U_1 = 0, \tag{10.192}$$

where $\omega_{b/a} = \sqrt{K_{eqv1}/M_{eqv1}}$ is the lowest resonance frequency of flexural vibration of the plates under boundary conditions that depend on geometry of the driver's cross sectional area. They are annular plates clamped on the inner (of radius b) and free on the outer boundary of radius a, and the plate of radius b_1 clamped on the boundary, as shown in Figure 10.36.

Vibration of the plates under these boundary conditions are considered in Ch. 4 (for the annular plate in Section 4.5.8.2) under the condition that elementary theory of bending is applicable. As our goal is to estimate the smallest acceptable thickness of the disk, we will assume that this approximation holds. Otherwise, the corrections for the finite thickness of the plates can be used that are introduced in Section 9.4.3. From Eqs. (10.190) and (10.191) will be obtained

$$U_1 = -\frac{M_{01}}{M_{eqv1}[1-(\omega_{b/a}/\omega)^2]}U_0 = -\frac{S_{av}}{S_{eff}[1-(\omega_{b/a}/\omega)^2]}U_0, \tag{10.193}$$

and the input impedance of the disk

$$Z_{in} = \frac{F_{dr}}{U_0} = j\omega M_d\left\{1-\frac{M_{01}^2}{M_d M_{eqv1}[1-(\omega_{b/a}/\omega)^2]}\right\} = $$
$$= j\omega M_d\left\{1-\frac{S_{av}^2}{S_d S_{eff}[1-(f_{b/a}/f)^2]}\right\}. \tag{10.194}$$

As we are interested in close to piston like vibration of the matching disk, the condition must be fulfilled that in expression (10.185) $|\xi_1| \ll \xi_0$. Let it be, for example,

$$|\xi_1| < 0.2\xi_0. \tag{10.195}$$

Then from relation (10.193) follows that at resonance frequency of the driver, f_{dr}, should be fulfilled relation

$$\frac{S_{av}}{S_{eff}[(f_{b/a}/f_{dr})^2-1]} \le 0.2, \tag{10.196}$$

or

$$f_{b/a} \ge f_{dr}\sqrt{5(S_{av}/S_{eff})+1}. \tag{10.197}$$

Expressions for the parameters $f_{b/a}$, S_{av} and S_{eff} of the annular plates vs. ratio b/a that are extracted from Table 4.5 are presented in Table 10.1.

Table 10.1. Parameters of the annular plates clamped on the inner diameter.

b/a	$S_{eff\,b/a}/\pi(a^2-b^2)$	$S_{av\,b/a}/\pi(a^2-b^2)$	$\lambda_{b/a}$
0.7	0.26	0.41	6.17
0.6	0.28	0.42	4.61
0. 5	0.29	0.44	3.69......

Note that

$$f_{b/a} = \lambda_{b/a}^2\frac{t}{2\pi\sqrt{12}\,a^2}\sqrt{\frac{Y}{\rho(1-\sigma^2)}}. \tag{10.198}$$

Thus, for example, if radius of the driver is $b = 0.5\,a$, than according to data from the Table $S_{av\,b/a} / S_{eff\,b/a} = 1.5$, and from relation (10.197) follows that it must be $f_{b/a} \geq 2.9 f_{dr}$. Using formula (10.198), where $\lambda = 3.68$, we arrive at the requirement for the geometry of the disk

$$\frac{t}{a^2}\sqrt{\frac{Y}{\rho(1-\sigma^2)}} \geq 4.6 f_{dr} \tag{10.199}$$

To get impression regarding an order of quantities involved consider an example. Let the operating frequency of the driver be = 10 kHz. Diameter of the matching disk according to our assumption is $2a = \lambda_w / 2 \approx$ 7.5 cm. Assuming that the disk is made of aluminum ($\sqrt{Y/\rho(1-\sigma^2)} = 5.4 \cdot 10^3$ m/s), from condition (10.199) will be obtained that $t \geq 1.2$ cm. Thus, at the dimensions a = 3.75 cm and $t \geq 1.2$ cm deviation of displacement from uniform over surface of the disk at frequency 10 kHz is less than 20%.

It is noteworthy that for the annular disk of these dimensions the thin plate theory can be not exact, and corrections to the expressions of the equivalent parameters must be required for the rotary inertia and shear deformations. Though the assumptions made for the considered example regarding the wave diameter of the radiating surface and acceptable nonuniformity of displacements are probably too demanding for practical applications. Thus, for example, if $2a = \lambda_w / 4 \approx 4$ cm, then at all other conditions equal it should be $t \geq 3.4$ mm, and the aspect ratio for the disk will be $a/t \approx 6$ vs. 3.3 in the previous example. The input impedance of the disk under the same condition (10.195) of uniformity of vibration, being obtained from (10.194) after substituting these values of the parameters, will be

$$Z_{in} = j\omega M_d \cdot 1.07\,. \tag{10.200}$$

An increase of the acceptable nonuniformity of displacement will result in reducing the thickness of the disk and in increasing deviation of value of input impedance from $Z_{in} = j\omega M_d$.

In the variant of driving the disk by a cylindrical piezoelement, as shown in Figure 10.36(b), the flexural vibrations may occur also of the central part of the disk at $r < b_{in}$. We will assume that vibration of this part of the disk can be considered, as of the plate clamped on its boundary at $r = b_{in}$. In this case the mode shape $\theta(r/b)$ in expression (10.185) is $\theta(r/b) = (1 - r^2/b_{in}^2)^2$, $S_{eff} = 0.18\pi b_{in}^2$, $S_{av} = 0.33\pi b_{in}^2$ (see Table 9.3), and in formula (10.198) for the resonance frequency a^2 must be replaced by b_{in}^2 and $\lambda_{b/a}$ by $\lambda_{b_{in}} = 3.2$.

Compare the resonance frequencies of the annular plate clamped at the radius b_o (denoted $\omega_{b_o/a}$) and of the inner plate clamped at radius b_{in} (denoted $\omega_{b_{in}}$). Obviously, under the condition that $\omega_{b_{in}} \gg \omega_{b_o/a}$ the nonuniformity of the surface displacement will be determined by displacements of the annular plate, which is already estimated for variant (a), and vibration of the central part of the disk can be considered as uniform. Given that $\omega_{b_o} < \omega_{b_{in}}$, the relation between the resonance frequencies $\omega_{b_o} \geq \omega_{b_o/a}$ will be stricter then required. Using formula (10.198) for both resonance frequencies and remembering that $\lambda_{b_o} = 3.2$ we arrive at relation

$$\left(\frac{b_o}{a}\right)^2 \cdot \frac{\lambda^2_{b_o/a}}{10.2} = \gamma \leq 1. \tag{10.201}$$

Assuming, for example, that $b_o / a = 0.6$ and 0.7, we obtain with values $\lambda_{b_o/a}$ taken from Table 10.1 that $\gamma = 0.7$ and 1.8, respectively. Thus, the condition (10.201) is satisfied at $b_o / a = 0.6$. And at values b_o / a somewhere between 0.6 and 0.7 the flexural vibration of the central part of the disk seem to start contributing to overall nonuniformity of the displacements. But at this point it is useful to keep in mind the following considerations.

For fare comparison with variant (a) it must be assumed that the cross section areas of the drivers in both cases are the same, i.e., the ratio $b_0^2 / 2b_{av}t$ should be kept constant. Following the previous example, in which case $S_d / S_{dr} = 4$, this means that thickness of the driver reduces with increase of b_o / a and at $b_o / a \approx 0.7$ reaches about $t \approx 0.2a$. At thicknesses of this order the assumption of clamped boundary conditions for both outer and inner parts of the disk become questionable. The assumption that the disk is simply supported on the average diameter of the driver may become more reasonable. The input impedance of a thin disk supported on the circle of radius $b < a$ is considered in Section 4.7.2.3. It is shown that if only the first mode of vibration $\theta_1 (r / a)$ is taken in calculation, then according to expression (4.696) the input impedance on the circle of radius b is

$$Z_{in} = j\omega M_d \left[1 - \frac{M_d}{M_{eqv1}} \cdot \frac{\theta_1^2 (b/a)}{\omega_1^2 / \omega^2 - 1}\right]^{-1}. \tag{10.202}$$

The first mode of vibration of free plate has a nodal line at $r / a = 0.68$. Thus, if a driver is acting on the line with radius $b = 0.68a$, $\theta_1(b / a) = 0$ and $Z_{in} = j\omega M_d$, i.e., the plate vibrates uniformly seemingly independent of frequency at least in vicinity of the first resonance frequency of the disk and even higher given that the resonance frequency of the next mode of vibration, which may contribute to nonuniformity of displacements, is $\omega_2 = 4.3\omega_1$. If to admit that contribution of the second mode may produce a visible nonuniformity of displacement at frequencies above $\omega_2 / 2$, then at the same operating frequency and diameter of the disk as in the previous example (f_{op} = 10 kHz, $2a$ = 7.5 cm) the acceptable thickness of the disk may be $t \approx 0.3$ cm.

Definitely, this result is approximate at least by the two reasons. Firstly, the driving force is applied not exactly on the line, but on the ring of a finite width, and simple supporting conditions are also not exact. Besides, as already mentioned, in the absence of the first vibrational mode contribution of the next vibrational mode $\theta_2(r/a)$, which was neglected, must be more accurately estimated. But these considerations show that by appropriate application of the acting force vibration of even relatively thin disk can be made close to uniform.

10.4.3 Transducers with Uniform over Length Cross Section Area

The nonuniform length expander transducer designs combined of parts having different cross section areas are typical for application at relatively low frequencies (in the range of about 2 kHz up to 50 kHz following Ref. 6-8). High-frequency transducers are normally made of single piezoelements (blocks of piezoceramics) and have uniform over the length cross section area. Typical schematic view of the transducers is shown in Figure 10.37.

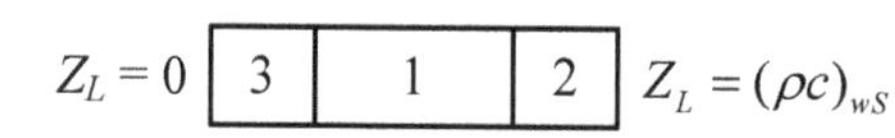

Figure 10.37: Schematic view of the transducer with uniform over length cross section area:1-piezoelement, 2 and 3 are metal blocks in variant of the sandwiched transducer and matching layers otherwise.

The half-wavelength piezoelements for operating in the frequency range about 50-200 kHz may appear to be not appropriately thick. For reducing the size, they can be sandwiched between two metal parts (2 and 3 in Figure 10.37) that perform mass loading of the piezoelements. Replacing the portions of piezoelement by passive parts occurs without loss (and even with some raise) of the effective coupling coefficient, and the electrical strength of the transducer is not compromised because of its significant reserves that are typical for the length expander transducers. In this case it is possible to excite the second mode of vibration. Also, the piezoelements may be cemented by two parts connected in parallel. In such design it is possible to excite vibration in the second mode by connecting the parts in antiphase, or both in the first and second modes by switching polarities of the parts. All these variants of transducers can be treated using the technique described in Section 10.2.3 in a straightforward way.

The lateral dimensions of the transducers may be significantly larger than their length (remember that the "length" is dimension in direction of vibration) from consideration of directionality required for their sonar applications. Therefore, the length to lateral dimensions aspect ratios of the piezoelements that may have configuration of stripes, plates and discs have to be taken into consideration for avoiding harmful effects of the coupled vibration in the piezoelements. The necessary information on these issues is presented in Section 10.3.

After the aspect ratio related conditions are fulfilled, all the equivalent circuits technique described in Section 10.2 can be employed in a straightforward way for calculating parameters of transducer. Effects of the matching layers may be considered as well by representing them as the passive T-networks.

Requirements for the matching layers and their design peculiarities were not considered so far, therefore we present some considerations regarding these issues following Ref. 10, where a comprehensive review can be found of different aspects of the high-frequency transducer designs. Usually, they offer less opportunities for optimizing acoustic load than Tonpilz like design. The functions of reducing mechanical Q of the transducers for achieving their acceptable efficiency and bandwidth are performed by the matching layers in the front and in the back of the piezoelement (blocks 2 and 3 in Figure 10.37). Employing the matching layers (blocks) is based on their properties to perform transformation of mechanical and acoustic loads (Z_L) applied to one of the ends according to the general formula (10.168).

$$Z_{in} = \frac{j(\rho c_{bc} S_{cs})_l \tan k_{bc} l + Z_L}{1 + j[Z_L / (\rho c_{bc} S_{cs})_l] \tan k_{bc} l}, \tag{10.203}$$

where subscript l stays for "layer". This formula is valid under the assumption that the layer vibrates in one dimensional (piston-like) mode. The value of the sound speed in the formula depends on whether the layers can be imagined as bars, strips, or plates. In other words, it depends on the boundary conditions for vibration of the layers. Therefore, the sound speed and wave number in the formula are used with the subscript "*bc*", which collectively reflects that the effective Young's moduli Y_{bc} in these cases have different values (Y for bars, Y_{st} and Y_{pl} by formulas (10.18) and (10.14) for the strips and plates) and $c_{bc} = \sqrt{Y_{bc} / \rho}$. Further the subscript will be omitted. Formula (10.203) does not account for mechanical losses in the layers for simplicity. In some cases, considering the losses may be essential, as it will be pointed out below.

For matching and backing purposes the quarter-wavelength ($l = \lambda / 4$ at the resonance frequency) and half-wavelength ($l = \lambda / 2$) layers are usually used. As it can be seen from formula (10.203), these layers perform transformations of loads that depend on frequency. At the resonance frequencies they are as follows. For the $\lambda / 4$ layer ($\tan kl = \infty$),

$$Z_{in} = (\rho c S)_l^2 / Z_L ; \tag{10.204}$$

for the $\lambda / 2$ layer ($\tan kl = 0$),

$$Z_{in} = Z_L . \tag{10.205}$$

In the common for underwater applications variant of loading shown in Figure 10.37 the impedances acting on the ends of piezoelement are as follows: with $\lambda / 4$ matching layers $Z_{in} = [(\rho c)_l^2 / (\rho c)_w] \cdot S$ on the front, and $Z_{in} = \infty$ on the back; with $\lambda / 2$ matching layers $Z_{in} = (\rho c)_w S$ on the front, and $Z_{in} = 0$ on the back.

Different combinations of the matching layers can be used depending on the requirements for transducer properties. For getting a loading that is near to optimal in terms of acoustic power radiated and ensures an acceptable bandwidth of the transducer operation (for the tradeoff between these properties see Section 3.1.5) the $\lambda / 4$ layer must be used in front of the piezoelement. In the situation that the bandwidth needs to be increased for improving conditions for signal processing this can be achieved for expense of efficiency of the transducer by backing the piezoelement with $\lambda / 2$ layer made of lossy material. Low loss $\lambda / 2$ layer in front of the piezoelement may serve as acoustic window and simultaneously as a protective layer that seals the transducer. Backing $\lambda / 4$ layer made of high-impedance material such as steel produces clamping effect ($Z_{in} = \infty$) on the end of the piezoelement, length of which must be reduced in this case to $\lambda / 4$ for retaining the same resonance frequency. The cross section that does not move can be used for mounting transducer within a case without (or with minimal) additional losses and interaction with the structure of the case.

Realization of the functions of the matching layers critically depends upon availability of materials with appropriate properties (sound speed, characteristic acoustic impedance $(\rho c)_l$ and coefficient of absorption). Review of the materials used for these applications can be found in Ref. 10. Most of them are composite materials synthesized from epoxy resins with the addition of proper fillers.

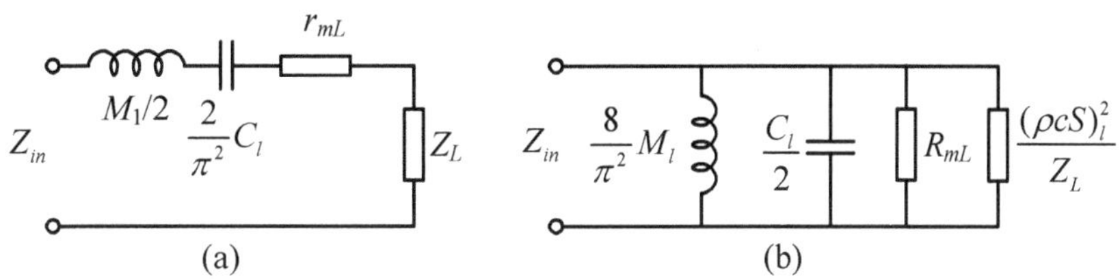

Figure 10.38: Simplified equivalent circuits in the frequency range around the resonance for the $\lambda/2$ (a) and $\lambda/4$ (b) layers loaded by impedance Z_L.

Quantitative estimations of input impedances of the matching layers made at resonance frequency subject to change with deviation of operating frequency. Variations of values of the input impedances in some frequency range around the resonance frequencies may be determined from approximate equivalent circuits with lump parameters that are presented in Ref. 10 and shown in Figure 10.38. These circuits were obtained as result of decomposition of the trigonometric functions in formula (10.203) into series in vicinity of the resonance frequencies. Resistances r_{mL} and R_{mL} are introduced in the circuits to account for losses of energy in the layers. In the equivalent circuits $M_l = \rho l S$ is the mass of the layer, $C_l = l/YS$ - static compliance of the layer, $r_{mL} = \pi^2/2\omega C_l \cdot Q_l$, and $R_{mL} = 2Q_l/\omega C_l$, where Q_l is the quality factor of the layer. It is shown that approximation of these circuits is valid within 3 dB bandwidth with an error less than 10% for the layers with Q-factor larger than 3, which is usually the case.

10.5 References

1. D. A. Berlincourt, D. R. Curran, and H. Jaffe, *Piezoelectric and Piezomagnetic Materials and their Function in Transducers*, in Physical Acoustics, Vol. I, Part A, edited by W. P. Mason (Academic, New York, 1964).
2. B. S. Aronov, C. L. Bachand, and D. A. Brown "Analytical modeling of piezoelectric ceramic transducers based on coupled vibration analysis with application to rectangular thickness poled plates," J. Acoust. Soc. Am. **126**(6), 2983–2990 (2009).
3. B. S. Aronov, "Dependency of resonance frequencies and effective coupling coefficients of piezoceramic rectangular bars and plates on aspect ratio (L)," J. Acoust. Soc. Am., **138**(3), 1481-1484 (2015).
4. B. S. Aronov and C. L. Bachand, "Piezoelectric cylindrical discs and solid rods: dependence of the resonance frequencies and effective coupling coefficients on aspect ratio," J. Acoust. Soc. Am., **140**(3), 2162-2172 (2016).
5. E. A. G. Shaw, "On the resonance vibration of thick barium titanate discs," J. Acoust. Soc. Am. **28**, 38–50 (1956).
6. R. S. Woollett, *Sonar Transducer Fundamentals*. (Naval Undersea Warfare Center, Newport, Rhode Island, undated).
7. D. Stansfield, *Underwater Electroacoustic Transducers* (Reprinted by Peninsula, Los Altos Hills, CA, 2003).
8. C. H. Sherman and J. L. Butler, *Transducers and Arrays for Underwater Sound* (Springer, New York, 2007).
9. Boris Aronov, "Application of the Tonpilz design for optimizing parameters of rod transducers," J. Acoust. Soc. Am., **105**, 1179 (1999) (A).
10. B. V. Smith and B. K. Gazey, "High-frequency sonar transducers: a review of current practice," IEEE Proc., **131**(3), 285-297 (1984).

Part IV

Some Aspects of Transducers Designing

CHAPTER 11

EFFECTS OF OPERATIONAL AND ENVIRONMENTAL CONDITIONS

11.1 Introduction

The preceding Parts I–III were concerned with solving the direct problem of transducers designing, that is calculating the transducer's output characteristics at specified input actions. It was assumed in the course of this treatment that operating and environmental conditions for the transducers are ideal (small signals for the projectors, room temperatures, negligible external pressure). Parameters of piezoceramic materials were assumed to be linear and independent of these conditions. Content of this Part is related to considering the reverse problems of the designing, which are the problems of achieving and optimizing transducer's operating characteristics at specified real environmental conditions and at as large as required levels of acoustic radiation including the maximum possible level for a given transducer type. Practical environmental and operating conditions result in external actions on the piezoceramic materials that influence their parameters. The actions that affect parameters of piezoceramics and hence properties of transducers include the following.

Large variable mechanical stress and electric fields, under which high power projectors may operate, that can reach values beyond the linearity of the piezoceramic properties and even cause mechanical failures or electrical breakdown of the transducer.

Large static mechanical stress that can arise both due to the hydrostatic pressure and due to structural features of the transducers (e.g., static compression bias stress on the ceramics). The static mechanical stress may induce instability of the piezoceramic properties and can result in the destruction of the transducer.

Large temperatures, which may cause instability in the piezoceramic properties. In the case of the high-power projectors, it is heating due to the internal energy losses that include dielectric and mechanical losses in the piezoceramics.

As the results the estimations of maximum available power radiated under the optimal load, and the value of the optimal load for projectors that were introduced in Ch. 3 by relations (3.126) and (3.127),

$$r_{opt\,W} = \frac{A_T E_P}{A_E T_P} - r_{mL} \approx \frac{A_T E_P}{A_E T_P}, \quad \dot{W}_{acm} \simeq \frac{E_P T_P}{A_E A_T}, \tag{11.1}$$

may change noticeably. Remember that coefficients A_E and A_T involve piezoelectric and elastic properties of ceramics (see Table 3.5), and the permissible electric field and mechanical stress, E_p and T_p, have values at which operation of a projector can be considered as linear and reliable for long term operation. Besides, the values of resonance frequency that is proportional to $1/\sqrt{s_{ii}^E}$ and clamped capacitance, $C_e^S \sim \varepsilon_{33}^T(1-k_{eff}^2)$, on which tuning the transducers drastically depend, subject to change. Increase of $\tan\delta_e$ may cause excessive heating of a transducer.

Properties of the receivers that are important for functioning of receive channel are the specific sensitivity and $\tan\delta_{tr}$. They characterize a receiver as the source of signal and of the internal noise, respectively. It follows from their expressions (3.181) and (3.175) that

$$\gamma_{sp} \sim k_{eff}\sqrt{s_{ii}^E} \text{ and } \tan\delta_{tr} = (1-k_{eff}^2)\tan\delta_e + \frac{k_{eff}^2}{Q_{m\Sigma}}. \tag{11.2}$$

Here k_{eff} is directly related to the coupling coefficient of the ceramics, k_m, the mechanical quality of transducer, $Q_{m\Sigma}$, mainly depends on the radiation resistance though includes contribution of the mechanical losses in ceramics.

So far, estimations of the listed transducer characteristics were made using parameters of piezoceramics determined at small signals, at atmospheric pressure and at room temperature. For the most usable PZT ceramic compositions they are presented in Table B.1 (Appendix B). For informed designing the underwater transducers dependences of the ceramic parameters on the operating and environmental conditions must be considered. The main results regarding the limiting factors for the transducers operation and dependence of parameters of ceramic compositions on the external actions were published in s Refs. 2-11. A brief review of these results is presented in the following sections. Before doing this some comments must be made about the basic data for the ceramic parameters at small signals.

11.2 Parameters of Ceramics at Small Signals

Data on the parameters of piezoelectric ceramics at small signals are related to Morgan standard original versions of the piezoelectric ceramics PZT-4, PZT-8, PZT-5A and PZT-5H (Navy types I. III, II. and VI, respectively). The same data are presented in Ref. 2 (Table VIII), in Ref. 1 (Table A5), and in Table B.1. As it is noted in Ref. 2, these low signal properties "...show statistical and systematic fluctuations from batch to batch or within a given batch due to slight chemical differences, variations in density, inhomogeneous chemical composition, variations in grain size, varying response to the poling treatment, and so forth. Measured values of coefficients mainly converge to those listed in typical table with variation of about 5% for elastic, 10% for piezoelectric, and 20% for dielectric constants".

Moreover, the same is true regarding different versions of these basic ceramic brands produced by different manufacturers. Thus, the results of calculations that employ values of ceramic parameters taken from Table B.1 cannot be fully accurate. To increase accuracy of predicting properties of a transducer design under development, parameters of the ceramics must be used that are obtained as a result of batch-to-batch incoming control of piezoelements provided by a particular manufacturer.

The variations that exist in the values of the small signal parameters of ceramics produced by different manufactures can be significantly magnified with increase of the electric field, dynamic stress and static stress, as it will be shown below. And these variations can be large even if the initial small signal properties of the samples were similar. This makes an accurate prediction of transducer parameters under the real operating conditions even more complicated. Therefore, developing methods and capabilities for physical testing transducers at conditions close to operating is of great importance.

It is noteworthy that presenting fixed values for deviating parameters of ceramics under high signals may be misleading without noting that these values are approximate and without referring to available results of investigating parameters of ceramics under different operating and environmental conditions.

11.3 Review of Published Results

11.3.1 Parameters of Ceramics at High Electric Fields

Investigating behavior of parameters of ceramics under strong electric field alone (without effects of high dynamic stress and/or hydrostatic pressure) has some reasons. One of the reasons (less important) is to make it possible to estimate the influence of other factors on change of the ceramic properties. More important reason is that even for projectors of moderate power and those intended for shallow water applications it may be required to operate under highest permissible electric field, E_p, in case that transducer is overloaded or operating at frequencies below resonance (the dynamic stress in this case may be relatively small, $T_d < T_p$). Such situation may be typical for the spherical and extensional ring projectors, and for the bender transducers at certain radiation loading conditions.

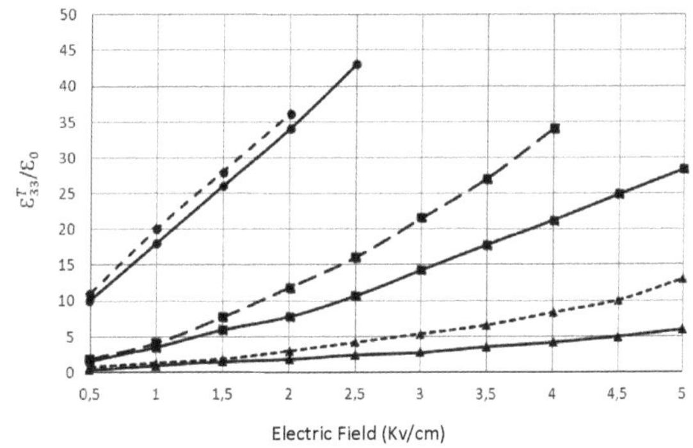

Figure 11.1: Increase in percents of dielectric constants $\varepsilon_{33}^T / \varepsilon_0$ vs. electric field. Solid lines - 25°C, dashed lines - 100°C; ● – PZT-5A, ■ – PZT-4, ▲ – PZT-8.

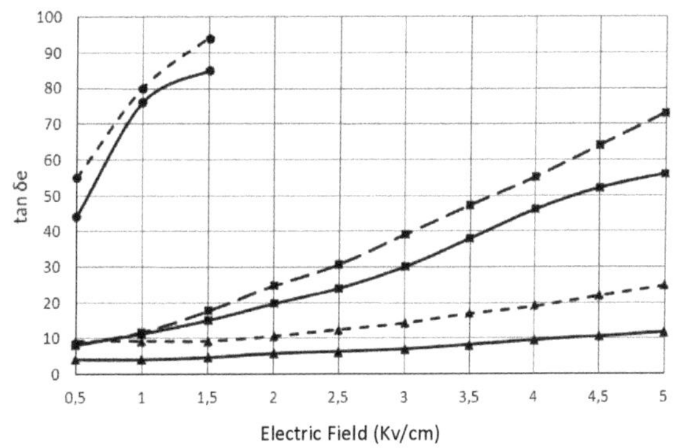

Figure 11.2: Increase in percents of dielectric loss factor $\tan\delta_e$ vs. electric field. Solid lines - 25°C, dashed lines - 100°C; ● – PZT-5A, ■ – PZT-4, ▲ – PZT-8.

In Figure 11.1 and Figure 11.2 dependencies are shown of increase in percents of the dielectric constants ε_{33}^T and dielectric loss factor $\tan\delta_e$ for 24 hour aged PZT ceramic compositions vs. electric field.[3] As operation of projectors may be accompanied by increase of temperature due to the dielectric loss, these dependencies are presented for temperatures 25°C and 100°C.

The advantage of PZT-8 composition is clearly seen especially at high temperature. It is noteworthy that with increasing time of aging the changes of parameters reduce. Therefore, it is important to know the history of the piezoelements that are used for transducers designing. It must be also kept in mind that variation in the dielectric properties of samples made of the same type of ceramic composition by different manufacturers may differ significantly under high-drive conditions though their initial values at small signals can be similar. This is illustrated by results of high-field measurements of PZT-4 samples produced by 3 manufacturers that are presented in Figure 11.3 according to Ref. 4.

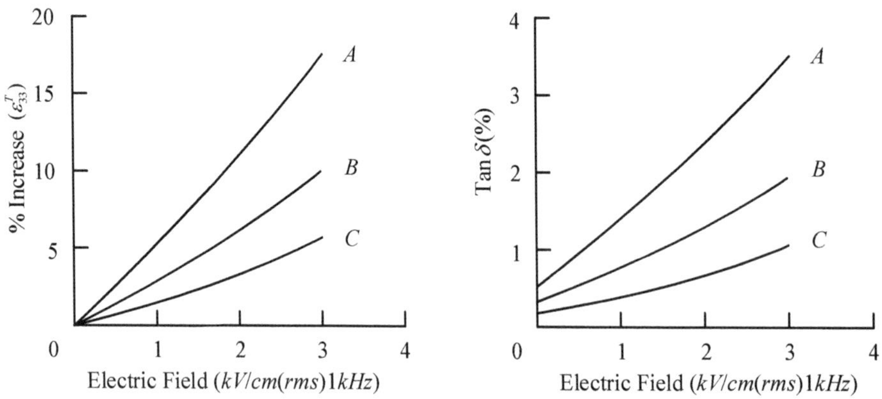

Figure 11.3: Increase in percents of differences in dielectric properties of PZT-4 type ceramic samples produced by different manufacturers revealed by high electric field.

11.3.2 Effect of Dynamic Stress on the Mechanical Quality Factor Q_m and Young Modulus ($Y_1^E = 1/s_{11}^E$)

The conditions of high dynamic stress at relatively small electric field take place in operating underloaded projectors in vicinity of their resonance frequency. In this case the dynamic stress may have the maximum permissible value $T_d = T_p$, whereas E_d may be much less than E_p. Such situation is typical for unmatched length expander bar projectors. Results of measuring the quality factor Q_m and Y_1^E vs. peak dynamic stress (Ref. 5) are presented in Figure 11.4. The measurements were made with well-aged (about 200 days) radially poled rings up to stress of their mechanical failure (about 3–4 kpsi or 25 MPa). It is noted in Ref. 5 that extent of the changes presented may be quite different for various modifications of the same ceramics.

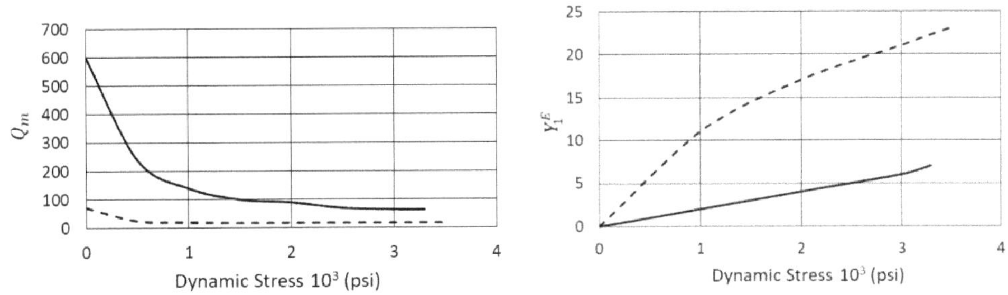

Figure 11.4: Dependencies of Q_m and percent of change of Y_1^E on dynamic stress perpendicular to polar axis; Solid lines – PZT-4, dashed lines – PZT-5A.

In case the projector is optimal acoustically loaded (well-matched) both the electric field and dynamic stress may reach their permissible values simultaneously, i. e., $E_d = E_p$ and $T_d = T_p$. No data exists on the combined effect of high electric field and dynamic stress on properties of piezoceramics.

11.3.3 Effects of Static Stress on the Properties of Piezoelectric Ceramics

At first, we consider effects of static stress, to which the mechanical systems of all the underwater transducers both the receivers and projectors are subjected due to action of the hydrostatic pressure or/and compression bias in a projector design. All the other limiting operating factors are typical for high power projectors only.

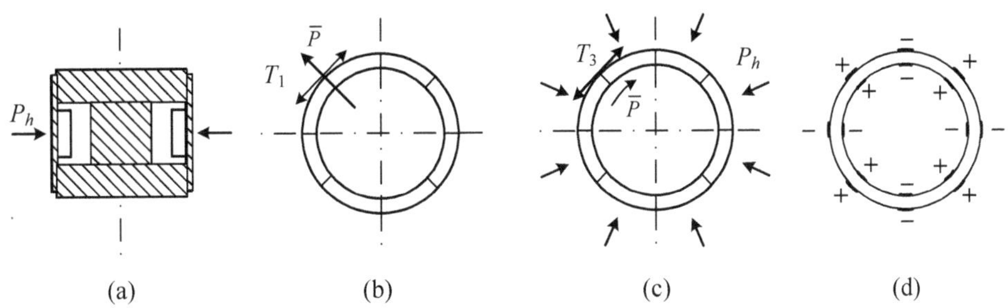

Figure 11.5: Air-backed cylindrical transducers: (a) schematic view, (b) perpendicular stress, (c) parallel stress, (d) stripe-electroded.

In terms of effects of hydrostatic pressure on the properties of piezoelectric ceramics the transducer designs can be divided into two groups: air-backed and hydrostatic pressure compensated. Schematic representations of these groups of designs are shown in Figure 11.5 through Figure 11.9. (The stresses are "perpendicular" and "parallel" to direction of polarization.)

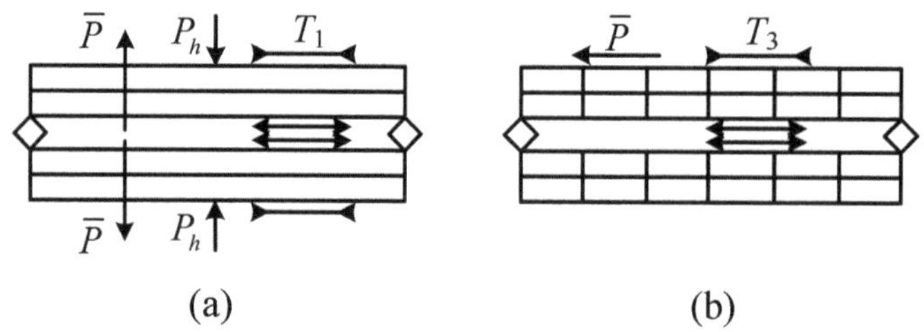

Figure 11.6: One-dimensional bender transducers: (a) perpendicular stress, (b) parallel stress.

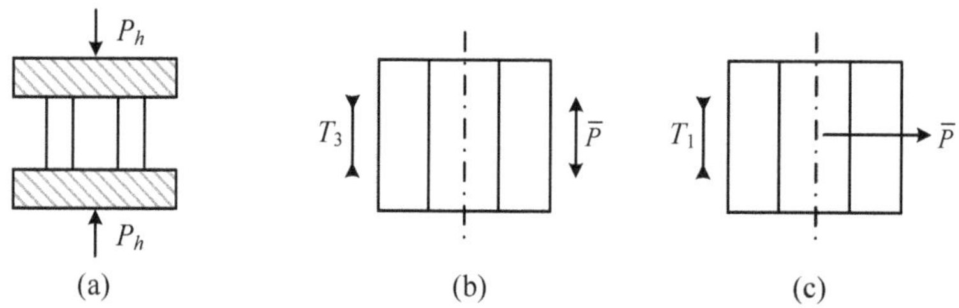

Figure 11.7: Mass loaded length expander transducer: (a) schematic view, (b) parallel stress, (c) perpendicular stress.

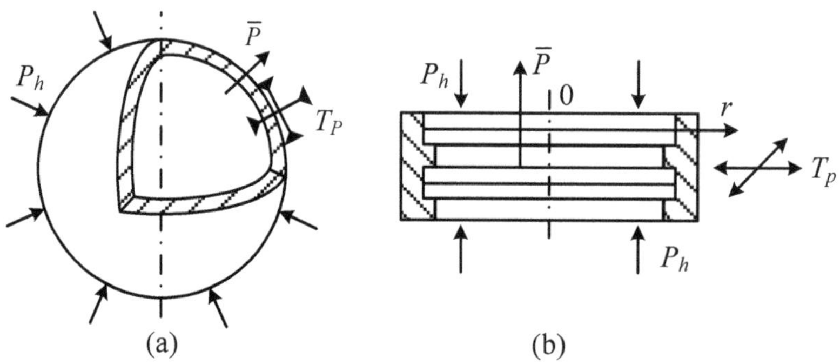

Figure 11.8: Two-dimensional transducers (planar stress): (a) Sphere, (b) Circular bender.

Besides of the hydrostatic pressure, the structural elements that produce compression bias in high power transducers to prevent mechanical failure of the transducers under dynamic stress in phase of tension may be a source of the mechanical stress in the transducer piezoelements. The variants of transducer designs with such structural elements are shown schematically in

Figure 11.10.

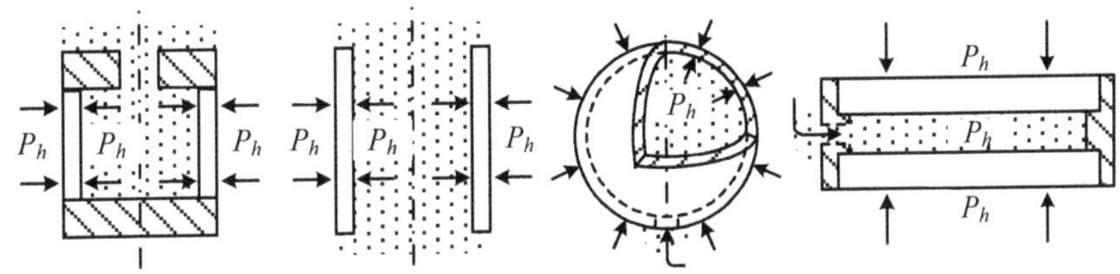

Figure 11.9: Hydrostatic pressure compensated transducers: (a) cylindrical transducers with acoustically isolated internal volume and (b) with open ends; (c) spherical and (d) bender transducers with acoustically isolated internal volume.

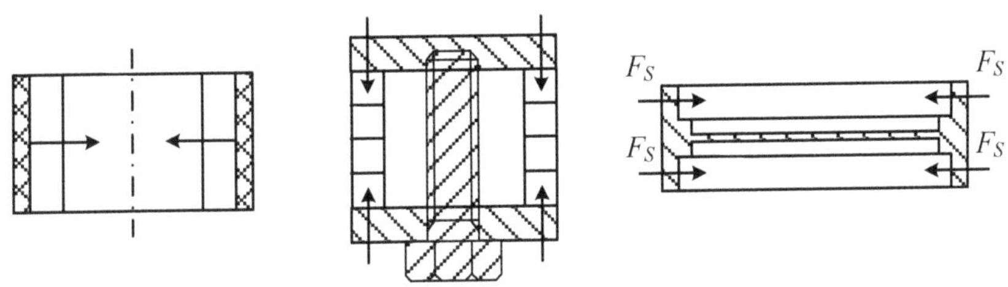

Figure 11.10: Schematic views of transducer designs with static stress bias imposed on piezoceramics: (a) fiber reinforced ring transducer, (b) Tonpilz transducer with central bolt, (c) rectangular bender with compressing frame.

Magnitude of stress in the mechanical systems of air-backed transducers that are developed under action of the hydrostatic pressure may be highly increased compared with magnitude of the pressure itself due to mechanical transformation performed by a mechanical system. Thus, the short rings (at $h/2a < 0.5$) with ends shielded (as shown in Figure 11.5) experience one-dimensional compression stress in the circumferential direction, T_i, which is

$$T_i = \frac{a}{t} P_h \,. \tag{11.3}$$

In the case of the radial polarization the stress is acting in perpendicular, and in case of circumferential polarization – in parallel to poling axis directions. Stress in a thin spherical shell is two-dimensional compression in the plane perpendicular to the poling axis, and

$$T_1 = T_2 = \frac{a}{t} P_h \,. \tag{11.4}$$

In the case of a simply supported rectangular beam transducer (Figure 11.6) magnitude of the stress changes through the thickness and over the length, and can be represented as

$$T_i = \left|T_{i\,\max}\right| \cdot (-2z/t) \cdot (16/5l)(x - 2x^3/l^2 + x^4/l^3), \tag{11.5}$$

where

$$\left|T_{i\,\max}\right| = 0.75 \frac{l^2}{t^2} P_h \,. \tag{11.6}$$

The stress is one-dimensional compression in the upper half of the beam (at $z > 0$) and tension in the lower. In the case of simply supported circular plates (Figure 11.8) the stress is two-dimensional (T_1 in radial and T_2 in circumferential directions). The maximum stress takes place in the center of the plate. At this point

$$T_1 = T_2 = -\frac{2z_0}{t} \cdot \frac{3(3+\sigma)}{8} \frac{a^2}{t^2} P_h \,.. \tag{11.7}$$

The stresses are compression in the upper half and tension in the lower half of the plate and change along the radius. The stresses remain approximately equal at the central part of the plate (up to $r/a \approx 0.5$), and after this they differ, though this difference is insignificant for practical estimation of effect of hydrostatic pressure on parameters of piezoceramics, of which the mechanical systems of the circular benders are built.

Given that in the practical designs it may be $a/t \approx 5 \div 10$ for cylindrical and spherical transducers, and for the benders l/t as well as a/t may be about 5, stress in the piezoelements can be significantly larger than the hydrostatic pressure.

For presenting the hydrostatic pressure and stress in the transducers besides the standard (SI) unit, which is Pa (N/m^2), often are used in literature: psi (lb/in^2), technical atmosphere atm (kg/cm^2) and Bar. The unit's conversions are: 1 psi = 6895 Pa; 1 atm = 101.3 kPa; 1 Bar = 105 Pa. And the depth correspondence is: 1 m $\rightarrow$ 10 kPa; 0.1 atm = 1.46 psi. Given that 1 ft $\approx$ 0.3 m, these relations may be presented accordingly. Thus, 1 ft $\rightarrow$ 0.44 psi.

Both SI and *psi* units are used throughout the chapter to make it easier comparison with previously reported data that are predominantly presented vs. pressure in psi. Operating and/or survival depth for transducers may be up to the full Ocean depth. Putting aside possible extreme conditions, the average depth of the Ocean can be taken about 4,300 m (14,000 ft) that corresponds to the hydrostatic pressure P_h = 43 MPa (6.2 kpsi).

Thus, the compression stress on the ceramics within mechanical systems of air-backed transducer designs, which are the most appreciable in terms of achieving the best electroacoustic characteristics, theoretically may reach values that are beyond the limits of mechanical strength of transducers. Moreover, parameters of piezoceramics are subject to changes that may result in significant degradation of transducers performance under compression stress that are below the limits of mechanical strength. Following Ref. 2 the allowable (that do not produce drastic change of ceramic parameters) compression bias stress and rated tensile stress are presented in Table 11.1 for several ceramic compositions.

Table 11.1: Allowable compression and rated tensile stress for piezoceramics.

	One-dimensional compression MPa (kpsi)		Tensile dynamic and static strength MPa (kpsi)	Hydrostatic pressure MPa (kpsi)
	$\parallel$	$\perp$		
PZT-4	84 (12)	56 (8)	24 (3.5)	350 (5)
PZT-8	84 (12)	56 (8)	35 (5.0)	350 (50)
PZT-5	21 (3)	14 (2)	–	140 (20)

Data presented in the Table must be considered as approximate. They were obtained with small samples of the materials. Actual values may depend upon configuration and quality of fabrication of piezoelements used in the mechanical systems of transducers. Besides, the data are given for temperatures 25°C and may change with rise of temperature. The data were obtained only with compression stress. As to the tensile stress, they are presented based on assumption that the dynamical tensile strength of small specimens is approximately equal to the static tensile strength.

In Ref. 6 the difference was reported between tensile strength of PZT ceramics and bending strength of a bar made of these ceramics. It was found that the bending strength is about 1.7 times larger than the tensile strength for both perpendicular and parallel directions of stress with respect to the poling axis. Possible reason for this may be the statistical effect leading to influence of stress distribution in the body of specimens (see Eq. (11.5)), which are different for both cases.

Thus, the allowed bending stress in the flexural type transducers may be increased approximately in factor of 1.7 compared with the tensile stress given in the Table. Though the values of stress in the Table are suggested as maximum allowable, the changes of ceramic parameters within this range of static stress may be significant enough to be taken in consideration in transducers designing. In operation the transducer is subjected to a maintained constant static stress (bias) and to a slowly variable compression stress due to changing of hydrostatic pressure. Properties of piezoceramics may change with time of being under maintained stress and due to variable stress in reversible and nonreversible way. These effects are dependent on orientation of stress with respect to the polar axis. And they are different for the "hard" and "soft" materials (PZT-4, PZT-8 and PZT-5, in particular). The most comprehensive analyses of the ceramics behavior under pressure were made in Refs. 7-12. Their results are summarized below. Note that values of stress will be given in kpsi, as this is done in the references. Effects of acting stress depend on their orientation with respect to the axis of polarization.

Noteworthy is that the most results of testing were obtained in the referenced works with ceramic specimens of small size (predominantly ½ inch cubes) under unidirectional action of static mechanical stress developed by a press. Applying the compression force and direct measurement of static charge and strain (the latter being produced by metal foil gauges bonded to surfaces of the specimens) required great precautions for achieving an acceptable accuracy of the results. While results obtained with small specimens may be appropriate for investigating physical properties of the piezoceramic materials per se, the accuracy of their application for predicting properties of transducers may be questionable. This is especially true regarding transducers, in mechanical systems of which distribution of hydrostatic stress is inherently nonuniform (e.g., in flexural type transducers). Besides, the effect of the static stress may depend on geometry (aspect ratio) of a piezoelement used. (To some extent different results were observed by authors of the referenced works even on small samples having different aspect ratios.) More appreciable in this sense would be performing testing under pressure the full size piezoelements in the real transducer designs. One of such possibilities is reported in Ref. 15, 16. A brief description of the experimental technique and results obtained therein are presented in Section 11.4.

11.3.3.1 Effects of Compression Stress Acting Parallel to the Polar Axis

Piezoceramics is subjected to compression stress parallel to the polar axis in the mechanical systems of circumferential poled rings and in outer half of longitudinally poled rectangular benders (item (c) in Figure 11.15 and item (b) in Figure 11.6). Note that the inner half of the bender is subjected to parallel tension, and behavior of parameters of ceramic under this action was not investigated. In the rings and rectangular benders with striped electrodes ceramics experiences combined action: parallel to polar axis stress in the main part of the volume and perpendicular stress in some parts. The behavior of parameters of ceramics in the piezoelements of this type was not previously investigated. Some results in this respect were reported in Ref. 16 and are presented in Section 11.4.1.

11.3.3.1.1. PZT-4

Typical effects of the reversible and non-reversible changes of ceramic parameters are illustrated with example of behavior the dielectric constant $\varepsilon_{33}^{T}/\varepsilon_0$ of PZT-4 vs. parallel compression stress acting in cycles up to 140 MPa (20 kpsi) are shown in Figure 11.11.

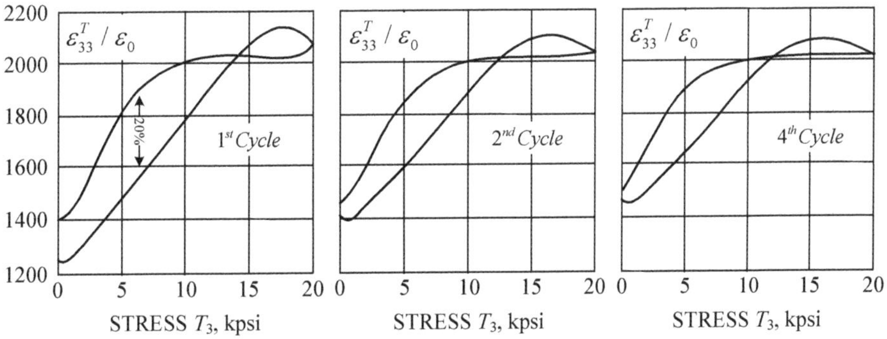

Figure 11.11: Dependencies of $\varepsilon_{33}^{T}/\varepsilon_0$ vs. parallel compression stress in course of the first, second and forth cycles of changing the stress.

During the first cycle of exposure to maximum stress a dramatic raise of about 80% takes place of initial value, which is indicated in specification for the ceramics. In the end of the cycle at zero stress $T_3 = 0$ the value of dielectric constant remains non-reversibly increased in about 13% of the initial value. The second and forth cycles of exposure to the pressure show the same kind of dielectric constant reversible behavior during the cycle, and gradual raise of nonreversible change of its value at $T_3 = 0$ (up to 22% in the end of the 4th cycle).

Such changes of the dielectric constant in the process of transducer operation are not acceptable. For the transmit channel they will result in dramatic detuning with a projector and loss of efficiency. Therefore, the allowable level of the parallel stress for PZT-4 in the Table 11.1 is restricted by 84 MPa (12 kpsi). (Strictly speaking, this should include allowable exposure to the survival depth.) In the range of stress below this level behavior of ceramics parameters vs. compression stress remains qualitatively similar, but quantitatively the values of parameters after maintained exposure to maximum stress (at $T_3 = 0$) and maximum values of reversible changes under variable stress (at variable submergence depth conditions) may be much more modest. Two dependencies are presented in Figure 11.12 of the relative values of dielectric constant $\varepsilon_{33}^T(T_3)/\varepsilon_{33}^T(0)$ for PZT-4 from maximum parallel compression stress ($\varepsilon_{33}^T(0)$ is the initial value before exposure to the stress). The lower curve represents the values after long exposure to the stress T_3 (value stabilized, or "aged" at this stress). It is shown that the finite (stabilized) value is reached after about 5-10 cycles of exposure to the maximum stress. The upper curve represents the value, to which the dielectric constant may be reversibly changed at the maximum stress.

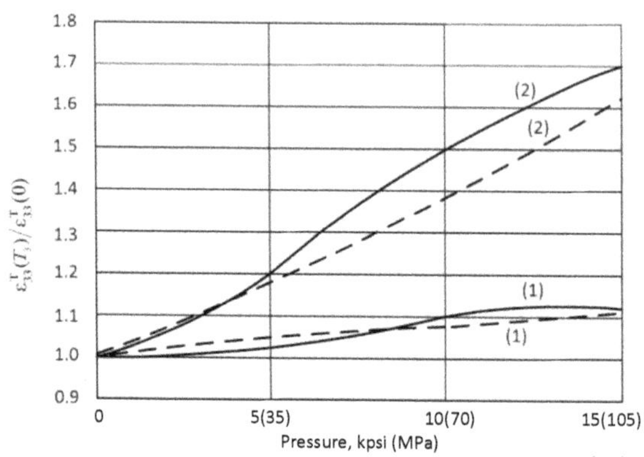

Figure 11.12: Dependencies of values of dielectric constant $\varepsilon_{33}^T(T_3)/\varepsilon_{33}^T(0)$ vs. parallel compression stress: (1) at $T_3 = 0$ after long exposure to the stress, (2) at the maintained stress T_3. Solid lines – for PZT-4, dashed lines – for PZT-8.

With stress changing from 0 to a specified maximum value of T_3 the dielectric constant will change in reversible way with hysteresis that is less than those shown in Figure 11.11

roughly proportional to ratio of maximum stresses. For example, for the maximum stress $T_3 \approx$ 35 Mpa (5 kpsi), which corresponds to operating depth about 700 m for the air-backed circumferentially poled ring having ratio $a/t = 5$, increase of dielectric constant between its value that it has close to surface and at the operating depth will be about 18%.

Analogous dependencies of the piezoelectric modulus d_{33} and $\tan\delta_e$ vs. compression stress are illustrated in the similar way in Figure 11.13(a) and (b), and Figure 11.14.

11.3.3.1.2. PZT-8

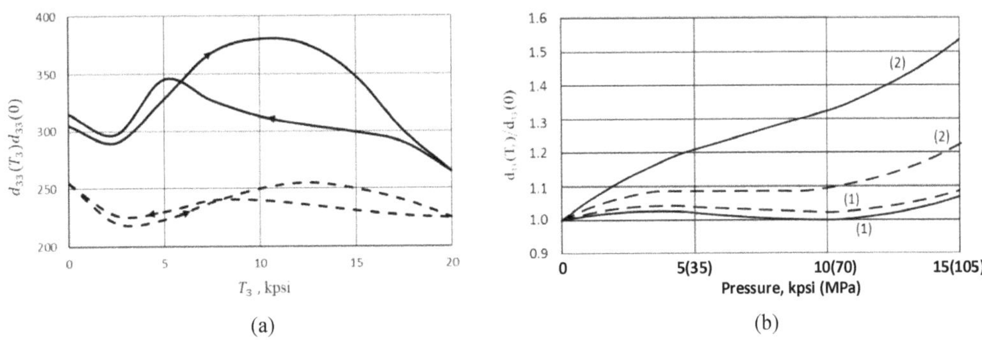

Figure 11.13: Dependencies of values of d_{33} vs. parallel compression stress: (a) in course of the fourth cycles of changing the stress for PZT-4, and in the 4th cycle for PZT-8; (b) at $T_3 = 0$ after long exposure to the stress (1), and at the maintained stress T_3 (2). Sold lines – for PZT-4, dashed lines – for PZT-8.

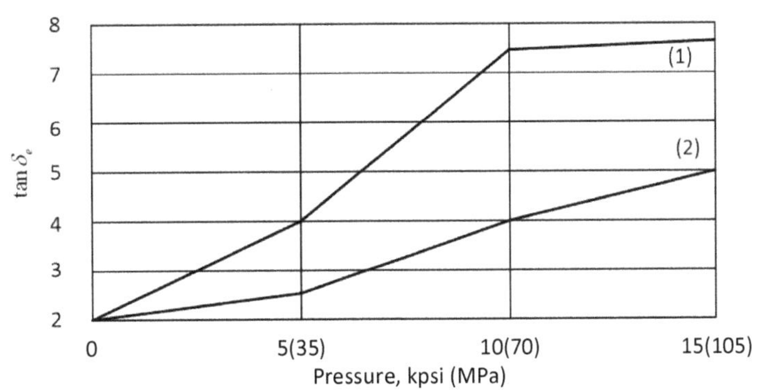

Figure 11.14: Dependencies of values of $\tan\delta_e$ for PZT-4 vs. parallel compression stress: (1) at the maintained stress T_3, (2) at $T_3 = 0$ after long exposure to the stress.

The behavior of PZT-8 ceramics under parallel compression stress is similar to that of PZT-4 but has a smaller range of change. This is illustrated by dependencies of the dielectric constant ε_{33}^{T} and piezoelectric modulus d_{33} vs. compression stress T_3, that are shown in Figure 11.12 and Figure 11.13 by the dashed lines.

But the main advantage of PZT-8 ceramics has under the compression bias at high drive conditions, which are typical for its application for projectors.

11.3.3.1.3. PZT-5A

The behavior of parameters of "soft" PZT-5 ceramics vs. parallel compression stress is quite different. Values of the dielectric constant ε_{33}^{T} and piezoelectric modulus d_{33} degrade sharply in course of the first cycle of exposure to the stress after its value reaches about 8kpsi.

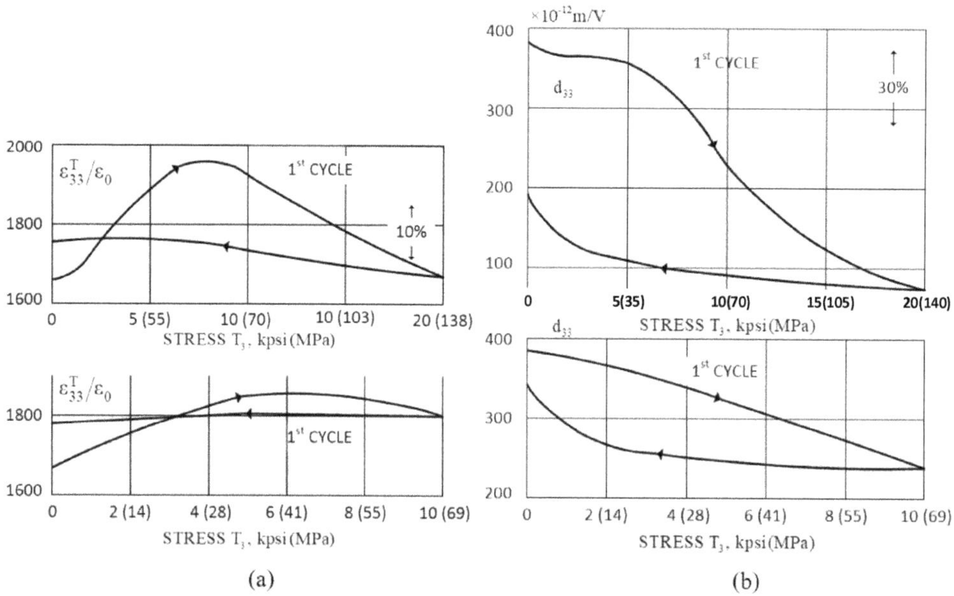

Figure 11.15: Dependences of parameters PZT-5A vs. compression stress T_3 in course of the first exposure to the stress: (a) dielectric constant $\varepsilon_{33}^{T} / \varepsilon_0$, (b) piezoelectric modulus d_{33}. Note that the lower plots are related to exposure to maximum stress 10 *kpsi* that is more typical for PZT-5.

This is illustrated with dependencies of these parameters vs. the stress shown in Figure 11.15(a) and (b). (By this reason the allowed stress in Table 11.1 is restricted by 8 kpsi.)

A favorable property of PZT-5 is that degradation of its parameters under maintained stress is less than for the "hard" materials. In Figure 11.16 comparison is made of relative changes of dielectric constants and piezoelectric moduli of PZT-4 and PZT-5 vs. time of exposure to compression stress $T_3 = 10$ kpsi.

Important peculiarity of PZT-5 (as of all the "soft" materials) is that no stabilization of its parameters after a certain number of cycles of exposure to operating compression stress occurs, unlike this is with "hard" ceramic materials, in which case stabilization of parameters takes place after few cycles of the exposure. This is illustrated with Figure 11.17, where behavior of d_{33} is shown vs. number of cycles of exposure to stress $T_3 = 10$ kpsi for PZT-5 and PZT-4.

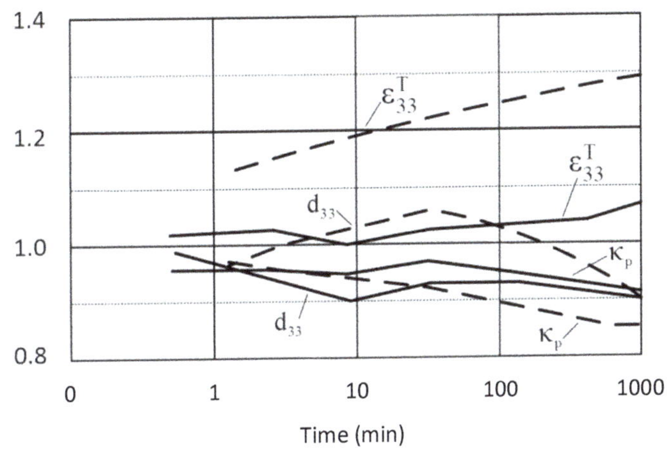

Figure 11.16: Behavior of the relative values of piezoelectric moduli d_{33} and dielectric constants ε_{33}^{T} under exposure to maintained stress $T_3 = 10$ kpsi: solid lines – for PZT-5A, dashed lines – for PZT-4.

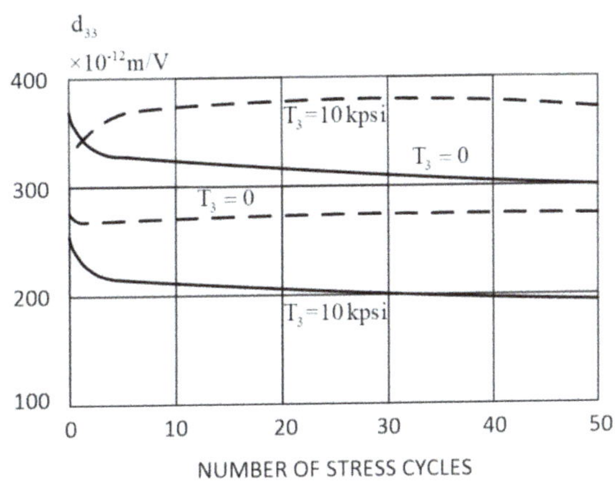

Figure 11.17: Values of piezoelectric modulus d_{33} vs. number of stress cycles up to stress $T_3 =$ 10 kpsi: solid lines – for PZT-5A, dashed lines – for PZT-4.

Thus, it can be recommended to avoid using PZT-5 ceramics for designing transducers that are intended for frequent submergences (exposure to cycling compression stress).

11.3.3.2 Compression Stress Perpendicular to the Polar Axis

Piezoceramics is subjected to compression stress perpendicular to the polar axis (or lateral stress) in the mechanical systems of radial poled rings and in outer half of transversely poled rectangular benders (items (b) in Figure 11.5 and item (a) in Figure 11.6). Note that the inner half of the bender is subjected to perpendicular tension, and behavior of parameters of ceramics under this action was not investigated. The main peculiarity in behavior of parameters of ceramics under the lateral stress is that piezoelectric modulus d_{31} reduces under lateral stress (T_1), whereas d_{32} rises for all the ceramic compositions. Piezoceramics become anisotropic in the plane perpendicular to the polar axis. This effect is large even for "hard" PZT-4 and PZT-8 piezoceramics and is especially pronounced for the "soft" PZT-5A and PZT-5H. Besides PZT-5 suffer remnant changes in ratio d_{31}/d_{32} after exposure to the stress. This is illustrated with plots in Figure 11.18 – Figure 11.20, where dependencies of d_{31} and d_{32} are shown in the 4th cycle of exposer to the lateral stress (with increase of number of the cycles the dependencies remain practically the same).

Ranges of stress for PZT-5A and PZT-5H are reduced because of stronger change of piezoelectric moduli vs. stress.

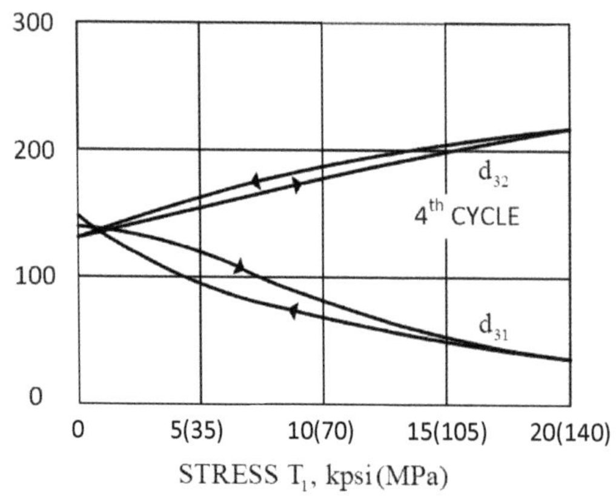

Figure 11.18: Dependencies of d_{31} and d_{32} vs. lateral stress T_1 in 4th cycle of exposure to the stress; PZT-4.

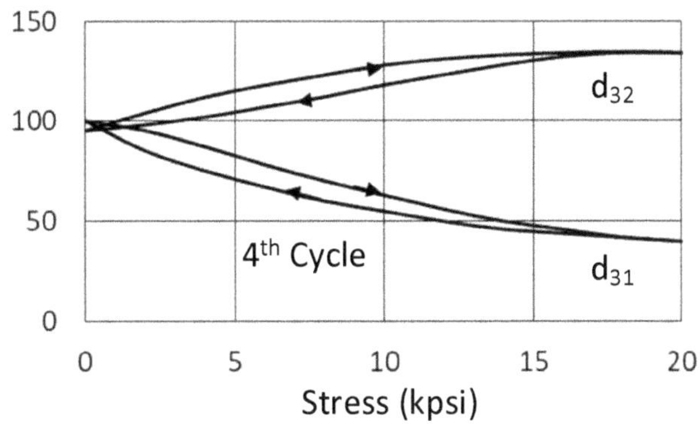

Figure 11.19: Dependencies of d_{31} and d_{32} vs. lateral stress T_1 in 4th cycle of exposure to the stress; PZT-8.

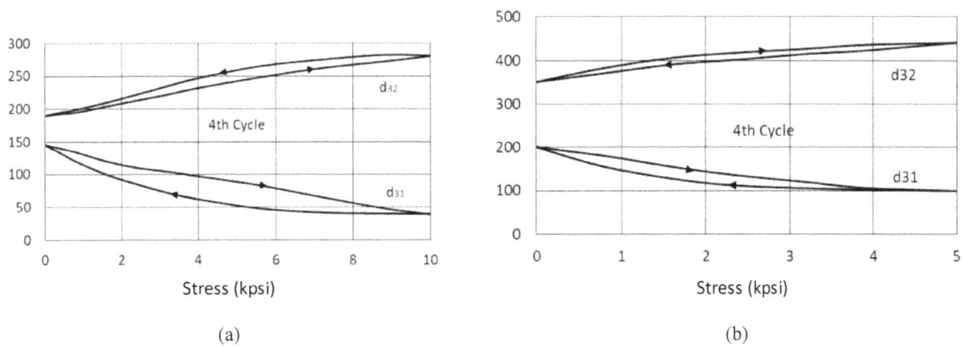

Figure 11.20: Dependencies of d_{31} and d_{32} vs. lateral stress T_1 in 4th cycle of exposure to the stress: (a) PZT-5A, (b) PZT-5H

As it was noted regarding the case of exposure to the parallel stress, in the range of stress below the maximum levels in cycles shown in Figure 11.18 – Figure 11.20 behavior of d_{31} and d_{32} vs. compression stress remains qualitatively similar, but quantitatively differences of their values at maximum operating stress T_1 and after removal of the maintained exposure to this stress (i.e., at $T_1 = 0$) are less pronounced. The plots are presented in Figure 11.21 and Figure 11.22 analogous to those shown in Figure 11.13 for making quantitative estimations of the values of piezoelectric moduli at different operating conditions. Dependencies are shown of the relative values of piezoelectric moduli $d_{3i}(T_1)/d_{31m}$ and $d_{3i}(T_1 = 0)/d_{31m}$ (where $i = 1,2$ and d_{31m} is the initial value before exposure to the compression stress). The lower curve represents the values after long exposure to the stress T_1 (stabilized, or "aged" at this stress). It follows that the finite (stabilized) values are reached after about 5-10 cycles of exposure to the maximum stress T_1 = 10 kpsi. The upper curve represents the level to which the value of may be reversibly changed at the maximum stress. At each maximum operating depth values of d_{3i} change vs. variable depth submergence qualitatively in the way as shown in Figure 11.19. For example, at the operating depth that corresponds to stress T_1 = 10 kpsi the changes occur between the values in points indicated as min and max.

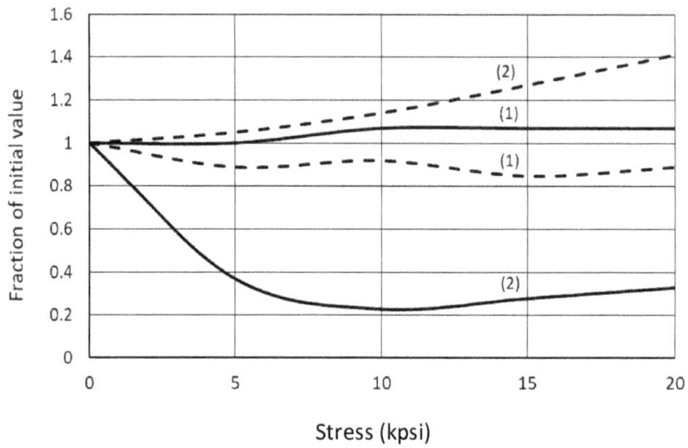

Figure 11.21: Dependencies of values of d_{31} and d_{32} for PZT-4 vs. lateral compression stress: (1) at $T_1 = 0$ after long exposure to the stress, (2) at the maintained stress T_1. Solid lines for d_{31}, dashed lines for d_{32}.

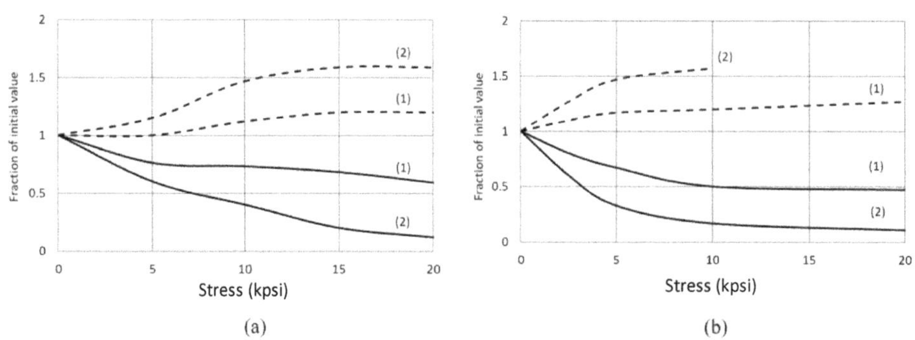

Figure 11.22: Dependencies of values of d_{31} and d_{32} vs. lateral compression stress: (1) at $T_1 = 0$ after long exposure to the stress, (2) at the maintained stress T_1. Sold lines for d_{31}, dashed lines for d_{32}. Figure (a) for PZT-5A, (b) for PZT-5H.

Plots in Figure 11.21 and Figure 11.22 show that "aging" of the piezoelectric moduli under the maintained lateral compression is not significant for "hard" piezoelectric ceramics and is rather severe for "soft" PZT-5A and PZT-5H ceramics. Remnant increase of d_{32} after prolonged exposure to compression stress $T_1 = 10kpsi$ is about in factor of 1.2. This property of PZT-5 materials theoretically can be used for improving hydrophones based on their employment. How this can be done is illustrated with Figure 11.23. It is noteworthy though that no data were available regarding behavior of thus improved transducers under compression stress T_2 until later investigation reported in Ref. 15. The results of this investigation are presented in Section 11.4.2.

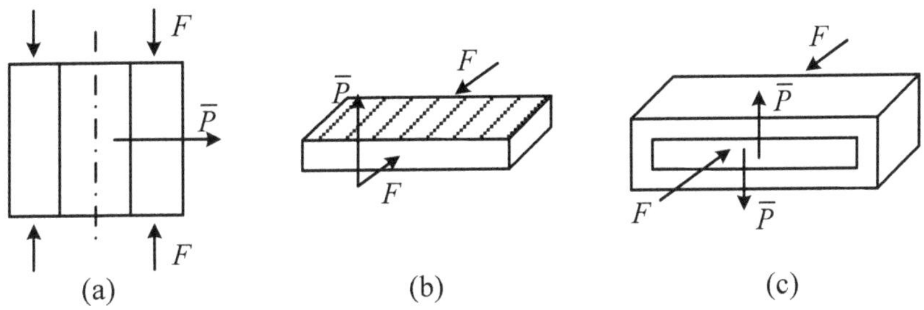

Figure 11.23: To the improving piezoelectric modulus in "working" direction of the piezoelements after subjecting them to prolonged compression in the direction perpendicular to the polar axis: (a) ring transducer, (b) length-expander side-electroded bar, (c) piezoelements for the rectangular bender transducer.

The effect of the lateral stress on the dielectric constants is relatively small. The nature of changing dielectric constants under varying stress is shown in Figure 11.24 for the "hard" and in Figure 11.25 for the "soft" ceramics with examples of the 4th cycle (for greater number of the cycles the dependencies remain qualitatively similar). In the same figures are also presented the dependencies that show correlation between the temporal and maximum reversible changes of the dielectric constants under stress that may correspond to different operating depths.

Knowing behavior of $\tan\delta_e$ vs. lateral compression stress is important for designing projectors that employ transverse piezoeffect. These dependencies measured at small electrical signals are shown in Figure 11.26 for PZT-4 and PZT-8. It is of note that no data is available in literature on the combined effect of the large electric fields and lateral stress on $\tan\delta_e$.

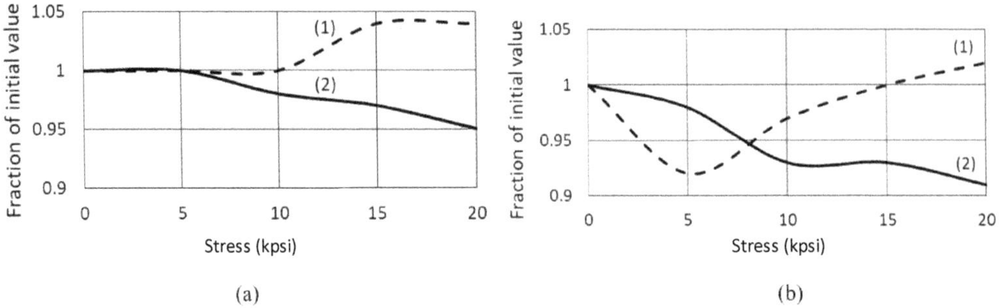

Figure 11.24: Dependencies of values of $\varepsilon_{33}^{T} / \varepsilon_0$ vs. lateral compression stress: (1) at $T_1 = 0$ after long exposure to the stress, (2) at the maintained stress T_1: (a) for PZT-4, (b) for PZT-8.

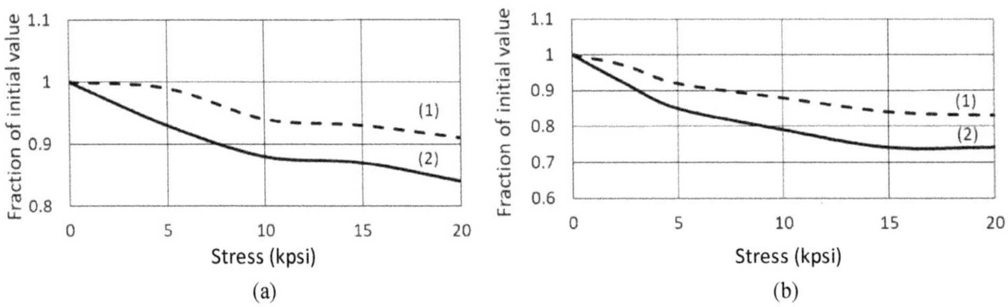

Figure 11.25. Dependencies of values of $\varepsilon_{33}^{T} / \varepsilon_0$ vs. lateral compression stress: (1) at $T_1 = 0$ after long exposure to the operating stress, (2) at the maintained stress T_1: (a) for PZT-5A, (b) for PZT-5H.

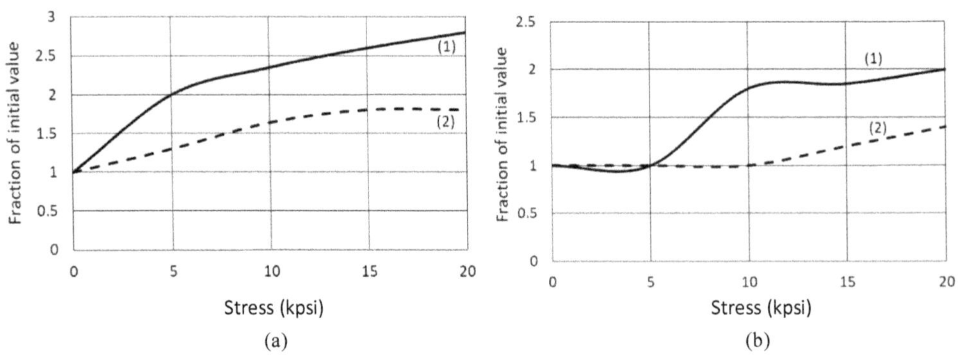

Figure 11.26: Dependences of $\tan \delta_e$ vs. lateral compression stress: (1) at $T_1 = 0$ after long exposure to the stress, (2) at the maintained stress T_1: (a) PZT-4, (b) PZT-8.

Analysis of the data regarding behavior of parameters of ceramics under the lateral stress leads to conclusion that the advantage of "soft" PZT-5 ceramics as material for designing hydrophones holds for shallow water applications. Judging by value of parameter $d_{31}/\sqrt{\varepsilon_{33}^T}$ that is related to the specific sensitivity γ_{sp} of a hydrophone (see (11.2)) even at moderate depths PZT-4 ceramic may successfully compete in this respect, especially in terms of stability under pressure.

11.3.3.3 Two-dimensional Compression Stress in Plane Perpendicular to the Polar Axis

These conditions exist in spherical transducers and with reasonable approximation in circular bender transducers shown in Figure 11.8. The most comprehensive data on behavior of ceramic parameters under the varying planar stress $T_p = T_1 = T_2$ up to T_p = 240 Mpa (35 kpsi) are presented in Ref. 11. Changing of the dielectric constants ε_{33}^T, elastic constants $s_p^E = 2(s_{11}^E + s_{12}^E)$, piezoelectric moduli d_{31} and coupling coefficients k_p were measured at small signals for PZT-4 and PZT-5 ceramics. The dielectric constant and dielectric loss factor $\tan\delta_e$ were measured also as functions of simultaneously acting driving electric field up to E = 3 kV/cm (rms) for PZT-4. The testing was conducted with spheres having diameter to thickness ratios $D/t = 20$ and $D/t = 12$ under the hydrostatic pressure up to 48 MPa (7 kpsi) that corresponds approximately to the full Ocean depth (on average). The results obtained are presented in Figure 11.27 – Figure 11.31.

Figure 11.27: Properties of PZT-4 under maintained planar stress.

Results presented in the figure were obtained in the first cycle of pressure with fresh samples. Measurements made in successive stress cycles showed that parameters of PZT-4 ceramics remained approximately independent of the number of the cycles.

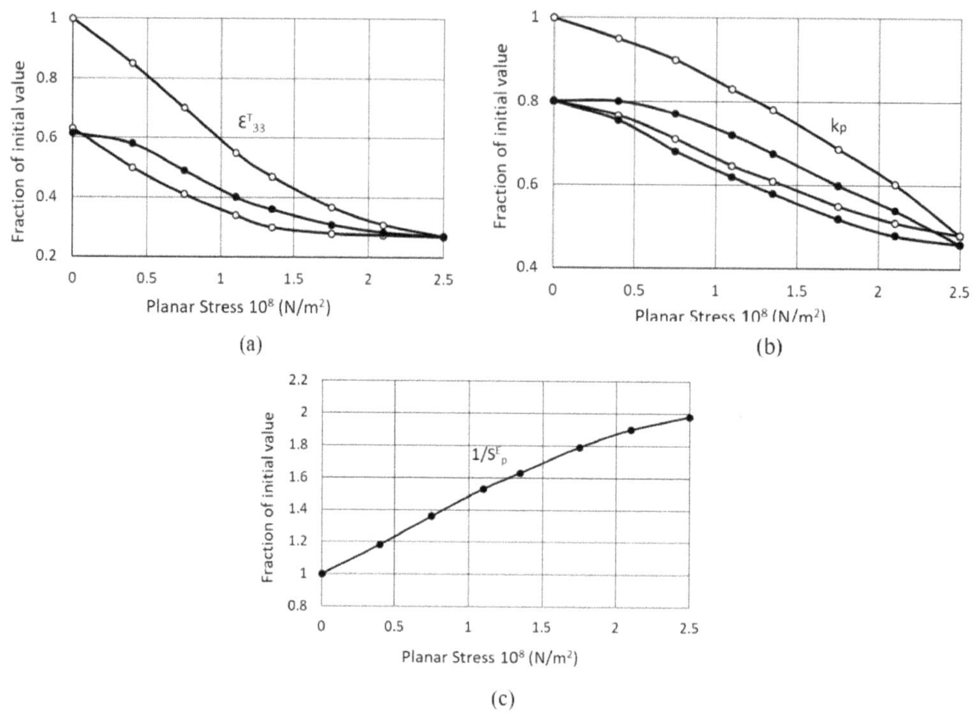

Figure 11.28: Properties of PZT-5 under maintained planar stress in the first two cycles on a fresh sample: ○ - first cycle, • - second cycle.

Unlike for PZT-4 the values of ε_{33}^{T} and d_{31} for PZT-5 ceramics depend on the number of cycles. They reach stable values after approximately 20 cycles of exposure to an operating pressure. This is shown in Figure 11.29.

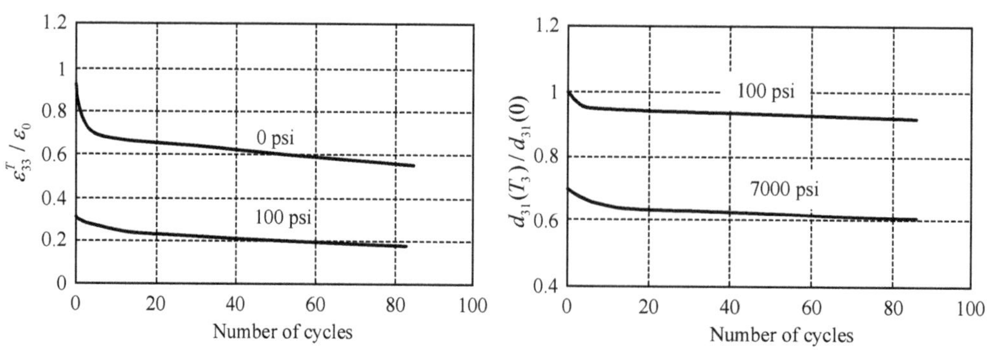

Figure 11.29: Effect of repeating pressure cycles on relative values of parameters of PZT-5 ceramics (in factions of unity).

Thus, parameters of transducers made of PZT-5 ceramics that experience planar static compression can be stabilized (though to the lower values) before they are used for applying multiple cycles of operating pressure. After such stabilization the properties of PZT-5 almost do not change in time under the operating pressure. In opposite the PZT-4 ceramics exhibits a significant aging in time under the same stress conditions, as this is shown in Figure 11.30 for the dielectric constant ε_{33}^{T}. This must be taken into consideration when designing spherical transducers for deep water applications.

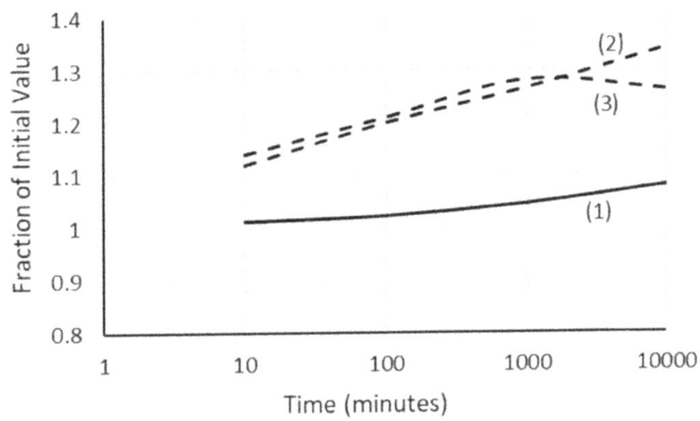

Figure 11.30: Aging of the dielectric constant ε_{33}^{T} of ceramics under the planar stress 250 MPa; solid line – PZT-5, dashed line – PZT-4.

Important for designing the spherical projectors made of PZT-4 ceramics are the data on behavior of the dielectric constant ε_{33}^{T} and dielectric loss factor $\tan\delta_e$ under combined action of the planar compression and electric field that are presented in Figure 11.31.

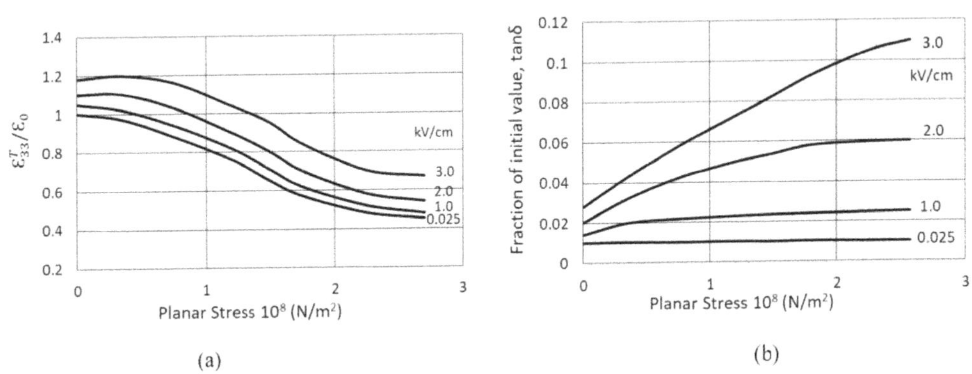

Figure 11.31: Dependence of the dielectric constant ε_{33}^{T} and dielectric loss factor $\tan\delta_e$ of PZT-4 from planar compression under different driving fields.

Comparison with analogous dependencies for the case of parallel one-dimensional compression at the same level of stress (see Figure 11.36) shows more severe raise of the loss factor.

11.3.3.4 Effects of the Hydrostatic Compression on Ceramics Parameters

In this case the piezoelements experience hydrostatic (uniform three-dimensional) compression in the pressure compensated transducer designs shown in Figure 11.19. There is no transformation of external pressure in the mechanical systems of these transducer designs. Therefore, the three-dimensional compression stresses do not exceed T_h = 43 Mpa (6.2 kpsi) at full (in average) Ocean depth. Such actions on the ceramics are relatively small compared with those that are experienced in the air-backed designs. The hydrostatic compression has relatively little effect on the domain's reorientation, and its effect on the ceramics properties is much less (up to one order of magnitude) than with other stress configurations, as this is noted and experimentally confirmed in Ref. 9 for piezoceramics PZT-4. Thus, the percentage change of the piezoelectric moduli d_{31} and d_{33} over their initial values at atmospheric pressure with increase of the hydrostatic pressure up to 10 *kpsi* was less than 7%, as it follows from Figure 11.32. The initial values of piezoelectric moduli were $d_{33} = 254 \cdot 10^{-12}$ m/V and $d_{31} = 124 \cdot 10^{-12}$ m/V with deviations between the tested samples about 2%.

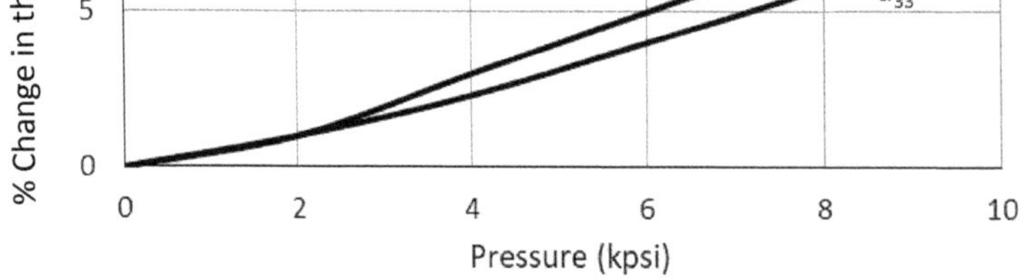

Figure 11.32: Variations of the piezoelectric moduli d_{31} and d_{32} with hydrostatic pressure.

Variations of the dielectric constant ε_{33}^{T} and dielectric loss factor $\tan\delta_e$ with hydrostatic pressure at different electric fields applied are shown in Figure 11.33.

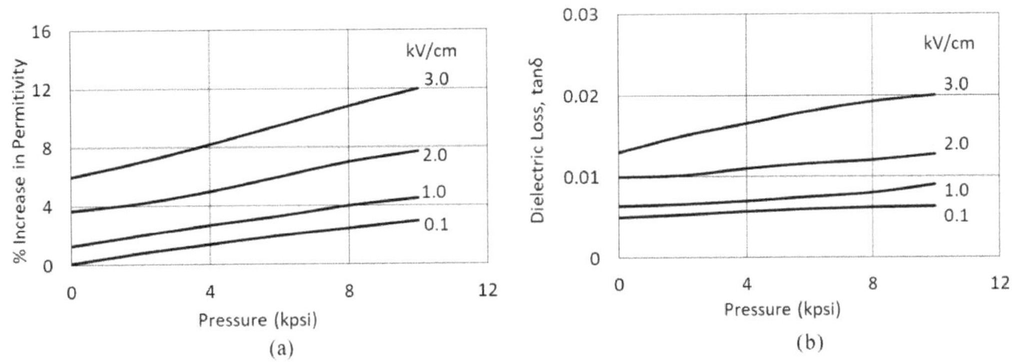

Figure 11.33: Variations of parameters with hydrostatic pressure at different electric fields: (a) ε_{33}^{T}, (b) $\tan\delta_e$.

The increase of the relative dielectric constant over its reference value at small signals and atmospheric pressure (that was $\varepsilon_{33}^{T}/\varepsilon_0 = 1190$) at E = 2 kV/cm and T_h = 10 kpsi is about 8%, whereas for PZT-4 ceramics with parallel stress T_h = 10 kpsi it was about 30%. It must be noted that for a ring transducer of air-backed design having radius to thickness ratio $a/t = 5$ this stress would correspond to operating depth about 1400 m, whereas for the pressure compensated (free flooded in particular) design – to about 7000 m. Dielectric loss factor changes not significantly under hydrostatic pressure compared with its increase vs. electric field at atmospheric pressure. It is noted in Ref. 10 that variations in the pressure and field dependences for samples from different batches or produced by different manufacturers may be significantly greater than it could be expected from deviation of the same parameters at low signals and atmospheric pressure.

The dielectric properties at high electric field reach their values shortly after application of the hydrostatic pressure and remain stable in time afterwards. Small changes of parameters under pressure and their time stability make transducers of pressure compensated design especially suitable for deep water applications. Though some complications regarding the frequency responses and directional factors of such transducers may arise due to additional acoustic effects that accompany designs of this kind. Influence of these effects on transducer parameters will be considered in Ch. 13 and 14 for the hydrophones and projectors respectively.

11.3.3.5 Combined Effect of the Compression Bias and Hydrostatic Pressure

This combination of the stress is typical for pressure compensated (free flooded in particular) projector designs shown in Figure 11.10, in which the compression one-dimensional stress bias is imposed on ceramics in order to increase dynamic strength of the transducers by avoiding tensile strain. Ceramics in the prestressed projectors is usually aged under the compression bias prior to subjecting to hydrostatic pressure. Effect of superposing the hydrostatic pressure and one-dimensional compression bias, under which the ceramics was aged, on the high-field dielectric properties of PZT-4 ceramics was considered in Ref. 10. Results of testing samples subjected to parallel compression bias of 2, 5 and 10 kpsi under the hydrostatic stress up to 10 kpsi showed that percentage change of the high-field dielectric properties (ε_{33}^{T} and $\tan\delta_e$) over their values aged under the bias is approximately the same, as under action of the hydrostatic stress alone. This is illustrated with plots in Figure 11.34 with 5 kpsi bias stress only, but effect of the magnitude of the bias stress on the dependence of parameters on the depth is small. It is noted also that the results did not depend on the number of pressure cycles.

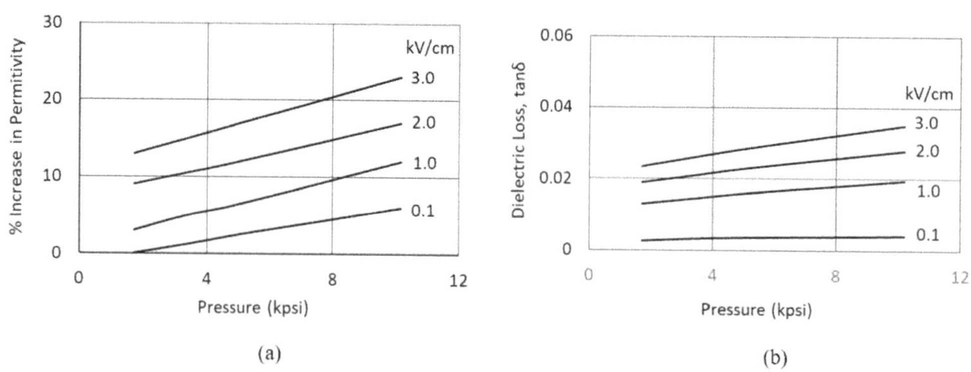

Figure 11.34: Variation of the (a) dielectric constant ε_{33}^{T} and (b) $\tan\delta_e$ with hydrostatic pressure for PZT-4 ceramics subjected to 5 kpsi parallel bias stress.

11.3.3.6 Combined Effect of Parallel Compression Stress and Ac Electric Field on Piezoceramics Parameters

The parallel compression bias is the most typical for the projectors. Therefore, the combined effect of the parallel stress and electric field on parameters of piezoceramics is of great interest. Dependencies of the dielectric constants and $\tan\delta_e$ vs. electric field at different compression stress are presented in Figure 11.35 and Figure 11.36.

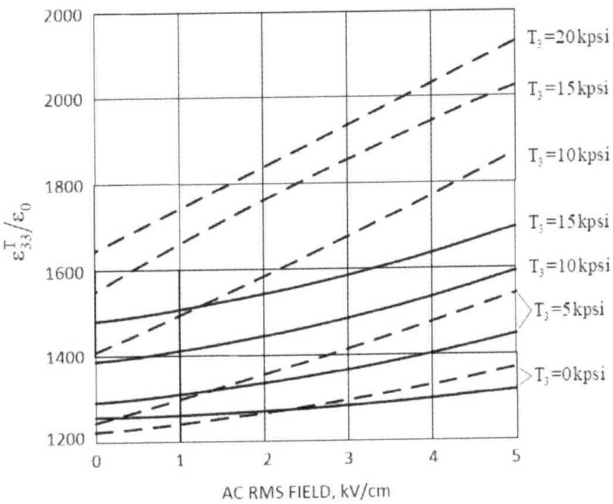

Figure 11.35: Dependencies of $\varepsilon_{33}^T / \varepsilon_0$ vs. electric field (in rms) at different values of compression stress: PZT-8 – solid lines, PZT-4 – dashed lines.

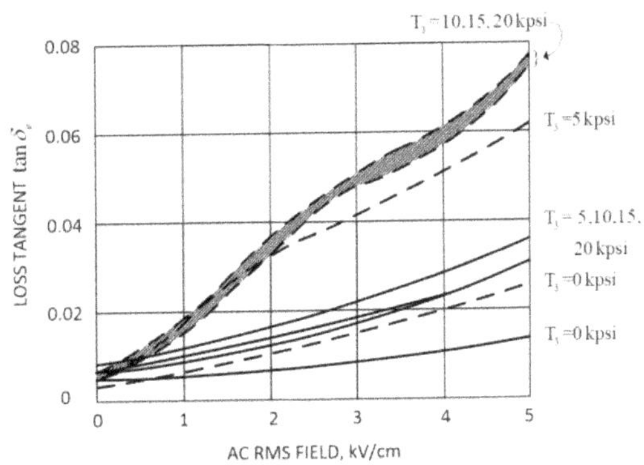

Figure 11.36: Dependencies of $\tan\delta_e$ vs. electric field (in rms) at different values of compression stress: PZT-8 – solid lines, PZT-4 – dashed lines.

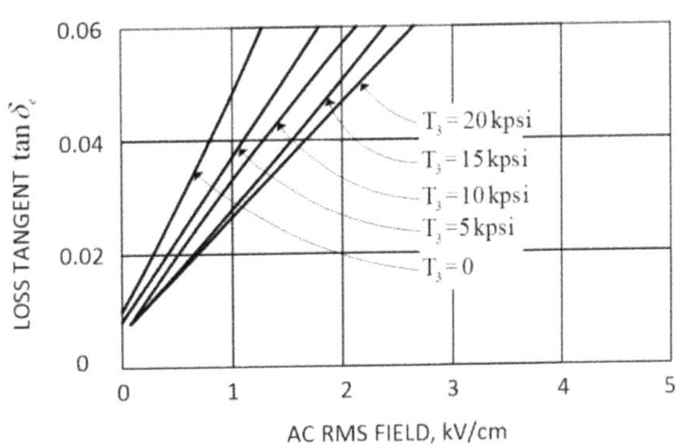

Figure 11.37: Dependencies of $\tan\delta_e$ vs. electric field (in rms) at different values of compression stress for PZT-5A.

They show clear advantages of PZT-8 ceramics for high drive applications. In Figure 11.37 dependencies of $\tan\delta_e$ vs electric field are shown for the "soft" PZT-5A ceramics that is not intended for application in high drive transducers due to large dielectric losses, particularly at high electric fields and under the static pressure.

11.4 Testing in High Pressure Air Chamber

The most natural way of investigating behavior of electromechanical parameters of piezoceramics in real piezoelements used in the transducer designs would be testing the transducers under changing hydrostatic pressure in the water filled pressure vessels (tanks). The most accurate measurements of the parameters in principle can be made by the dynamic resonance-antiresonance method. But this method cannot be used in a water filled tank because of acoustic loading of the transducers and possible tank resonances.

Thus, for example, when investigating parameters of ceramics under the planar stress with spheres subjected to hydrostatic pressure in water filled vessel, for getting results authors of Ref. 10 were forced to use strain gauges and the reciprocity method for determining the low frequency sensitivity of the transducer. This made the measurement procedure quite complicated and significantly reduced accuracy of the results.

The above problems can be overcome, if to perform testing in the air pressure vessel (or small chamber) having sufficient pressure multiplied by volume ($PS\cdot\tilde{V}$) capabilities. Equipment for such purposes is not commercially available and has to be custom made. But this is a solvable engineering problem and once being built equipment of this kind may allow a broad spectrum of investigations of transducer parameters under pressure using the most accurate dynamical measurements. Examples of performing testing the piezoelements under pressure in the air chamber are presented in Ref. 15 and 16. All the experiments were performed in a custom air pressure chamber with air pressure capabilities up to about 10 MPa ($\approx$ 1500 psi) that corresponds to 1000 m of operating depth for a transducer. Parameters of the piezoelements were measured by an Impedance Analyzer. Two subjects were considered: (a) effects of static circumferential stress on the piezoelectric properties of tangentially polarized stripe-electroded cylinders, which could not be predicted based on the available dependencies of parameters under the parallel and lateral stress due to nonuniformity of the electric field of polarization; (b) possibility of increasing piezoelectric effect in radially polarized soft PZT cylinders by pressure treating that was predicted in earlier work[7] but never practically implemented. The results obtained are presented in the following sections.

11.4.1 Effects of Static Circumferential Stress on the Piezoelectric Properties of Tangentially Polarized Stripe-Electroded Cylinders

The electric field of polarization in the tangentially polarized (stripe-electroded) piezoelements is inherently non uniform. Schematic representation of the electric field geometry that was suggested and used for analysis of electromechanical properties of the piezoelements in Ref. 17 is shown in Figure 11.38. In the stripe-electroded cylinders the stress acting in the circumferential direction produces predominantly longitudinal piezoeffect in region *A* and predominantly transverse piezoeffect in the region *B*. As a result, the effective piezoelectric modulus of the tangentially polarized piezoelement reduces in comparison with those under regular

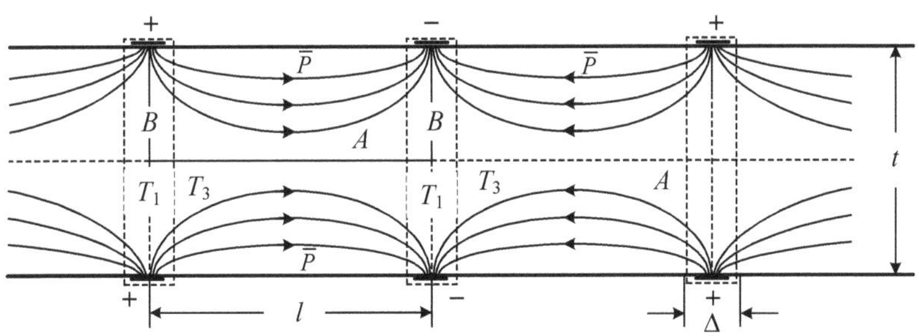

Figure 11.38: Representation of the electric field in the tangentially polarized (stripe-electroded) piezoelement

longitudinal piezoeffect, which is quantitatively estimated in Section 7.2.2.1.2. Qualitatively the value of effective piezoelectric modulus of the tangentially polarized piezoelement can be represented as

$$d_{33eff} = Ad_{33} - B|d_{31}|, \qquad (11.8)$$

where coefficients A and B depend on the geometry of the piezoelement and electrodes (thickness t, l and Δ, in particular). Thus, it can be expected that sensitivity of the effective electromechanical parameters of the tangentially polarized piezoelements to compression circumferential stress can be different from those under the stress parallel to polar axis considered above.

The more so, as behavior of the parameters under the parallel and perpendicular to polar axis compression stress is different. For verifying this assumption and for getting information that was not available from previous works, dependencies of the effective electromechanical parameters of the stripe-electroded piezoelements from the compression stress produced by the hydrostatic pressure were investigated.

The testing was performed using a custom air pressure chamber capable of producing pressure up to P = 10 MPa (≈ 1500 psi) with samples of the stripe-electroded cylinders made of commercial hard PZT-4 like piezoceramics (further referred to as samples *I*) and commercial soft PZT-5A like piezoceramics (further referred to as samples *II*). Dimensions of the cylinders (mean diameter $2a$ = 35 mm, height h = 13 mm, thickness t = 3.15 mm) insured close to uniform circumferential stress distribution under the external pressure. The samples had N = 12 striped electrodes of Δ = 1.5 mm width. The cylinders were installed in the fixture shown in Figure 11.39 that imitated the air-backed transducer design. The pressure P in the chamber and the circumferential stress T_c in the cylinders were related as $T_c = P \cdot (a/t) = 5.5P$.

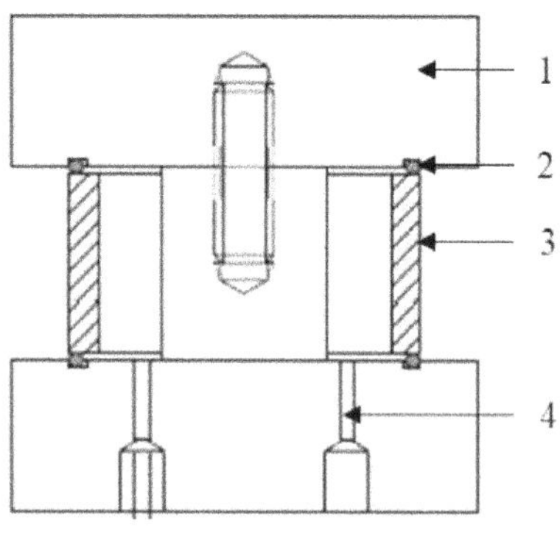

Figure 11.39: Schematic view of the fixture for exposing the samples to the external pressure; 1 – end caps with supporting stud, 2 – O-ring, 3 – tangentially polarized piezoelectric cylinder, 4 – leads from the internal volume

In process of the testing the samples were subjected to increasing and decreasing stress cycles ranging from 0 up to 8250 psi with 825 psi increments. At each pressure step all the parameters (f_T, f_{ar}, C_{Lf}) were measured with an Impedance Analyzer. The effective electromechanical parameters of the stripe-electroded cylinders were calculated from results of the measurements by the following formulas (see Section 7.2.2.1.2):

the *effective coupling coefficient*,

$$k_{33eff}^2 = 1 - (f_r / f_{ar})^2 ; \tag{11.9}$$

the *effective dielectric constant*,

$$\varepsilon_{33eff}^T = \frac{C_{Lf}}{1 - k_{eff}^2} \frac{th}{l} N , \tag{11.10}$$

where N is the number of stripes and l is the separation between the stripes:

the *effective elastic constant*,

$$s_{33eff}^E = 1 / f_r^2 (2\pi a)^2 \rho \tag{11.11}$$

the effective piezoelectric modulus,

$$d_{33eff} = k_{33eff} \sqrt{\varepsilon_{33eff}^T s_{33eff}^E} \, . \tag{11.12}$$

In Figure 11.40 -Figure 11.43 are presented results of the 4-th circle of the testing. After the first cycle of testing some remnant changes were observed in results of measurements, and after the second circle they practically stabilized.

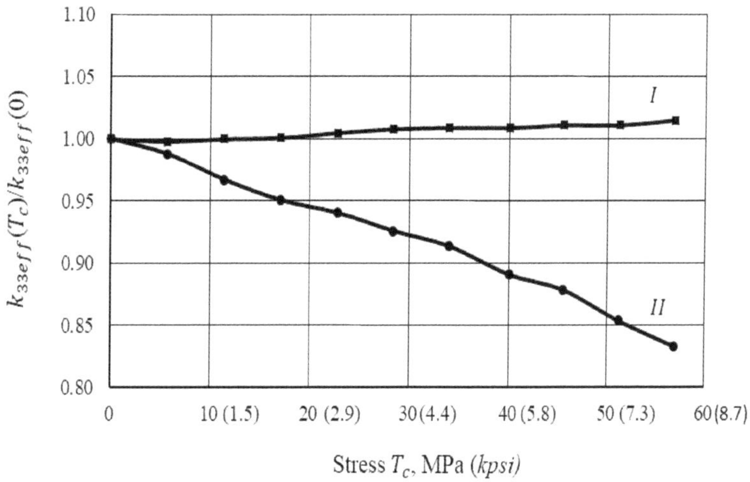

Figure 11.40: Dependence of the relative change of effective coupling coefficient $k_{33eff}(T_c)/k_{33eff}(0)$ of ceramics *I* and *II* under compressive circumferential stress T_c.

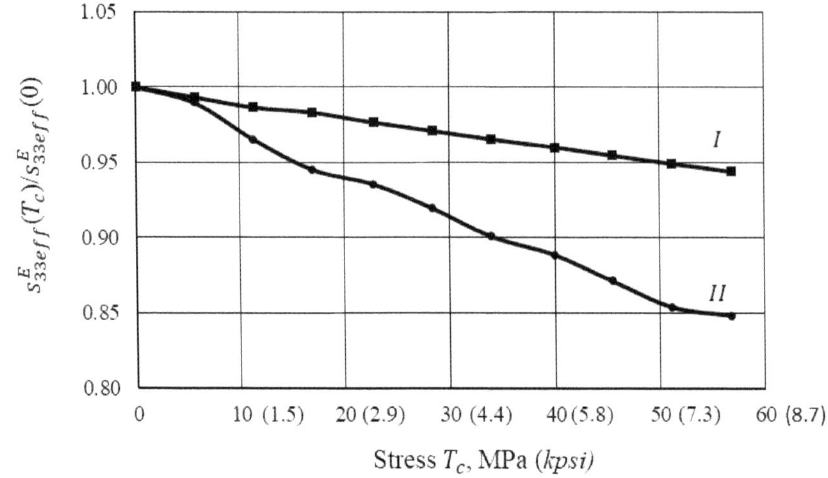

Figure 11.41: Dependence of relative change of the effective elastic constant $s^E_{33eff}(T_c)/s^E_{33eff}(0)$ of ceramics *I* and *II* under compressive circumferential stress T_c.

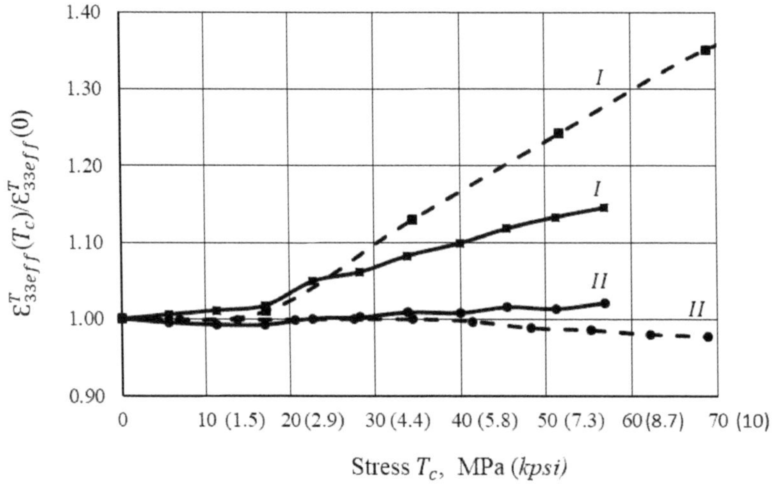

Figure 11.42: Dependence of relative change of the effective dielectric constant $\varepsilon^T_{33etf}(T_c)/\varepsilon^T_{33eff}(0)$ of ceramics *I* and *II* under compressive circumferential stress T_c.(solid lines). Dashed lines – results from Figure 11.12 for $\varepsilon^T_{33}(T_3)/\varepsilon^T_{33}(0)$ at the stress T_3 up to 10 *kpsi*)

As follows from the plots in Figure 11.42 and Figure 11.43, the relative change of the effective dielectric constants and piezoelectric moduli of the tangential polarized piezoelements made both from the hard and soft ceramics under the circumferential compression stress (shown by the solid lines *I* and *II*) are less pronounced than for the PZT-4 and PZT-5A ceramics under compression stress parallel to the polar axis, as was reported in Ref. 8 (shown by the dashed lines *I* and *II*). It is noteworthy that analogous comparison for the elastic constants and coupling coefficients cannot be made, because data on their sensitivities to the parallel compression stress for regularly polarized ceramics are not available from literature.

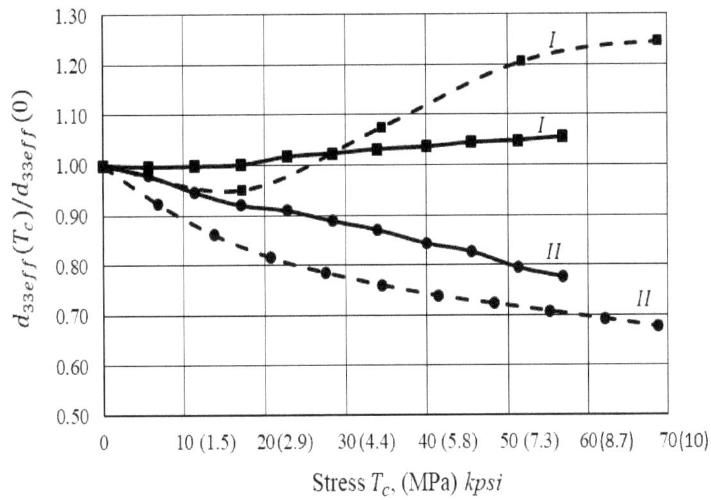

Figure 11.43: Dependence of the relative change of effective piezoelectric modulus $d^T_{33eff}(T_c)/d^T_{33eff}(0)$ of ceramics *I* and *II* under compressive circumferential stress T_c (solid lines). Dashed lines - results from Figure 11.13 (b) for $d^T_{33}(T_3)/d^T_{33}(0)$ at stress T_3 up to 10 *kpsi*)

11.4.2 Increasing Piezoelectric Effect in Radially Polarized Cylinders Made of Soft PZT Ceramics by Pressure Treating

It was shown in previous investigations (Ref. 7) that piezoceramics became anisotropic in the plane perpendicular to the polar axis being subjected to the lateral (perpendicular to the axis of polarization) compression stress. Namely, the piezoelectric modulus d_{31} reduces under lateral stress, whereas d_{32} rises for all the piezoceramic compositions. The changes in ratio d_{31}/d_{32} for PZT-5A and -5H piezoceramics became remnant after exposure to the first cycle of the stress and then remain practically the same after the subsequent cycles. This is illustrated with plots in Figure 11.20 (a) and (b), where dependencies of d_{31} and d_{32} are shown in the 4th cycles of exposure to the lateral stress for PZT-5A and -5H piezoceramics. The effect of uniaxial compression prestress on the properties of soft PZT piezoceramics was also investigated recently in Ref. 18, and the results obtained agree with previously predicted.

This property of the soft piezoceramics can potentially be used for improving parameters of transducers after the pressure treating of their piezoelements by applying the compression force as is illustrated in Figure 11.23 with examples of side-electroded length expander bar, hollow cylinder, and rectangular flexural plate. Without doubt these improvements can be used in the transducers intended for applications in air or in shallow water. But the question arises regarding behavior of parameters of the pressure treated piezoelements under significant operating hydrostatic pressure. Obviously, in the air-backed designs of the transducers operating hydrostatic pressure must produce a reverse effect on the piezoelectric moduli, and it can be expected that at some level of the pressure their values may return to original before treatment. Investigation of the possible applications in the transducer designs of the pressure treated piezoelements in the shape of cylinders was reported in Ref. 15. The results obtained are summarized below.

All the testing was performed in the same air pressure chamber that was referred to in the previous section. Samples of the radial poled cylinders with mean diameter $2a = 35$ mm, height $h = 13$ mm, and thickness $t = 3.15$ mm made of commercial soft PZT piezoceramics (further referred to as ceramic A) were used for the experimenting. The aspect ratio $h/2a \approx 0.35$ guaranteed that deformation of cylinders was one-dimensional both in axial and circumferential directions. The experiments were performed with three prototype configurations that are schematically illustrated with sketches in Figure 11.44 (a), (b) and (c). They insured different actions of stress in the cylinders depending on the goal of the experiment. Analogous cylinders made of Navy Type-I piezoceramics (further referred to as ceramic B) were tested simultaneously in the same way for comparison.

In the case (a) the cylinders were exposed to axial stress $T_1 = (S_{cap} / S_{csc}) \cdot P$, where $S_{cap} = \pi r_{cap}^2$ with $r_{cap} = 27.5$ mm and $S_{csc} = 2\pi at$ is the cross-section area of the cylinder. Thus, $T_2 = (a/t) \cdot P = 6.6 \cdot P$. In the case (b) surface of the cylinder with ends mechanically isolated from the caps was exposed to the radial pressure that imitated stress conditions in the air-backed transducer design. Correlation between the pressure applied and circumferential stress T_2 in ceramic was $T_2 = (a/t) \cdot P = 5.5 \cdot P$. In the case (c) the bare cylinders were subjected to pressure over all the surfaces that imitated hydrostatic stress conditions in the pressure equalized (liquid filled) transducer design.

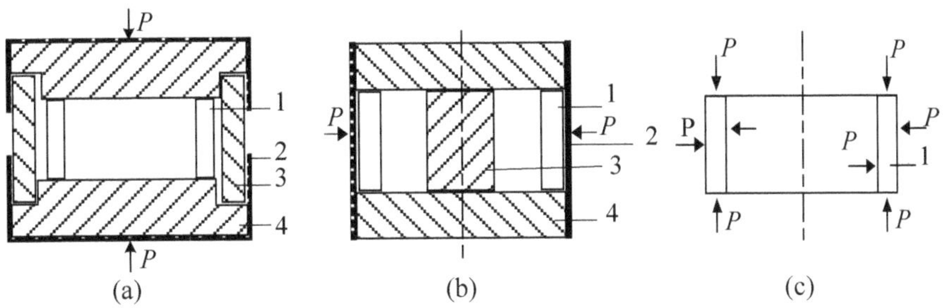

Figure 11.44: Schematics of the prototypes designed for testing cylindrical piezoelements under the external actions: (a) in course of the pressure treating; (b) under condition of use in the air-backed design; (c) for application in the pressure equalized (liquid filled) design. 1 – piezoceramic cylinder, 2 – polyurethane boot, 3 – mid-section/central support, 4 – end cap

All the prototypes were subjected to increasing and decreasing cycles of pressure ranging from 0 up to 1500 psi in 150 psi increments. The pressure was held for 10 minutes at each increment step and was maintained for 30 minutes at 1500 psi. Electromechanical parameters of the samples were determined from results of measuring with Impedance Analyzer by standard formulas from Ref. 19.

11.4.2.1 Results of the Pressure Treating

Only the capacitance and conductance of the cylinders could be accurately measured in the course of testing under the axial pressure because the resonance and antiresonance frequencies were subjected to change due to attachment of the caps used for application of the pressure. Changes in all the other parameters were determined after the pressure was released.

Dependence of the dielectric constant and dielectric loss factor on axial stress T_1 up to $T_1 \approx 10$ kpsi (≈ 70 MPa) that corresponds to maximum pressure in the chamber $P = 1500$ psi are presented in Figure 11.45. These data can be used in case that the axial vibration of the radial polarized cylindrical piezoelement is employed (e.g., in the Tonpilz like design).

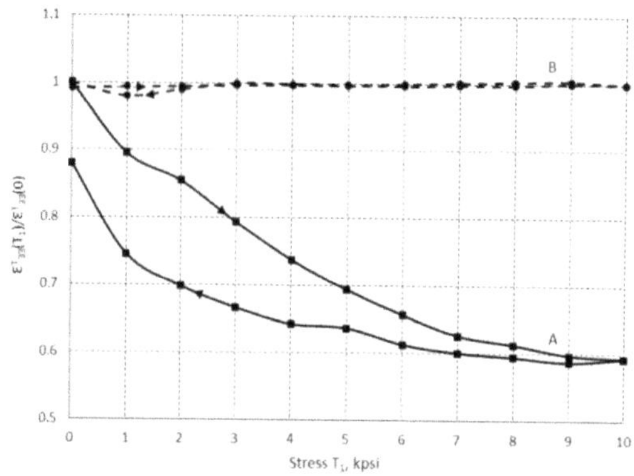

(a)

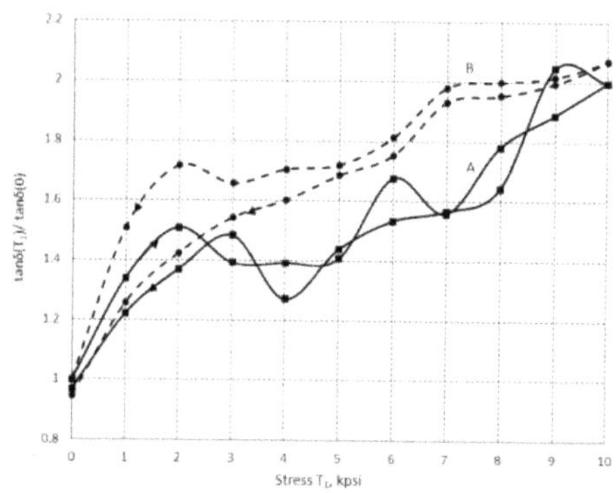

(b)

Figure 11.45: Dependences of the relative change of (a) dielectric constant $\varepsilon_{33}^T(T_1)/\varepsilon_{33}^T(0)$ and (b) dielectric loss factor $\tan\delta_e(T_1)/\tan\delta_e(0)$ from axial stress T_1. For ceramic A – solid line, for ceramic B – dashed line.

Table 11.2: Summary of parameters of piezoceramics before and after treating by stress T_1.

Parameters	$\frac{k_{32}(T_2)}{k'_{32}}$	$\frac{\varepsilon_{33}^T(T_2)}{\varepsilon_{33}^{T\,\prime}}$	$\frac{s_{22}^E(T_2)}{s_{22}^{E\,\prime}}$	$\frac{d_{32}(T_2)}{d'_{32}}$	$\frac{\tan\delta(T_1)}{\tan\delta'}$
Values	1.21	0.88	1.04	1.16	0.950

Relative change of parameters of the pressure treated piezoelements (marked with T_2 in parenthesis) in comparison with their values before the treating (marked with primes) are summarized in Table 11.2. Results of the testing of five samples of cylinders A were in close agreement, therefore the averaged data are presented in the Table. These results proved to be stable with time. Presumably, greater increase of d_{32} and k_{32} could be achieved for cylinders made of PZT-5H ceramics, as it follows from comparison between plots in Figure 11.20 (a) and (b).

It is noteworthy that the pressure treating of the cylinders could be performed under the static pressure in a water filled chamber, as well as under pressure mechanically applied to the caps, because in this case an additional axial loading would not influence the results of measuring capacitance and $\tan\delta_e$ under pressure anyway. The air pressure chamber was used for this purpose for standardizing the experimental technique.

11.4.2.2 Results of Testing in the Air-Backed Design

After it was confirmed that the effect of treating cylinders under the axial stress is significant and stable, the question arises regarding possible underwater acoustic transducer applications of the cylinders with the improved piezoelectric properties. The most common application of the radial poled cylinders is in the air-backed transducer designs, $T_2 = (a/t)\cdot P$, in which case the static pressure that is acting on the cylinder's surface transforms in the circumferential stress $T_2 = (a/t)\cdot P$. Judging by the previously obtained data[1] that are presented in Figure 11.20(a), it may be expected that under stress T_2 the piezoelectric modulus d_{32} and hence coupling coefficient k_{32} should reduce in an irreversible manner. To verify the factual behavior of the pressure treated cylinders in the air-backed design, prototypes shown in Figure 11.42 (b) were tested in accordance with the accepted pressure cycle. All the parameters of the cylinders were measured in the process of changing the pressure. Results of the 1st cycle of exposure to the radial pressure are presented in Figure 11.46 - Figure 11.48, where the relative change of the coupling coefficient, $k_{32}(T_2)/k'_{32}$, dielectric constant, $\varepsilon_{33}^T(T_2)/\varepsilon_{33}^{T\prime}$, and elastic constant, $s_{22}^E(T_2)/s_{22}^{E\prime}$, are depicted.

In the same Figures the dependences on the stress are shown of the analogous relative quantities for cylinders made of ceramics *B*.

Plots in the figures have two scales: one for the stress in ceramics (kpsi) another that corresponds with operating depth (*m*) in water for the transducers that employ cylinders used in the testing. The relative changes of the piezoelectric modulus $d_{32}(T_2)/d'_{32}$ vs. stress T_2 were calculated from formula $d_{32} = k_{32}\sqrt{\varepsilon_{33}^T s_{22}^E}$ after dependencies of other parameters became known. They are presented in Figure 11.49 for the first cycle of exposure to the stress.

Noticeable remnant reduction of ceramic parameters takes place after the 1st cycle of exposing pressure treated cylinders to stress T_2. Then under subsequent cycles of testing under pressure parameters of ceramic were gradually reducing and stabilized after the 4th cycle at the level that is below the original level before the treatment.

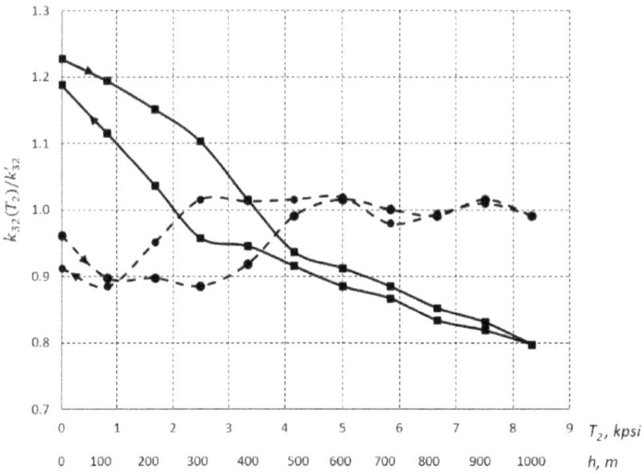

Figure 11.46: Dependence of the relative change of the coupling coefficients, $k_{32}(T_2)/k'_{32}$, from the circumferential stress T_2 (from the operating depth h): solid line – ceramics A, dashed line – ceramics B. The 1st cycle of exposure to the stress.

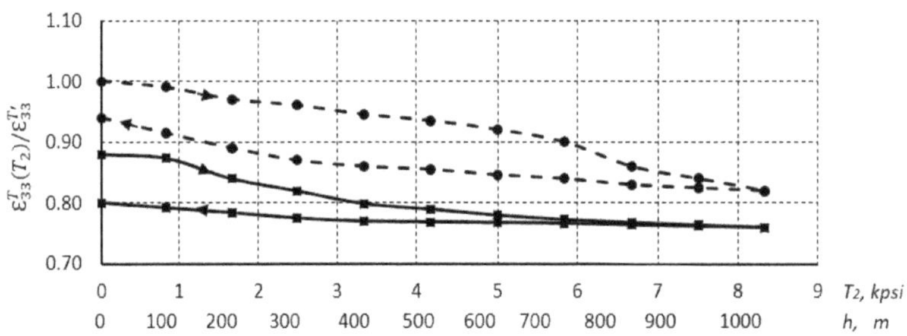

Figure 11.47: Dependence of the relative change of $\varepsilon^T_{33}(T_2)/\varepsilon^{T'}_{33}$ from the circumferential stress T_2 (from the operating depth *h*): solid line – ceramics *A*, dashed line – ceramics *B* (the 1st cycle of exposure to the stress)

The values of parameters of the treated ceramic after the 4th circle of exposure to the circumferential stress are presented in Table 11.3. (Note that data in the Table and in the Figures may be somewhat different, because in the Table they are averaged, and in the Figures, they are taken for a particular sample). Dependences of the coupling coefficient and piezoelectric modulus vs. circumferential stress occur in the reversible way after the 4th stress cycle for the treated and after the 2nd cycle for untreated ceramics.

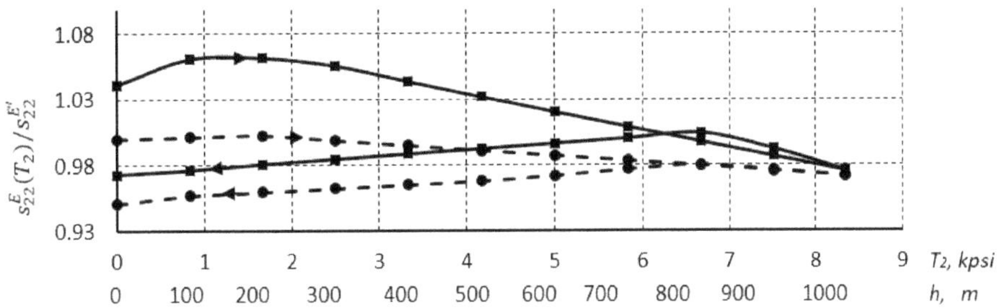

Figure 11.48: Dependence of the relative change of the elastic constant $s_{22}^{E}(T_2)/s_{22}^{E'}$ from the circumferential stress T_2 (from the operating depth *h*): solid line –ceramic *A*, dashed line – ceramic *B*. The 1st cycle of exposure to the stress.

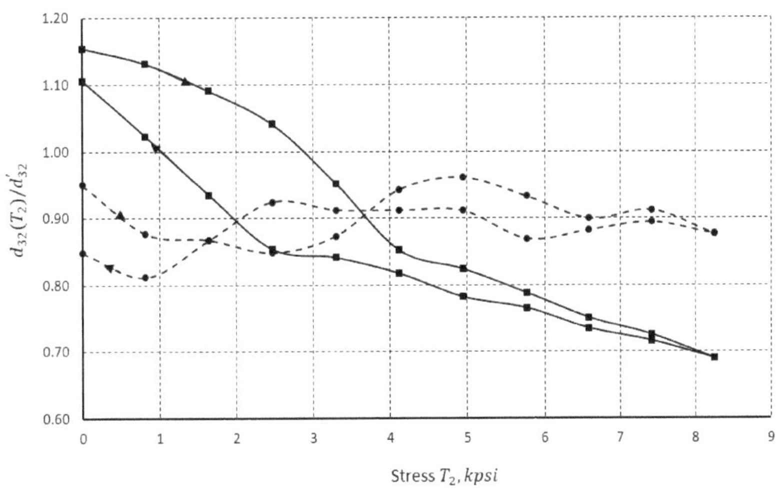

Figure 11.49: Dependence of the relative change of the piezoelectric modulus, $d_{32}(T_2)/d'_{32}$, from the circumferential stress T_2: solid line – ceramic A, dashed line – ceramics B.

Table 11.3: Changing parameters of ceramic after the 1st and 4th cycles of exposure to stress T_2

Parameter # cycles		$k_{32}(T_2)/k'_{32}$	$\varepsilon_{33}^{T}(T_2)/\varepsilon_{33}^{T'}$	$s_{22}^{E}(T_2)/s_{22}^{E'}$	$d_{32}(T_2)/d'_{32}$
1	start	1.21	0.88	1.04	1.16
	end	1.17	0.92	0.97	1.11
5	start	0.96	0.89	0.90	0.85
	end	0.96	0.88	0.90	0.85

Results of testing the cylinders up to pressure equivalent to 1000 m. depth being interesting from the point of view of the physical processes in the piezoceramics do not characterize a possibility of their practical applications in the air-backed transducer designs operating at a reasonable depth.

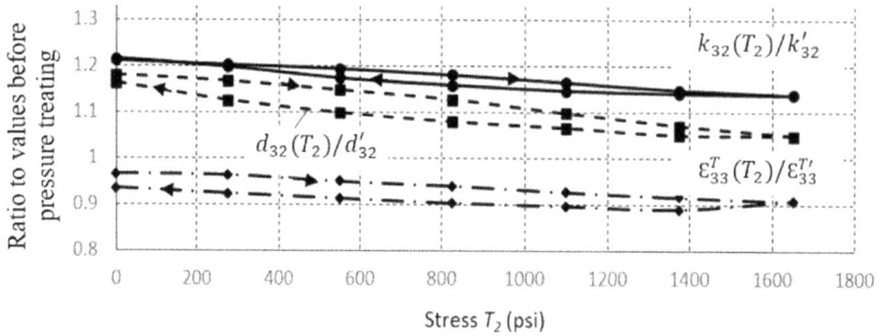

Figure 11.50: Dependences of the coupling coefficient $k_{32}(T_2)/k'_{32}$, $d_{32}(T_2)/d'_{32}$ and $\varepsilon^T_{33}(T_2)/\varepsilon^{T'}_{33}$ of the pressure treated cylinders in the range of circumferential stress up to T_2 = 1600 psi (equivalent to about 200 m depth).

Usually, transducers made of soft PZT ceramics are intended for relatively shallow water applications because of significant drop of their parameters under great stress. For more realistic estimation in Figure 11.50 is shown, how parameters of the pressure treated cylinders change in the range of circumferential stress up to $T_2 \approx 1600$ psi (equivalent to about 200 m depth for the cylinders tested). Changes of the values of parameters are reversible (within accuracy of the measurement).

Thus, it can be concluded that pressure treated piezoceramic cylinders with improved piezoelectric parameters can be used in the air-backed transducer designs operating at the relatively shallow depth (with equivalent circumferential stress of about $T_2 \approx 1600$ psi). Changing parameters of ceramics vs. pressure within this range of the stress is reversible and takes place in the typical for soft PZT materials way. The possibility of application of the pressure treated piezoelements to the greater depth is considered in the next section.

11.4.2.3 Results of Testing in the Pressure Equalized Design

In order to determine, how parameters of the pressure equalized (free flooded) transducer designs that employ the pressure treated cylinders may depend on the operating depth, the bare cylinders were tested in the air chamber up to pressure P = 1500 psi that imitated submergence of the transducers to about 1000 m depth. The treated cylinders made of ceramics A were tested together with the cylinders made of ceramics B for comparison. Results of the testing are presented in Figure 11.51 in the form of the plots normalized to original values of the parameters before the pressure treating. They show that the pressure treated soft PZT piezoceramics with improved parameters can be successfully used in the pressure equalized transducer designs.

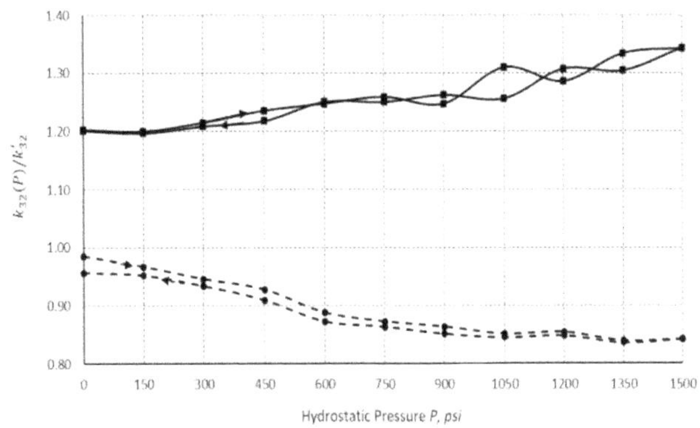

(a)

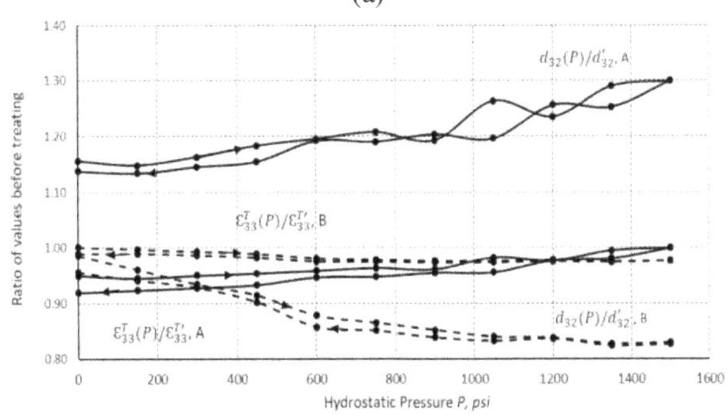

(b)

Figure 11.51: Dependences of the relative changing parameters of the pressure treated cylinders vs. hydrostatic pressure *P*: (a) coupling coefficients $k_{32}(T_2)/k_{32}'$, (b) piezoelectric moduli, $d_{32}(T_2)/d_{32}'$, and dielectric constants, $\varepsilon_{33}^T(T_2)/\varepsilon_{33}^{T\prime}$. Solid lines – ceramics *A*, dashed lines – ceramics *B*.

It is noteworthy that results obtained for the pressure treated ceramics *B* differ qualitatively from those previously presented in Ref. 10 for untreated PZT-4 ceramics (see Figure 11.35 and Figure 11.36). Whereas in the latter case at the hydrostatic pressure about 1.5 kpsi both d_{31} and

ε_{33}^{T} slightly raise, in the current testing they drop noticeably. Hardly possible is that the reason for this may be in difference of properties of the ceramics used (commercially available materials in both cases, but with interval of more than 50 years). Probably the main difference needs to be attributed to peculiarity of properties of the treated vs. untreated ceramics

11.4.2.4 Concluding Remarks for the Section

It is shown that electromechanical properties of the cylinders made of soft PZT piezoceramics can be improved by prolonged applying the compression stress of about 8 kpsi in the axial direction, as it was predicted in Ref. 9 based on results of investigations performed with small samples of PZT-5 piezoceramics. The effect of improvement proved to be stable in time and can be realized in underwater transducer designs. It is shown that in the air-backed design, in which case the ceramics is subjected to unidirectional compression stress in the circumferential direction, $T_2 = (a/t)P_h$, the improved properties remain reversible up to an operating depth that corresponds to the stress about 1600 psi (this corresponded to depth about 200 m for the cylinders used). With variation of the stress within 1600 psi range the improved parameters of treated cylinders behave in the same way as for the untreated cylinders. In the course of gradual increase of the compression stress the remnant values of the improved parameters drop and at $T_2 \approx 8$ kpsi they come to a level that is slightly lower than they were before the pressure treatment.

Investigation of the pressure treated cylinders was performed also under the conditions that imitated their application in the hydrostatic pressure equalized (free flooded) transducer designs. It is shown that at least up to the static pressure P_h = 1500 psi (the maximum pressure available in the air pressure chamber used), which imitated submergence of the transducers to about 1000 m depth, the improved parameters remained reversible. With increase of pressure the dielectric constant practically did not change, while the piezoelectric module and coupling coefficient rose by about 10%. Thus, application of the pressure treated cylinders in the pressure equalized designs can be highly recommended. Especially advantages can be using the pressure treated piezoelements made of soft PZT piezoceramics for designing sensors that do not experience intensive external actions.

On the separate issue, the conclusion can be made that the technique of experimenting involving employment of air pressure chamber in combination with an impedance analyzer, which allows dynamic measurement parameters of transducers in process of their subjecting to action of pressure, is an extremely useful tool for investigating transducers performance under their operating environmental conditions. The apparatus used allowed testing up to pressure P = 1500 psi (≈ 10 MPa) equivalent to 1000 m depth. This is sufficient for most practical transducers applications. In case that it is necessary to expend the pressure capability for investigating the physical properties of piezoceramic materials and/or behavior of transduces intended for greater depth of operation, the requirements for safety of the air pressure chamber may seriously complicate its design and increase its cost. These complications can be avoided by placing an air chamber inside of a water filled high pressure vessel, as it is shown schematically in Figure 11.52, and synchronizing the pressure inside and outside the chamber in course of testing.

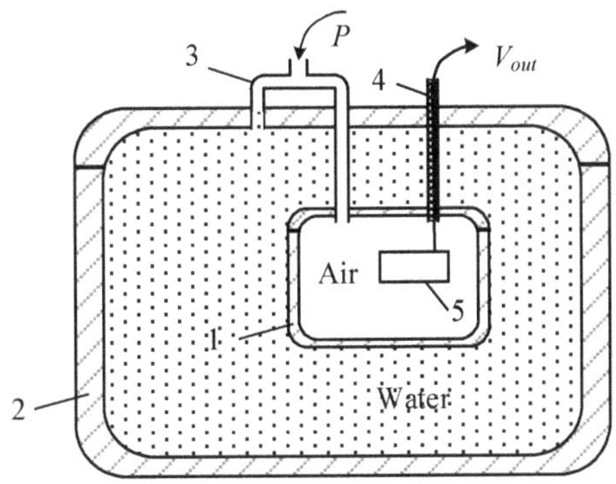

Figure 11.52: Arrangement for compensating pressure on the walls of an air chamber: 1 - air chamber, 2 – water filled high pressure vessel, 3 - pressure input, 4 - electrical output, 5-prototype under testing.

Requirements for the strength of the air-chamber can even be reduced in this case. Noteworthy is that with such an arrangement it is easy to perform a combined transducer testing under air pressure and under elevated temperature by heating water in the vessel.

11.5 Concluding Remarks for the Chapter

Results of experimental investigations presented in Refs. 6-1112 and reproduced here are in agreement with concept of the domain structure of piezoelectric ceramic materials and could be qualitatively predicted. But in terms of practical transducers designing the data obtained must be considered more as the tendencies in the parameters of piezoceramics behavior. Many uncertainties in values of ceramics parameters were revealed by these investigations that do not allow accurate prediction of maximum power that can be reliably radiated by a projector. Great variations in maximum achievable values of electric field, dynamic stress and resulting acoustic power may exist between practical projector designs depending on conditions of operation. This is illustrated by results of study presented in Ref. 13, where information was examined on the mechanical and electrical limits for transducers obtained from literature, from transducer designers and ceramics manufacturers. One of conclusions of this study is that permissible values of electric field and stress for transducers are lower than those that ceramics alone may withstand. At the same time, it was revealed that up to about 70-100 MPa of compression and up to 4-6 kV/cm electric field are achievable in projectors at certain operating conditions. But it is noted that these are single-point results.

In the conclusion it may be acknowledged that Woollett's assertion made in year 1968 (Ref. 14) remains intact today: "No general agreement as yet exists on the values of power limits of piezoceramic materials that should be used in design. The reasons for this include inadequate knowledge of the nonlinear properties of the materials, uncertain reliability of the high-voltage insolation techniques employed, and variability of the mechanical strength of ceramics in production lots and between manufacturers". And the same problems formulated in Ref. 14 remain actual: determining more precisely the power limits of the piezoelectric ceramics, increasing uniformity of the piezoelectric ceramics in production lots and between manufacturers, considering the problem of stability after reliability of operation has been achieved.

It seems to be wise to concentrate on improving the existing materials in terms of solving the above listed problems. Resulting improving of the traditional transducers capabilities may bring great economic effect with much less funding than searching for "silver bullet" new materials, implementation of which for underwater sonar techniques may encounter the same (if not greater) problems.

Remembering that most of the transducers users would rather have reliability than extra power achievable by calculation, it seems to be a right approach on the first stages of a transducer designing to use more conservative values of permissible (allowable) values for purpose of comparing different transducer types and for prototyping. Further the prototypes made of ceramics produced by a certain vendor (or by different vendors for comparison) have to be tested for determining the upper limits of a safe transducer operation under the specified operating conditions. If it appears as result of the testing that the transducer design has excessive reserves of safety for achieving the set goals, the design can be optimized by reducing these reserves.

In our farther estimations the following conservative values of permissible dynamic stress and electric fields will be used for PZT-4 and PZT-8 like ceramics: electric field E_p = 2kV/cm (5.1 mil); dynamic compression one-dimensional stress T_{pc} = 84 MPa and tensile stress T_{pc} = 24 MPa for the transverse and for the longitudinal piezoeffect with solid piezoelements; dynamic bending compression, T_{pc} = 84×1.7 = 142.8 MPa, and tensile, T_{pbt} = 24×1.7 = 40.8 MPa, stress. In case that the longitudinal piezoeffect with segmented piezoelements is used, it is assumed that the piezoelements are pre-compressed up to values that exclude tension in the bonding.

11.6 References

1. C. H. Sherman, and J. L. Butler, *Transducers and Arrays for Underwater Sound*, Springer Science + Business Media, LLC, New York, 2007.
2. D. A. Berlincourt, D. R. Currant and H. Juffe, *Piezoelectric and Piezomagnetic Materials and Their Function in Transducers*, in Physical Acoustics edited by W.P. Mason, V. I, part A, Academic Press, New York.
3. D. A. Berlincourt, and H. H. A. Krueger, "Behavior of Piezoelectric Ceramics under Various Environmental and Operating Conditions of Radiating Sonar Transducers", Technical Publication -228, Morgan Electro Ceramics, Bedford, Ohio, 44146.
4. R. S. Woollett and C. L. LeBlanc, "Ferroelectric Nonlinearities in Transducer ceramics", IEEE Transection on Sonics and Ultrasonics, Vol. SU-20, No. 1, January 1973
5. R. Gerson, "Dependence of Mechanical Q and Young's modulus of Ferroelectric Ceramics on Stress Amplitude", J. Acoust. Soc. Am., 32 (10), 1297-1301, (1960)
6. R. Gerson, S. R. Burlage, and Don Berlincourt, "Dynamic Tensile Strength of a Ferroelectric Ceramic", J. Acoust. Soc. Am., 33 (11), 1483-1485, (1961)
7. H. H. A. Krueger and Don Berlincourt, "Effects of High Static Stress on the Piezoelectric Properties of Transducer Materials", J. Acoust. Soc. Am., 33 (10), 1339-1344, (1961)
8. H. H. A. Krueger, "Stress Sensitivity of Piezoelectric Ceramics: Part 1. Sensitivity to Compression Stress Parallel to the Polar Axis", J. Acoust. Soc. Am., 42 (3), 636-645, (1967)
9. H. H. A. Krueger, "Stress Sensitivity of Piezoelectric Ceramics: Part 3. Sensitivity to Compression Stress Perpendicular to the Polar Axis", J. Acoust. Soc. Am., 43 (3), 583-591, (1968)
10. R. Y. Nishi and R. F. Brown, "Behavior of Piezoceramic Projector Material under Hydrostatic Pressure", J. Acoust. Soc. Am., 36 (7), 1292-1296, (1964)
11. R. F. Brown and G. W. McMagon, "Properties of Transducer Ceramics under Maintained Planar Stress", J. Acoust. Soc. Am., 38, 570-575, (1965)
12. R.Y. Nishi, "Effects of One-Dimensional Pressure on the Properties of Several Transducer Ceramics", J. Acoust. Soc. Am., 40 (2), 486-495, (1966)
13. J. L. Butler, K. D. Rolt, and F. A. Tito, "Piezoelectric ceramic mechanical and electrical stress study", J. Acoust. Soc. Am., 96 (3), 1914-1917, (1994)
14. R. S. Woollett, "Power Limitations of Sonic Transducers", IEEE Trans. on Sonic and Ultrasonics, Vol. SU-15, no. 4, pp. 218-229, (1968)
15. B. S. Aronov, E. K. Atkins, and D. A. Brown, "Increasing piezoelectric effect in radially polarized cylinders made of soft PZT ceramics by pressure treating and its practical application", J. Acoust. Soc. Am., 147 (6), 4145-4152, (2020)
16. E. K. Atkins, B. S. Aronov and D. A. Brown, "Effects of circumferential stress on tangentially polarized piezoelectric cylinders (L)", J. Acoust. Soc. Am., 148 (5), November 2020

17. B. S. Aronov, C. Bachand, and D. A. Brown, "Analysis of the piezoelectric properties of tangentially polarized, stripe-electroded cylinders," J. Acoust. Soc. Am., **129** (5), 2960-2967, (2011).

18. Rattikorn Yimnirun et al, "Effect of uniaxial compression prestress on ferroelectric properties of soft PZT ceramics", J. Appl. Phys. 39 759, 2006

19. IEE Standard on Piezoelectricity, ANSI/IEEE Standard No 176-1987, 1987, sect. 6.4.5.

CHAPTER 12

HYDROSTATIC PRESSURE EQUALIZATION

12.1 Introduction

Harmful effects of the hydrostatic pressure in the air-backed transducer designs (Figure 11.5) both in terms of static mechanical strength and changing parameters of ceramic materials are avoided in the pressure equalized (pressure compensated) designs presented schematically in Figure 11.9. In these designs the ceramic material experiences overall pressure equal to the ambient hydrostatic pressure. As shown in Sec. 11.3.3.4, the influence of the hydrostatic pressure on the ceramic parameters and their stability is insignificant. But the pressure equalization does not come for free. It requires including additional structural elements in the transducer designs, which may influence operating characteristics of the transducers. These elements are in the shape of cavities (compliant chambers) connected with outside space through a narrow neck or through an opening. The least influencing the transducer characteristics is the pressure equalization by means of pressurized gas; however, it requires special equipment for providing pressurized gas supply and its control. The size and complexity of this equipment depends on compliance of mechanical system of the transducer and on its internal cavity volume. Seemingly this volume can be made sufficiently small due to large compressibility of the gas, but in reality, it may be restricted because of reducing compressibility of the gas under pressure at a great depth. For low frequency transducers that employ mechanical systems vibrating in flexure and therefore having large compliance the air compensation is, probably, the only way to extend their application to great depth.

The most widely used with transducer types vibrating in the extensional modes and benders operating at moderate frequencies is pressure equalizing by means of liquids (or rubber like materials) filled cavities connected with outside space. Some of them may play role of the low frequency filters that pass signals with very low frequencies $\omega < \omega_L$ (including hydrostatic pressure at $\omega = 0$), where ω_L is below the operating range of a transducer, and acoustically isolates the volume of cavity from the external acoustic field at higher frequencies (usually used with hydrophones). Some of them may be used as Helmholtz resonators having resonance frequencies close to operating resonance frequency of the transducer (presumably, may be used with benders and slotted ring projectors). In some cases, the internal volume may freely radiate, as for example from open ends of a cylindrical transducer. Accordingly, the following issues will be considered related to problem of the hydrostatic pressure equalization: acoustic elements of transducer designs (compliance of the internal volume, acoustic mass and resistance of the neck and/or opening, Helmholtz resonator) (Sections 12.2 and 12.3); properties of the fluids (gasses and liquids) vs. pressure (Section 12.4); input impedances of the internal liquid filled volumes (including coupled vibration with internal Helmholtz resonator), and contribution of vibrating internal volumes to acoustic radiation of a transducer (these issues will be considered in Section 13.4.2.)

12.2 Acoustic Elements of Transducer Designs

12.2.1 Lumped Acoustic Elements

Consider input impedance of a column of fluid having density ρ and sound speed c that is confined within a tube of length l with rigid walls surrounded by the same fluid (note that under the term "fluid" we will mean both liquid and gas unless it will be necessary to consider their properties separately). At one end of the column, we imagine a piston having area of the column cross-section S_{cs} that can freely move along the walls under action of applied force F_m. The other end of the tube is closed by an imaginary cap loaded by impedance Z_L, as shown in Figure 12.1. All the dimensions of the tube are supposed to be small with respect to wavelength.

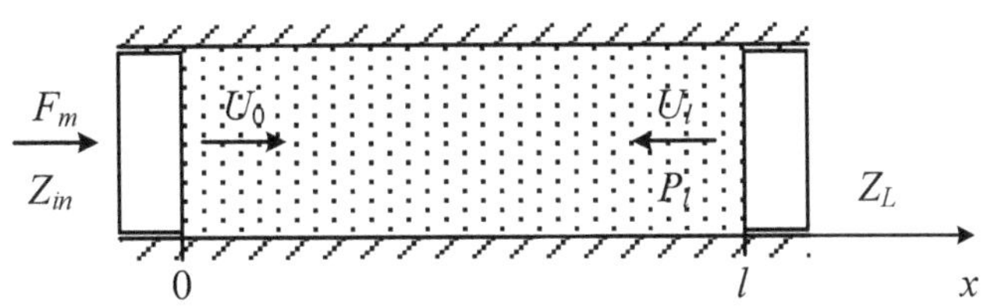

Figure 12.1: Column of fluid within tube with rigid walls loaded at the end by an impedance Z_L.

Since the wave motion within the tube obeys the same one-dimensional wave equation as the longitudinal vibrations in an elastic bar (see Eq. (4.90)), the mechanical input impedance of the column can be determined by the same formula (4.101)

$$Z_{inm} = \frac{F_m}{U_0} = \frac{j\rho c S_{cs} \tan kl + Z_L}{1 + j\dfrac{Z_L}{\rho c S_{cs}} \tan kl}, \tag{12.1}$$

which corresponds with the equivalent circuit of Figure 12.2 (see the original circuit in Figure 4.6) loaded by impedance Z_L.

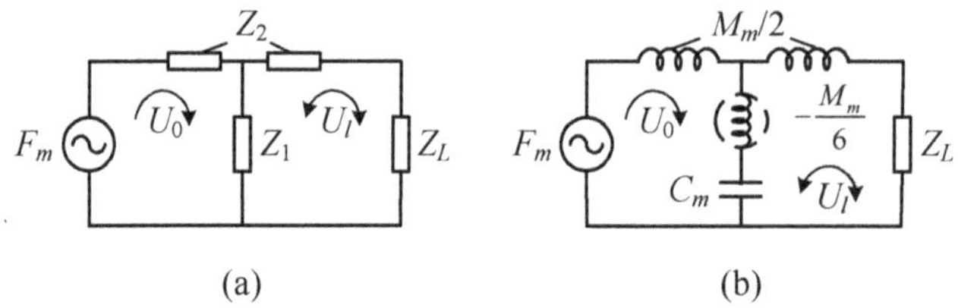

Figure 12.2: Equivalent mechanical circuit of the liquid column vibrating within tube with rigid walls loaded by impedance Z_L: (a) general view ($Z_1 = -j\rho c S_{c.s} / \sin kl$, $Z_2 = j\rho c S_{c.s} \tan(kl/2)$) (b) modification at $kl < 0.6$.

In the case that the end is clamped $Z_L = \infty$, $U_L = 0$. In the case that the end is open (acoustic short circuit) $Z_L = 0$, $P_l = 0$. The more realistic case is that the end of the tube radiates in the surrounding space. The radiation from the end can be approximated, as being produced by one-sided piston of cross-section S_{cs}. Assuming that the tube has circular cross-section of radius *a*, and following the expressions for nondimensional coefficients of radiation impedance of one-sided circular piston at small *ka* presented by formulas (see expressions (6.324) and (6.325)) $\alpha = (ka)^2/4$, $\beta = 2ka/\pi$, we arrive at expression

$$Z_L = \rho c \pi a^2 \frac{(ka)^2}{4} + j\rho c \cdot 2ka^3. \tag{12.2}$$

Consider input impedance of the column of fluid under the assumption that dimensions of the tube are small with respect to wavelength of sound, e.g., $kl < 0.6$. After replacing $\sin kl \approx kl - (kl)^3/6$ and $\tan(kl/2) \approx kl/2$ the equivalent circuit Figure 12.2 (a) will be simplified to the circuit in Figure 12.2 (b), where

$$M_m = \rho l S_{cs} \text{ and } C_m = \frac{l}{\rho c^2 S_{cs}} \tag{12.3}$$

are the mechanical mass and compliance of the volume (note that for elastic bar of the same size $C_m = l/YS_{cs}$ as $c^2 = Y/\rho$).

When considering the acoustic elements of transducer designs it is more convenient to use acoustic generalized quantities rather than mechanical, in which case the generalized force is the acoustic pressure P, and the generalized velocity is the volume velocity $U_{\tilde{V}}$. In our case the volume velocity of the piston is $U_{\tilde{V}} = U_0 S_{cs}$. Accordingly, the acoustic input impedance of the column is related to mechanical impedance as

$$Z_{in\,ac} = \frac{P}{U_{\tilde{V}}} = \frac{(F_m / S_{cs})}{U_0 S_{cs}} = \frac{Z_{in\,m}}{S_{cs}^2}. \tag{12.4}$$

Thus, the acoustic analogs of mechanical elements (12.3) are the acoustic mass and acoustic compliance,

$$M_{ac} = \rho l / S_{cs} \text{ and } C_{ac} = \frac{\tilde{V}}{\rho c^2}. \tag{12.5}$$

The equivalent mechanical circuit of Figure 12.2 (b) can be represented in the equivalent acoustic form shown in Figure 12.3. Impedance of the load in the acoustic circuit must be replaced by its acoustic analog $Z_{L\,ac} = Z_L / S_{cs}^2$.

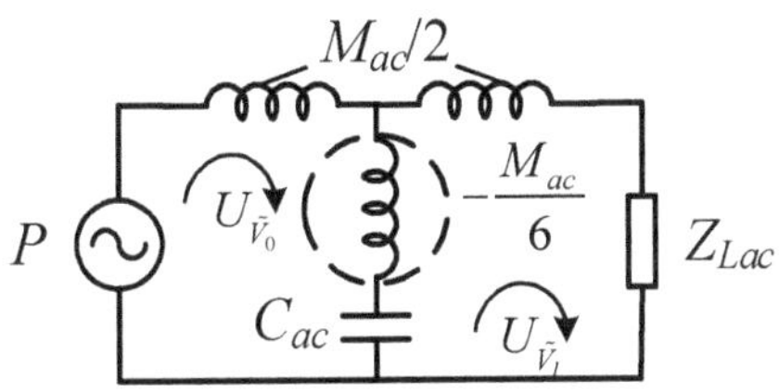

Figure 12.3: Equivalent acoustic circuit of the column of fluid vibrating within tube with rigid walls having small wave size.

In the case that the end of the tube is closed ($Z_{L\,ac} = \infty$, output of the circuit is open) we will find that

$$Z_{in\,ac} = \frac{1}{j\omega C_{ac}} + j\omega \frac{M_{ac}}{3}. \tag{12.6}$$

The mass term at $kl < 0.6$ can be neglected. Indeed, it can be shown using expressions (12.5) that

$$|Z_{in\,ac}| = \frac{1}{\omega C_{ac}} \sqrt{1 + (kl)^4 / 9}, \tag{12.7}$$

and

$$Z_{in\,ac} \approx \frac{1}{j\omega C_{ac}} = \frac{K_{ac}}{j\omega} = \frac{\rho c^2}{j\omega \tilde{V}}. \tag{12.8}$$

Thus, the input impedance of the column of fluid within closed tube having rigid walls is that of compliance (rigidity K_{ac}). This conclusion can be generalized to volumes of fluid confined within closed cavities with rigid walls of any configuration provided their dimensions are small with respect to length of the acoustic wave, practically, when they are less than $\lambda/6$. In this case the rigidity and potential energy of the volume are

$$K_{ac} = \frac{1}{C_{ac}} = \frac{\rho c^2}{\tilde{V}},\ W_{pot} = \frac{\xi_{\tilde{V}}^2}{2} K_{ac}, \tag{12.9}$$

where $\xi_{\tilde{V}}$ is the volume displacement at the input of the cavity. If the energy is supplied through a part of its surface, on which a nonuniform distribution of displacements takes place, then $\xi_{\tilde{V}} = \xi_o S_{av}$, where ξ_o is displacement of a reference point on the surface. From the obvious equality

$$W_{pot} = \frac{\xi_o^2}{2} K_m = \frac{\xi_{\tilde{V}}^2}{2} K_{ac} \tag{12.10}$$

follows that

$$K_{ac} = K_m / S_{av}^2,\ C_{ac} = C_m S_{av}^2. \tag{12.11}$$

In the case that the end of the tube is open ($Z_{L\,ac} = 0$), output of the circuit in Figure 12.3 is short circuited) we obtain that

$$Z_{in\,ac} \approx j\omega M_{ac}, \tag{12.12}$$

where according to formulas (12.5)

$$M_{ac} = \rho l / S_{cs} = \rho \tilde{V} / S_{cs}^2 = M_{\tilde{V}} / S_{cs}^2. \tag{12.13}$$

This expression, where $M_{\tilde{V}}$ is the total mass of fluid in the tube, can be generalized to volumes of fluid confined within open ducts of small wave size with rigid walls. The kinetic energy of an open volume of fluid is

$$W_{kin} = \frac{\dot{\xi}_{\tilde{V}}^2}{2} M_{ac}. \tag{12.14}$$

If the energy is supplied by nonuniform vibration of the input surface of a duct, then $\dot{\xi}_{\tilde{V}} = \dot{\xi}_o S_{av}$, and it can be shown analogous to the previous case that in general

$$M_{ac} = M_{\tilde{V}} / S_{av}^2 . \tag{12.15}$$

Consider the effect of radiation from the end of the tube under assumption that the radiation takes place in the media having the same properties as the fluid inside of the resonator (otherwise ρ and c below will be different). After substituting impedance of radiation from the end of a tube (12.2) in the form of acoustic impedance

$$Z_{L\,ac} = Z_L / (\pi a^2)^2 = \rho c \frac{\pi}{\lambda^2} + j\rho\omega \cdot \frac{2}{\pi^2 a} \tag{12.16}$$

into equivalent circuit Figure 12.3, we will obtain that

$$Z_{in\,ac} = j\omega M_{ac} + Z_{La} = \rho c \frac{\pi}{\lambda^2} + j\omega\rho \frac{1}{\pi a^2}\left(l + \frac{2}{\pi} a \right). \tag{12.17}$$

(Note that according to (12.21), $M_{ac} = \rho l / \pi a^2$). The active radiation resistance characterizes energy losses, while the presence of the reactance results in the equivalent increase of mass of the fluid that can be accounted for by introducing the effective length of the tube $l_{eff} = l + 0.64a$.

12.2.2 Effects of Finite Viscosity of the Fluids

So far, the vibration of fluid was supposed to be lossless. Due to finite viscosity of the real fluids, losses of energy take place in course of vibration of fluids inside the ducts with rigid walls. Predominantly they occur in the thin boundary layers near the walls (see Figure 13.18 and formula (13.109)), $l_b = \sqrt{2\mu / \rho\omega}$), within which the transverse viscous wave develops that rapidly attenuates with distance from the wall. Therefore, the effects of viscosity become essential in case of propagating sound in narrow long tubes and slots, where separation between the walls is comparable with length of the shear wave. The related issues are considered in detail in Refs. 1–3. As viscosity of fluids plays an essential role in determining properties of the acoustic elements, some information regarding this quantity that is available from literature is summarized here.

The dynamic (absolute) viscosity, μ , that is used in expressions for the viscose resistances is by definition (Newton's Law of friction) $\mu = \tau / (du / dn)$, where τ is the shearing stress (tangential force per unit area between layers of flow), and (du / dn) is the rate of change of velocity of flow in direction of normal to the layers The dynamic viscosity units are: in system SI - N·s/m^2 or Pa·s and smaller unit mPa·s (millipascal-second); in system CGS - *poise* (P) (dyn·s/cm^2), and *centipoise* (cP), 1 cP = 0.01 P. Correlations between the units are: 1 P = 0.1 Pa·s; 1 cP = 1 mPa·s. The viscosity μ is known to be independent of pressure (except for extremely high pressure that is beyond practical range of interest for transducer applications). However, viscosity depends on temperature: its values tend to decrease for liquids and increase for gases as temperature rises. Therefore, the reference temperature must be quoted each time together with value of viscosity. The range of operating temperatures in underwater applications for hydrophones (receivers in general) is as an ambient temperature in ocean and can be estimated following Ref. 4 as -2°C < T < 35°C. Temperatures in the projector designs can be higher due to heat generation in course of operation. Values of the dynamic viscosity of several fluids are presented in Table 12.1 at temperature 20°C. Note that modifications of the silicon oils are available having substantially different viscosities. Low viscosity modification is chosen for example. It must be remembered tat the viscosities significantly depend on temperature.

Table 12.1: Values of the dynamic viscosity of fluids at 20°C.

	air	water	Motor oil, SAE-10	Castor oil	Silicon oil (low viscosity)
μ, Pa·s $\Delta P_{tr} \approx P_h \dfrac{C_{\tilde{V}}}{C_{\tilde{V}} + C_{ad}}$	$2.0 \cdot 10^{-5}$	$1.0 \cdot 10^{-3}$	0.06	1.0 - 1.5	0.01
$\rho \cdot 10^{-3}$, kg/m^3	$1.2 \cdot 10^{-3}$	1.0	0.95	0.95	0.96
$c \cdot 10^{-3}$, m/s	0.33	1.5	1.75	1.54	1.35

As shown in Ref. 1, the averaged input mechanical impedance of viscous fluid in the circular tube is,

$$Z'_{in} = j\omega\rho l S_{cs} / \left[1 - \frac{2J_1(k'a)}{k'aJ_0(k'a)} \right], \tag{12.18}$$

where

$$k' = \sqrt{\frac{j\omega\rho}{\mu}} = \frac{2\pi}{\lambda'}(1+j)\text{, and } \lambda' = 2\pi\sqrt{2\mu/\rho\omega} \tag{12.19}$$

is the wavelength of the viscous wave. For example, in case of air ($\mu = 2\cdot 10^{-5}$, N·s/m², ρ = 1.2 kg/m³) at frequencies f > 100 Hz λ' < 1.4 mm. In case of water ($\mu = 10^{-3}$ N·s/m², ρ = 1000 kg/m³) at f > 100 Hz λ' < 0.35 mm.

If the radius of a tube is small with respect to the wavelength λ' (at $|k'a| < 2$), then the input impedance (12.18) may be approximated as

$$Z'_{in} = S_{cs}(8\mu l / a^2 + j\omega\cdot 4\rho l/3)\,. \tag{12.20}$$

Thus, the effect of viscosity results in introducing resistance of losses,

$$R' = 8\mu l S_{cs} / a^2\,, \tag{12.21}$$

and in increase of the effective mass of vibrating fluid to value

$$M' = \frac{4}{3}M_{\tilde{V}}\,, \tag{12.22}$$

where $M_{\tilde{V}}$ is the actual mass of the fluid. The attached mass due to the effect of viscosity is 1/3 of the actual mass of the fluid.

In the case of sufficiently narrow tube and low frequencies the resistive term of the impedance may become dominant ($R'/\omega M' \approx \mu / a^2 f\rho$). The tube in this case will present a pure acoustic resistance,

$$R'_{ac} = 8\mu l / \pi a^4\,. \tag{12.23}$$

With raising the values of a/λ' (at $|k'a| > 10$) the following expression for the impedance can be obtained from the general formula (12.18)

$$Z'_{in} \approx S_{cs}\frac{1}{a}\sqrt{2\rho\mu\omega} + j\omega\rho l S_{cs}\left(1 + \frac{1}{k'a}\right). \tag{12.24}$$

In case $|1/k'a| < 0.1$ and given that this quantity drops with increase of radius and frequency, the term in parenthesis may be neglected, and input impedance of tube with not extremely small diameter can be presented in the form,

$$Z'_{in} = R' + j\omega M_{\tilde{V}}\,, \tag{12.25}$$

where

$$R' = \pi a\sqrt{2\rho\mu\omega}\ ,\ \text{or}\ R'_{ac} = \frac{1}{\pi a^3}\sqrt{2\rho\mu\omega} \tag{12.26}$$

is the resistance of losses due to finite viscosity of the fluid. The resistance is small compared with the inertia term, indeed,

$$\frac{R'}{\omega M_{\tilde{V}}} = \frac{1}{a}\sqrt{\frac{2\mu}{\rho\omega}} = \frac{1}{|k'a|}, \tag{12.27}$$

but this term cannot be neglected, as it is the only source of losses.

Expression for the input impedance of fluid within a thin rectangular slit with rigid walls (Figure 12.4) is also presented in Ref. 1 (theoretically under the assumption that the slit is

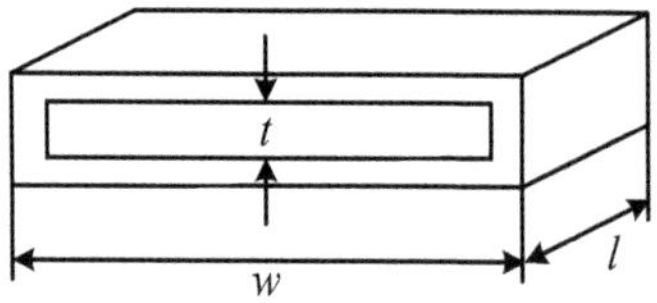

Figure 12.4: Rectangular slit with rigid walls.

infinitely wide). It is shown that at $|k't| < 1$, where $k' = \sqrt{\omega\rho / j\mu}$,

$$Z'_{in} = R' + j\omega M' \approx \frac{12\mu l}{t^2} S_{cs} + j\omega\frac{6}{5}\rho l S_{cs}\,. \tag{12.28}$$

Thus, the resistance of losses due to finite viscosity of the fluid is

$$R' = \frac{12\mu l}{t^2} S_{cs}\,,\ \text{or}\ R'_{ac} = \frac{12\mu l}{wt^3}\,. \tag{12.29}$$

Another effect of the viscosity is in addition to the actual mass of fluid inside the slit.

If to take into consideration combined effect of radiation from the end (Z_{Lac} by formula (12.17)) and viscosity of the fluid ($Z'_{in\,ac}$ obtained from formula (12.25)), then the input impedance of the tube will be

$$Z_{in\,ac} = \left(\frac{1}{\pi a^3}\sqrt{2\rho\mu\omega} + \rho c\frac{\pi}{\lambda^2}\right) + j\omega\frac{\rho l}{\pi a^2}(1 + \frac{2}{\pi}a) = R_{loss} + j\omega M_{ac\,eff}\,. \tag{12.30}$$

Remember that if the tube is especially thin the input impedance is determined by formula (12.20).

12.3 Acoustic Elements in Transducers

12.3.1 Helmholtz Resonators

Acoustic structures that are used in transducers represent a combination of the considered components: cavities, in which the fluid behaves as a rigidity; tubes (slots), in which the fluid behaves as a vibrating mass; and resistances that account for the viscous losses and losses due to radiation from inside of a cavity. Two variants of these combinations and the equivalent circuits, which can be used to determine, how the structures may influence transducer parameters, are shown in Figure 12.5.

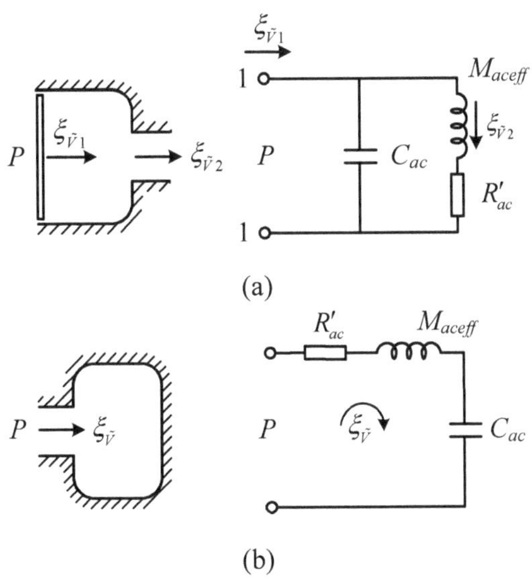

Figure 12.5: Equivalent circuits of the acoustic resonators.

In the variant (a) the sound pressure P is generated by vibration of a part of the cavity wall with volume displacement $\xi_{\tilde V1}$. The volume displacement at the output of the cavity is denoted as $\xi_{\tilde V2}$. The energies involved in the process being expressed through the volume displacements as the generalized coordinates are

$$W_{pot} = K_{ac}(\xi_{\tilde V1} - \xi_{\tilde V2})^2 / 2\,,\ W_{kin} = M_{aceff}\dot{\xi}^2_{\tilde V2} / 2\,,\ W_e = P\xi_{\tilde V1}\,. \tag{12.31}$$

It is easy to verify that the Lagrange's equations for this system are equivalent to relations for the parallel contour shown in Figure 12.5 (a), where resistance of losses, R'_{ac}, that accompanies the acoustic mass is introduced by observation. The acoustic mass, M_{aceff}, and resistance of losses that are due to effects of viscosity and radiation from the opening can be determined from expression (12.30).

In the variant (b) the outside pressure P generates the volume displacement $\xi_{\tilde{V}}$ at the neck input. Expressions for the energies involved are, as follows:

$$W_{pot} = K_{ac}(\xi_{\tilde{V}1} - \xi_{\tilde{V}2})^2/2\,,\; W_{kin} = M_{ac\,eff}\dot{\xi}^2_{\tilde{V}2}/2\,,\; W_e = P\xi_{\tilde{V}1}\,. \tag{12.32}$$

and the equivalent circuit corresponding to the balance of these energies is the series contour shown in Figure 12.5 (b). Both structures represent the Helmholtz resonators. Assuming that the neck is a tube, parameters of the resonators are according to relation (12.30):

$$C_{ac} = \frac{\tilde{V}}{\rho c^2}\,,\; M_{ac\,eff} = \frac{\rho l}{\pi a^2}(1+\frac{2}{\pi}a)\,, \tag{12.33}$$

And

$$R'_{ac} = \frac{1}{\pi a^3}\sqrt{2\rho\mu\omega} + \rho c\frac{\pi}{\lambda^2}\,. \tag{12.34}$$

If the opening is in the form of especially thin tube or slit (the later may be typical for application with the rectangular bender designs), then $M_{ac\,eff}$ and R'_{ac} must be determined from expressions (12.23) and (12.29), respectively. We will assume further that the neck is cylindrical and parameters have values by formulas (12.33) and (12.34) unless noted differently. The resonators have resonance frequency

$$\omega_H = \frac{1}{\sqrt{C_{ac}M_{ac\,eff}}} = c\sqrt{\frac{\pi a^2}{\tilde{V}l(1+0.64a)}}\,. \tag{12.35}$$

The quality factor of the resonators is

$$Q = \frac{\omega_H M_{ac\,eff}}{R'_{ac}} = \frac{1}{R'_{ac}}\sqrt{\frac{M_{ac\,eff}}{C_{ac}}} = \frac{1}{R'_{ac}}\sqrt{\frac{l(1+0.64a)}{\pi a^2\tilde{V}}}\,. \tag{12.36}$$

12.3.2 Transducer with Helmholtz Resonator as Hydrostatic Pressure Equalizing System

If to assume that the vibrating part of the cavity wall in the case shown in Figure 12.65 (a) is mechanical system of a transducer, then operation of the entire system can be described by means of the equivalent electro-mechano-acoustic circuit presented in Figure 12.6, in which the mechanical system of the transducer is loaded with input of the cavity. Being commonly described as electromechanical with equivalent parameters n, $C^E_{eqv}, M_{eqv}, r_{mL}$ and Z_s (radiation impedance), the equivalent circuit of the transducer must be converted to electroacoustic, i.e., presented in respect to the volume velocity $\dot{\xi}_{\tilde{V}} = U_{\tilde{V}}$.

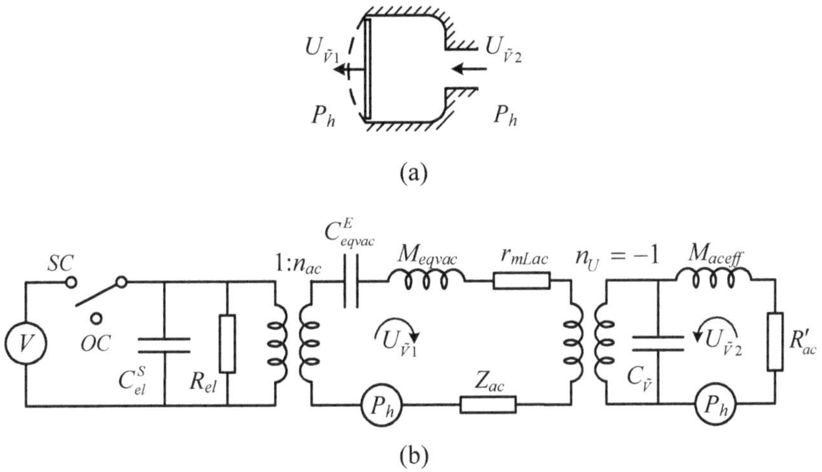

Figure 12.6: To equalizing the hydrostatic pressure on the transducer mechanical system: (a) schematic view of the arrangement, (b) acoustic equivalent circuit of the transducer with resonator.

The generalized displacement in Figure 12.6 (a) is $\xi_{\tilde{V}1} = \xi_0 S_{av}$, where ξ_0 is displacement of the reference point and S_{av} is the average surface of the mechanical system of the transducer. Accordingly, the values of equivalent mechanical parameters must be converted into the corresponding acoustic values, namely,

$$n_{ac} = n / S_{av}, C^E_{eqvac} = C^E_{eqv} S^2_{av}, M_{eqvac} = M_{eqv} / S^2_{av}, r_{mLac} = r_{mL} / S^2_{av}, Z_{sac} = Z_s / S^2_{av}. \qquad (12.37)$$

Acting on the mechanical system of the transducer external pressure that is denoted P_h includes a variable component (actually, $P_h = P(\omega)$). It is) introduced in the circuit directly as acoustic equivalent of F_{eqv}. Given that in terms of equalizing the hydrostatic pressure we are interested in functioning of the system at low frequencies close to $\omega = 0$, we assume that the same external pressure is acting on the input of the cavity neck and can be included in the circuit of the resonator. It is noteworthy that in operating range of the transducer these sound pressures can be different, and radiations from the transducer and through the neck may interact. But these issues are out of the scope of this Section.

Contours that correspond to vibration of the mechanical system of the transducer and resonator are connected by means of the ideal transformer with transformation coefficient $n_U = -1$ that converts direction of the volume velocity. The sign (–) reflects the fact that the volume velocity of the inner surface of transducer is in anti-phase with velocity of outer surface (the conditionally positive being direction of radiation into external medium). This feature is not essential unless radiation from the neck is considered, and the transformer can be removed from the circuit for simplicity. One more simplifying assumption is made regarding the equivalent circuit. We assume that the volume velocities of the outer and inner surface of the transducer mechanical system are the same. While this is true for transducers with flat mechanical system (such as rectangular bender), it can be not exact for transducers with cylindrical mechanical system (such as slotted ring).

Analysis of the mechano-acoustic circuit may be performed using technique of the coupled vibrations considered in Section 4.6 for the pure mechanical systems. The general outline of applying this technique can be introduced here, though a more complete analysis requires knowing properties of transducer designs that employ the pressure equalizing system. This will be done in Chapter 13.

The energies associated with operation of the system under the assumption that the electrical side of a transducer is short circuited are

$$W_{pot} = \frac{\xi_{\tilde{V}1}^2}{2C_{eqv\,ac}^E} + \frac{(\xi_{\tilde{V}1} - \xi_{\tilde{V}2})^2}{2C_{\tilde{V}}} = \frac{\xi_{\tilde{V}1}^2}{2}(K_{eqvac}^E + K_{\tilde{V}}) - \xi_{\tilde{V}1}\xi_{\tilde{V}2}K_{\tilde{V}} + \frac{\xi_{\tilde{V}2}^2}{2}K_{\tilde{V}}, \tag{12.38}$$

$$W_{kin} = \frac{M_{eqvac}\dot{\xi}_{\tilde{V}1}^2}{2} + \frac{M_{ac\,eff}\dot{\xi}_{\tilde{V}2}^2}{2}, \tag{12.39}$$

$$W_e = P_h(\xi_{\tilde{V}1} - \xi_{\tilde{V}2}). \tag{12.40}$$

The energies of losses including the radiation impedance are omitted, because they are not essential for illustrating the approach.

The expressions for the potential and kinetic energies are analogous to expressions (4.520) and (4.521) that characterize coupled vibrations of two partial systems. In our case one of the partial systems (that remains at $\xi_{\tilde{V}2} = 0$) has combined rigidity $K_{eqvac}^E + K_{\tilde{V}}$, mass M_{eqvac} and its vibration is generated by external generalized force P_h. Another partial system (that remains at $\xi_{\tilde{V}1} = 0$) has rigidity $K_{\tilde{V}}$ and mass $M_{ac\,eff}$, i.e., represents the Helmholtz resonator shown in Figure 12.5(b). Vibration in the system is generated by pressure P_h. Coupling between the partial systems is due to mutual rigidity $K_{\tilde{V}}$ (compliance $C_{\tilde{V}}$).

Using expressions (12.38)-(12.40) the following Lagrange's equations in complex form may be obtained that describe vibrations in this coupled system

$$[(K^E_{eqv\,ac} + K_{\tilde{V}}) - \omega^2 M_{eqv\,ac}]U_{\tilde{V}1} + K_{\tilde{V}}U_{\tilde{V}2} = P_h, \tag{12.41}$$

$$K_{\tilde{V}}U_{\tilde{V}1} + (K_{\tilde{V}}U_{\tilde{V}2} - \omega^2 M_{ac\,eff})U_{\tilde{V}2} = -P_h. \tag{12.42}$$

These equations may be accounted for by the equivalent circuit shown in Figure 12.6 (c) with the terms that represent losses of energy and electrical side of the transducer included. It must be noted that so far, the electrical side of transducer was supposed to be short circuited, and therefore did not influence parameters of the corresponding partial system. The same condition holds if the transducer operates in the transmit mode. In case that the electrical side is open circuited, which is typical for the hydrophones, reaction of the electrical side (reactance introduced into the mechanical contour) must be taken into consideration. Analysis of operating characteristics of the coupled system can be made using general procedures of theory of coupled vibrations described in Section 4.6. Thus, expressions for the partial resonance frequencies of the system, f_{tp} for the transducer and f_{rp} for the resonator, are

$$f_{tp} = \frac{1}{2\pi}\sqrt{\frac{K^E_{eqv\,ac} + K_{\tilde{V}}}{M_{eqv\,ac}}}, \quad f_{rp} = \frac{1}{2\pi}\sqrt{\frac{K_{\tilde{V}}}{M_{ac\,eff}}}, \tag{12.43}$$

and coefficient of the "compliant" coupling is

$$\gamma_C = \frac{K_{\tilde{V}}}{\sqrt{(K^E_{eqv\,ac} + K_{\tilde{V}})K_{\tilde{V}}}} = \frac{1}{\sqrt{1 + C_{\tilde{V}} / C^E_{eqv\,ac}}}. \tag{12.44}$$

The frequency equation for determining the resonance frequencies of the system is

$$[1 - (f_{rp} / f)^2][1 - (f_{tp} / f)^2] - \gamma_C^2 = 0. \tag{12.45}$$

Further analysis requires specifying the transducer design and will be considered in Section 13.4.3 with examples of the rectangular bender and slotted ring projectors.

12.3.2.1 Transient Process in the Helmholtz Resonator

Under maintained action of the hydrostatic pressure, which can be considered as the extreme case of the sound pressure at $\omega \to 0$, the inertial impedances in the circuit of Figure 12.6 (c) can be neglected, and it can be transformed to the circuit in Figure 12.7 (a) (remember that the electrical side is assumed to be short circuited). The force that causes deformation of the

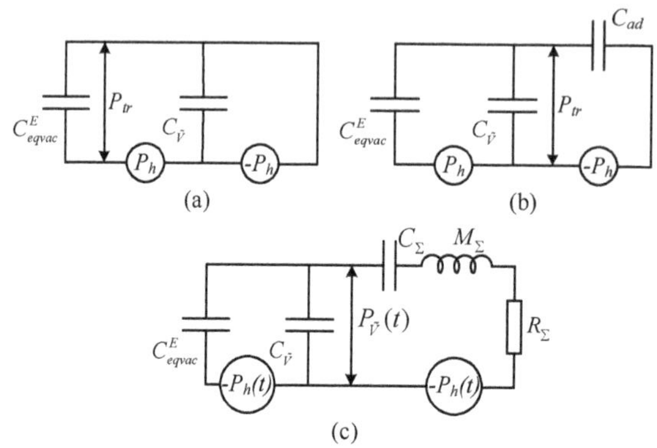

Figure 12.7: Equivalent circuits of the coupled system transducer-resonator: (a) and (b) at frequencies $\omega \to 0$ under steady state conditions, (b) with additional rigidity related term included; (c) under rapidly changing hydrostatic pressure (M_Σ and R_Σ are the combined masses and resistances in the contour of the resonator).

mechanical system of transducer is proportional to difference ΔP_{tr} between the external pressure $P_h(t)$ and pressure within the cavity $P_{\tilde{V}}(t)$,

$$\Delta P_{tr} = P_h(t) - P_{\tilde{V}}(t) . \tag{12.46}$$

At the steady state $P_{\tilde{V}}(t) = P_h(t)$ and $\Delta P_{tr} = 0$. Thus, the arrangement shown in Figure 12.6 performs the function of the static pressure equalizer. In real transducer designs the interior of the fluid filled volume of the resonator must be separated from the outer medium (water) with aid of some sealing member that possesses its own rigidity and mass. Their acoustic values will be denoted as K_{ad} and M_{ad}. In the steady state the corresponding compliance, $C_{ad} = 1/K_{ad}$, must be included in the equivalent circuit of the resonator, as this is shown in Figure 12.7 (b). This results in reducing $|P_{\tilde{V}}|$. Namely, $|P_{\tilde{V}}| \approx P_h \cdot C_{ad}/(C_{ad} + C_{\tilde{V}})$, and the mismatch between the outer and inner pressure on the mechanical system of a transducer becomes

$$\Delta P_{tr} \approx P_h \frac{C_{\tilde{V}}}{C_{\tilde{V}} + C_{ad}} . \tag{12.47}$$

In a rational design of the transducer with pressure compensation the ratio $C^E_{eqv\,ac} / C_{\tilde{V}}$ must be considerably less than unity and $C_{ad} \gg C_{\tilde{V}}$ for minimizing the mismatch. In this case the mismatch will be

$$\Delta P_{tr} \approx P_h \cdot C_{\tilde{V}} / C_{ad} \,. \tag{12.48}$$

Maximum value of ΔP_{tr} must not exceed some value that is permissible in terms of strength of mechanical system of the transducer and/or changing properties of piezoceramic. This quantity can be called allowed accuracy of the pressure equalization, $\Delta P_{tr\,a}$. Its value depends on a transducer design.

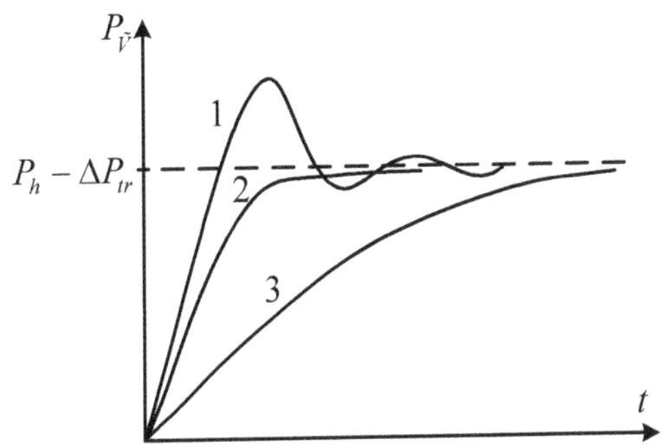

Figure 12.8: Step response of the resonator at different values of the damping coefficient ζ: (1) $\zeta < 1$ underdamped resonator, (2) $\zeta = 1$ -critical damping; (3) $\zeta > 1$ -overdamped resonator.

The condition of maintained (steady state) hydrostatic pressure is ideal in terms of operating the pressure equalizing system. In real operation the hydrostatic pressure may change in time according to deviations of the depth of transducer submergence. The most demanding condition is that of a rapid submergence of a vehicle, on which the transducer is installed. In this case the final value of the pressure $P_{\tilde{V}}(t)$ within the cavity is achieved as result of a transient process. In course of the transient process all the elements of the resonator must be taken into consideration in the equivalent circuit of the equalization system, as is shown in Figure 12.7 (c). Reaction of this circuit to changing input pressure depends on the rate of change of hydrostatic pressure. Thus, if the pressure changed in the step like way, the response would be as shown qualitatively in Figure 12.8.

The rate of changing value of pressure $P_{\tilde{V}}$ depends on the damping coefficient of the resonator,

$$\zeta = \frac{R_\Sigma}{2}\sqrt{\frac{M_\Sigma}{C_\Sigma}}, \tag{12.49}$$

where for brevity the combined reactive and active parameters of the resonator denoted as C_Σ, M_Σ and R_Σ. The pressure reaches save for transducer level $P_{\tilde{V}}(t) = P_h(t) - \Delta P_{tr}$ sooner or later, but in all the cases at the first moment all the hydrostatic pressure is acting on the mechanical system of transducer.

In reality the rate of changing hydrostatic pressure (rate of submergence), dP_h / dt, is finite. At the beginning of submergence, the flow of fluid in the neck (assumed to be in the shape of a tube) is proportional to $P_h(t)$. Due to viscous resistance of fluid inside the tube a drop of pressure, ΔP, on the length of the tube takes place. For approximate estimating the value of the drop the Hagen-Poiseuille law (Ref. 3) can be used that establishes relation

$$\Delta P = Q\frac{8\mu l}{\pi a^4} \tag{12.50}$$

between pressure drop (ΔP), resistance of fluid ($R_{ac} = 8\mu l / \pi a^4$) and rate of laminar flow through the tube (Q). (It is analogous to Ohm's law for electrical circuits). As noted, $Q = (dP_h / dt)$. Equating $\Delta P = \Delta P_{tp}$, geometry of the tube can be found from Eq. (12.50) for a particular rate of transducer submergence. For achieving a higher accuracy of hydrostatic pressure equalization parameter of resonator a^4 / l must be increased.

It is noteworthy that relation (12.50) is valid, strictly speaking, for the long thin tubes. Therefore, the results obtained must be considered more as tendencies rather than accurate values. The same is true regarding all the calculations that are related to processes of propagating through the tubes of finite size, moreover, even properties of the fluids in real transducer designs may be known not to a great accuracy.

Characteristics of the resonator, as a partial system that accomplishes the hydrostatic pressure compensation, influence the operating parameters of transducer, as this can be quantitatively analyzed using the equivalent circuit in Figure 12.6 (b). Therefore, designing the transducer and resonator, as the parts of the coupled system, must be produced in accord. Two variants must be considered differently in this respect: operating of the transducer in the receive mode in a broad frequency range that includes low frequencies and operating in the transmit mode in vicinity of the resonance frequency of the projector. These situations will be illustrated with examples of transduces in Chapter 13 (for projectors) and in Chapter 14 (for receivers).

12.4 Properties of Fluids under Pressure

A common peculiarity of transducers with hydrostatic pressure equalization systems is that their parameters (sensitivity for receivers, resonance frequencies for projectors) may change to some extent due to dependence of properties of used fluids from environmental conditions. The most vulnerable in this respect are the mechanical systems of low frequency transducers of the flexural type. Properties of the fluids, of which parameters of resonators depend, according to formulas (12.9), (12.13), and (12.21), (12.26) are: $\rho c^2 \sim K_{ac}$, $\rho \sim M_{ac}$, $\mu \sim R'_{ac}$ (for thin tubes), and $\sqrt{\rho\mu} \sim R'_{ac}$ (for tubes of moderate diameter). A brief information regarding properties of the fluids that are essential for determining variations of parameters of resonators is presented below.

For a fluid

$$c = 1/\sqrt{K\rho} \text{ and } \rho c^2 = 1/K , \tag{12.51}$$

where K is compressibility of the fluid. For liquids that have great heat conductivity the isothermal compressibility, K_T, must be used in the formulas. More appropriate for a gas is the adiabatic compressibility, K_S, (except for high frequencies that are beyond the range of underwater acoustics). Correlation $K_S = K_T / \gamma$ exists, where $\gamma = C_P / C_{\tilde{V}}$ is ratio of the specific heats of the gas at constant pressure and at constant volume. Thus, $K_{ac} \sim K_T$ for liquids, and $K_{ac} \sim K_T / \gamma$ for gasses. By definition

$$K_T = -\frac{i}{\tilde{V}}\left(\frac{\partial \tilde{V}}{\partial P}\right)_T = \frac{1}{\rho}\left(\frac{\partial \rho}{\partial P}\right)_T . \tag{12.52}$$

From equation of state for a perfect gas (in the range of operating temperature and hydrostatic pressure conditions all the gases behave approximately as a perfect gas),

$$P\tilde{V} = nRT , \tag{12.53}$$

where n is number of moles (mass equal to molecular weight in grams) in the volume, $R = 8.3 \cdot 10^3$ joules/mole and T is the absolute temperature, follows that

$$K_T = \frac{1}{P} \text{ , and } K_S = \frac{1}{\gamma P} . \tag{12.54}$$

For the liquids the compressibility is

$$K = \frac{1}{B} , \tag{12.55}$$

where $B = \rho c^2$ is the bulk modulus, $B = \Delta P / (\Delta \tilde{V} / \tilde{V})$. (The bulk modulus B is analogous to the Young's modulus Y for solids). Approximately linear relationship exists between the volume of a liquid and pressure (between the density and pressure) up to pressures at least about 1000 atm (this corresponds to the pressure at the deepest point in the Ocean, Mariana Trench). Thus, compressibility of seawater with $B \approx 23 \cdot 10^3$ atm changes at this depth by about 5% (the density changes as well), and therefore compressibility of the liquids can be considered practically independent of pressure in underwater applications. Data on the bulk moduli of several liquids are presented in Table A.2

Table 12.2: Bulk moduli of liquids

Liquid	Water	Seawater	Castor oil	Motor oil SAE-30	Hydraulic fluid ISO 32	Silicon oil
B, GPa	2.15	2.34	2.1	1.5	1.8	1.1

According to formulas (12.9) for acoustic rigidity of a cavity, and (12.54), (12.55) for compressibility of the gasses and liquids, the ratio of rigidities of cavities having equal volumes filled with commonly used castor oil and air under pressure is

$$\frac{K_{ac\,air}}{K_{ac\,oil}} = \frac{B_{oil}}{\gamma P} = 1.25 \cdot 10^9 \cdot \frac{1}{P_{(Pa)}} . \tag{12.56}$$

(For air $\gamma = 1.4$). Thus, at the average depth of Ocean ($\approx$ 4200 m) this ratio is about 30, i.e., at equal rigidities the volume of cavity filled with air can be made accordingly smaller. Especially vulnerable to problems of reducing the size of the pressure compensating systems and their influence on the transducer performance are the low-frequency projectors that employ mechanical systems of flexural type (benders, slotted ring and flextensional type transducers). These problems and possible ways of their overcoming are briefly discussed in Ref. 6. Using the pressurized gas would be the best choice in this respect, but its employing encounters significant technical problems with reliable pressurized gas supply. Using compressible liquids such as the silicon oil may allow reducing volume of cavity approximately in factor of two. For transducers operating at moderate depth, it is possible to increase compressibility by inserting compliant metal or plastic sealed tubes inside the volume.

In the case of widely used pressure compensated cylindrical transducers the main concern may be to reduce effect of input impedance of the internal volume on the transducer performance. For most practical applications it is advantageous to use polyurethane rather than liquids for filling the interior of transducers. This simplifies the transducer designs and increases their reliability. Physical properties of the polyurethane modifications differ, but they can be characterized in average for approximate estimating parameters of the internal volume. Useful information regarding mechanical and acoustic properties of the polyurethanes can be found in Ref.7. For example, the data are presented in Table 12.3 for two commercial polyurethane (PU) rubbers: DeSoto PR1547 and GS960PU of Gallaher Corporation.

Table 12.3: Properties of the polyurethanes.

Property		ρ, kg/m^3	C, m/s	B, GPa	G, MPa
PR1547	4°C	1.05	1650	2.9	6
	34°C		1500	2.3	4
GS960PU, 20°C		1.08	1700	3.3	1.2

The properties of PU to some extent depend on temperature, as is illustrated in the table, and on frequency. The dynamic bulk modulus of elasticity for different PU compositions may change approximately within ±35% of the mean value of 2.9 GPa. Therefore, it may be desirable to increase compressibility of the PU filled internal volume of a transducer. For limited depth applications, the compliant inserts can be used. Thus, in case that transducer is intended to operate up to relatively small depth, but must withstand much deeper submergence, using corprene for this purpose can be recommended. For example, 3 mm thick sheet of corprene DC-100 being encapsulated in PU proved to be effective up to about 250 meters before its compliance dropped, but the original properties were restored after exposure to pressure equivalent to 1000 m depth.

12.5 References

1. I. B. Crendall, *Theory of Vibrating Systems and Sound*, New York: D. Van Nostrand, 1926.
2. S. N. Rschevkin, *A Course of Lectures on the Theory of Sound*, Pergamon Press Ltd, Oxford, England, 1963.
3. P. M. Morse, K. U. Ingard, *Theoretical Acoustics*, (McGraw-Hill, New York, 1968)
4. R. J. Urik, *Principles of Underwater Sound*, Third Edition, Peninsula Publishing, Los Altos, CA, (1983).
5. A. D. Pierce, *Acoustics*, Published by Acoustic Society of America, 1989.
6. R.S. Woollett, "Basic problems caused by depth and size constraints in low-frequency underwater transducers," J. Acoust. Soc. Am., 68 (4), 1031-1037, (1980)
7. P. H. Motta and C. M. Roland, "Acoustic and dynamic mechanical properties of a polyurethane rubber," J. Acoust. Soc. Am., 111 (4), 1782-1790, (2002)

CHAPTER 13

PROJECTORS DESIGNING RELATED ISSUES

13.1 Introduction

The theoretical considerations presented in the preceding chapters allow calculating the electroacoustic characteristics of transducers regardless of their application. But an optimal designing the real transduces (especially high-power projectors) requires many skills and engineering intuition besides the ability to calculate parameters of the transducers analytically or numerically. Therefore, attempts to describe detail of designs of the particular transducer types without specifying requirements for their electroacoustic characteristics and operating and environmental conditions hardly can be successful. It may be useful and possible instead to show the tendencies that can be pursued for achieving certain goals. In terms of meeting requirements for the electroacoustic parameters, the most challenging are the projector designs due to limitations of acoustic power radiated predominantly by the dynamic stress and electric field in the active material. These limitations are influenced by the environmental conditions, especially by the hydrostatic pressure that may produce reversible and permanent changing parameters of piezoceramics (see Chapter 11). In this Chapter several aspects of the projector designing are considered related to optimizing their operation.

For making decisions regarding optimal matching transducers of different type with acoustic field the concept of reserves of mechanical and electrical strength of a transducer that was introduced in Section 3.1.3.2 is used. It is shown how reducing the excessive reserves of strength by changing transducer design may lead to optimizing transducer parameters: to increasing power radiated, or to making the design more balanced and therefore more reliable. This is done in Section 13.2.

Mechanical strength of the load bearing (air-backed) transducers of different type under action of the hydrostatic pressure is considered in Section 13.3. The notions are introduced in this regard of the survival depth of submergence and of the maximum operating depth that is limited by a tolerable changing of ceramics parameters under the pressure.

Insuring the static mechanical strength and stable parameters of the transducer design under the hydrostatic pressure is one of important issues in their practical realization. Especially demanding in this respect are the low frequency flexural type projectors, as their mechanical systems inherently have relatively large compliance and hence vulnerability to static loads. Solving this problem requires using the hydrostatic pressure equalizing arrangements. The questions arise as to the influence of these arrangements on the electromechanical parameters of the projectors. The related issues are considered in Section 13.4.

13.2 Using Concept of Reserves of Strength

13.2.1 The Optimum Projector Loading and Coefficients of Reserves of Strength

Summary is presented below of expressions that characterize the maximum acoustic power $\dot{W}_{ac\,m}$ that can be radiated by a transducer under the condition that radiation resistance has optimum for this transducer value $r_{opt\,W}$, or a real for the transducer value r_{ac}, which were introduced in Section 3.1.3 under certain assumptions. Thus, for example, the mechanical losses in the transducers are neglected to the first approximation in order to simplify relations that illustrate the essence of the matter without loss of generality, as it is explained in Section 3.1.3. Therefore, all the following relations must be considered as approximate. When using the relations, it has to be remembered that numerical values of optimum acoustic load and maximum power radiated depend significantly on the values of permissible electric field and dynamic mechanical stress, which may change under particular operating conditions (temperature regime, duty cycle). Therefore, the numerical data must be considered as illustrative for the methodical purpose rather than as exact values.

The optimum radiation resistance is determined by formula (3.126) as

$$r_{opt\,W} = \frac{A_T E_P}{A_E T_P}. \tag{13.1}$$

Here the quantities A_E and A_T for a particular transducer type must be determined using relations (3.121) and (3.123), respectively, where

$$E_p = 2\cdot 10^5 \text{ V/m},\ T_p = 2\cdot 10^7 \text{ N/m}^2,\ \text{and } E_p/T_p = 10^{-2} \text{ Vm/N} \tag{13.2}$$

for the extensional type transducers, and

$$E_p = 2 \cdot 10^5 \text{ V/m}, \ T_p = 2 \cdot 1.7 \cdot 10^7 = 3.4 \cdot 10^7 \text{ N/m}^2, \text{ and } E_p / T_p = 0.6 \cdot 10^{-2} \text{ Vm/N} \quad (13.3)$$

for transducers of the flexural type. It is noteworthy that the same values E_p = 2 kV/cm and T_p = 3000 psi (about 21 MPa) were used as permissible for estimating the extensional type transducers in Ref. 5. Regarding the permissible values of electric field and mechanical stress must be remembered the following.

Magnitude of the permissible electric field may be limited by a smaller value depending on the temperature regime of a particular transducer design. The maximum power of electrical losses that is generated in volume of a transducer at resonance frequency ω_0 at permissible electric field E_p can be estimated as

$$\dot{W}_{eL} = V^2 \omega_0 C_e^S \tan\delta = E_p^2 t_e^2 \omega_0 C_e^S \tan\delta, \quad (13.4)$$

where t_e is separation between electrodes. This quantity can be calculated for a particular transducer type. Thus, for example, for the spherical transducer made of PZT-4 ceramics:. $C_e^S = \varepsilon_{33}^T (1 - k_p^2) \cdot 4\pi a^2 / t_e$., where $t_e = t$ is the thickness of the sphere; $\tan\delta = 0.02$ at $E_p = 2 \cdot 10^5$ V/m, and $\omega_0 = (1/a)\sqrt{2/\rho s_{11}^E (1 - \sigma_1^E)}$. After substituting these data in formula (13.4) we arrive at $\dot{W}_{eL} = 43at$, where *a* and *t* are in *cm*. Or being normalized to the radiating surface area $S_\Sigma = 4\pi a^2$,

$$\frac{\dot{W}_{eL}}{S_\Sigma} = 3.4 \frac{t}{a} \text{ w/cm}^2. \quad (13.5)$$

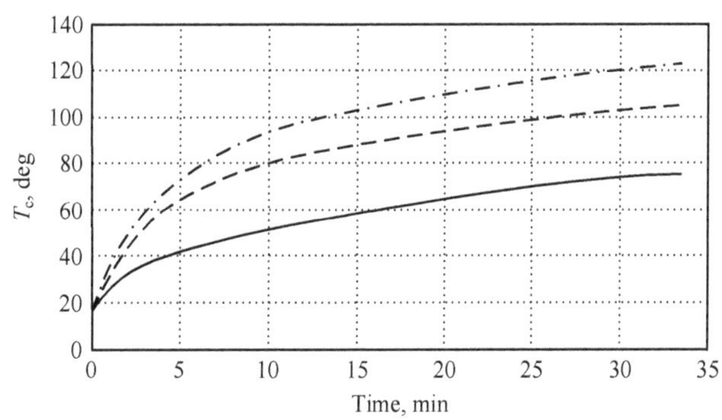

Figure 13.1: Temperature dependencies for sphere with $a = 3.3$ cm and $t = 0.3$ cm that vibrates in water at room temperature in *cw* mode at $E_p = 2 \cdot 10^5$ V/m. Solid line for bare sphere, dash and dot-dash lines for the sphere incapsulated by 3 and 5 mm thick layers of polyurethane, respectively.

If the raise of temperature due to dissipating this amount of energy in a piezoelement exceeds an acceptable level, then the permissible value of the electric field must be reduced accordingly to ensure a safe level of temperature and radiating power.

This new permissible value of the electric field must be taken in estimation of the optimum loading. Evidentially, this can be done only after considering particularities of the transducer design and conditions of its operation. Example of the temperature dependences for the spherical transducer with dimensions $a = 3.3$ cm and $t = 0.3$ cm made of PZT-4 ceramics that vibrates in water at room temperature in *cw* mode at $E_p = 2\cdot 10^5$ V/m, are presented in Figure 13.1. The cases considered include the sphere being bare or encapsulated with layers of polyurethane of varying thicknesses.

For illustrating general approach to estimations of the optimum loading we will keep the permissible value of electric field $E_p = 2\cdot 10^5$ V/m until specified otherwise. The permissible value of the mechanical stress is related to the solid piezoelements. If the piezoelements are cemented of parts, this value should be significantly reduced unless the piezoelement is pre-compressed to a value that excludes tension in the bonds.

Real radiation resistance, r_{ac}, may differ from the optimum, r_{opt}, if no special measures are taken. The difference can be characterized by the coefficient of mismatch,

$$r_{ac} / r_{opt\,W} = m_W . \tag{13.6}$$

At the real radiation resistance, the maximum power radiated can be limited by the electric field ($\dot{W}_{mE}$), or by the mechanical stress ($\dot{W}_{mT}$) that are expressed by formulas

$$\dot{W}_{mE} \approx \frac{E_P^2}{A_E^2}\frac{1}{r_{ac}}, \quad \dot{W}_{mT} = \frac{T_P^2}{A_T^2} r_{ac} . \tag{13.7}$$

In the expression for $\dot{W}_{mE}$ resistance of mechanical losses is neglected. This is especially justified in a situation where the radiation resistance is close to the optimum. Ratio of these limiting values of power is related to the coefficient of mismatch, as

$$\frac{\dot{W}_{mT}}{\dot{W}_{mE}} \approx \left(\frac{A_E}{A_T}\right)^2 \cdot \left(\frac{T_P}{E_P}\right)^2 = \frac{r_{ac}^2}{r_{optW}^2} = m_w^2 . \tag{13.8}$$

The ratio (A_E / A_T) is called in Section 3.1.3 the design factor, because it is determined by the type and mode of vibration of the transducer mechanical system and by the electromechanical properties of the piezoelectric ceramic material used.

The ratio (T_p / E_p) is called the technological factor because it depends on the existing average technological level of fabrication of the piezoelements and of the finished transducers that determines the permissible values of electric field and mechanical stress. The mismatch coefficient m_w can be considered as the acoustic load factor. In the case that $m_w < 1$ the transducer is underloaded, and the maximum power radiated is mechanical stress limited. At $m_w > 1$ the transducer is overloaded, and the maximum power radiated is electric field limited. At $m_w = 1$ the transducer is optimal loaded, and $\dot{W}_{mT} = \dot{W}_{mE}$. These situations were illustrated in Table 3.5 with examples of the basic designs of the uniform rod and circular ring transducer under the loads that are real for their application in array of a large size and in a single cylindrical transducer of a big height. It is seen that under these "natural" loading conditions they represent typical examples of underloaded and overloaded transducer types, respectively. Results of the analogous estimations that are made for broader range of the transducer types are presented in Table 13.1.

The estimations in the Table 13.1 are made under the following assumptions:

The radiation resistances for the baffled cylinder, sphere, and mass-loaded bar (assuming that its head is a round piston that has wave size $D / \lambda = 0.5$) are determined from the plots in Figures 6.10, 6.28 and 6.38 at values ka that correspond to the resonance frequencies of the transducers made of PZT-4 ceramics.

The radiation resistances of the flexural type transducers are calculated by formulas presented in Section 9.4: (9.283) for the single transducer units, (9.288) for the column like rectangular benders and slotted rings, (9.294) for the benders in arrays of a large size. It is assumed that the single rectangular plate and slotted ring transducer units have dimensions $w = l$ and $h = 2a$, accordingly. The column like transducers have the height that exceeds the wavelength. The wave sizes of the single flexural transducers are assumed to be $D / \lambda \approx t / a$,were for the numerical estimations is used $t / a = 0.2$. Convention is that for the rectangular benders $D = 2a = w = l$.

All the values of the mismatch coefficients m_w are determined for the transducers made of PZT-4 ceramics. In the case that they depend on the relative dimensions of the piezoelements, the values of ratios $a / t = 5$ and $l / t = 5$ are used for numerical estimations.

Table 13.1. Estimations of the optimum acoustic loads for different projector types and their mismatch with "real" acoustic loads.

Transducer Type	i	A_E	A_T	$\frac{r_{opt}}{S_\Sigma}\cdot 10^2$	$\frac{r_{ac}}{S_\Sigma}$	m_w
Sphere	1	$\frac{s_{11}^E(1-\sigma_1^E)}{2d_{31}S_\Sigma}\frac{a}{t}$	$(\rho c_1^E)\sqrt{\frac{1-\sigma_1^E}{2}}$	$(\rho c_1^E)\frac{d_{31}}{s_{11}^E}$ $\times\sqrt{\frac{2}{1-\sigma_1^E}\frac{t}{a}}$	$0.98(\rho c)_w$ at $ka = 3.3$	$0.35\frac{a}{t}$
Baffled Cylinder	1	$\frac{s_{ii}^E}{d_{3i}S_\Sigma}\frac{a}{t}$	(ρc_i^E)	$(\rho c_i^E)\frac{d_{3i}}{s_{ii}^E}\frac{t}{a}$	$0.46(\rho c)_w$ at $ka = 2.2$	$0.28\frac{a}{t}$
	3					$0.17\frac{a}{t}$
Mass-Loaded Bar	1	$\frac{s_{ii}^E}{2d_{3i}S_{\Sigma c}}$	$(\rho c_i^E)\sqrt{\frac{2M_h}{m_c}}$	$2(\rho c_i^E)\frac{d_{3i}}{s_{ii}^E}$ $\times\frac{S_{\Sigma c}}{S_{\Sigma h}}\sqrt{\frac{2M_h}{m_c}}$	$0.8(\rho c)_w$ at $ka = 1.7$	$0.24K_m^*$
	3					$0.13K_m^*$
Single Circular Plate Unit	1	$1.3\frac{s_{11}^E(1-\sigma_1^E)}{d_{31}S_\Sigma}$ $\times\left(\frac{a}{t}\right)^2$	$0.86(\rho c_i^E)$	$0.4(\rho c_1^E)$ $\times\frac{d_{31}}{s_{11}^E(1-\sigma_1^E)}$ $\times\left(\frac{t}{a}\right)^2$	$1.1(\rho c)_w\left(\frac{t}{a}\right)^2$	0.64
Single Rectangular Plate Unit	1	$\frac{2}{\pi}\frac{s_{ii}^E}{d_{3i}S_\Sigma}\left(\frac{l}{t}\right)^2$	$1.7(\rho c_i^E)$	**) $\frac{S_\Sigma}{dx\,dy}$	$2.5(\rho c)_w\left(\frac{t}{l}\right)^2$	0.93
	3					0.5
Column-like Rectangular Bender	1			$1.22\frac{S_\Sigma}{dx\,dy}$	$1.3(\rho c)_w\left(\frac{t}{l}\right)^2$	$0.5\frac{l}{t}$
	3			$2.7(\rho c_i^E)$		$0.27\frac{l}{t}$
Rectangular Bender in Array	1			$\times\frac{d_{3i}}{s_{ii}^E}\left(\frac{t}{l}\right)^2$		$2.3K_s$ **)
	3					$1.22K_s$ **

Transducer Type	i	A_E	A_T	$\frac{r_{opt}}{S_\Sigma}\cdot 10^2$	$\frac{r_{ac}}{S_\Sigma}$	m_w
Slotted Ring Single Unit	1	$9.1\frac{s_{ii}^E}{d_{3i}S_\Sigma}\left(\frac{a}{t}\right)^2$	$25(\rho c_i^E)$	$1.6(\rho c_i^E)\times\frac{d_{3i}}{s_{ii}^E}\left(\frac{t}{a}\right)^2$	$3.3\cdot 10^{-2}\times(\rho c)_w\left(\frac{t}{a}\right)^2$	$1.2\cdot 10^{-2}$
	3					$0.7\cdot 10^{-2}$
Slotted Ring Column-Like Transducer	1				$0.2(\rho c)_w\frac{t}{a}$	$7.5\cdot 10^{-2}\times(a/t)$
	3					$4.3\cdot 10^{-2}\times(a/t)$

*) $K_m = \frac{S_{\Sigma h}}{S_\Sigma}\cdot\sqrt{\frac{m_c}{2M_h}} = \frac{1}{\sqrt{2}}\sqrt{\frac{\rho_c h_c a_h^2}{\rho_h h_h a_c^2}}$, where and a_c are the radiuses of the head and piezoelement, is the matching coefficient. **) $K_s = S_\Sigma / dx\cdot dy$ is the space factor of an array.

For the mass-loaded bar transducer the numerical values depend on the transducer dimensions through the coefficient $K_m = (S_{\Sigma h}/S_\Sigma)\cdot\sqrt{m_c/2M_h}$ that must be calculated for a particular transducer design. Loading for the rectangular benders in an array can be changed by changing the space factor $K_s = S_\Sigma / dx\cdot dy$, where S_Σ is the transducer surface area, and dx, dy are the dimensions of the place that is occupied by the transducer in the array.

Several conclusions can be reached following the data presented in Table 13.1. Example of the mass-loaded bar transducer, which is a particular case of the Tonpilz design, illustrates a way for matching the underloaded equivalent uniform bar transducer with acoustic field (regarding the equivalent mutual transforming between the uniform and nonuniform bars in a general case see Sec. 10.4). Close to optimum loading can be achieved by proper choosing the matching coefficient K_m. It is noteworthy that due to cementing the parts of transducer design with piezoelement and segments of the piezoelement in case that longitudinal piezoeffect is employed prestress a_h of the whole structure is required. Introducing a prestressing member of the design requires replacing a part of ceramic volume by a passive material, and thus results in some changes of coefficients A_E and A_T. Therefore, the example in the table is numerically not exact, but it illustrates the way of better matching the transducer with acoustic field. How the conditions of optimum matching can be achieved in a transducer design will be considered in the next section.

Example of the baffled cylindrical transducer shows much better loading in comparison with omni radiating transducer. Given that *a/t* aspect ratio may be slightly changed the loading of baffled transducer can be considered almost optimum especially for the case that longitudinal piezoeffect is employed. In the last case a prestressing the segmented piezoelement is needed, but it does not require any replacement of the active material. Though the coefficient A_E may be reduced due to exposure of the ceramics to the maintained compression stress in direction of polarization. One of the possible ways of prestressing the cylindrical piezoelement without a significant changing its mechanical properties is by wrapping with a fiber (for example, glass fiber) under tension in combination with epoxy compound, as illustrated in Figure 13.2.

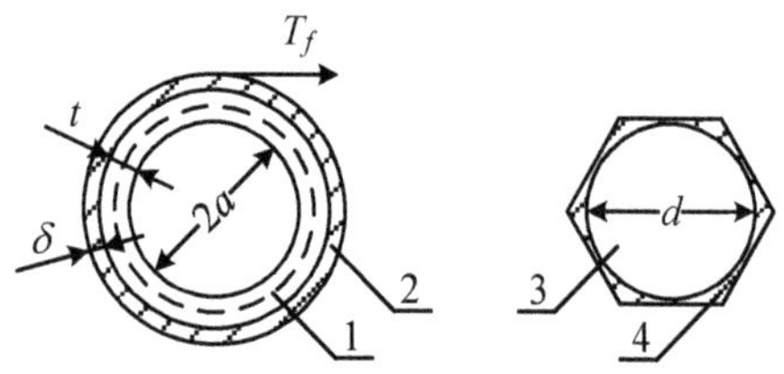

(a) (b)

Figure 13.2: To prestressing the cylindrical piezoelement. (a) General view, (b) cross section of the single element of the winding.: 1 - ring, 2 - winding, 3 - fiber, 4 - epoxy.

Correlation between necessary ceramics prestressing, T_c, and tension in the fiber, T_f, can be determined as follows. Assuming that thickness of the winding is much less than thickness of the ring, $\delta \ll t$, it can be concluded that,

$$T_c t + T_f \delta \cdot k_f = 0, \qquad (13.9)$$

where k_f is the space factor for the winding, $k_f = S_f / S_\Sigma$ (S_f is cross-section area of the fiber, and S_Σ is the total cross-section area of the element). From geometry considerations for the cross-section of a single element of the winding shown in Figure 13.2 (b) follows that $k_f = 0.9$. Thickness δ of the winding will be found from this relation with known maximum permissible value of tension T_f in the fiber. For estimating effect of the winding on parameters of a ring the resulting equivalent rigidity and mass of the cylinder with winding, K_Σ and M_Σ, must be determined. Approximately they are (subscript *c* stays for "ceramics")

$$K_\Sigma = K_c + K_f \approx K_c[1 + Y_w \delta / Y_c t], \; M_\Sigma = M_c + M_f \approx M_c[1 + \rho_w \delta / \rho_c t]. \qquad (13.10)$$

Here $Y_w = 0.9Y_f + 0.1Y_{ep}$, $\rho_w = 0.9\rho_f + 0.1\rho_{ep}$; Y_f, ρ_f and Y_{ep}, ρ_{ep}, are the Young's moduli and densities of the fiber and epoxy, respectively. After these corrections due to the winding are determined, calculating changes of the ring parameters becomes straightforward. Given that $\delta \ll t$, $Y_w \ll Y_c$, and $\rho_w < \rho_c$ all the changes of parameters most probable are negligible.

It is noteworthy that though this method of reinforcing the piezoceramic rings is widely used, it is not known for how long and to what extent the originally established stress is maintained due to possible stress relief in the process of the fiber relaxation.

Baffling can be considered as the way of matching the cylindrical transducers with acoustic field in case that they are used as unidirectional. Otherwise, the only way of matching is by changing the ratio a/t. Using the cylindrical transducers as unidirectional is typical for their employment in arrays. The range of their effective operating vs. depth depends significantly on properties of the baffles and on their behavior under the hydrostatic pressure. We considered in our estimations in Table 13.1 the ideal (rigid or compliant) baffle representations for revealing their influence on transducer performance. In practical applications, baffles made of encapsulated corprene can be used for operation in shallow water (up to about 200 m in depth). Designing the baffles intended for operating at great depth and at low frequencies (below approximately 10 kHz) is a complicated engineering problem. Description of several examples of baffles for underwater transducers and methods of their calculating can be found in Ref. 9.

The spherical transducer is overloaded, if the piezoelement is solid. This means that the power radiated is limited by the electric field. The degree of overloading can be reduced by increasing the relative thickness of the sphere. Although usually the piezoceramic spherical shells are cemented out of hemispheres, and a prestress of the bonding is hardly possible. In this case the permissible stress will be limited by the strength of bonding the hemispheres, and the transducer may become underloaded, i.e., limited by the mechanical strength.

The single circular and rectangular plate benders at transverse piezoeffect are almost optimally loaded. The single rectangular plate benders that employ the longitudinal piezoeffect are underloaded. But it must be kept in mind that the segmented rectangular beam requires prestress, and a part of the active material can be used for achieving this goal. As a result, the loading conditions may become closer to optimum. It is of note that the single flexural transducer units are seldom used for radiation. The circular plate benders are not capable of radiating significant power (remember that condition of optimum loading, $m_w = 1$, means only that

$\dot{W}_{mE} = \dot{W}_{mT}$, and does not say about their absolute values), and rectangular benders are usually used in columns, or as the members of the large arrays. The column like rectangular benders are noticeably overloaded (electric field limited), if the transverse piezoeffect is used, and are almost optimally loaded, if the longitudinal piezoeffect is employed (though the prestress is required in this case). Loading of the rectangular benders in the large arrays can be tailored by changing the "space factor" K_s. With transverse piezoeffect used the benders are overloaded at reasonable (that does not contradict overall amount of power radiated) value of the space factor, and with longitudinal effect used the optimum loading may be achieved at $K_s \approx 0.5$, or a little less given that the prestress is required.

The slotted ring transducers are dramatically underloaded (mechanical stress limited) as the single units and in the column like transducers. This is especially true regarding the transducers that employ longitudinal piezoeffect, moreover, they require a prestress.

It must be kept in mind that the numerical estimations are presented in Table 13.1 without considering real environmental conditions. Thus, the possible effects of hydrostatic pressure on peculiarities of the transducer designs and on their parameters are not taken into consideration.

13.2.2 About Using Concept of Reserves of Strength for Optimizing Projector Designs

Reserves of the electrical and mechanical strength of the projector regarding the maximum values of power limited by permissible electric field and mechanical stress are characterized by the coefficients (see relations (3.130) and (3.131))

$$k_E = \frac{E_P}{E_{op}} = \sqrt{\frac{\dot{W}_{mE}}{\dot{W}_{op}}} \text{ and } k_T = \frac{T_P}{T_{op}} = \sqrt{\frac{\dot{W}_{mT}}{\dot{W}_{op}}} . \tag{13.11}$$

Here the operating acoustic power of a projector is denoted by $\dot{W}_{op}$, and the corresponding electrical field and mechanical stress for the projector are denoted by E_{op} and T_{op}. According to these definitions for the reserves of strength coefficients (they may be also considered as safety factors), it must be $k_E \geq 1$ and $k_T \geq 1$. If k_E and/or k_T exceeds the unity, it means that the projector has an excessive reserve of strength. In general, the projector design may be regarded as rational in the case that it does not have excessive reserves of either electrical or mechanical strength, i.e., $k_E = k_T = 1$ under the operating loading conditions. This situation corresponds to the maximum power available from the projector for a particular application. If a moderate acoustic power is required, then it will be $k_E = k_T > 1$, which means that the projector acquires excessive reserves and its design can be made simpler, cheaper, and even more reliable for expense of reducing these reserves (for example, the amount of active material can be reduced, or the prestressing arrangements can be simplified, etc.). Note, that the resulting increase of E_{op} and T_{op} up to their permissible values E_p and T_p should not compromise the reliability of a projector, because the permissible values by their definitions must insure a long term reliability of a transducer. This assertion is seemingly unusual, and most of the users may prefer to have excessive reserves of strength to be on a save side, but the concept of possibility of optimizing projector design by reducing its reserves of the strength rests on this assumption.

13.2.2.1 Length Expander Bar

To illustrate how the concept of reserves of strength may be realized in way of optimizing a projector design, consider an example of transducer that employs piezoceramic length expander bar, as the typical representative of underloaded transducer type. The concept of equivalent transformation between the uniform and nonuniform (composite) bar representations that was introduced in Section 10.4.2 can be used with this goal. For simplicity we consider symmetrical composite transducer. Geometry of both composite and equivalent uniform bars is shown in Figure 13.3.

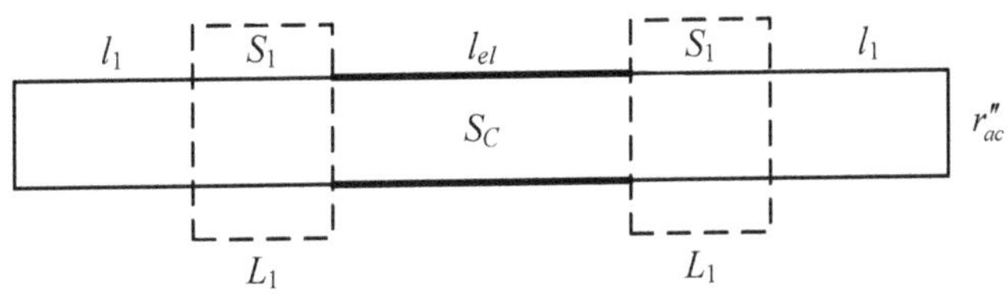

Figure 13.3: To the equivalent conversion of the uniform bar into composite transducer. By the solid line is shown imaginary uniform bar, by the doted lines are shown passive parts of composite transducer that are equivalent to their ceramic counterparts in terms of the input impedances.

As it was shown in Section 10.4, dimensions of the composite bar transducer can be related to parameters of the equivalent uniform bar that has a certain resonance frequency, f_r, area of the end surface, S_c, and is loaded by arbitrary radiation resistance r''_a. They must satisfy relations (10.177) - (10.179) that are replicated below as

$$l = l_{el} + 2l_1 = c_c^E / 2f_r, \tag{13.12}$$

$$(\rho c S)_1 \tan(kL)_1 = (\rho c^E S)_c \tan k_c^E \frac{l - l_{el}}{2}, \tag{13.13}$$

$$\cos^2(kL)_1 = (\rho c)_w \frac{S_1}{r''_{ac}} \cos^2 k_c^E \frac{l - l_{el}}{2}. \tag{13.14}$$

Given that parameters of ceramics (ρ_c , c_c^E) are known beforehand, for determining parameters ρ_1, c_1, L_1, l_{el}, and S_c (S_1 should be known as the area that is occupied by the transducer in array) we have the two first equations. Eq. (13.14) can be specified for the optimum matching with acoustic field by a proper choice of the radiation resistance r''_{ac}. We will assume that at optimum matching the mechanical and electrical safety factors must be equal ($k_E = k_T = 1$ that correspond to the maximum power radiated by the transducer). This condition can be formulated as follows. Expression for the electromechanical transformation coefficient of a uniform bar transducer in the case that only a central part of ceramics is active is

$$n(l_{el}) = n(l) \sin(\pi l_{el} / 2l). \tag{13.15}$$

Accordingly, at the same radiated power the electric field E'_{op} for a transducer with partially active ceramics as compared to E_{op} for the fully active piezoelement is

$$E'_{op} = E_{op}\, n(l) / n(l_{el}) \tag{13.16}$$

and

$$A'_E = A_E \cdot n(l) / n(l_{el}) = A_E \cdot \sin(\pi l_{el} / 2l). \tag{13.17}$$

Now from the condition $k_E = k_T$, i.e., $\dot{W}_{mE} = \dot{W}_{mT}$, making use of formula (13.1) we obtain

$$r''_{ac} = \frac{E_P}{T_P} \cdot \frac{A_T}{A'_E} = \frac{E_P}{T_P} \frac{A_T}{A_E} \cdot \frac{1}{\sin(\pi l_{el} / 2l)} = r_{opt\,W} \frac{1}{\sin(\pi l_{el} / 2l)}. \tag{13.18}$$

Upon substituting this expression into Eq. (13.14) we arrive at the condition

$$\cos^2(kL)_1 = \frac{(\rho c)_w S_1}{r_{opt}} \cos^2 k_c^E \frac{l - l_{el}}{2} \cdot \sin(\pi l_{el} / 2l). \tag{13.19}$$

It must be kept in mind that reducing the amount of active material for achieving better matching with acoustic field comes for expense of the electromechanical force (reducing the electromechanical transformation coefficient). At the same time, it is desirable to retain (or at least not to reduce significantly) the value of the effective coupling coefficient in order not to compromise the bandwidth of matching the transducer with power amplifier. These conditions can be not contradictory until $l_{el}/l \geq 0.5$, as it was previously illustrated with Figure 10.33.

For maximizing the overall bandwidth of the transmit channel and increasing with this goal the bandwidth of mechanical system (reducing its Q_{mch}) predominantly light passive materials must be used for replacing ceramics such, for example, as aluminum alloys and titanium. This imposes an additional condition for determining values that characterize the composite transducer design.

13.2.2.2 Rectangular Bender

Consider the rectangular bender projector employing the longitudinal piezoeffect as one more example of achieving a close to balanced acoustic loading. Projectors of this type are the candidates for covering frequency range below 3-4 kHz. As this is seen from Table 13.1, the mismatch coefficient of loading the rectangular bender in the large array is $m_w = 2.2K_s$, and it can be tailored by changing the "space factor" $K_s = S_\Sigma / dxdy$. Thus, seemingly close to optimum loading could be achieved at $K_s \approx 0.5$, but it must be taken into consideration that a prestressing of ceramic may be required, if the goal of radiating maximum possible power is pursued. Therefore, an element that produces the prestressing must be included in the transducer design. In the case of the composite length expander transducer, it was the central bolt stretched between its head and tail. Introducing the bolt could be considered in terms of influencing transducer electromechanical parameters as replacing some part of ceramic volume by passive material. This could be tolerated because the length expander transducer had excessive reserve of electrical strength. Specifics of the rectangular bender as a member of an array is that it is overloaded and introducing an additional structural element ideally must not result in reducing the electromechanical transformation coefficient of transducer, or this reduction must be as small as possible. At the same time, the changes in transducer unit design that are followed by reducing its space factor (area occupied by the active element in array) can be made for expense of existing reserve of the mechanical strength. For illustration two possible variants of the rectangular bender design with elements that produce precompression are shown in Figure 13.4 and Figure 13.5.

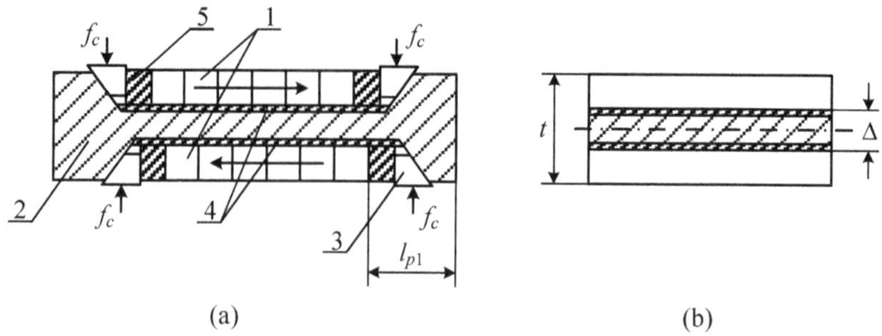

Figure 13.4: (a) Design with the compression member that constitutes a part of the bender plate: 1- segmented piezoceramic blocks, 2 - enforcing member, 3 - wedges for transforming the applied force, 4 - dielectric substrate, 5 - metal end parts. Arrows indicate directions of polarization. (b) Cross-section of the plate.

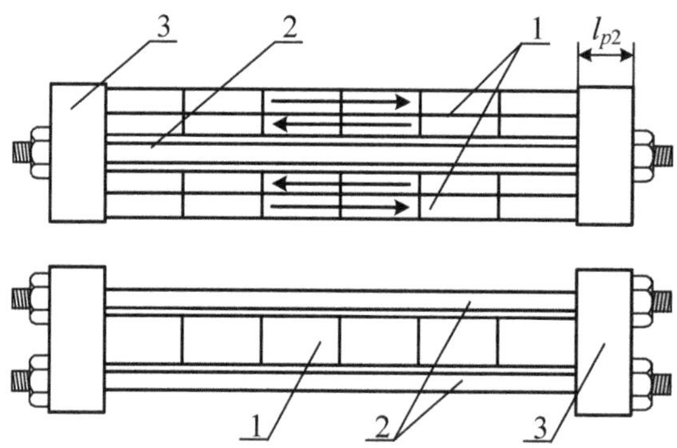

Figure 13.5: Design with the compression member outside of the bender active volume: 1– segmented piezoceramic blocks, 2 – compression bolts, 3 – the end parts. Arrows indicate directions of polarization.

In the variant of Figure 13.4 precompression of ceramic blocks is produced with help of the parts of transducer mechanical system. This must be done in process of cementing the mechanical system under action of compressive forces, f_c, applied to the wedges. The design requires replacing parts of the active volume at the ends (with length l_{p1}) and through the thickness (the latter includes dielectric substrates for isolating electrodes form the central metal layer). In the variant of Figure 13.5 the compression members (bolts) are located outside of the active mechanical system. They are positioned between extensions of the passive end parts of the mechanical system (with length l_{p2}) in the neutral plane of the bender.

The compression can be produced after the mechanical system of the bender is assembled.

Compare these variants in terms of minimal reducing reserves of electrical strength under the condition that dimensions of the radiating surfaces are the same. In both cases replacement of a part of the active material by passive (presumably metal) is required. This results in reducing electromechanical force ($F_{em} = n_{em}V$) according to plots presented in Figures 9.12 and 9.13 that are reproduced here as Figure 13.6 and Figure 13.7.

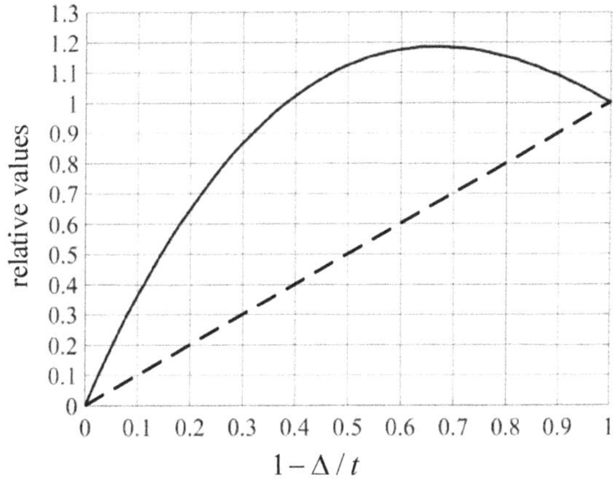

Figure 13.6: Dependence of relative values of the transformation coefficient and effective coupling coefficient from the relative thickness of the passive substrate Δ / t. Dashed line $n(\Delta / t) / n(0)$, solid line $k_{eff}(\Delta / t) / k_{eff}(0)$

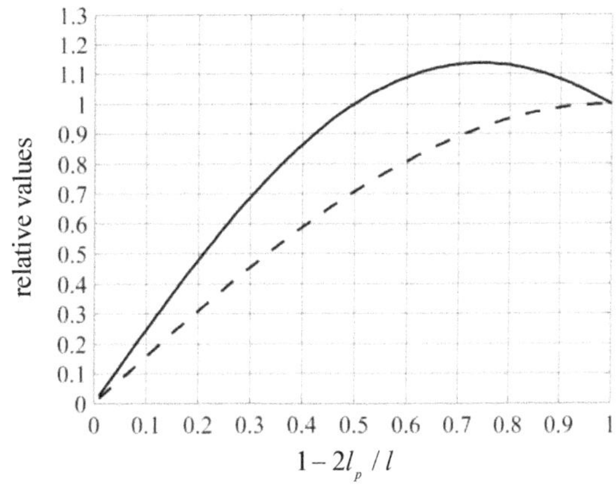

Figure 13.7: Dependence of relative values of the transformation coefficient and effective coupling coefficient from the relative length of passive parts l_p / l. Dashed line $n(l_p / l) / n(0)$, solid line $k_{eff}(l_p / l) / k_{eff}(0)$.

For the qualitative estimations we use the plots obtained under the assumption that values of Young's moduli of the passive material and piezoceramics are close. Although in the first variant the metal parts may be made of metal having noticeably larger Young's modulus, because they experience significant tensile stress. More accurate estimations can be made in this case using the results of Section 9.2.1.2.

As can be concluded from the plots, in both cases replacing a part of volume by a passive material is followed by loss of the transformation coefficient, i.e., by reducing reserves of electrical strength. Especially large drop of electromechanical transformation takes place in the first variant, in which case the required volume of passive material is much greater than in the second variant. This is because the stress-bearing central layer must have a certain thickness, and the passive parts at the ends must be more massive ($l_{p1} > l_{p2}$) to arrange for developing the compressive force.

Therefore, this variant of design can be recommended for projectors of a moderate power in an array, or for the single unit transducers that are underloaded (have reserves of electrical strength). On the positive side is that reducing amount of active material is followed by increase of the effective coupling coefficient according to plots in Figure 13.6 and Figure 13.7.

For the more powerful transducers the second variant is preferable in terms of preserving the level of electromechanical force. But while the prestressing bolts do not influence the electromechanical force, they may reduce to some extent the effective coupling coefficient of a transducer even though their axes are in the neutral plane due to a finite diameter of the bolts. The diameter of a bolt must be determined from the condition that the tensile stress in the bolt, T_{tb}, is related to compression stress in the ceramics, T_{cc}, by relation

$$2\cdot\pi a_b^2\cdot T_{tb} = S_{csc}T_{cc}, \tag{13.20}$$

where a_b is the radius of the bolt, and S_{csc} is the cross-section area of the ceramic plate. Thus,

$$2\cdot\pi a_b^2\cdot T_{tb} = S_{csc}T_{cc}. \tag{13.21}$$

Suppose that the bolts are made from beryllium copper, as was suggested in Ref. 1 for the center bolt of a Tonpilz transducer. This material has the tolerable stress T_{tb} = 700 MPa, $Y_b \simeq$ 200 GPa, $\rho_b = 8\cdot 10^3$ kg/m^3. it is noteworthy that the bolts do not experience a dynamic stress in course of vibration, because of their location in the neutral plane, therefore the static stress is the only factor that must be taken into consideration, when determining their mechanical strength. Thus, for reaching the suggested permissible value of the tensile stress in solid ceramics it must be $T_{cc} = T_{pt} = 34$ MPa (see Eq. (13.3)), and

$$a_b = \sqrt{8 \cdot 10^{-3} S_{csc}} \,. \tag{13.22}$$

To estimate effects of the bolts on the equivalent mechanical parameters of the transducer, consider the energies associated with their deformation in course of flexural vibration of the mechanical system of the transducer. If the bolts have the same length, and the same boundary conditions as the mechanical system of the transducer, its first mode of vibration can be represented in the same way as for the piezoelement, i.e.,

$$\xi_b(x) = \xi_{ob} \sin(\pi x / l) \,, \tag{13.23}$$

and given that the slops at the ends, $\left((d\xi / dx)_{x=0,l}\right)$, are the same, $\xi_{ob} = \xi_o$ -the maximum displacement of the piezoelement.

The potential energy of the bolts is,

$$W_{pot} = 2 \cdot \frac{1}{2} Y_b J_y \int_0^l \left(\frac{d^2 \xi}{dx^2} \right)^2 dx = K_{eqvb} \xi_o^2 \,, \tag{13.24}$$

where the inertia moment of the circular cross-section in respect to y axis is $J_y = \pi a_b^4 / 4$, and K_{eqvb} is the equivalent rigidity of the single bolt. Thus,

$$K_{eqvb} = \frac{\pi^5 a_b^4}{8 l^3} Y_b \,. \tag{13.25}$$

The kinetic energy of the bolts is,

$$W_{pot} = 2 \cdot \frac{1}{2} \rho_b \pi a_b^2 \int_0^l \dot{\xi}^2 dx = M_{eqvb} \dot{\xi}_o^2 \,, \tag{13.26}$$

where M_{eqvb} is the equivalent mass of the single bolt. Thus,

$$M_{eqvb} = \rho_b \pi a_b^2 l / 2 = M_b / 2 \,. \tag{13.27}$$

The validity of representing vibration of the bolts in the operating range of the bender by the first mode can be estimated by comparing the resonance frequencies of the bolts and of the bender itself. Based on the expressions for the equivalent parameters of the bolts and mechanical system of the bender and using formula $\omega_r = \sqrt{K_{eqv} / M_{eqv}}$ may be obtained that the first resonance frequency for the bolt is about two times lower than for the bender. Keeping in mind that the next (third) mode resonance frequency of the bolt is 9 times higher, representing vibration of the bolts by the first mode in the frequency range of the bender operation can be considered appropriate.

Thus, the total equivalent rigidity and the mass of the piezoelement and the bolts made of beryllium copper is,

$$K_\Sigma = K_{eqv}^E(1+2K_{eqvb} / K_{eqv}^E) = K_{eqv}^E\left(1+1.2\cdot 10^{-3}\frac{w}{t}\frac{Y_b}{Y_3^E}\right), \tag{13.28}$$

$$M_\Sigma = M_{eqv}(1+2M_{eqvb} / M_{eqv}) = M_{eqv}\left(1+5\cdot 10^{-2}\rho_b / \rho_c\right). \tag{13.29}$$

Here $K_{eqv}^E = \pi^4 wt^3 Y_3^E / 24l^3$ and $M_{eqv} = \rho_c wtl / 2$. With PZT ceramics used and at reasonable value of ratio w/t the second terms in parenthesis are negligible and $K_\Sigma \approx K_{eqv}^E$, $M_\Sigma \approx M_{eqv}$. Thus, the compression bolts do not influence equivalent mechanical parameters of the piezoelement to the first approximation.

Another question is whether the compressive stress acting along the piezoceramic beam influences its elastic property. It is shown in Ref. 7 that relative change of the resonance frequency of a beam due to compressive stress, T_c, acting along its length can be estimated by formula

$$\frac{\Delta f_r}{f_r} = \sqrt{1-\frac{12}{\pi^2}s_{33}^E T_c\left(\frac{l}{t}\right)^2}. \tag{13.30}$$

Thus, if T_c = 40 MPa and l/t = 5, then for PZT-4 ceramics $\Delta f_r / f_r = 0.99$. This shift of the resonance frequency is negligible, but for a larger compression and relatively thinner beams this value can become noticeable. For example, it was assumed in Ref. 4 that T_c = 70 MPa and l/t = 10. The shift of resonance frequency under this assumption $\Delta f_r / f_r = 0.93$ is big enough to be considered in the transducer designing.

13.3 Static Mechanical Strength of Transducers

The typical environmental conditions for underwater transducers involve the action of the hydrostatic pressure. Therefore, it must be estimated how the resulting stress and measures for insuring the static mechanical strength of projectors of the load-bearing design may influence the above considerations regarding maximum possible power radiated and optimum loading. The severity of acting the hydrostatic pressure on the transducers operation is quite different for the designs that realize the extensional vibrations and those vibrating in flexure. The latter are especially vulnerable in this respect as their mechanical systems inherently have relatively large compliance and hence larger deformations under static loads. Transducers of these types will be considered separately.

13.3.1 The Extensional Type Transducers

The static compression stress in the piezoelements of the extensional type transducers such as spheres and circular rings (planar for spheres and one dimensional for rings) under the hydrostatic pressure can be determined by formula

$$T_{sc} = \frac{a}{t} p_h, \tag{13.31}$$

so far as they are relatively thin and distribution of stress over the thickness can be neglected, which is usually the case. Assuming that compression static strength of ceramics that is typically used for projectors (PZT-4 and PZT-8) is $T_{sc\,\max} = 350$ MPa and the aspect ratio of the piezoelements may be approximately estimated as $a/t \approx 5$, from formula (13.31) follows that in terms of strength the pure piezoelements can withstand hydrostatic pressure up to $p_h = 70$ MPa. This corresponds to depth 7000 m, which exceeds the average depth of ocean (4300 m, and $p_h = 43$ MPa). (Definitely, this does not mean that transducer designs that employ such piezoelements may have equal mechanical strength). To estimate a real operating range of depth for application of the piezoelements, dependences of their electromechanical parameters on the compression stress that are presented in Section 11.3.3 must be taken into consideration. Thus, under the two-dimensional (planar) stress in the spherical shell properties of PZT-4 ceramics may change as it is shown by plots in Figure 11.27.

Under the one-dimensional stress in the circumferentially poled (segmented) rings a prestress in the circumferential direction is usually required. In the extreme case the prestress up to $T_{ps} = 20$ MPa must be imposed to prevent developing the dynamic tensile stress in the ring.

The situation regarding the stress acting in the ceramics in the circumferential direction in this case is illustrated with plots in Figure 13.8. (Remember that the compressive stress is negative according to the sign convention.) It follows that for the transducers intended to operate below 400 m the prestress is not needed for avoiding the tensile dynamic stress.

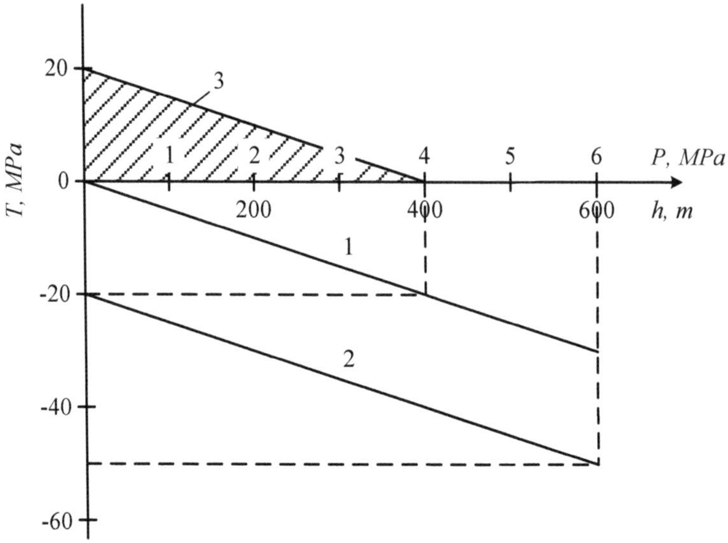

Figure 13.8: Dependencies of the stress in circumferential direction in a ring vs. hydrostatic pressure (operating depth): 1 - hydrostatic compression stress in the ring having aspect ratio $a/t = 5$; 2 – compression stress in the same ring that is prestressed up to 20 MPa; 3 – dynamic tensile stress in the ring without prestress.

It must be noted that the static and dynamic strengths are usually determined in peak values of stress, whereas in calculating power radiated the *rms* values are used. Thus, if the permissible value $T_p = 20$ MPa is accepted for calculating the power, the prestress for avoiding the tensile stress in ceramics should be, strictly speaking, $T_{ps} \approx 28$ MPa. However, we will retain for simplicity in approximate estimations the *rms* value. Moreover, some level of self-strength of bonding exists, at small depth radiation of ultimate power is not possible because of cavitation and at a greater depth an additional compression in ceramics occurs due to the hydrostatic pressure. Dependencies of the dielectric constants and piezoelectric moduli of PZT-4 and PZT-8 ceramics from the static stress parallel to direction of polarization are presented in Section 11.3.3.1 (Figures 11.12 and 11.13).

13.3.2 The Flexural Type Transducers

The stresses in the mechanical system of a bender depend significantly on the boundary conditions. We will assume that the circular plates and rectangular beams employed in the benders are predominantly simply supported. The reasons behind these are that simply supported piezoelements proved to be the most effective electromechanically, and these conditions are the most easily achievable, particularly, in the double-sided symmetrical bender transducer designs.

13.3.2.1 Rectangular Beams4.4/6

The stresses in the rectangular beam are (see Sections 2.6.1 and 4.5.6)

$$T_1^E = -zY_a^E \xi_0 \frac{d^2\theta}{dx^2}. \tag{13.32}$$

The displacement ξ_0 under action of the hydrostatic pressure p_h on the transducer surface can be found as

$$\xi_0 = p_h S_{av} / K_m^E. \tag{13.33}$$

Thus, for the maximum stress that takes place at $z = t/2$ in the middle section of the beam

$$T_{\max} = p_h \frac{tS_{av} Y_1^E}{2K_m^E} \frac{d^2\theta}{dx^2}\bigg|_{x=l/2}. \tag{13.34}$$

At simply supported boundary conditions with $\theta(x) = (16/51)(x - 2x^3/l^2 + x^4/l^3)$

$$K_m^E = 4\frac{wt^3}{l^3 s_{11}^E} \tag{13.35}$$

and $S_{av} = 0.64S = 0.64wl$ (see (2.128) and (2.130)). After substituting expressions for K_m^E and $\theta(x)$ into Eq. (13.34) we obtain

$$T_{\max} = 0.78\frac{l^2}{t^2} P_h, \tag{13.36}$$

For estimating the limiting survival depth of the rectangular bender, we will use values of the static bending tensile and compression stress T_{ps} = 40 MPa and 500 MPa, respectively, as permissible in terms of mechanical strength for solid piezoelements. (They are the tensile and compression strength for the extensional deformation taken from Table 11.1 and multiplied by factor of 1.5–1.7). The permissible tensile stresses for the segmented piezoelements are significantly smaller and they depend on quality of cementing the segments. Such piezoelements are subject to prestress in the transducer designs for avoiding tension. Here we will assume that transducers are made from solid piezoelements that may include the stripe electroded piezoelements employing the longitudinal piezoeffect, or from the segmented piezoelements prestressed up to excluding the tension, as it was assumed for the rings in the previous article.

When using the numerical estimations made in this section, it should be kept in mind that the value of stress that characterizes the bending tensile strength of ceramics must be considered as conditional. It has to be admitted that no reliable data regarding the tensile strength (both static and dynamic) exists. Significantly different results of determining the tensile strength can be encountered in literature (in Ref. 2 this value is about 60-70 MPa, in Ref. 10 this value is estimated as peak dynamic stress about 90 MPa). Probably, $T_{ps} \approx 40$ MPa accepted here is somewhat underestimated value. The subsequent numerical results can be easily recalculated if more trustworthy estimations for the bending strength of ceramics become available.

In determining the maximum depth to which the benders can be used, the notion of the maximum survival and operating depth must be distinguished. Additional limitations on the operating depth may arise due to changing parameters of ceramics under static stress, and due to combined effect of the static and dynamic tensile stress on the strength of the beams. At first, we will consider the survival depth without considering these effects. Estimation of the maximum survival depth that follows from relation (13.35) at value for the static tensile strength T_{ps} = 40 MPa is

$$h_{\max} = 5 \cdot 10^3 \frac{t^2}{l^2} (m) . \tag{13.37}$$

Thus, the thickness to length aspect ratio of a beam is related to the depth as

$$(t/l) \geq 1.4 \cdot 10^{-2} \sqrt{h_{(\mathrm{m})}} . \tag{13.38}$$

The theoretical assumption for applicability of the elementary theory of bending is that $t/l \ll 1$. But the results obtained under this assumption can be used for benders of finite thickness with corrections that are introduced in Section 9.2.2.1. The corrections are negligible for practical application up to values $t/l < 1/5$. Increase of the thickness to length aspect ratio not only affects the electromechanical parameters of the piezoelements. It results also in an increase of the shear stress under the bending that has maximum values in the midplane and close to the ends of a beam. Therefore, the shear strength of epoxy that bonds the parts of the beam must be taken into consideration, when determining a reasonable increase of the aspect ratio. As an example, the shear strength for metal-to-metal bonding is about $T_{p\,sh} \approx 13$ MPa according to the data from manufacturer for a particular epoxy used. Following (9.145) expression for the shear stress is

$$T_5 = \frac{Y_a^E}{2}\xi_o\left(\frac{t^2}{4} - z^2\right)\frac{d^3\theta}{dx^3}. \tag{13.39}$$

The shear stress T_5 reaches its maximum value at $z = 0$ and at $x = 0, l$ (for the simply supported beam $\theta(x) = \sin(\pi x / l)$), and after substituting expression (13.33) for ξ_0 the maximum value will be obtained as

$$T_{m\,sh} = 0.6\frac{l}{t}p_h. \tag{13.40}$$

Given that $Y_a^E = 1/s_{11}^E = 81$ GPa and the permissible shear stress is $T_{psh} = 13$ MPa, the depth limit due to the shear strength is about

$$h_{m\,sh} \approx 2.2\cdot 10^3\frac{t}{l}\ (m). \tag{13.41}$$

If $t/l < 1/5$, the depth limited by shear strength of the bonding between ceramics is $h = 440$ m. The depth limited by the tensile strength according to formula (13.37) is 200 m. Thus, the tensile strength remains the limiting factor and the geometry of the beam vs. required survival depth must be determined using formula (13.37) and its modification

$$l \geq 6.3\cdot 10^{-3}\frac{c_c}{f_r}\sqrt{h_{(\mathrm{m})}} \tag{13.42}$$

that relates the length of a beam to its resonance frequency and to the depth. For example, the length of the fully active beam that has resonance frequency $f_r = 1.5$ kHz at the aspect ratio $t/l = 1/5$ and must withstand maximum depth 200 m is $l \approx 20$ cm. This makes the overall dimensions of the beam 4 cm × 20 cm. Beam of such a large size cannot typically be produced as a solid piezoelement, and must be glued out of smaller parts, in which case the tensile strength of the whole structure can be reduced dramatically. In order to illustrate the correlation between more realistic geometry of a beam and its limiting survival depth, the dependences of the length of a beam having aspect ratio $t/l = 1/5$ from the depth for different resonance frequencies are presented in Figure 13.9 for the case that PZT–4 piezoceramics is used. For different aspect ratios the maximum depth must be determined from formula (13.41). This is noteworthy that the range of the operating depths for the benders as self-sustaining structures was estimated in Ref. 11 as up to 200 m approximately. In our estimation this corresponds to the transducer with resonance frequency $f_r = 1.5\ kHz$.

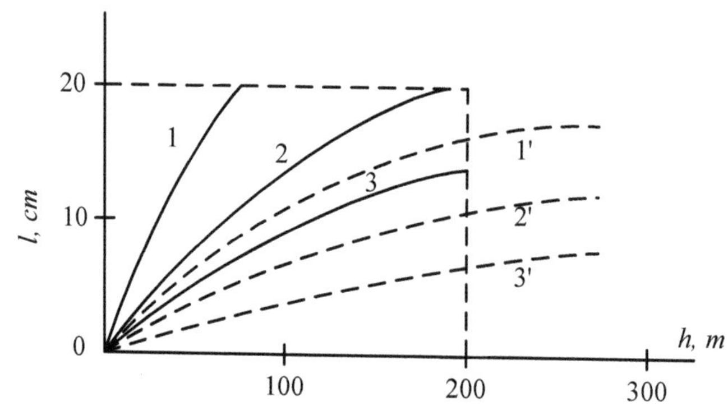

Figure 13.9: Dependences of the length of the beam vs. the maximum survival depth for different resonance frequencies: 1,1'- 1.0 kHz; 2, 2' – 1.5 kHz; 3, 3' – 2.0 kHz. Curves 1, 2, 3 for the fully active beams; 1', 2', 3' for the half passive beams. Piezoceramics is PZT-4, passive material aluminum. Aspect ratio of the beams is t/l=0.2.

For increasing the survival depth and still retaining reasonably compact dimensions of the flexural beam transducers, the half passive beam designs are used, in which case the part of the beam that experiences tension is made of a metal that has a much higher yield tensile strength than piezoceramics. For example, if aluminum alloy AA2011 with T_{pt} = 330 MPa is used as a passive material in combination with PZT-4 ceramics, then formulas (13.38) and (13.42) must be replaced by relations

$$\frac{t}{l} > 0.5 \cdot 10^{-2} \sqrt{h_{(m)}} \text{, and } l \geq 2.2 \cdot 10^{-3} \frac{c_c}{f_r} \sqrt{h_{(m)}} \qquad (13.43)$$

respectively. The plots of dependencies of the length of half passive beam vs. maximum operating depth for different resonance frequencies are presented in Figure 13.9 by the dash lines. Increase of the maximum survival depth of a bender comes for expense of reducing the electric field limited acoustic power. If the beams are prestressed for expanding their dynamic range of operation, then the corresponding operating depth is also increased.

13.3.2.2 Circular Plates

The issues related to the strength of the circular plates under action of the hydrostatic pressure can be considered in the same way as for the rectangular beams. The difference is that the stresses in the circular plates are two dimensional: $T_{rr} = T_1$ in the radial direction, and $T_{\varphi\varphi} = T_2$ in the circumferential direction. The maximum stress takes place in the center of the plate. At this point $T_{rr} = T_{\varphi\varphi}$. Expression for the radial stress in the center of a circular plate is

$$T_{1\max} = -z\frac{Y_1^E}{[1-(\sigma_1^E)^2]}\xi_0 \left.\frac{d^2\theta}{dr^2}\right|_{r=0} = P_h \frac{tS_{av}Y_1^E}{2K_m^E[1-(\sigma_1^E)^2]}\left.\frac{d^2\theta}{dr^2}\right|_{r=0}, \tag{13.44}$$

where the static deflection curve for the simply supported boundary conditions is given by formula

$$\theta(r/a) \approx \left(1-\frac{r^2}{a^2}\right)\left(1-\frac{r^2}{4a^2}\right). \tag{13.45}$$

According to (2.151) the rigidity, K_m^E, and the average surface area, S_{av}, are

$$K_{eqv}^E = \frac{0.6\pi t^3}{a^2 s_{11}^E(1-\sigma_1^{E2})}, \quad S_{av} = 0.46\pi a^2. \tag{13.46}$$

Substituting these quantities into Eq. (13.44) results in

$$T_{1\max} = 0.92\frac{a^2}{t^2}P_h. \tag{13.47}$$

Estimation of the maximum operating depth that follows from this relation at value of the permissible static tensile stress for the flexural deformation T_{ps} = 40 MPa is

$$h_{\max} = 4.3\cdot 10^3 (t/a)^2 \ (m). \tag{13.48}$$

It is noteworthy that for the two-dimensional case the permissible static tensile stress under the flexural deformation must be greater than that for the one dimensional system, but numerically it is not certain and therefore the same value T_{ps} = 40 MPa is used. Thus, the thickness to radius ratio for a circular plate is

$$(t/a) \geq 1.5 \cdot 10^{-2} \sqrt{h_{(m)}}\,. \tag{13.49}$$

On the other hand, we consider that $t/a \leq 1/5$ is the maximum aspect ratio, at which the corrections for the finite thickness of a plate still remain small. At $t/a = 1/5$ the correlation between the radius of the plate for a certain resonance frequency, $f_r = 0.23(t/a^2)c_c^E$ (see (9.200)), and maximum operating depth can be estimated as

$$a \geq 3.6 \cdot 10^{-3} \frac{c_c^E}{f_r} \sqrt{h_{(m)}}\,, \tag{13.50}$$

where $h_{(m)}$ is taken in meters. From this relation the radius of a plate that insures its static strength at a required depth and for a certain resonance frequency ca be determined.

The maximum theoretically achievable operating depth being obtained from relation (13.48) at $t/a = 1/5$ is about 170 m. Thus, for example, if the resonance frequency of transducer is $f_r = 1.5\ kHz$ the radius of plate for this depth must be $a \geq 10$ cm, which makes the overall dimensions of the plate $D \geq 20$ cm, $t \geq 2$ cm. A transducer of this size is too massive. Besides, the plates of this size cannot be manufactured as solid piezoelements and being glued out of number of smaller parts they will lose their tensile strength dramatically. More realistic correlation between the plate radius and its maximum operating depth is illustrated for several resonance frequencies with plots in Figure 13.10 following Eq. (13.50).

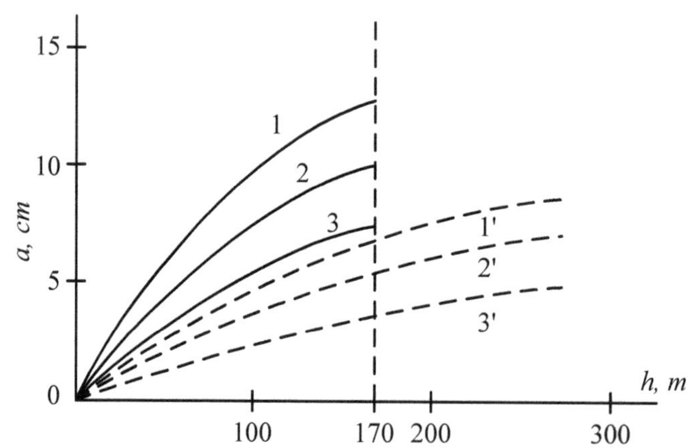

Figure 13.10: Dependences of the radius of the plate at the maximum operating depth for different resonance frequencies: 1, 1' – 1.0 kHz; 2, 2' – 1.5 kHz; 3, 3' – 2.0 kHz. Curves 1, 2, 3 for the fully active plates; 1', 2', 3' for the active- passive plates.

For increasing the operating depth of the piezoelements having realistic geometries, a half passive plate design can be used, in which material of the lower half plate is replaced by a metal. It must be kept in mind that not only the permissible stress is changed, when calculating the active-passive plates (beam), but in general the Young's moduli Y_a^E and sound speeds c_c in the due formulas must be changed to Y_{eqv} and c_{eqv} determined for the active–passive designs, as this is shown in Sections 9.2.3 and 9.3.2. The changes are small, if the values of Young's moduli of the ceramics and metal are close, as it is in case that aluminum alloy AA2011 is used. Plots for the half passive ceramics-aluminum plates are shown in Figure 13.10 by the dash lines.

As it was already noted, using the circular plates for projectors is less common compared with the rectangular beams. The half passive circular plate designs are typical for hydrophones.

13.3.2.3 Slotted Rings

The circumferential stress in the slotted ring shown in Figure 13.11 is according to Eqs. (4.500) and (4.513),

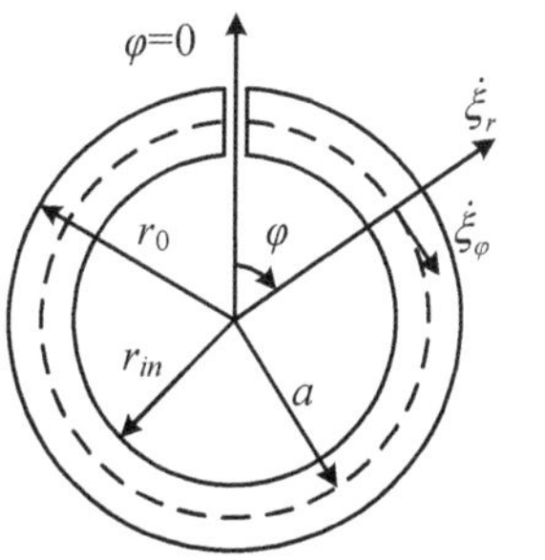

Figure 13.11: Slotted ring geometry.

$$T_\varphi = \frac{1}{s_{ii}^E} S_\varphi = -\frac{z}{s_{ii}^E a^2}\left(\frac{\partial^2 \xi_r}{\partial \varphi^2} + \xi_r\right) = -\frac{z}{s_{ii}^E a^2} \cdot 0.44(1-\cos\varphi)\xi_{ro}\,, \tag{13.51}$$

where ξ_{ro} is the maximum radial displacement at $\varphi = 0$. The maximum stress $T_{\max}$ is developed at $|z| = t/2$ and $\varphi = \pi$. In the outer layer ($z = t/2$)

$$T_{\max} = -0.44\frac{t}{s_{ii}^E a^2}\xi_{ro}\,. \tag{13.52}$$

Formula (13.52) can be used for determining the static stress, if the displacement ξ_{ro} under action of the hydrostatic pressure is known. This displacement can be found from relation

$|\xi_{ro}| = F_{eqv} / K_m^E$, where the equivalent force produced by the hydrostatic pressure P_h is $F_{eqv} = S_{av} P_h$ and

$$S_{av} = 0.62 S_{outer} = 0.62 \cdot 2\pi[a + (t/2)]h \tag{13.53}$$

following expression (7.289). The rigidity K_m^E for the short circuited ring is (see (7.250))

$$K_m^E = \frac{\pi}{20} \frac{t^3 h}{a^3 s_{11}^E}. \tag{13.54}$$

Note that here and in formula (13.53) h denotes the height of the slotted ring, whereas $h_{(m)}$ denotes the depth in meters.

As the result we obtain

$$\xi_{ro} = -25.0 \cdot \frac{a^4}{t^3} s_{ii}^E \left(1 + \frac{t}{2a}\right) P_h. \tag{13.55}$$

Sign (-) shows that the displacement is directed to the center of a ring, whereas positive is direction outside. Now from formula (13.52) follows that the maximum tension in the outer layer is

$$T_{\max} = 11 \cdot \frac{a^2}{t^2} \left(1 + \frac{t}{2a}\right) P_h, \tag{13.56}$$

or given that the resonance frequency of the ring is $f_r = 0.02(t/a^2)c_c$,

$$T_{\max} = 0.22 \frac{c_c}{f_r t} \left(1 + \frac{t}{2a}\right) P_h. \tag{13.57}$$

Thus, for example, for a ring made of PZT-4 ceramics having dimensions

$$D_o = 5.2 \cdot 10^{-2}\ m,\ t = 1.6 \cdot 10^{-2}\ m\ \ (f_r = 1.5\ kHz) \tag{13.58}$$

the limiting survival depth may be determined from relation

$$h_{(m)} \leq 2.5 \cdot T_{p(MPa)}. \tag{13.59}$$

With T_p = 40 MPa it will be $h_{(m)} \leq 100\ m$. This estimation is valid for the fully active transverse poled and for the tangential poled (stripe-electroded) rings. For the segmented longitudinal polarized rings, the permissible static tension is smaller and so is the survival depth of the transducer.

The limiting depth can be significantly extended in the half passive transducer design, in which case the outer layer of ceramics is replaced with metal (for example, aluminum). Thus, the tension in the ceramics will be avoided at the expense of reducing the electromechanical force by the factor of two.

13.3.2.4 On the Correlation Between the Operating and Survival Depth of the Transducers

So far, the maximum survival depth for the transducers was considered. In determining the maximum allowable operating depth, the additional limitations must be considered that arise due to changing parameters of ceramics under static stress, and due to combined effect of the static and dynamic tensile stress on the strength of the transducers. This issue will be considered with example of the flexural type transducers that are the most vulnerable in this respect. For estimating the operating depth an assumption must be made regarding the permissible values of the dynamic stress, T_d, in presence of the biasing static stress, T_{st}. We will assume that the maximum permissible stress remains the same for the total stress T_Σ, as this was suggested in Ref. 5.

The maximum biasing stress due to the hydrostatic pressure is that of tension on the inner surface, compression on the outer surface of the plates (beams), and vice versa for the slotted rings. Accordingly, the operating (dynamic) tensile stress becomes biased by the static compression stress on the inner surface for the plates (beams), as

$$T_{\Sigma i} = T_d + T_{st}, \tag{13.60}$$

and on the outer surface, as

$$T_{\Sigma o} = T_d - T_{st}. \tag{13.61}$$

Remember that according to the accepted sign convention the tensile stress is considered as positive and the compressive - as negative. For a slotted ring these relations must be changed places.

Further an example of a beam as a part of the rectangular bender transducer will be used for illustrating. We introduce the following notations. The survival depth limited by tension that corresponds to permissible tensile stress T_{pt} is denoted h_{st}, limited by compression that corresponds to permissible compressive stress T_{pc} is denoted h_{sc}. The operating depth that corresponds to permissible stress at this depth $T_{op}(h)$ is denoted h_{op}

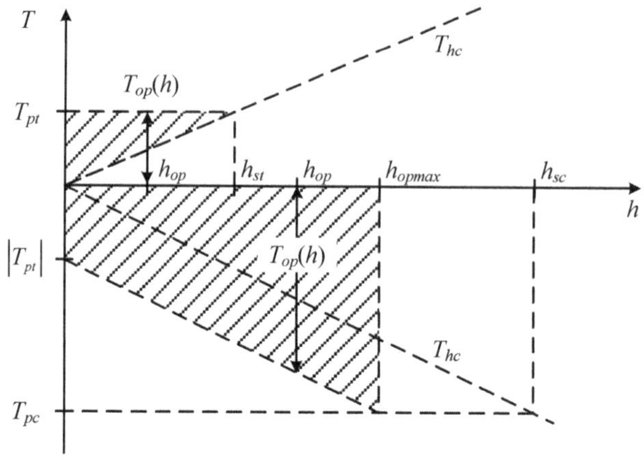

Figure 13.12: Illustration of the operating and survival depths determined for the fully active bender plates on the lower and upper halves of the mechanical system. Values of the permissible operating stresses for a particular depth are within the shaded areas. The survival depth limited by tension is h_{st}, survival depth limited by compression is h_{sc}.

The correlations between the survival and operating depths that are obtained with help of relations (13.60) and (13.61) for the fully active bender plates on the lower and upper halves of the mechanical system are qualitatively illustrated with diagrams in Figure 13.12.

The dashed areas show the limits for values of permissible operating stress $T_{op}(h)$ vs. depth. Obviously, at $h = 0$ $T_{op}(0) = T_d = T_{pt}$. On the inner surface the permissible operating stress comes to zero at $h = h_{st}$. On the outer surface the permissible operating stress raises with depth until it reaches the survival depth $h = h_{sc}$. Thus, the value of stress limited power radiated by the bender is restricted by stress conditions in the lover half of the mechanical system, and the larger the required power the smaller is the operating depth. Radiating by the fully active bender at the survival depth is impossible. Maximum power can be radiated close to the surface. Though, it must be noted that close to the surface the radiated power may be limited by cavitation rather than by tensile strength.

The situation is quite different for the half passive design with the inner half of ceramics replaced by a metal having tensile strength comparable to compressive strength of the ceramics. Much larger survival depth that is determined by permissible compressive static stress, T_{ps}, of ceramics, and operating depth for radiating much larger stress-limited power is now allowed by strength of the upper part. But this advantage comes for expense of reduction of the bending moment, or in other words, by reduction of acoustic power limited by the electric field. Another factor that must be taken into consideration regarding the half passive design is related to possible changing parameters of ceramics under the compressive stress generated by the hydrostatic pressure. This circumstance is not important for the fully active design, because its strength is limited by the relatively small tensile strength of the ceramics. In the half passive design the compressive stress in ceramics may reach its permissible value of about $T_{pc} \approx 350$ MPa, which may produce changes in the ceramic parameters. Note that the permissible compressive bending stress can be even as large as about 500 MPa, but no data exists regarding changing parameters of ceramics under the bending stress. Approximate estimations can be made following the data presented for extensional deformations in Section 11.3.3.

Alternative to the half passive design in terms of increasing operating depth and dynamic range of a transducer is precompression of the ceramics for reducing tension to a safe level in case of the transverse effect in the fully active design, and for avoiding tension in the segmented beams. The precompression may be required also in the half passive designs in case that the bias due to hydrostatic pressure is not sufficient for insuring dynamic strength of the ceramics.

13.4 Hydrostatic Pressure Equalized Designs

In this Section the cylindrical and flexural type (bender) transducer designs with internal volume filled with liquid are considered. The intended purpose of these designs is to equalize hydrostatic pressure and thus to avoid its harmful effect on the strength and electromechanical parameters of the transducers. Radiation from the internal volume of the transducers in this case must be regarded as parasitic. Considering the liquid filled transducer designs intended for radiation from the internal volume, such as cylindrical free flooded ("squirter") and Helmholtz resonator transducers, are out of scope of this treatment. Some information on the performance characteristics of transducers of this kind can be found in Refs. 12-18.

13.4.1 Liquid Filled Cylindrical Shell of Infinite Length

At first, we consider input impedance, Z_{in}, of internal volume of infinitely long liquid filled cylindrical shell in the two modifications shown in Figure 13.13: (a) the volume is free of other parts, and (b) a rigid cylindrical post is installed coaxially inside the volume. Though this case is not an exact representation of a real transducer design, it can serve as a prerequisite to considering the finite size cylindrical transducers, and the results obtained even may be applicable to the transducers having height comparable to wavelength.

The assumptions are that vibration of the shell is radial with velocity uniform along z-axis and the shell is thin enough for velocities of the outer and inner surfaces of the shell, U_{out} and U_{in}, being considered equal. Thus, the radiation problem is two dimensional due to symmetry, and is equivalent to radiation into internal volume of a ring that is confined between two rigid planes perpendicular to axis of the shell as is shown in the Figure 13.13 (b).

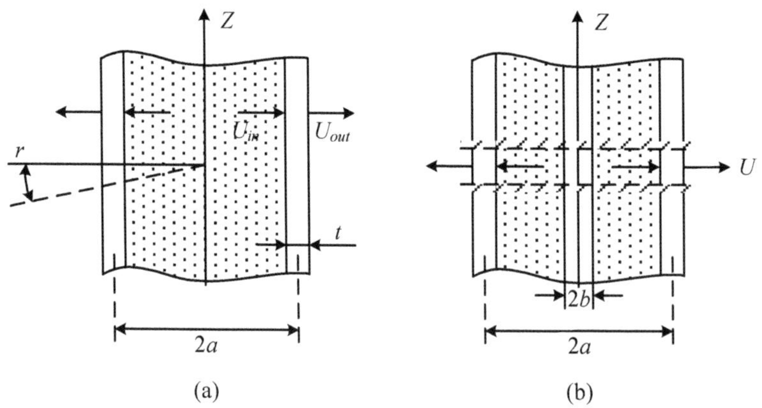

Figure 13.13: Infinite liquid filled cylindrical shell: (a) with internal volume free of other parts, (b) with a coaxially installed cylinder.

The input impedance per unit length of internal volume for different modes of radial vibration of the shell,

$$U(\varphi,t)\big|_{r=a} = U_n e^{j\omega t}\cos n\varphi\,, \tag{13.62}$$

will be determined from relation

$$\bar{\dot{W}}_{ac\tilde{V}} = a\int_0^{2\pi} U_n^* P(a,\varphi)\cos n\varphi d\varphi = Z_{in}\left|U_n\right|^2. \tag{13.63}$$

The general solution for different modes of sound pressure in the internal volume with time dependent factor omitted is (see Section 6.3.1.1)

$$P_n(r,\varphi) = \cos n\varphi[A_n J_n(kr) + B_n N_n(kr)]\,, \tag{13.64}$$

and the boundary conditions are:

for the case (a)

$$\text{value of } P\big|_{r=0} \text{ is limited, and } \left.\frac{\partial P}{\partial r}\right|_{r=a} = j\omega\rho U\big|_{r=a}\,; \tag{13.65}$$

for the case (b)

$$\left.\frac{\partial P}{\partial r}\right|_{r=a} = j\omega\rho U\big|_{r=a}\,, \quad \left.\frac{\partial P}{\partial r}\right|_{r=b} = 0\,. \tag{13.66}$$

The note must be made regarding the signs in the boundary conditions. Generally, the boundary conditions are derived from description of the sound field in some volume using relations (6.45) or (6.47), where sign (-) in the relation

$$\bar{u} = -\,grad\,\Phi \tag{13.67}$$

corresponds to the case that velocity of the boundary vibration is directed inside the volume. In the situation shown in Figure 13.13 the velocity of the surface at $r = a$ is directed outside the internal volume. Therefore, the signs in relations (13.65) and (133.66) are changed when formulating condition at $r = a$.

In the case (a) it must be $B_n = 0$ in Eq. (13.64) due to the condition that sound pressure in the center has a limited value, while $N_n(kr)\big|_{r\to 0} \to \infty$. Thus,

$$P_n(r,\varphi) = A_n J_n(kr)\cos n\varphi\,. \tag{13.68}$$

13.4.1.1 The Zero Mode of Vibration

At the zero mode of vibration $U(a,\varphi) = U_0$ and after using boundary condition (13.65) will be obtained

$$A_0 = -j\omega\rho\frac{1}{kJ_1(ka)}U_o \text{ and } P_0(r) = -j\rho c\frac{J_0(kr)}{J_1(ka)}U_o\,. \tag{13.69}$$

Considering relation (13.63) we arrive at

$$Z_{in0} = -j(\rho c)_l\cdot 2\pi a\frac{J_0(k_l a)}{J_1(k_l a)}\,. \tag{13.70}$$

Here, the subscript l represents 'liquid', which fills the internal volume.

The internal volume has resonances and antiresonances at $Z_{in0} = 0$ ($J_0(k_l a) = 0$) and at

$Z_{in0} = \infty$ ($J_1(k_l a) = 0$), respectively. The corresponding values of $k_l a$ are:

for the resonances

$$(k_l a)_r = 2.4,\ 5.5,\ ...;\ ; \tag{13.71}$$

for the antiresonances

$$(k_l a)_{ar} = 3.8,\ 7.0,\ ...\ . \tag{13.72}$$

Plot of the function $-jZ_{in0}$ is shown qualitatively in Figure 13.14 by solid lines.

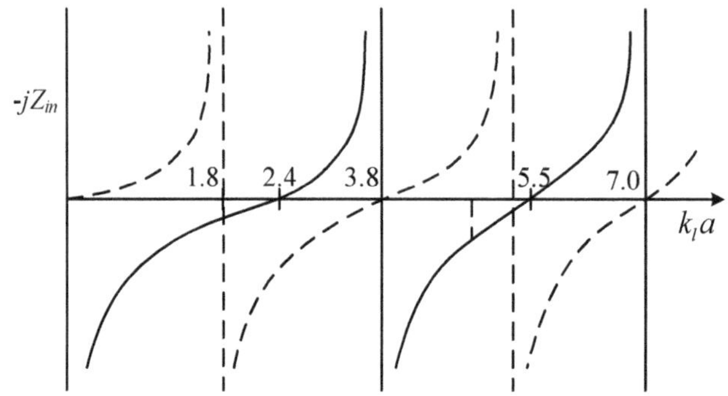

Figure 13.14: Input impedance per unit length of the internal volume of infinitely long cylindrical shell: vibrating in the zero mode-solid lines, vibrating in the first mode-dash lines.

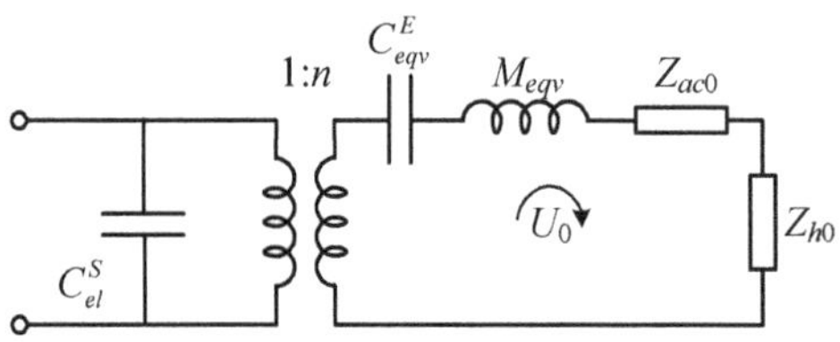

Figure 13.15: Equivalent electromechanical circuit of the ring transducer loaded with the input impedance of the liquid filled internal volume of cylindrical shell.

Let us suppose that a cylindrical transducer is composed of short rings (generally with aspect ratio $h/2a \le 0.5$) made of PZT piezoceramics and it is long enough to be approximated in terms of radiation by infinitely long cylindrical shell. The common equivalent electromechanical circuit of a ring transducer that is loaded by the input impedance of the internal volume may be represented as is shown in Figure 13.15. The input impedance per ring of height h is denoted as

$$Z_{h0} = Z_{in0} \cdot h = -j(\rho c)_l \cdot 2\pi a h \frac{J_0(k_l a)}{J_1(k_l a)}\ . \tag{13.73}$$

The wave size of a ring at the resonance frequency of the zero mode without reaction of the internal volume is $k_w a \approx 2.2$ in case that it is made of PZT-4 ceramics (subscript w stays for water). For the internal volume at the same frequency $k_l a = (c_w / c_l)(k_w a)$. If for example, c_l = 1500 m/s, then $k_l a = 2.2$ and input impedance of internal volume is that of rigidity, as it follows from the plot in Figure 13.14. Therefore, the resonance frequency of the transducer with internal volume filled with liquid will go up. Besides, the presence of an additional rigidity will result in some reducing the effective coupling coefficient of the transducer and notches in its frequency response can be expected at antiresonance frequencies of the internal volume. At low frequencies, at $k_l a \to 0$, $J_0(k_l a) \to 1$ and $J_1(k_l a) \to k_l a / 2$. Thus,

$$Z_{h0}\Big|_{k_l a \to 0} = 4\pi h \frac{(\rho c^2)_l}{j\omega} = \frac{1}{j\omega C_{h0}}, \tag{13.74}$$

where $C_{h0} = 1/4\pi h(\rho c^2)_l$ is the mechanical compliance of the internal volume of the ring that is confined between two rigid planes. The acoustic compliance of the volume with vibrating surface area $S = 2\pi a h$ is

$$C_{ac0} = C_{h0} \cdot S^2 = \frac{\pi a^2 h}{(\rho c^2)_l} = \frac{\tilde{V}}{(\rho c^2)_l}, \tag{13.75}$$

which coincides with relation (12.5) obtained in Chapter 12 for the small volumes.

13.4.1.2 The First Mode of Vibration

The first mode of vibration $U(a,\varphi) = U_1 \cos\varphi$ is typical for cylindrical transducers described in Section7.2.3. In this case $P_1(r,\varphi) = A_1\, J_1(k_l r) \cos\varphi$. Using the boundary condition (13.65) and remembering that

$$J_1'(kr) = kJ_0(kr) - J_1(kr)/r \tag{13.76}$$

(see Appendix C) will be obtained

$$P_1(r,\varphi) = j(\rho c)_l \frac{J_1(k_l r)}{J_0(k_l a) - J_1(k_l a)/k_l a} U_1 \cos\varphi. \tag{13.77}$$

After substituting P_1 into relation (13.63) we obtain expression for the input impedance per unit length of the shell in the form of

$$Z_{in1} = j(\rho c)_l \pi a \frac{J_1(k_l a)}{J_0(k_l a) - J_1(k_l a)/k_l a}. \tag{13.78}$$

The resonance frequencies of the internal volume take place at $Z_{in1} = 0$, i.e., at $J_1(k_l a) = 0$. Thus,

$$(k_l a)_r = 0,\ 3.8,\ 7.0, \ldots. \tag{13.79}$$

The antiresonance frequencies (at $Z_{in1} = \infty$) take place at $J_0(k_l a) - J_1(k_l a)/k_l a = 0$, where from

$$(k_l a)_{ar} = 1.8,\ 5.3, \ldots. \tag{13.80}$$

At low frequencies, at $k_l a \to 0$, the input impedance of internal volume per unit length of the shell is

$$Z_{in1} = j\omega\rho_l \pi a^2 = j\omega M_{in1} \quad , \tag{13.81}$$

where M_{in1} is the mass of the liquid per unit length. Function $-jZ_{in1}$ is presented in Figure 13.14 by the dashed lines.

In this case the equivalent circuit of Figure 13.15 is applicable with replacement the subscript *0* by *1*. The loading impedance for a ring is

$$Z_{h1} = Z_{in1} \cdot h = j(\rho c)_l \cdot \pi a h \frac{J_0(k_l a)}{J_0(k_l a) - J_1(k_l a)/k_l a}. \tag{13.82}$$

The wave size of a ring made of PZT-4 ceramics at resonance frequency of the first mode of vibration without reaction of the internal volume is $k_w a \approx 3.1$. This corresponds to the same value $k_l a \approx 3.1$ of the wave size of the internal volume filled with liquid, for which c_l = 1500 m/s. As it follows from the plot in Figure 13.14, the input impedance at this wave size is that of rigidity, and the resonance frequency of the ring goes up, as well as for the ring vibrating in the zero mode. Quantitatively the relation between the resonance frequencies of a ring for the zero and first modes, which is $f_1 / f_0 \approx \sqrt{2}$, may change due to the additional loading by the impedances of the internal volume. This may result in changing the frequency responses of transducers that employ 0 ± 1 modes of ring vibration (see Section 7.2.4).

13.4.1.3 The Second Mode of Vibration

The second mode of vibration $U(a,\varphi) = U_2 \cos 2\varphi$ is typical for low frequency transducers that employ flexural vibration of the complete rings that are described in Section 7.3.1. Using the same procedure as in the previous cases, for this mode of vibration will be obtained expression for the input impedance per unit length of the internal volume

$$Z_{in2} = j(\rho c)_l \pi a \frac{J_2(k_l a)}{J_1(k_l a) - 2J_2(k_l a)/k_l a}. \tag{13.83}$$

Resonances of the internal volume are at frequencies that correspond to $J_2(k_l a) = 0$, where from

$$k_l a = 0,\ 5.13, \ldots. \tag{13.84}$$

The antiresonance frequencies obtained from condition $J_1(k_l a) - 2J_2(k_l a)/k_l a = 0$ correspond to

$$k_l a = 0,\ 5.13, \ldots \tag{13.85}$$

From analysis made in Chapter 7 follows that ratio of the resonance frequency of the flexural (see formula (7.230)) and zero mode radial vibrations of a ring is $(f_{fl}/f_{rad}) = 0.75(t/a)$. Assuming that maximum aspect ratio for the flexural ring may be $t/a \approx 1/5$, this makes $f_{fl}/f_{rad} \approx 0.15$. Thus, the value of $k_l a$ at the resonance frequency of flexural vibration is $k_l a \approx 0.35$, which is much below the first antiresonance frequency of the internal volume. At small $k_l a$, $J_1(k_l a) \approx (k_l a)/2$, and $Z_{in2} = j\omega\rho_l \pi a^2/2$. In the equivalent circuit of Figure 13.15 for a single ring the inertia reactance must be introduced

$$Z_{h2} = Z_{in2} \cdot h = j\omega\rho_l \pi a^2 h/2. \tag{13.86}$$

This will result in some lowering of the resonance frequency of the transducer.

13.4.1.4 Liquid Filled Shell with Coaxial Rigid Cylindrical Post

The boundary conditions for the case that a rigid cylinder of radius $r = b$ is coaxially installed inside of the internal volume are

$$\left.\frac{\partial P}{\partial r}\right|_{r=a} = j\omega\rho U\big|_{r=a},\quad \left.\frac{\partial P}{\partial r}\right|_{r=b} = 0. \tag{13.87}$$

We will consider the zero mode of the shell vibration. From expression (13.64) at $n = 0$,

$$P_0(r) = A_0 J_0(k_l r) + B_0 N_0(k_l r) A_0. \tag{13.88}$$

Using the boundary conditions (13.87) we obtain expressions for the constants and B_0 as

$$A_0 = -j(\rho c)_l U_0 \frac{N_1(k_l b)}{\Delta}, \; B_0 = j(\rho c)_l U_0 \frac{J_1(k_l b)}{\Delta}, \tag{13.89}$$

where

$$\Delta = J_1(k_l a)N_1(k_l b) - J_1(k_l b)N_1(k_l a)\,. \tag{13.90}$$

Thus,

$$P_0(r) = -j(\rho c)_l U_0 \frac{1}{\Delta}[J_0(k_l r)N_1(k_l b) - N_0(k_l r)J_1(k_l b)]\,. \tag{13.91}$$

After applying formula (13.63) we arrive at the expression for the input impedance per unit length of the shell

$$Z_{in\tilde{V}} = -j(\rho c)_l 2\pi a \frac{1}{\Delta}[J_0(k_l a)N_1(k_l b) - N_0(k_l a)J_1(k_l b)] \tag{13.92}$$

Conditions for determining the resonance and antiresonance frequencies of the input impedance are:

$$J_0(x)/J_1(mx) = N_0(x)/N_1(mx)\,; \tag{13.93}$$

and

$$J_1(x)/J_1(mx) = N_1(x)/N_1(mx)k_l a\,. \tag{13.94}$$

Here $x = k_l a$ and $m = b/a$. Solutions for Eqs. (13.93) and (13.94) that correspond to the resonance and antiresonance frequencies are presented in Table 13.2.

Table 13.2: Values of at the resonance and antiresonance frequencies.

m	0	0.2	0.4	0.6	0.8
$(k_l a)_r$	2.40	2.57	3.11	4.34	8.21
$(k_l a)_{ar}$	3.83	4.24	5.39	7.93	15.74

At low frequencies (at $k_l a \to 0$) from expression (13.92) follows that the input impedance per single ring of the height h is

$$Z_{h\tilde{V}} = \frac{4\pi a^2 h(\rho c^2)_l}{j\omega(a^2 - b^2)} = \frac{1}{j\omega C_{h\tilde{V}}}, \tag{13.95}$$

i.e., that of the rigidity. After converting the mechanical compliance $C_{h\tilde{V}}$ to acoustic value, we obtain

$$C_{h\tilde{V}ac} = C_{h\tilde{V}} \cdot 4\pi^2 a^2 h^2 = \frac{\pi(a^2 - b^2)h}{(\rho c^2)_l} = \frac{\tilde{V}_{\text{int}}}{(\rho c^2)_l}, \tag{13.96}$$

that is the common expression for the acoustic compliance of a small volume with vibrating surface area $S = 2\pi ah$, which is confined within rigid walls.

13.4.2 Input Impedance of Internal Volume of the Liquid Filled Finite Height Cylindrical Transducer

Real cylindrical projectors are usually composed of rings with aspect ratio $(h/2a) < 0.5$. At number of rings in the transducer greater than two a nonuniform distribution of vibration over the transducer height may take place due to acoustic interaction between the comprising rings both in the external and internal acoustic fields, if special measures are not taken for their equalizing (for example, by the series connection of the rings). In this treatment we will assume that distribution of velocity on the surface of transducer is uniform. One more assumption will be made regarding accounting for losses in the internal acoustic field. For simplicity we will consider equations for determining acoustic field inside the volume as lossless. And then, after velocity distributions in the field become known, the viscose losses of energy due to shear in the regions close to boundaries (in the boundary layers) will be introduced. This approach seems to be appropriate, because the shear losses in the boundary layers contribute the most to dissipation of energy inside the internal volume. We will assume further that the liquid filled volume is closed at the ends by the floating disks that separate the internal volume from outside environment, as shown in Figure 13.16. The axisymmetric rings vibration will be considered only. Due to symmetry solution for the problem will be the same as for the half-length transducer closed on the bottom by the absolutely rigid baffle, as shown in Figure 13.16 (b). We will assume that velocities of vibration of the transducer and disks, $U(r)\big|_{r=a} = U_0$ and $U(z)\big|_{z=\pm L} = U_L$, are directed outside the volume to make impedance of the internal volume compatible with radiation impedance of the transducer.

The boundary conditions for the problem are formulated as follows:

$$\left.\frac{\partial P}{\partial r}\right|_{r=a} = j\omega\rho U_0 \tag{13.97}$$

at the internal surface of the transducer, and

$$\left.\frac{\partial P}{\partial z}\right|_{z=\pm L} = j\omega\rho U_L \tag{13.98}$$

at the surfaces of the disks.

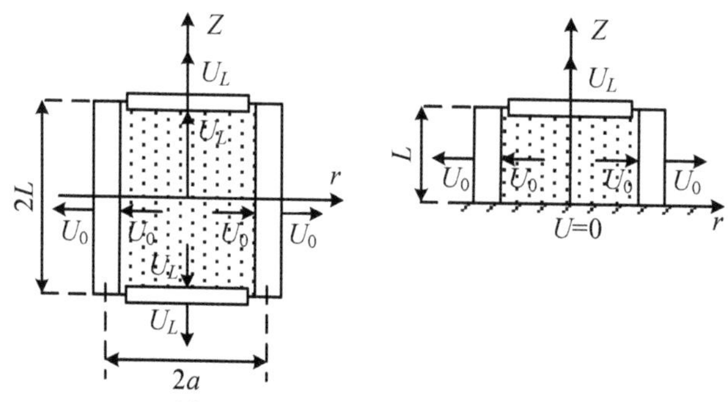

Figure 13.16: (a) Geometry of the finite height cylindrical transducer with internal volume filled by a liquid and with floating disks at the ends. (b) The half-length transducer with imaginary rigid foundation

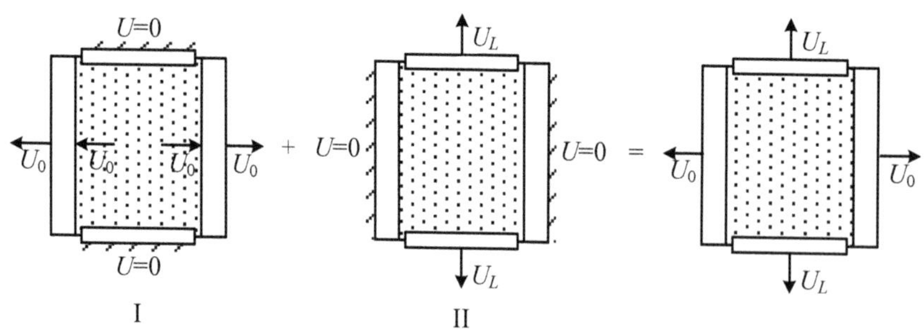

Figure 13.17: Illustration of superposition of the boundary conditions.

Solution to the problem can be represented as superposition of solutions *I* and *II* that correspond to the following combinations of boundary conditions:

$$\left.\frac{\partial P_I}{\partial r}\right|_{r=a} = j\omega\rho_l U_0\,,\ \left.\frac{\partial P_I}{\partial z}\right|_{z=\pm L} = 0\,; \tag{13.99}$$

$$\left.\frac{\partial P_{II}}{\partial z}\right|_{z=\pm L} = j\omega\rho_l U_L , \quad \left.\frac{\partial P_{II}}{\partial r}\right|_{r=a} = 0 . \tag{13.100}$$

The superposition of boundary conditions is illustrated with Figure 13.17.

The set I of the boundary conditions is the same as for the infinitely long cylindrical shell, due to symmetry. Therefore,

$$P_I(r) = -j(\rho c)_l \frac{J_0(k_l r)}{J_1(k_l a)} U_o \tag{13.101}$$

according to expression (13.69), and velocity $U_I(r)$ is

$$U_I(r) = \frac{J_1(k_l r)}{J_1(k_l a)} U_0 . \tag{13.102}$$

Solution to the wave equation that corresponds to set II of boundary conditions is (the liquid is supposed to be ideal)

$$P_{II}(z) = D\cos kz + C\sin kz . \tag{13.103}$$

Due to symmetry in respect to plane $z = 0$ it should be $C = 0$. Thus, $P_{II} = D\cos kz$ and after applying boundary condition (13.100) at $z = L$ will be obtained

$$P_{II}(z) = -j(\rho c)_l \frac{\cos k_l z}{\sin k_l L} U_L , \tag{13.104}$$

$$U_{II}(z) = \frac{\sin k_l z}{\sin k_l L} U_L . \tag{13.105}$$

Now the characteristics of the acoustic field in the internal volume can be represented as

$$P(r,z) = -j(\rho c)_l \left[\frac{J_0(k_l r)}{J_1(k_l a)} U_o + \frac{\cos k_l z}{\sin k_l L} U_L \right], \tag{13.106}$$

$$U(r) = \frac{J_1(k_l r)}{J_1(k_l a)} U_0 , \quad U(z) = \frac{\sin k_l z}{\sin k_l L} U_L . \tag{13.107}$$

After the expressions for velocity distributions are obtained, the losses of energy due to viscosity of the liquid can be considered. The conditions $\partial U(z)/\partial r = \partial U(r)/\partial z = 0$ hold. Therefore, the viscous losses inside the volume do not exist under the approximations accepted. The energy of losses and corresponding resistances that must be included into the equivalent circuit of the transducer may be attributed to dissipation of energy in the boundary layers near the side surfaces of the cylinder and disks. The dissipation of energy in the boundary layer can be introduced as follows (Ref. 19).

Due to finite viscosity of a real fluid, a thin layer exists near a boundary, in which the shear stress can develop, if the tangential to the boundary movement of the fluid takes place, as it is shown in Figure 13.18. The shear stress is defined as

$$T_{xy} = \mu \frac{\partial v_y}{\partial x}, \tag{13.108}$$

where the coefficient μ is called the coefficient of viscosity. Thickness l_b of the boundary layer is

$$l_b = \sqrt{2\mu / \rho\omega}. \tag{13.109}$$

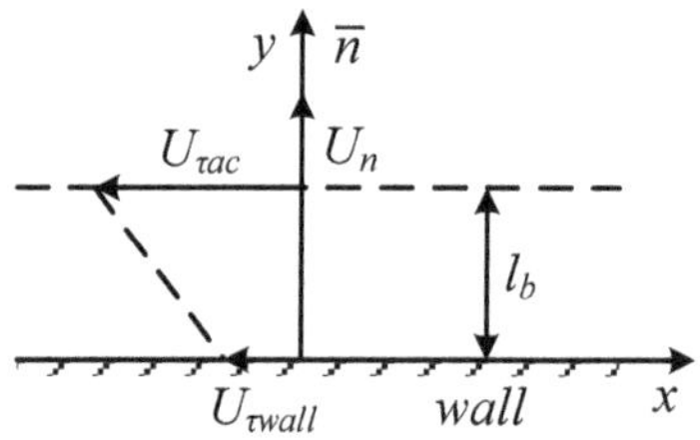

Figure 13.18: The boundary layer and distribution of the tangential velocity.

The layer is very thin compared with acoustic wavelength. Thus (see Chapter 12), for water $\mu = 10^{-3}$ Pa·s and $l_b \simeq 0.2 \cdot 10^{-2} / \sqrt{f_{\mathrm{kHz}}}$ cm, for the castor oil $\mu \simeq 1.0$ Pa·s and $l_b \simeq 6.0 \cdot 10^{-2} / \sqrt{f_{\mathrm{kHz}}}$ cm. Within the boundary layer the propagating wave itself cannot fit the boundary condition, and the shear wave in the tangential to the vibrating surface direction must be considered also. This wave is dying down rapidly in the normal to the surface direction. In Figure 13.18 $U_{shear} = (U_{\tau wall} - U_{\tau ac})e^{-(1-j)x/l_b}$, where U_{shear}, $U_{\tau wall}$ and $U_{\tau ac}$ are the velocity in the shear wave, velocity of vibration of the transducer surface in the tangential direction, and the tangential component of the acoustic wave velocity, respectively. The shear wave does not influence the propagating wave. The effect of the shear wave is in producing the loss of energy within the boundary layer. The rate of energy loss per unit area of the transducer surface can be estimated by formula (Ref. 19).

$$\dot{W}_{loss} \simeq \frac{1}{2} \rho\omega l_b \left| U_{\tau wall} - U_{\tau ac} \right|^2. \tag{13.110}$$

According to this expression the energy of losses in the boundary layer on the inner surface of the cylinder due to velocity distribution $U(z)$ along the surface will be

$$\bar{\dot{W}}_{loss\,L} = \frac{1}{2}\rho_l \omega l_b 2\pi a \int_{-L}^{L} |U(z)|^2\, dz = \tilde{r}_{lL} U_L^2, \tag{13.111}$$

where $\tilde{r}_{lL}$ is the equivalent resistance of losses on the inner surface attributed to velocity U_L; l_b is the thickness of the boundary layer that is determined by formula (13.109). After integrating in expression (13.111) will be obtained that

$$\tilde{r}_{lL} = \rho_l \omega l_b 4\pi a L^2 \cdot \left(\frac{1}{\sin^2 k_l L} - \frac{1}{k_l L \tan k_l L} \right). \tag{13.112}$$

In the analogous way on the surface of the disk

$$\bar{\dot{W}}_{loss\,a} = \frac{1}{2}\rho_l \omega l_b 2\pi \int_{0}^{a} |U(r)|^2\, r dr = \tilde{r}_{la} U_0^2, \tag{13.113}$$

and

$$\tilde{r}_{la} = \frac{1}{2}\rho_l \omega l_b \pi a^2 \left[1 - \frac{J_0(k_l a) J_2(k_l a)}{J_1^2(k_l a)} \right]. \tag{13.114}$$

Vibration of the disks is generated by vibration of the transducer and is influenced by a reaction of the outside sound field. Consider the movement of the disks under these actions assuming that the disks vibrate in the piston like mode, and their thickness is much smaller than wavelength in their material. In this case the input impedance of the disk itself can be represented as $Z_d = j\omega M_d$. The external reaction strictly speaking may be due to self-radiation of the disk and due to interaction with radiation produced by the outer side surface of the transducer. The effect of the interaction may be neglected if the height of the cylinder is not too small. The self-impedance will be introduced as radiation impedance, $Z_{ac\,d}$, of the disk vibrating on one side (see Section 6.5.7). Besides, the resistance of losses $\tilde{r}_{lL}$ related to velocity U_L must be included. Thus, the total input impedance of the disk will be

$$Z_{in\,d} = j\omega M_d + Z_{ac\,d} + \tilde{r}_{lL}. \tag{13.115}$$

The force acting on the surface of the disk that is produced by the internal acoustic field is

$$F_d = 2\pi \int_0^a P(r,z)\big|_{z=L} rdr . \qquad (13.116)$$

After substituting expression (13.106) for $P(r,z)$ and integrating will be obtained

$$F_d = -j(\rho c)_l \left[\frac{2\pi a}{k_l} U_0 + \frac{\pi a^2}{\tan k_l L} U_L \right]. \qquad (13.117)$$

Using relation $F_d = Z_{ind} U_L$ we arrive at

$$U_L = -j(\rho c)_l \frac{2\pi a / k_l}{Z_{ind} + j(\rho c)_l \pi a^2 / \tan k_l L} U_0 . \qquad (13.118)$$

Note that at $Z_{ind} = 0$ (liquid filled cylinder with ends closed by imaginary thin films vibrating in air)

$$\frac{U_L}{U_0} = -\frac{2}{k_l a} \tan k_l L . \qquad (13.119)$$

At low frequencies (at $\omega \to 0$)

$$\frac{U_L}{U_0} = -\frac{2L}{a} . \qquad (13.120)$$

The same result can be obtained from consideration that the volume of incompressible liquid must remain constant during vibration, i.e., $U_0 4\pi a L + U_L \pi a^2 = 0$ and $U_L / U_0 = -2L / a$.

After substituting relation (13.118) for velocity U_L in formulas (13.106) and (13.107) we arrive at expressions for the field characteristics inside the volume attributed to velocity U_0 of the transducer vibration:

$$P(r,z) = -j(\rho c)_l \left[\frac{J_0(k_l r)}{J_1(k_l a)} - j(\rho c)_l \frac{2\pi a}{k_l} \frac{1}{Z_{ind} + j(\rho c)_l \pi a^2 / \tan k_l L} \cdot \frac{\cos k_l z}{\sin k_l L} \right] U_0 , \qquad (13.121)$$

$$U(r) = \frac{J_1(k_l r)}{J_1(k_l a)} U_0 , \qquad (13.122)$$

$$U(z) = -j(\rho c)_l \frac{2\pi a / k_l}{Z_{ind} + j(\rho c)_l \pi a^2 / \tan k_l L} \frac{\sin k_l z}{\sin k_l L} U_0 . \qquad (13.123)$$

The resistance of losses $\tilde{r}_{lL}$ also must be recalculated to velocity U_0 . The transformed value, which we denote r_{iL}, will be found from equality $r_{iL}|U_0|^2 = \tilde{r}_{iL}|U_L|^2$ as

$$r_{IL} = \rho_l \omega l_b 4\pi a L \left| (\rho c)_l \frac{\pi a}{k_l} \frac{1}{Z_{ind} + j(\rho c)_l \pi a^2 / \tan k_l L} \right|^2 \cdot \left(\frac{1}{\sin^2 k_l L} - \frac{1}{k_l L \tan k_l L} \right). \quad (13.124)$$

Now the input impedance of the internal volume per the entire transducer that must be included in its equivalent circuit, as this is shown in Figure 13.15, can be determined from relation

$$\bar{\tilde{W}}_{in\tilde{V}} = 2\pi a U_0^* \int_{-L}^{L} P(a,z)dz + (r_{IL} + 2\tilde{r}_{la})|U_0|^2 = Z_{in\tilde{V}}|U_0|^2, \quad (13.125)$$

which represents the total energy spent by the transducer for generating acoustic field in the internal volume. After performing the integration, the following expressions will be obtained for the input impedance ant its components:

$$Z_{in\tilde{V}} = Z_{ina}(k_l a) + Z_{inL}(k_l L), \quad (13.126)$$

where $$Z_{inL}(k_l L) = -(\rho c)_l 4\pi a L \left[\frac{2}{k_l a \cdot k_l L} \cdot \frac{(\rho c)_l \pi a^2}{Z_{ind} + j(\rho c)_l \pi a^2 / \tan k_l L} \right] + r_{IL}, \quad (13.127)$$

$$Z_{ina}(k_l a) = -j(\rho c)_l 4\pi a L \left\{ \frac{J_0(k_l a)}{J_1(k_l a)} + j \frac{\rho_l \omega l_b}{(\rho c)_l} \cdot \frac{a}{4L} \cdot \left[1 - \frac{J_0(k_l a) J_2(k_l a)}{J_1^2(k_l a)} \right] \right\}. \quad (13.128)$$

The values of the impedances are average for the entire transducer, whereas input impedances per comprising rings may be somewhat different, because the sound pressure inside the volume is not uniform over the height.

Due to symmetry the solutions obtained are valid for the transducer design with internal volume closed on the bottom by a rigid (theoretically absolutely rigid) element, as shown in Figure 13.16 (b). In this case the impedances of the internal volume will have half the value of those determined for the full size transducer.

The reactions of the internal volume influence the operating characteristics of a cylindrical transducer. The reactive loading for the transducer at frequencies around antiresonances are the largest by magnitude and change signs. This may produce shift of the resonance frequency of the transducer in case that it is close to the frequency of the antiresonance. Introducing large impedances in the equivalent circuit of a transducer in vicinity of the antiresonance frequencies may result in appearance of notches in frequency response of the transducer. Besides, the resistances of losses are introduced in the equivalent circuit that represent dissipation of energy in the internal volume due to viscosity of the liquid. The combined effect of all the sources of energy dissipation results in reducing the seemingly infinite values of the input impedances at the antiresonance frequencies.

It must be admitted that calculations made for the frequency regions around the antiresonances of the input impedances should be considered as qualitative, because of the assumptions made. For estimating the main features of the impedance of internal volume and its influence on the operating characteristics of a transducer consider the following example. Suppose that the transducer vibrates in air, the liquid is lossless, and the internal volume is confined between thin disks with negligible mass. In this case $Z_{ind} = 0$. (Note that analogous problem was considered at such approximation in Ref. 9 in a different way).

Putting $Z_{ind} = 0$ and neglecting other losses related terms in formulas (13.127) and (13.128), we obtain expressions for the input reactance:

$$Z_{ina} = -j(\rho c)_l 4\pi a L \frac{J_0(k_l a)}{J_1(k_l a)}, \tag{13.129}$$

$$Z_{inL} = j(\rho c)_l \frac{8\pi L^2}{k_l L} \cdot \frac{\tan k_l L}{k_l L}. \tag{13.130}$$

Impedance Z_{ina} represents the impedance per unit length of the internal volume of the infinitely long cylindrical shell multiplied by the length of the finite height cylinder $2L$. The corresponding set of the antiresonance frequencies was obtained from condition $J_1(k_l a) = 0$ at

$$k_l a = 3.8;\ 7.0;\dots. \tag{13.131}$$

The first antiresonance frequency that is close to the resonance frequency of the pulsating cylindrical transducer (at $k_l a \approx 2.2$) is

$$f_{ara} \approx 1.7 f_{r0} \tag{13.132}$$

in case that the sound speeds in the liquid is $c_l = 1.5 \cdot 10^3\ m/s$. As it was noted in Section 13.4.1.1, the input reactance Z_{ina} produces increase of the resonance frequency of the transducer. Another set of antiresonance frequencies will be obtained from condition $\tan k_l L = \infty$, i.e., from

$$k_l L = \frac{\pi}{2}(2m-1), \quad m = 1, 2, \ldots . \tag{13.133}$$

This set of the antiresonance frequencies is

$$f_{ar\,L} = \frac{c_l}{4L}(2m-1), \quad m = 1, 2, \ldots . \tag{13.134}$$

i.e., the set of resonance frequencies of longitudinal vibration of the column of liquid with free ends having height $2L$. Being expressed in terms of the resonance frequency of the cylindrical transducer, these frequencies are

$$f_{ar\,L} = 0.35\frac{a}{L} f_{r0}(2m-1), \quad m = 1, 2, \ldots \tag{13.135}$$

Thus, the number of the antiresonances below the resonance frequency of the transducer increases with increase of aspect ratio L/a, the lowest antiresonance frequency being

$$f_{ar\,L} = 0.35\frac{a}{L} f_{r0} . \tag{13.136}$$

The input reactance $Z_{in\,L}$ may produce notches in the frequency response of the transducer. It is noteworthy that according to formula (13.112) the resistance of losses $\tilde{r}_{lL}$ increases with L. This reduces the depths of the notches.

Internal volume has resonances. The resonance frequencies may be determined from condition

$$Jm[Z_{in\tilde{V}}] = Jm[Z_{in\,a}(k_l a) + Z_{in\,L}(k_l L)] = 0 . \tag{13.137}$$

Approximately this condition can be represented as

$$-\frac{\tan k_l L}{k_l L} - \frac{k_l a}{2}\frac{J_0(k_l a)}{J_1(k_l a)} = 0 , \tag{13.138}$$

following the expressions (13.129) and (13.130) for the input impedances. At the resonance frequencies only the resistances of losses are introduced in the equivalent circuit of the transducer. Their effect on the frequency response of the transducer is straightforward. At the same time, radiation from the transducer ends may increase due to raise of magnitude of vibration inside the internal volume in axial direction.

Both the antiresonance and resonance frequencies can be changed to some extent by changing dimensions of the disks on the ends of the volume.

In the general case of operating transducer in water the active energy of radiation from the ends and its relation to the active energy radiated from the side surface of the transducer can be estimated in the following way. The energy radiated by the disks is

$$\mathrm{Re}[\bar{\dot{W}}_{ac\,d}] = 2R_{ac\,d}|U_l|^2 = 2R_{ac\,d}\left|(\rho c)_l \frac{2\pi a / k_l}{Z_{ind} + j(\rho c)_l \pi a^2 / \tan k_l L}\right|^2 |U_0|^2 . \quad (13.139)$$

The energy of radiation from the side surface of the transducer having radiation resistance $R_{ac\,tr}$ is

$$\mathrm{Re}[\bar{\dot{W}}_{ac\,tr}] = R_{ac\,tr}|U_o|^2 . \quad (13.140)$$

If transducer is made of PZT-4 ceramics and operates in vicinity of its resonance frequency, then $ka \approx 2.2$. At this value of ka $R_{ac\,d} \approx (\rho c)_w \pi a^2$. For the cylindrical transducer of the length comparable with wavelength (approximately at $(2L/\lambda) > 0.8$) the radiation resistance is $R_{ac\,tr} \approx (\rho c)_w 4\pi a L$. (For the radiation resistances of the cylindrical transducer and disc radiating from one side see Sections 6.3.1.1 and 6.3.3.7). For the approximate estimation we assume that $(\rho c)_l \approx (\rho c)_w$, and the mass of the disk, which can be considered as imaginary, is neglected. With help of expressions (13.139) and (13.140) will be obtained

$$\frac{\mathrm{Re}[\bar{\dot{W}}_{ac\,d}]}{\mathrm{Re}[\bar{\dot{W}}_{ac\,tr}]} \approx \frac{a}{2L}\tan^2 kL . \quad (13.141)$$

Given that for this case $kL \approx 2\pi$, the radiation from the ends is negligible.

Although the above analysis was not intended for calculating acoustic field due to combined radiation from the side surface and from the ends of the liquid filled transducer, this can be done within the accepted approximations in the following way. Using relation (13.112) between velocities of the uniformly vibrating side surface and of the ends of transducer (or of the one end, if another end rests on a rigid foundation), U_o and U_L, the sound pressure may be found, as radiation of the finite size cylinder that is described in Section 6.3.1.3. The results that can be obtained in this way may be less accurate for cylinders of small height, because the acoustic interaction between radiations of the side and end surfaces was not considered when determining relation between their velocities.

For rough estimation of peculiarities of the directional factor of the cylindrical transducer the combined radiation from the side surface and from the ends can be represented as their superposition that is illustrated with Figure 13.19. The resulting expression for the sound pressure in the far field is

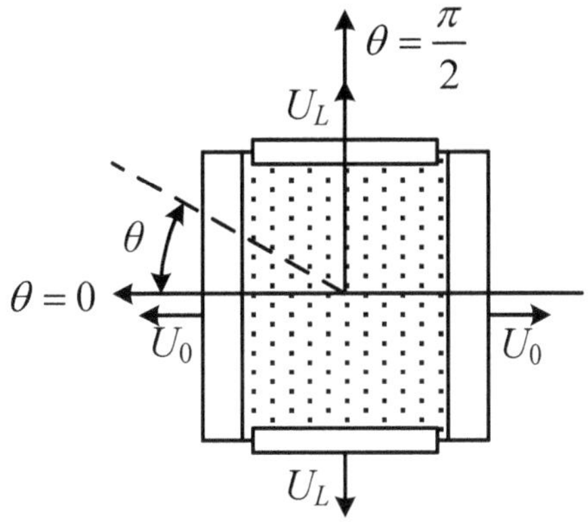

Figure 13.19: Liquid filled cylindrical projector radiating from the side and end surfaces.

$$P(\theta) = P_{side}(\theta) + 2P_{end}[\theta - (\pi/2)], \qquad (13.142)$$

where according to expressions (2.67) and (6.327)

$$P_{side}(\theta) = -(\rho c)_w 2LU_o \frac{1}{\pi \cos H_1^{(2)}(ka\cos\theta)} \cdot \frac{sin[(2kL\sin\theta)/2]}{(2kL\sin\theta)/2} \cdot \frac{e^{-jkR}}{R}, \qquad (13.143)$$

$$P_{disk}[\theta - (\pi/2)] =$$
$$= j(\rho c)_w U_L \frac{ka^2}{4}\left[\frac{2J_1(ka\cos\theta)}{ka\cos\theta} + \frac{2}{(ka)^2} F(ka)\cdot E[ka,(\theta - \pi/2)]\right] \cdot \frac{e^{-jkR}}{R}. \qquad (13.144)$$

The values of functions $F(ka)$ and $E(ka,\theta)$ can be found from Table 6.3 and expression (6.315).

Note that some additional error in determining the combined acoustic field exists because the acoustic centers of radiation of the cylinder (which is in the center of coordinates on the axis of the cylinder) and of the disks do not coincide.

13.4.3 Flexural Type Projectors

Specifics of the flexural type projectors (rectangular benders, slotted rings) in comparison with the above considered cylindrical transducers in terms of influencing the pressure equalization arrangements on their operating characteristics is due to much greater compliance of their mechanical systems and smaller wave size. Their fluid filled internal volumes represent acoustic chambers with lumped acoustic parameters, and their influence on the transducer operation is more pronounced. The acoustic properties of the fluid filled chambers are considered in Chapter 12. Requirements for properties of the pressure equalizing system differ depending on the frequency response of a transducer. Regarding the hydrophones operating in a broad low frequency range these issues are considered in Chapter 14.

Using the liquid filled internal volumes for hydrostatic pressure equalizing in low frequency projectors that operate in vicinity of resonance frequency is more questionable. Impedance of the internal volume may influence dramatically on the properties of the projector. The related problems are discussed in Ref. 11. In particular, using compliant inserts (metal or plastic tubes) into the liquid filled chamber is considered as one of the measures for reducing its rigidity. But this restricts the operating depth of projector by the strength of the insert. In terms of expending the range of pressure equalizing for the low frequency projectors, attractive looks idea to use the internal volume as Helmholtz resonator having resonance frequency close to the resonance frequency of the projector. The problem in this case is that influencing parameters of the resonator on the operating characteristics of the projector is not that obvious. In principle, the operating characteristics of a projector with Helmholtz resonator may be investigated quantitively using the equivalent circuit presented in Figure 12.6. Execution of this task can be greatly facilitated by a preliminary qualitative analysis of the main peculiarities of this approach. A detailed consideration of related to the problem issues is out of scope of this treatment. As to the approximate qualitative analysis, applying technique of the coupled vibration can be recommended for this purpose. Employing this technique is illustrated in the following sections with an example of the slotted ring projector.

13.4.3.1 Coupled Vibration Analysis of the Pressure Equalizing System

Requirements for the pressure equalizing system of the low frequency projectors may be formulated based on analysis of the equivalent circuit of a transducer operating in vicinity of its resonance frequency that includes acoustic elements of this system. The corresponding electro-mechano-acoustic equivalent circuit was introduced in Chapter 12 (Figure 12.6). The slightly modified version of this circuit is shown in Figure 13.20.

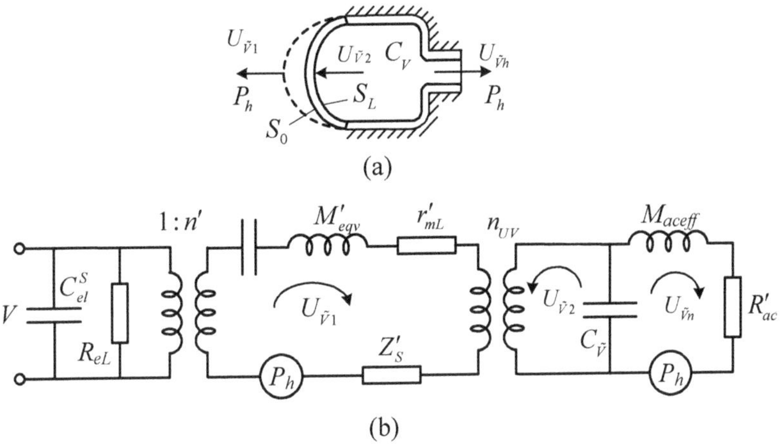

Figure 13.20: (a) Schematic representation and (b) equivalent electro-mechano-acoustic circuit of a projector of the flexural type with hydrostatic pressure equalizing system in the shape of Helmholtz resonator.

Namely, a possible transformation of the volume velocity between the outer and inner surfaces of a transducer with cylindrical mechanical system is introduced in the form of imaginary transformer with coefficient of transformation $n_{U_{\tilde{V}}} = -U_{\tilde{V}2} / U_{\tilde{V}1}$. Sign (-) indicates change of direction of the velocity, given that conditionally positive is the direction into the volume, in which the radiation takes place. This is direction outside a projector. The conditionally positive directions of velocity in contours of the equivalent circuit are indicated by arrows pointing in the clockwise direction (these are velocities $U_{\tilde{V}1}$ and $U_{\tilde{V}n}$). The electromechanical parameters of the circuit are described in acoustic generalized coordinates, which are the volume displacements. The volume displacement of the mechanical system is denoted $\xi_{\tilde{V}} = \xi_o S_{av}$, where ξ_o is displacement of the reference point and S_{av} is the average surface of the mechanical system. The corresponding acoustic values of the parameters are

$$\begin{aligned} n_{ac} = n / S_{av}, C^E_{eqvac} = C^E_{eqv} S^2_{av}, M_{eqvac} = M_{eqv} / S^2_{av}, \\ r_{mLac} = r_{mL} / S^2_{av}, Z_{sac} = Z_s / S^2_{av} \end{aligned}, \tag{13.145}$$

For brevity they are marked by primes ($n_{ac} = n'$, $C^E_{eqvac} = C'^E_{eqv}$, $M_{eqvac} = M'_{eqv}$, $r_{mLac} = r'_{mL}$, $Z_{sac} = Z'_s$).

The energies associated with vibration of the coupled mechano-acoustic system under voltage applied to the electrical input are

$$W_{pot} = \frac{\xi_{\tilde{V}1}^2}{2C'^E_{eqv}} + \frac{(\xi_{\tilde{V}2} - \xi_{\tilde{V}n})^2}{2C_{\tilde{V}}} = \frac{\xi_{\tilde{V}1}^2}{2}(K'^E_{eqv} + n^2_{U_{\tilde{V}}} K_{\tilde{V}}) - \xi_{\tilde{V}1}\xi_{\tilde{V}n} n_{U_{\tilde{V}}} K_{\tilde{V}} + \frac{\xi_{\tilde{V}n}^2}{2} K_{\tilde{V}}, \tag{13.146}$$

$$W_{kin} = \frac{M'_{eqv}\dot{\xi}_{\tilde{V}1}^2}{2} + \frac{M_{ac\,eff}\dot{\xi}_{\tilde{V}n}^2}{2}, \tag{13.147}$$

$$W_{ext} = Vn'\xi_{\tilde{V}1}. \tag{13.148}$$

Here $\xi_{\tilde{V}n}$ is the generalized displacement in the neck, $\xi_{\tilde{V}2} = \xi_{\tilde{V}1} \cdot n_{U_{\tilde{V}}}$, and $\xi_{\tilde{V}2} - \xi_{\tilde{V}n}$ is the generalized deformation of the internal volume of the resonator. Effects of losses and acoustic radiation impedance are omitted (both for transducer and resonator) when considering the coupled vibrations. The expressions for the potential and kinetic energies are analogous to expressions (4.520) and (4.521) that characterize the coupled vibrations of two partial mechanical systems. In our case one of the partial systems that remains, if to put $\xi_{\tilde{V}n} = 0$, has combined rigidity $K'^E_{eqv} + n^2_{U_{\tilde{V}}} K_{\tilde{V}}$, mass M'_{eqv} and its vibration is generated by voltage applied. Another partial system that remains at $\xi_{\tilde{V}1} = 0$ is the resonator with rigid walls that has rigidity $K_{\tilde{V}}$ and mass $M_{ac\,eff}$. Coupling between the partial systems is due to the mutual rigidity $K_{12} = \left| n_{U_{\tilde{V}}} K_{\tilde{V}} \right|$. Analysis of operating characteristics of this coupled system can be made using general procedures of theory of coupled vibrations described in Section 4.6. The starting considerations almost do not differ from those presented by expressions (12.37) - (12.45).

Expressions for the partial resonance frequencies of the system, f_{1p} for the transducer and f_{2p} for the resonator, are

$$f_{1p} = \frac{1}{2\pi}\sqrt{\frac{K'^E_{eqv} + n^2_{U_{\tilde{V}}} K_{\tilde{V}}}{M'_{eqv}}}, \quad f_{2p} = \frac{1}{2\pi}\sqrt{\frac{K_{\tilde{V}}}{M_{ac\,eff}}}, \tag{13.149}$$

and coefficient of the elastic coupling is

$$\gamma^2 = \frac{n^2_{U_{\tilde{V}}} K^2_{\tilde{V}}}{(K'^E_{eqv} + n^2_{U_{\tilde{V}}} K_{\tilde{V}})K_{\tilde{V}}} = \frac{1}{1 + K'^E_{eqv} / n^2_{U_{\tilde{V}}} K_{\tilde{V}}}. \tag{13.150}$$

Using expressions (13.136) - (13.138) the following Lagrange's equations in complex form may be obtained that describe free vibrations in this coupled system

$$[(K'^E_{eqv} + n^2_{U_{\tilde{V}}} K_{\tilde{V}}) - \omega^2 M'_{eqv}]U_{\tilde{V}1} - n_{U_{\tilde{V}}} K_{\tilde{V}} U_{\tilde{V}n} = Vn', \tag{13.151}$$

$$-n_{U_{\tilde{V}}} K_{\tilde{V}} U_{\tilde{V}1} + (K_{\tilde{V}} - \omega^2 M_{ac\,eff}) U_{\tilde{V}n} = 0 . \quad (13.152)$$

Due to relations $|U_{\tilde{V}1} / S_{avo}| = |U_{\tilde{V}2} / S_{avi}| = U_o$,

$$|n_{U_{\tilde{V}}}| = |U_{\tilde{V}2} / U_{\tilde{V}1}| = S_{avi} / S_{avo}, \quad (13.153)$$

where S_{avi} and S_{avo} are the average inner and outer surface areas of the mechanical system of a transducer.

Using expressions for the partial resonance frequencies and coefficient of coupling, the frequency equation for determining resonance frequencies of the system will be obtained in the form

$$[1-(f / f_{1p})^2][1-(f / f_{2p})^2] - \gamma^2 = 0 . \quad (13.154)$$

For convenience of the general analysis of the coupled vibrations in the system the detuning factor,

$$\beta = f_{1p} / f_{2p}, \quad (13.155)$$

and the normalized frequency factor,

$$\Omega = f^2 / f_{1p}^2, \quad (13.156)$$

will be introduced. Using these factors, Eq. (13.154) can be represented in the form

$$\Omega^2 - [1 + (1/\beta^2)]\Omega + (1-\gamma^2)/\beta^2 = 0\Omega_1(\beta) \quad . \quad (13.157)$$

Two branches and $\Omega_2(\beta)$ of normalized resonance frequencies of the coupled system will be found as solutions for this equation. The convention will be used that $\Omega_1(\beta) < \Omega_2(\beta)$. We will assume further that parameters of the mechanical system of transducer do not change, and all the changes of resonance frequencies of the coupled system are due to alternations of the resonator. For further manipulations denote

$$\frac{n_{U_{\tilde{V}}}^2 K_{\tilde{V}}}{K'^{E}_{eqv}} = \psi . \quad (13.158)$$

Then parameters of the coupled system will be expressed as

$$f_{1p} = \frac{1}{2\pi} \sqrt{\frac{K'^{E}_{eqv}}{M'_{eqv}}} \sqrt{1+\psi} = f_{tr} \sqrt{1+\psi} , \quad (13.159)$$

$$\gamma^2 = \psi / (1+\psi), \tag{13.160}$$

$$\beta = (f_{tr} / f_{2p})\sqrt{1+\psi} = \beta'\sqrt{1+\psi}\ ,\ \beta' = f_{tr} / f_{2p} \tag{13.161}$$

$$\Omega = (f^2 / f_{tr}^2) / (1+\psi). \tag{13.162}$$

From Eq. (13.157) follows that

$$\Omega_{1,2} = \frac{1+\beta^2}{2\beta^2}\left[1 \pm \sqrt{1 - \frac{4\beta^2(1-\gamma^2)}{(1+\beta^2)^2}}\right]. \tag{13.163}$$

After substituting expressions (13.160) and (13.161) for parameters γ and β, we arrive at equation for determining resonance frequencies of the coupled system, as functions of parameter $\beta' = f_{tr} / f_{2p}$ at different values of constant ψ

$$\Omega_{1,2} = \frac{1/\beta'^2 + (1+\psi)}{2(1+\psi)}\left\{1 \pm \sqrt{1 - \frac{4\beta'^2}{[1+\beta'^2(1+\psi)]^2}}\right\}. \tag{13.164}$$

After the values of Ω_i are determined, ratios of the volume velocities $\left(U_{\tilde{V}n} / U_{\tilde{V}1}\right)_i$ at resonance frequencies of the system can be found from either of equations (13.151) or (13.152). For example, from Eq. (13.152) we obtain

$$\left(\frac{U_{\tilde{V}n}}{U_{\tilde{V}1}}\right)_{i=1,2} = n_{U_{\tilde{V}}} \cdot \frac{1+\psi}{\psi} \cdot (1-\Omega_i). \tag{13.165}$$

This quantity characterizes the contribution of unintended radiation from the neck in comparison with radiation from the transducer surface at corresponding resonance frequencies together with corresponding radiation impedances.

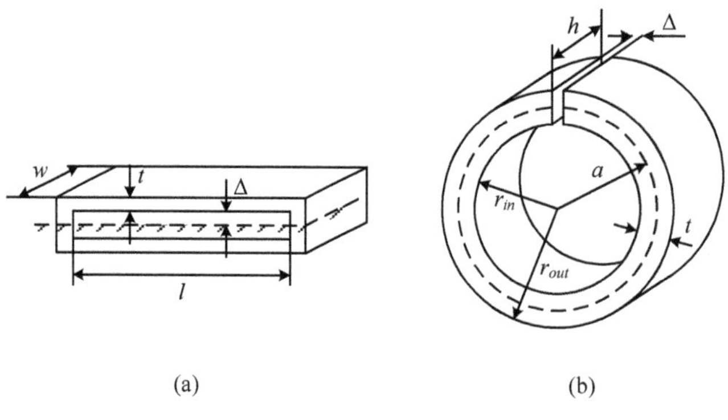

Figure 13.21: Geometry of the rectangular bender (a) and of the slotted ring (b) with resonators.

Constant ψ depends on the geometry and elastic properties of the transducer mechanical system and resonator. We will specify values of this quantity and subsequent calculations for the rectangular bender and for the slotted ring projector. Their schematic representations with Helmholtz resonator chambers are shown in Figure 13.21. Remember that the walls of the chambers as partial systems are considered to be rigid.

13.4.3.2 Rectangular Benders with Helmholtz Resonator

Due to symmetry of the design one half of it can be considered under the condition that the plane of symmetry is absolutely rigid. For the rectangular plate bender $n_{U_{\tilde{V}}} = -1$.

Acoustic rigidities of the rectangular simply supported plate and of the chamber of resonator are

$$K'^{E}_{eqv} = \frac{\pi^4 w t^3 Y_i^E}{24 l^3} \cdot \frac{1}{(0.64wl)^2}, \tag{13.166}$$

$$K_{\tilde{V}} = \frac{(\rho c^2)_l}{\tilde{V}_{av}} = \frac{(\rho c^2)_l}{0.64wl\Delta}. \tag{13.167}$$

Thus,

$$\psi = \frac{n^2_{U_{\tilde{V}}} K_{\tilde{V}}}{K'^{E}_{eqv}} = 0.16 \frac{(\rho c^2)_l}{Y_i^E} \frac{l^4}{t^4} \frac{t}{\Delta}. \tag{13.168}$$

For approximate estimation we will assume that the internal volume is filled with liquid having parameters of water, $(\rho c^2)_w$ = 2.25 GPa and the benders are made of piezoceramics having parameters of PZT-4: $Y_1^E = 81$ GPa and $Y_3^E = 64$ GPa for the transverse and longitudinal piezoeffect, respectively. Assuming that the maximum length to thickness ratio is l/t = 5, we obtain

$$\psi \approx 2.8 \frac{t}{\Delta} \text{ and } \psi \approx 3.5 \frac{t}{\Delta}. \tag{13.169}$$

For a particular value of ratio t/Δ (i.e., ψ) and resonance frequency of the mechanical system of transducer in air, f_{tr}, dependencies can be obtained of the frequency factor, Ω, and of the "mode shape" coefficients, $U_{\tilde{V}n}/U_{\tilde{V}1}$, by formulas (13.158) and (13.159) from parameter $\beta' = f_{tr}/f_{2p}$. Calculating these quantities will be considered with the example of a slotted ring.

13.4.3.3 Slotted Ring Projector

The average inner surface of a slotted ring is following expression (7.289)

$$S_{avin} = 1.34\pi ha[1-(t/2a)], \tag{13.170}$$

and coefficient of velocity transformation, is

$$n_{U_{\tilde{V}}} = -\frac{1-(t/2a)}{1+(t/2a)}. \tag{13.171}$$

The acoustic rigidies of a ring (after expressions (7.250) and (13.166)) and of a resonator are

$$K'^{E}_{eqv} = \frac{\pi t^3 h Y_i^E}{20a^3} \cdot \frac{1}{(1.34\pi ha)^2[1-(t/2a)]^2}, \tag{13.172}$$

$$K_{\tilde{V}} = \frac{(\rho c^2)_l}{\pi a^2 h[1-(t/2a)]^2}, \tag{13.173}$$

$$\psi = \frac{n^2_{U_{\tilde{V}}} K_{\tilde{V}}}{K'^{E}_{eqv}} = 10\left[\frac{1-(t/2a)}{1+(t/2a)}\right]^2 \frac{(\rho c^2)_l}{Y_i^E}\frac{a^3}{t^3}. \tag{13.174}$$

Consider an example of slotted ring made of PZT-4 ceramics with the realistic dimensions: a = 3.5 cm, t = 2 cm, Δ = 2 mm. Let it be filled with liquid having parameters of water, i.e., $c_l = c_w$ = 1500 m/s and $(\rho c^2)_l$ = 2.25 GPa. Under these assumptions ψ = 0.46 for the transverse and ψ = 0.58 for the longitudinal piezoeffect.

Using formulas for the resonance frequencies of the resonator (f_{2p} by formula (13.149)) and of the ring, $f_{ring} = 0.02(t/a^2)c_c^E$, ($c_c^E$ is the sound speed in ceramics) the value of coefficient β' will be obtained as

$$\beta' = \frac{f_{ring}}{f_{2p}} = 0.22\frac{c_c^E}{c_l}\frac{1-(t/2a)}{a}\sqrt{\frac{t^3}{\Delta}}. \tag{13.175}$$

Values of the coefficients ψ, β' and all the non-dimensional resonance frequencies for this example are presented in Table 13.3. Note that relation holds between the resonance frequencies of the coupled system and partial frequencies, $f_1 < f_{1p} < f_{2p} < f_2$, that is known from the general theory of the coupled vibrations (see Section 4.6).

Important characteristics of the coupled system are the relations between the volume velocities of the transducer surface and surface of the neck (the "mode shape" coefficients), $\left(U_{\tilde{V}n}/U_{\tilde{V}1}\right)_i$, which are determined from Eq. (13.165). Their values for the example considered are presented in Table 13.3. As it follows from the data, at the low (operating) resonance frequency of the transducer the volume velocity of the neck is distractive (in antiphase) and smaller than velocity of the ring surface. At high resonance frequency (resonance frequency of the resonator operating as projector) it is in phase and larger. Noteworthy is that outside the resonance frequency region $U_{\tilde{V}1}$ and $U_{\tilde{V}n}$ are in quadrature.

Table 13.3: Quantities that characterize the coupled ring-resonator system.

	ψ	β'	Ω_1	Ω_2	f_1	f_2	f_{1p}	f_{2p}	$\left(\frac{U_{\tilde{V}n}}{U_{\tilde{V}1}}\right)_{f_1}$	$\left(\frac{U_{\tilde{V}n}}{U_{\tilde{V}1}}\right)_{f_2}$
Eq.	(13.174)	(13.175)	(13.164)		(13.162)		(13.159)	(13.161)	(13.165)	
Transverse effect	0.46	0.62	0.55	2.2	$0.9f_{ring}$	$1.8f_{ring}$	$1.2f_{ring}$	$1.6f_{ring}$	-0.80	2.2
Longitudinal effect	0.58	0.55	0.51	2.6	$0.8f_{ring}$	$2.0f_{ring}$	$1.26f_{ring}$	$1.8f_{ring}$	-0.64	1.2

By changing parameters ψ and β' that are related to properties of transducer and resonator the most favorable characteristics of the coupled system can be determined. After this is done, a detailed calculation of operating characteristics of the transducer must be produced using the equivalent circuit of Figure 13.20, which includes radiation impedances and resistances of losses for the transducer and for the resonator.

13.5 References

1. D. Stansfield, *Underwater Electroacoustic Transducers* (Reprinted by Peninsula, Los Altos Hills, CA, 2003).
2. T. Fett, D. Munz, G. Thun, “Tensile and bending strength of piezoelectric ceramics,” J. Mater. Sci. Letters, 18(1999), 1899 – 1902
3. Fett, T., Munz, D. and Thun, G., “Strength of a soft PZT ceramic under a transversal electric field,” J. Mater. Sci. Letters, 2000, 19, 1921–1924
4. T. Fett, D. Munz, G. Thun, “Bending strength of a PZT ceramic under electric fields,” Journal of the European Ceramic Society 23 (2003) 195–202.
5. Ralf S. Woollett, “Power Limitations of Sonic Transducers,” IEEE Transections on Sonics and Ultrasonics, Vol. SU-15, No. 4, Oktober, 1968.
6. D. A. Berlincourt, D. R. Curran, and H. Jaffe, *Piezoelectric and Piezomagnetic Materials and their Function in Transducers*, in Physical Acoustics, Vol. I, Part A, edited by W. P. Mason (Academic, New York, 1964).
7. S. P. Timoshenko, *Vibration Problems in Engineering*, Second Edition, New York, D. Van Nostrand Company, Inc.,1937.
8. Boris Aronov, “Application of the Tonpilz design for optimizing parameters of rod transducers,” J. Acoust. Soc. Am. 105, 1179, (A), (1999)
9. V. E. Glazanov, *Acoustic baffles for underwater transducers and arrays*, Sankt-Petersburg, Electrotechnical University “LETI”,2013, 176 p. (in Russian).
10. R. Gerson, S. R. Burlage, and Don Berlincourt, “Dynamic Tensile Strength of a Ferroelectric Ceramic”, J. Acoust. Soc. Am., 33 (11), 1483-1485, (1961)
11. R.S. Woollett, “Basic problems caused by depth and size constrains in low-frequency underwater transducers,” J. Acoust. Soc. Am., **68** (4), 1031-1037, (1980)
12. D.H. Robey, “On the radiation Impedance of the Liquid-Filled Squirting Cylinder,” J. Acoust. Soc. Am. 27 (4), 711- 714, (1955).
13. G. W. McMagon, “Performance of Open Ferroelectric Ceramic Cylinders in Underwater Transducers,” J. Acoust. Soc. Am. 36 (3), 528-533, (1964)
14. M.C. Junger, “Design parameters of free-flooding cylindrical transducers,” Cambridge Acoustic Associates Inc. MA, Technical Report U-308-210, Feb. 1969.
15. P. H. Rogers, “Mathematical model for a free flooded piezoelectric cylinder transducer,” J. Acoust. Soc. Am. 80 (1), 14, July 1986.
16. C. H. Sherman and J. L. Butler, *Transducers and Arrays for Underwater Sound*, Springer Science & Business Media LLC, 2007, pp. 91-95
17. R. S. Woollett, “Underwater Helmholtz-Resonator Transducers,” General Design Principles”, NUSC, New London, Technical Report 5633, July 1977

18. E. L. Shenderov, *Radiation and Scattering of Sound*, pp. 109-127, Sudostroenie Press, Leningrad, 1989, 302 pp, in Russian.

19. P. M. Morse, K. U. Ingard, *Theoretical Acoustics*, Section 6.4, (McGraw-Hill, New York, 1968).

D.H. Robey, "On the Contribution of a Contained Viscous Liquid to the Acoustic Impedance of a Radially Vibrating Tube," J. Acoust. Soc. Am. 27 (1), 22- 25, (1955).

CHAPTER 14

SENSORS DESIGNING RELATED ISSUES

14.1 Introduction

General characterization of properties of the sensors (hydrophones and accelerometers) as members of a receive channel was produced in Section 3.3. Namely, their properties as a source of signal and internal noise, and their immunity to unwanted actions were concerned. Methods of analyses and results obtained for particular transducer types that allow calculating characteristics of the sensors in detail were addressed in the related chapters of Part III. Here some general properties of the sensors of different type that allow their comparison for operating under real environmental conditions will be considered, and recommendations on improving these properties will be discussed. The most widely used in underwater applications are hydrophones. Using accelerometers per se is less common. Therefore, the main body of the chapter is devoted to the hydrophones.

Requirements for the hydrophones can differ depending on their applications. Thus, for the hydrophones intended for measuring characteristics of the acoustic fields (pressure and pressure gradient) their immunity to unwanted actions may be especially important. Hydrophones intended for populating arrays of passive sonars are the most demanding in terms of their sensitivity under the operating environmental conditions. As to the reversible transducers that are used interchangeably in the transmit and receive modes of operation, requirements for their properties are usually dictated by the transmit mode and there is no need to consider them as hydrophones.

Considerations regarding optimizing properties of the hydrophones operating in a broad band under hydrostatic pressure are addressed in Section 14.3. Harmful effects of unwanted actions on characteristics of hydrophones and some recommendations regarding improving their immunity to the unwanted actions are considered in Section 14.4. Most of the results that are presented in this chapter were reported in Ref. 1. More information regarding the first order sensor designs and related literature can be found in Ref. 2 and Ref. 3.

14.2 Zero and First Order Hydrophones

14.2.1 Ideal Zero Order (Pressure) and First Order (Pressure gradient) Hydrophones

Harmonic sound fields are characterized by the sound pressure P (a scalar quantity) and by the particle velocity $\bar{U}$, or by the pressure gradient $\bar{\nabla}P$ (vector quantity). The velocity and pressure are related by Euler's equation

$$\bar{\nabla}P = -j\omega\rho_0\bar{U} . \tag{14.1}$$

This is noteworthy that there is no real means for direct measuring of the velocity, if not consider hypothetical mechanisms of converting movement of particles into electrical signals. All the real devices (especially those employing piezoceramics) produce the effect of measuring, because of deformation or motion of solid bodies under action of forces related to pressure gradient in the acoustic fields. Measurement of sound pressure and pressure gradient requires acoustic hydrophones of zero order (omnidirectional, or of the monopole type) and of the first order (having figure of eight directionality, or of the dipole type).

An ideal sensor of zero order can be considered as a device with dimensions much smaller than the acoustic wavelength λ ($D/\lambda \ll 1$, where D is its maximum dimension of the hydrophone), and having electrical response that is omnidirectional and proportional to the sound pressure only, as it is illustrated in Figure 14.1 (a). The ideal hydrophone must be insensitive to motion regardless of its origin, especially if it is produced by acoustic field.

An ideal sensor of the first order or of the "dipole type" can be realized by a pair of identical ideal monopoles closely spaced (at distance d, where $d/\lambda \ll 1$) and electrically connected in opposite phase, as illustrated in Figure 14.1 (b.1). Ideally the dipole possesses a figure of eight directionality with the acoustic axis directed along the line connecting the monopoles. Zero sensitivity in the lateral direction is equivalent to insensitivity of the dipole to the sound pressure. In the case that the monopoles in Figure 14.1 (b.1) are identical (with equal sensitivities $\gamma_{1p} = \gamma_{1p} = \gamma$), the output of the dipole hydrophone in the incident plane wave with magnitude P propagating in the direction of axis $\varphi = 0$ is

$$V = P_2\gamma_{2p} - P_1\gamma_{1p} = \gamma d\frac{P_2 - P_1}{d} \approx \gamma d \cdot gradP , \tag{14.2}$$

and electrical output V of the device is proportional to the pressure gradient only (is insensitive to the sound pressure). For the wave propagating in direction of angle $\varphi = 90°$ the output voltage is $V = 0$.

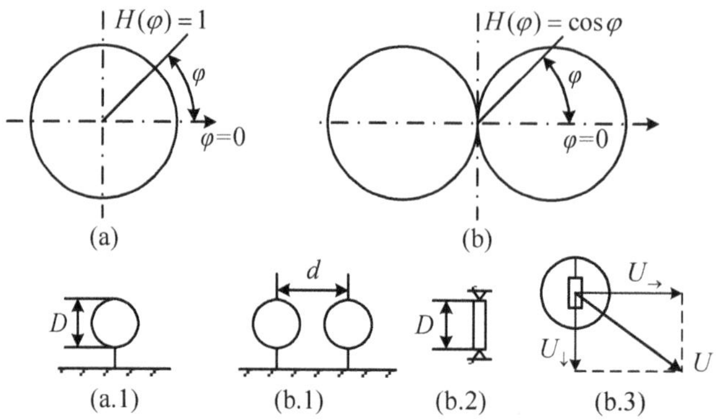

Figure 14.1: Various types of ideal hydrophones and their directivities: (a) ideal zero order (sound pressure) hydrophone; (b) ideal first order (pressure gradient) hydrophones. Diffraction (fixed) type pressure gradient hydrophones: dipole (b.1) and flexural plate (b.2). Motion type pressure gradient hydrophone with an accelerometer (b.3).

Note that the ideal figure of eight directional response of the device with zero response at $\varphi = 90°$ can be observed only under the assumption that there is no acoustic noise (or the signal to noise ratio in the sound field is infinitely large) and the sensitivities of the monopoles are infinitely large in order not to add electrical noise from the receiving system. Otherwise, the "nulls" of the directional characteristic will be limited by the environmental and receiving system noise.

Another example of the ideal first order hydrophone represents bimorph flexural plate supported on the circumference by foundation that does not move (massive enough), as is shown in Figure 14.1 (b.2). Under the condition that force acting in plane of the plate (denoted as $F_{\uparrow}$) does not produce an output effect (halves of bimorph are identical) the output voltage of the hydrophone is proportional to the force $F_{\rightarrow}$ only. Assuming that dimensions of the plate are much smaller than wavelength, in direction $\varphi = 0$ of the wave propagation the output voltage of the hydrophone will be

$$V = \gamma(P_2 - P_1) \approx \gamma t \cdot gradP , \qquad (14.3)$$

where γ is sensitivity of the bimorph plate. The approximation sign is used because in principle a correction must be made due to diffraction of the sound wave on the plate despite its small dimensions. The first order hydrophone of this kind will be conditionally referred to as the diffraction type hydrophone.

An ideal hydrophone of the first order can also be realized with a small body containing an accelerometer, which can move under the action of acoustic field in the direction of wave propagation without any restriction, as shown in Figure 14.1 (b.3). We denote sensitivity of the accelerometer to acceleration in the direction of motion as $\gamma_{\dot{U}\rightarrow}$, where $\dot{U}$ denotes the derivative of velocity or acceleration and the arrow denotes motion in the direction of acoustic wave propagation. Likewise, sensitivity of the accelerometer to motion in the perpendicular direction will be denoted as $\gamma_{\dot{U}\uparrow}$. Ideally sensitivity to this motion must be zero, i.e., $\gamma_{\dot{U}\uparrow}=0$. In this case the device has the directional response of the first order, $H(\varphi)=\cos\varphi$. Hydrophone of this kind will be referred to as the motion type hydrophone. The ideal device of this type must be insensitive to deformations of the body under the action of sound pressure. By the same reason, as in the previous case of the dipole type hydrophone, the signal to noise ratio in the acoustic field and the sensitivity of the accelerometer must be infinitely large in order to get an ideal first order directional response of the device. Due to the inherent directional figure of eight property that allows in principle resolving direction of the wave propagation the first order hydrophones are also called the vector hydrophones.

Thus, for obtaining the ideal responses of the zero and first order, the hydrophones must meet certain unrealistic requirements. They must be immune to unwanted actions. In the case of sound pressure hydrophones, they must be immune to the actions of the pressure gradient that exists in the sound field, and in the case of pressure gradient hydrophones they must be immune to sound pressure. The sound pressure hydrophones used in the dipole type device must also be absolutely identical. Accelerometers used in the motion type hydrophones must be insensitive to vibrations in the perpendicular direction. And lastly, the sensitivities of the sound pressure hydrophones and accelerometers used for obtaining an ideal first order response must be infinitely large. As these requirements cannot be completely met, the responses of actual hydrophones are not ideal and will always be compromised to some degree. Real hydrophones always have a finite sensitivity to signal and are sensitive to unwanted actions to some extent. The goal of rational designing is to keep these sensitivities within acceptable limits that must be established for each sensor type and its intended application.

14.2.2 Real Zero and First Order Hydrophones

A real hydrophone placed in the acoustic field can be represented as an elastic body involving a mechanism of mechano-electrical transduction (Figure 14.2 (a). The body with surface denoted Σ vibrates under the action of sound wave. Related movement and/or deformation of the body are transformed into electrical output. The motion of the body can be imagined as the superposition of its vibration as a "clamped" body (as if it is rigid) allowed by the compliance of the mounting conditions, and the elastic deformations of the body under consideration that the body is held fixed (does not move). These conditions are illustrated in Figure 14.2.

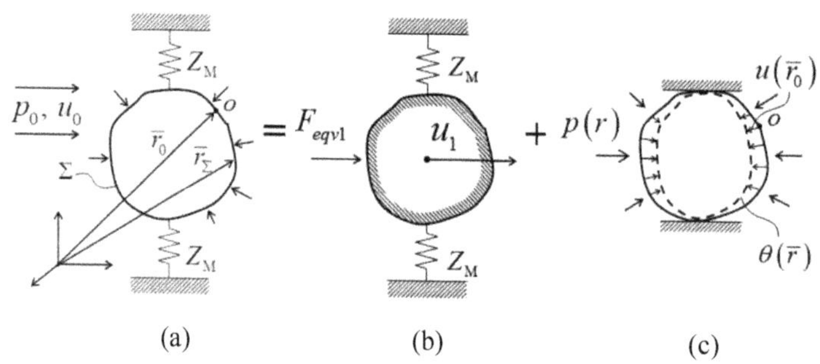

Figure 14.2: Representation of motion of a real body (a) by superposition of its motion as the rigid body (b) and elastic deformation of the body that is fixed (c).

For introducing the forces that generate these components of motion the concept of diffraction coefficients must be used (see Section 6.1.3.1.4). For this purpose, consider the mechanical power W_{am} delivered by the acoustic field that is associated with motion of a rigid body and deformation of the body that does not move.

In the case of movement of the rigid body all the surface elements $d\Sigma$ vibrate with the same velocity U_1 in the direction of the wave propagation, as shown in Figure 14.3(a). As the sound pressure is acting by the normal to a surface element, we have

$$\bar{\dot{W}}_{am} = \int_{\Sigma} P(\bar{r}_\Sigma) U_1^* \cos\varphi d\Sigma = F_{eqvm} U_1^*, \tag{14.4}$$

where (*) denotes the complex conjugate. The equivalent force F_{eqvm} can be represented using definition for the diffraction coefficient k_{dif}, as

$$F_{eqvm} = P_0 S_\Sigma k_{dif\,m}, \tag{14.5}$$

where P_0 is sound pressure in the free acoustic field at the point, where the acoustic center of the body is located, S_Σ is the total surface of the body and

$$k_{dif\,m} = \frac{1}{P_0 S_\Sigma} \int_\Sigma P(\bar{r}_\Sigma) \cos\varphi \cdot d\Sigma. \tag{14.6}$$

In formulas (14.4) and (14.6) $P(\bar{r}_\Sigma)$ is the sound pressure on the "clamped" body surface. The diffraction coefficient depends on configuration of the body and on the acoustic wavelength.

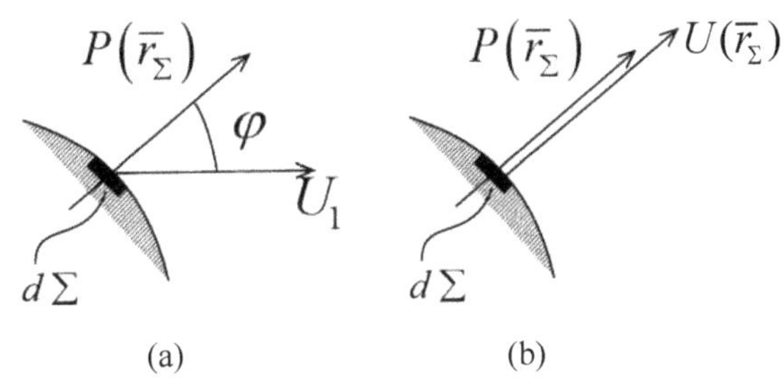

Figure 14.3: Surface elements vibrating with the same velocity u_1 : (a) in the direction of the wave propagation, (b) under the action of sound pressure.

In the case of deformation of the body that does not move as a whole (Figure 14.2 (c)), we have for the mechanical power

$$\bar{\bar{W}}_{am} = \int_\Sigma P(\bar{r}_\Sigma) U^*(\bar{r}_\Sigma) d\Sigma, \tag{14.7}$$

where $U(\bar{r}_\Sigma)$ is the distribution of velocity of vibration in the normal direction over the body surface (Figure 14.3 (b)) and $P(\bar{r}_\Sigma)$ is the sound pressure on the surface determined under the condition that the surface is "clamped" (the same sound pressure distribution, as in the previous case). Under assumption that the body has single mechanical degree of freedom, which is usually the case in the frequency range below its first resonance frequency, distribution of velocity may be represented in the form $U(\bar{r}_\Sigma) = U(\bar{r}_o)\theta(\bar{r}_\Sigma)$, where $U(\bar{r}_o)$ is the velocity of the reference point and $\theta(\bar{r}_\Sigma)$ is the mode of vibration of the mechanical system. Thus, expression (14.7) for the acoustomechanical power can be presented as $\bar{\dot{W}}_{am} = F_{eqvd}U^*(\bar{r}_o)$. The equivalent force F_{eqvd} that produces deformation of the body may be expressed in the form analogous to formula (14.5), as

$$F_{eqvd} = P_0 S_\Sigma k_{dif\,d}, \tag{14.8}$$

the diffraction coefficient is

$$k_{dif\,d} = \frac{1}{P_0 S_\Sigma} \int_\Sigma P(\bar{r}_\Sigma)\theta(\bar{r}_\Sigma) d\Sigma . \tag{14.9}$$

The hydrophones, in which output effect is associated with motion of their bodies, will be referred to as the motion type hydrophones. The hydrophones, in which effect is associated with deformation of their bodies, will be referred to as the diffraction (or fixed) type hydrophones. In both cases the output effect is proportional to the same property of acoustic field through the diffraction coefficients $k_{dif\,d}$ and $k_{dif\,m}$ (the subscripts *d* and *m* in the diffraction coefficients are introduced for distinguishing between the diffraction and motion type situations). For different hydrophone types the coefficients may have the same values, as this is illustrated with examples (b) and (c) shown in Figure 14.4 (in case (b) the mode shape is $\theta(\bar{r}_\Sigma) = \cos\varphi$).

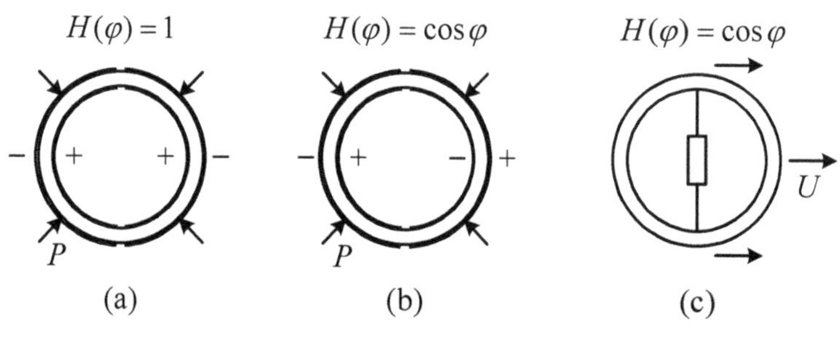

Figure 14.4: Examples of spherical mechanical systems of hydrophones of (a) zero order and of (b) the first order hydrophones of diffraction and (c) motion types. In the cases (a) and (b) the spherical shells do not move and realize the zero and first modes of deformation (diffraction type hydrophones). In the case (c) the spherical body moves in the direction of wave propagating.

Under the condition that wave size of the spherical shell is small the following values of the diffraction coefficients will be obtained from expressions (14.6) and (14.9) for the hydrophone that realizes zero mode of deformation

$$k_{dif\,0} \approx 1 . \tag{14.10}$$

For the hydrophone of diffraction type that realizes the first mode of the spherical shell deformation, and for the hydrophone of motion type that employs movement of spherical shell the diffraction coefficients have the same values. From expression (6.250) follows that value of the diffraction coefficient for spheres of small wave size at $\varphi = 0$ is $k_{dif} \approx -jka/2$.

Thus,

$$k_{dif\,d}(\varphi) = k_{dif\,m}(\varphi) \approx -j(ka/2)\cos\varphi. \tag{14.11}$$

Given that in the plane wave propagating in direction $\boldsymbol{r}$ under angle φ relative to the horizontal plane (to x axis) $P = P_0 e^{j(\omega t - \boldsymbol{k}\cdot\boldsymbol{r})}$,

$$gradP_x = (-jk\cos\varphi)\cdot P, \tag{14.12}$$

i. e., the diffraction coefficients (14.11) are proportional to the pressure gradient in acoustic field. With increase of the wave size of the hydrophones the values of diffraction coefficients become frequency dependent by modulus and by phase (for hydrophones that employ the spherical shells this is shown in Figure 14.5), and the correspondences between their outputs and values of sound pressure and pressure gradient in the acoustic field become not accurate. Although in terms of the directional characteristics the hydrophones remain being of the zero and first order.

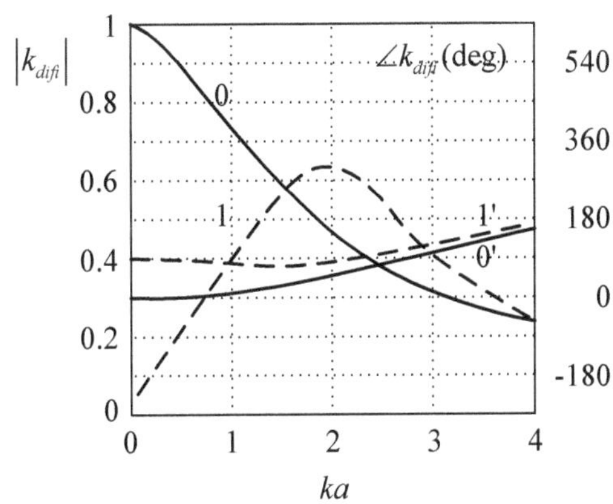

Figure 14.5: Diffraction coefficients for spherical hydrophones of the zero (solid lines) and first (dashed lines) order. The plots for $|k_{dif}|$ are marked 0 and 1, the plots for $\arg k_{dif}$ are marked 0' and 1'.

The analogous situation exists regarding the hydrophones that employ cylindrical shells vibrating in the zero and first modes. Dependence of the diffraction coefficients from ka for this case is shown in Figure 14.6 by modulus and phase.

For the oscillating cylinder of small wave diameter (at $ka < 0.3$)

$$k_{dif} = -jka \cdot \cos\varphi, \tag{14.13}$$

as it follows from formula (6.119). For the oscillating disk of small wave size (at ka <0.5 as follows from (6.322))

$$k_{dif} = -j\frac{4}{3\pi}ka \cdot \cos\varphi. \tag{14.14}$$

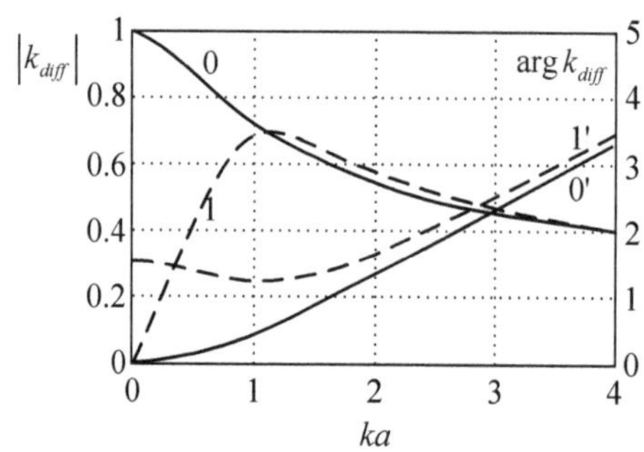

Figure 14.6: Diffraction coefficients for the cylindrical hydrophones of the zero (solid line) and first (dash line) order. The plots for $|k_{dif}|$ are marked 0 and 1, the plots for $\arg k_{dif}$ are marked 0' and 1'.

The diffraction coefficients are the only quantities that correlate outputs of hydrophones with properties of acoustic fields. All other characteristics of the hydrophones depend on the mechanisms of mechanoelectrical conversion employed and on the frequency responses of their mechanical systems. They can be changed by a receive channel and presented in a form of the frequency dependence that is convenient for a particular application. Thus, the output effect of the first order hydrophone in plane harmonic wave can be presented in the units of the pressure gradient, of velocity of vibration, of the displacement, and even in units of sound pressure proceeding from the fact that all these quantities are proportional to value of sound pressure.

The fact that the modulus and phase of diffraction coefficients deviate from their real values for the sound pressure and pressure gradient with increase of the hydrophones wave size does not compromise ability of different calibration of the output, so far as the hydrophones are used as the single units. But in the case of combined use of the zero and first order hydrophones (for example, for forming cardioid response, or measuring intensity of sound field) the additional magnitude and phase differences that arise from frequency dependences of the diffraction coefficients (analogous to those shown in Figure 14.5 and Figure 14.6) must be taken into consideration.

As result of the above discussion the following classification of the hydrophones that is based on their directional characteristics and on actions that they experience in the acoustic field may be considered as appropriate: all the hydrophones of zero order may be called "acoustic pressure hydrophones" and all the hydrophones of the first order may be called "pressure gradient hydrophones" or "vector hydrophones." The pressure gradient hydrophones may be subcategorized as either of motion or fixed type. And further the fixed type hydrophones may be characterized as either differential (dipole), or diffraction types. The pressure gradient hydrophones are often interchangeably called in literature as "velocity hydrophones", though this may produce a misleading impression that the corresponding hydrophones possess some magical acoustic field related properties other than the pressure gradient hydrophones have.

In reality the bodies of motion type hydrophones are subjected to deformation under action of acoustic field, and bodies of the fixed type hydrophones move under unwanted action of the structural vibrations of their supporting elements and under pressure gradient in acoustic field. All the real hydrophones are sensitive to some extent to these unwanted actions, and output effect produced by unwanted actions may compromise the zero and first order characteristics. Distortions of the assumed characteristics depend on the environmental conditions, and they may be very significant, as the example considered in Section 3.2.4 shows. These issues will be addressed in Sec. 14.4. Besides, the ideal zero and first order characteristics of the hydrophones may be distorted due contribution of noise of the receive channel if their sensitivity to the intended signals is insufficient.

First, we consider issues related to properties of hydrophones as converters of the intended actions (signals).

14.3 Hydrophone as Source of Signals

14.3.1 Parameters of Hydrophone as Source of Signals

It was noted in Section 14.1 that we are considering the low frequency broad band hydrophones. As it has been established in Section 3.2.2, in a broad band below the first resonance frequency a hydrophone can be represented as equivalent generator shown in Figure 14.57 with electromotive force $E = \gamma P_0$ and internal impedance $Z_{in} \approx j\omega C_{Lf}$, where according to expressions (3.169) and (3.170)

$$C_{Lf} = C_e^S / (1 - k_{eff}^2), \qquad (14.15)$$

$$\gamma(\omega) = \frac{C_{eqv}^E}{C_{Lf}} \cdot nS_\Sigma k_{dif}(\omega). \qquad (14.16)$$

(Here is omitted resistance of losses that is responsible for the internal noise of a hydrophone (see Section 3.2.2).

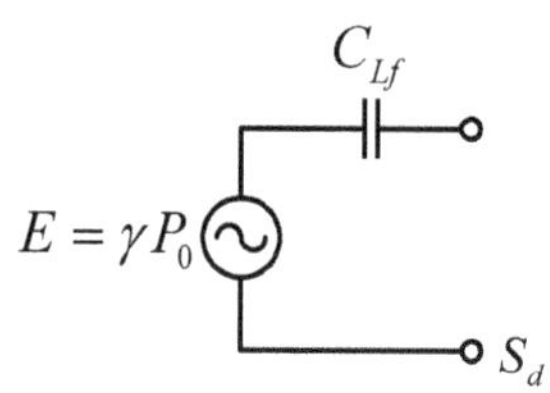

Figure 14.7: Representation of a hydrophone as equivalent generator of signal for a preamplifier.

Together with these parameters, which fully characterize a hydrophone as source of energy of signal, important characterization of the hydrophone is the specific sensitivity

$$\gamma_{sp} = \gamma\sqrt{C_{Lf}}. \qquad (14.17)$$

This figure of merit allows comparison of potential quality of the hydrophones having different parameters γ and C_{Lf}. It has to do with the rated power of a hydrophone as source of energy by formula (see (3.182))

$$\dot{W}_{rt} = \omega\frac{\gamma_{sp}^2}{4}|P_0|^2. \qquad (14.18)$$

In case the surface occupied by the hydrophone in an array (that will be denoted as - "dimensional area") is limited, the more objective figure of merit may be helpful, which will be called the "reduced sensitivity", γ_{rd},

$$\gamma_{rd} = \gamma_{sp} / \sqrt{S_d}\,. \tag{14.19}$$

The reduced sensitivity has to do with efficiency of acoustoelectrical conversion of energy of acoustic wave, $\dot{W}_{ac} = |P_o|^2 S_d / (\rho c)_w$, performed by the hydrophone per unit of its dimensional area. Namely,

$$\frac{\dot{W}_{rt}}{\dot{W}_{ac}} = \frac{1}{4}\omega(\rho c)_w \frac{\gamma_{sp}^2}{S_d} = \frac{1}{4}\omega(\rho c)_w \gamma_{rd}^2\,. \tag{14.20}$$

After using relations (14.15) and (14.16) for γ and C and definition for k_{eff} (2.93), the expression for γ_{rd} becomes

$$\gamma_{rd} = k_{eff} S_\Sigma k_{dif} \sqrt{C_{eqv}^E / S_d}\,. \tag{14.21}$$

Considering that

$$C_{eqv}^E = \frac{1}{\omega_r^2 M_{eqv}} = \frac{1-k_{eff}^2}{\omega_{me}^2 M_{eqv}}, \tag{14.22}$$

where ω_r and ω_{em} are the frequencies of the mechanical and electromechanical resonances of a mechanical system, expression (14.21) may be presented in the form

$$\gamma_{rd} \cdot f_{em} = k_{dif} \cdot k_{eff} \sqrt{1-k_{eff}^2} \cdot \sqrt{\frac{S_\Sigma^2}{4\pi^2 M_{eqv} S_d}}\,. \tag{14.23}$$

Given that in general, $M_{eqv} = S_{eff} t \rho$, where t is the thickness of mechanical system, we obtain the relation

$$\gamma_{rd} \cdot f_{em} = k_{dif} \cdot k_{eff} \sqrt{1-k_{eff}^2} \cdot \sqrt{\frac{S_\Sigma^2}{4\pi^2 S_{eff} S_d}} \cdot \frac{1}{\sqrt{\rho t}}\,. \tag{14.24}$$

For all the hydrophones of the same transducer type the coefficient

$$Ff = k_{dif} \cdot \sqrt{\frac{S_\Sigma^2}{4\pi^2 S_{eff} S_d}}, \tag{14.25}$$

which can be called the "form factor", has approximately the same value. Therefore, the product of the reduced sensitivity and bandwidth that is characterized by the first resonance frequency can be estimated for a particular hydrophone transducer type, as

$$\gamma_{rd} \cdot f_{em} = Ff \cdot k_{eff}\sqrt{1-k_{eff}^2} \cdot \frac{1}{\sqrt{\rho t}} \sim k_{eff}\sqrt{1-k_{eff}^2} \cdot \frac{1}{\sqrt{\rho t}} . \qquad (14.26)$$

14.3.2 Variants of the Sensor Designs

Designs of the pressure and pressure gradient hydrophones employ electromechanical transducers of different type. The most widely used of them are schematically depicted in Figure 14.8.

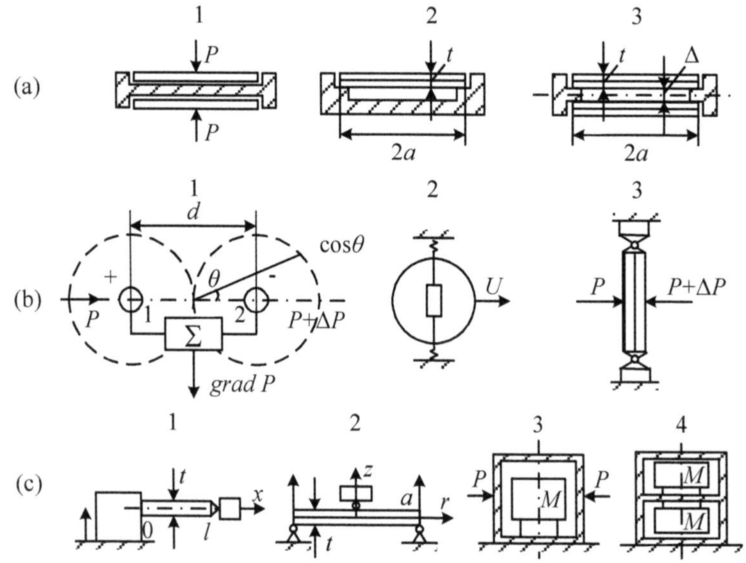

Figure 14.8: Variants of the sensors and electromechanical transducers involved in their designs. (a) Sound pressure hydrophones (1-cylindrical, 2-single plate bender, 3-double plate bender). (b) Pressure gradient hydrophones. (c) Accelerometers for the pressure gradient hydrophones of the motion type (c.1, c.2) and for measuring vibration of structures (c.3, c.4).

Accordingly, the issues will be considered related to designing: the sound pressure hydrophones, pressure gradient hydrophones of the dipole and diffraction type, and accelerometers as a part of pressure gradient transducers of the motion type and for measuring vibration of structures.

14.3.3 Sound Pressure Hydrophones

Data necessary for estimations by formula (14.26) for pressure hydrophone types that are shown in Figure 14.8 (a) are presented in previous chapters. When using this formula for the pressure hydrophones of the air-backed design that belong to group (a) in Figure 14.8, it must be verified whether the thickness of their mechanical system insures sufficient strength at a required operating depth. Thus, the maximum static mechanical stress in the spherical and cylindrical shells at fixed resonance frequency is

$$T = P_h \frac{a}{t} = P_h \cdot \frac{1}{t} \cdot \frac{c_c}{2\pi f_r} = P_h \cdot \frac{1}{t} \cdot \frac{c_c \sqrt{1-k_{eff}^2}}{2\pi f_{em}}, \tag{14.27}$$

where $c_c = 1/\sqrt{s_{11}^E \rho}$ and $k_{eff} = k_{31}$ for radial polled cylinder with aspect ratio $(h/2a) < 0.5$; $c_c = \sqrt{2}/\sqrt{s_{11}^E \rho (1-\sigma_1^E)}$ and $k_{eff}^2 = k_p$ for sphere.

From relation (14.27) follows that the minimal thickness of the shells may be expressed as

$$t = \frac{P_h}{T_p} \cdot \frac{c_c \sqrt{1-k_{eff}^2}}{2\pi f_{em}}, \tag{14.28}$$

where T_p is the value of stress in the piezoelement that is allowed from consideration of its mechanical strength or of changing parameters under hydrostatic pressure according to the data presented in Chapter 11. After substituting this value of the minimal thickness in formula (14.26) we obtain expression

$$\gamma_{rd} \cdot \sqrt{f_{em}} \cdot \sqrt{P_h} = \sqrt{T_p} \cdot Ff \cdot \frac{2.5}{\sqrt{\rho c_c}} \cdot k_{eff} \cdot (1-k_{eff}^2)^{1/4}, \tag{14.29}$$

Analogous expression can be obtained for the circular flexural plate hydrophone. In the case that the plate is simply supported,

$$T \approx 1.25 P_h \left(\frac{a}{t}\right)^2 = 0.56 P_h \cdot \frac{1}{t} \cdot \frac{c_c \sqrt{1-k_{eff}^2}}{f_{em}}, \tag{14.30}$$

and the minimal thickness is

$$t = 0.56 \frac{P_h}{T_p} \cdot \frac{c_c \sqrt{1-k_{eff}^2}}{f_{em}}. \tag{14.31}$$

Accordingly,

$$\gamma_{rd} \cdot \sqrt{f_{em}} \cdot \sqrt{P_h} = \sqrt{T_p} \cdot Ff \cdot \frac{1.3}{\sqrt{\rho c_c}} \cdot k_{eff} \cdot (1 - k_{eff}^2)^{1/4}, \tag{14.32}$$

The inherent difference between variants of transducers of extensional type (formula (14.29)) and flexural bimorph type (formula (14.32)) is that in the first case T_p is permissible compression stress, whereas in the second – permissible tensile stress that is significantly smaller, so far as a fully active plate design is concerned.

Given that maximum operating frequency band of a hydrophone determined from the condition of linearity of its frequency response may be estimated approximately as $\Delta f_{op} \approx 0.7 f_{em}$, it follows from formulas (14.29) and (14.32) that requirements for a large sensitivity per some restricted volume of the hydrophone, operating in a broad frequency range and at large depth are contradictory. Product of quantities that characterize these properties of a hydrophone is numerically constant for a given transducer type.

Thus, the broader operating frequency range and larger depth of operation, the smaller sensitivity can be achieved per unit area occupied by the hydrophone by normal to direction of wave propagation. Note that the spherical and cylindrical hydrophones have much higher resonance frequency (much broader frequency range of operation) compared with the flexural plate sensors of a comparable size. Therefore, they significantly exceed the plate sensors in the static mechanical strength (operating depth), although they have smaller reduced sensitivity.

The factors that allow increasing this product for a particular transducer of air-backed design type are: increasing the effective coupling coefficient and/or increasing of permissible hydrostatic pressure. As the tensile strength of a circular piezoceramic plate is much less than the compressive, the overall static strength of this transducer can be increased, if to replace the internal half of ceramic plate by the one made of a metal. Though the increase of the static strength in this case is achieved for expense of the reducing its effective coupling coefficient in factor of about $\sqrt{2}$, this measure may result in increase of the sensitivity at great operating depths.

In terms of a possible increase of the effective coupling coefficients it must be noted that there is no means for increase of the coupling coefficients of the spherical and cylindrical hydrophones except for using an appropriate piezoceramic composition because of uniform distribution of dynamic stress in their mechanical systems. As to the flexural plate designs that have essentially nonuniform distribution of stress, possibilities for optimizing the effective coupling coefficients were considered in Chapter 9. Here a summary of the results obtained is presented.

14.3.3.1 Regarding Optimizing the Effective Coupling Coefficient of the Flexural Plate Hydrophone

In the case of a simply supported bilaminar circular plate the effective coupling coefficient is $k_{eff} \approx 0.8k_p$. The effective coupling coefficient may be increased by combining active and passive materials in the flexural plate design as described in Section 9.3. The relative changes of k_{eff} vs. active circular plate dimensions are shown in Figure 14.9. It is assumed that the mechanical parameters of the active and passive portions of the plate are approximately the same. As it follows from the Figure, the optimum thickness δ of the active layer is $\delta = t/3$, and the gain of the coupling coefficient is approximately 20%.

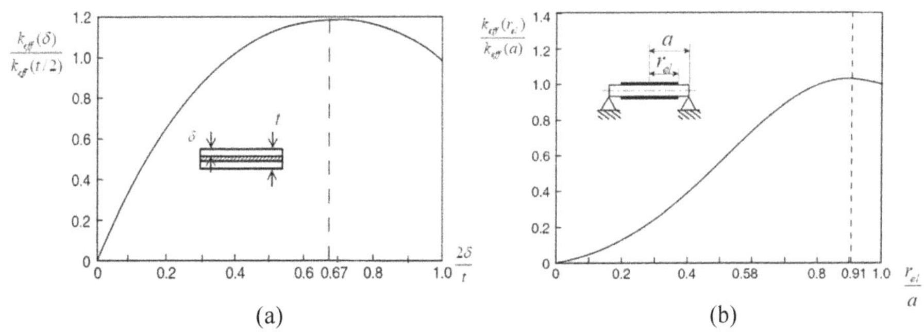

Figure 14.9: The effective coupling k_{eff} as functions of (a) normalized thickness and (b) radius of the active layer.

The optimum radius of the active layer appears to be $r \approx 0.9a$. As the gain of the coupling coefficient k_{eff} is insignificant, it seems to be more advantageous to reduce radius of the active plate to $r \approx 0.8a$. While the specific sensitivity in this case remains approximately the same, it is achieved with smaller capacitance, which makes matching with a preamplifier easier. Besides, the boundary conditions in real designs may differ from the assumed simply supported, and this measure may reduce a distractive effect of non-ideal boundary condition on the output

of transducer. The optimized version of the hydrophone with potentially higher sensitivity is realized by reducing the volume of active material as is illustrated in Figure 14.10. However, the major limitation of sensitivity of the flexural type hydrophone designs for underwater applications is due to hydrostatic pressure that produces compression on the outer surface and tension on the inner surface of the plate. Ceramic materials fail under tensile stress of approximately $T_t \approx 40\ MPa$ whereas the permissible compression is $T_c \approx 350\ MPa$,
as is adopted in Chapter 14. In different references and for different materials these values may differ, but their relation remains about $T_c / T_t \approx 10$. Thus, the operating depth may be increased in this proportion for the same hydrophone dimensions if the inner ceramic half plate is replaced by a metal. It can be, for example, titanium or an aluminum alloy. The modified hydrophone design for deeper water operation is illustrated in Figure 14.11. The specific sensitivity of this half passive design is smaller in factor of $\sqrt{2}$ compared with the original bimorph design due to reducing the volume of active material. Therefore, the projected increase of the product in formula (14.32) may be estimated approximately in factor of 2.5.

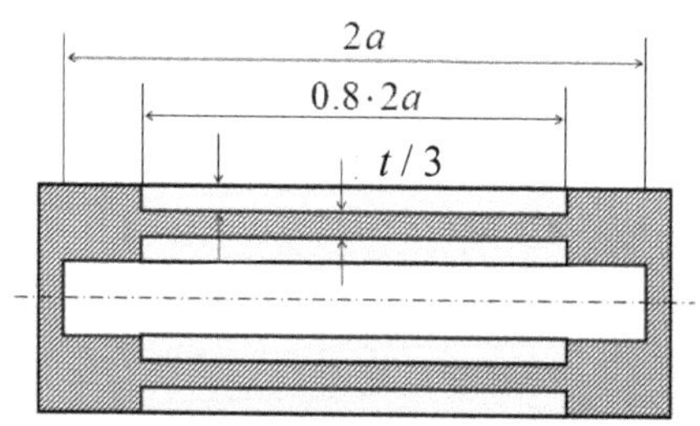

Figure 14.10: The hydrophone design with sensitivity optimized by using combination of active and passive materials.

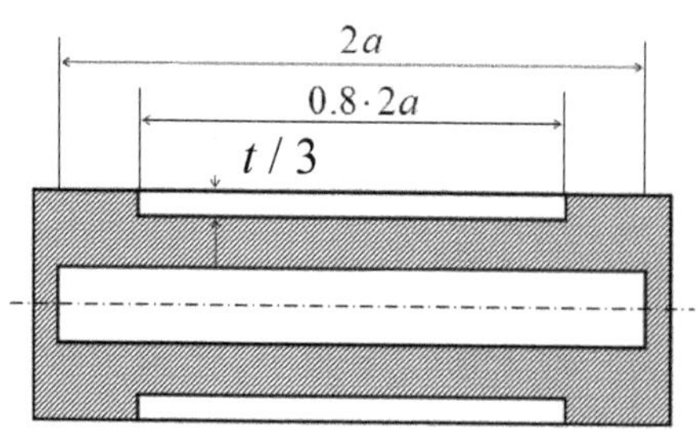

Figure 14.11: Example of the "half passive" flexural plate hydrophone design for deep water operation.

It must be kept in mind when using formulas (14.29) and (14.32) for estimating properties of the hydrophones that parameters of piezoceramics can change under stress due to hydrostatic pressure in both reversible and irreversible ways, as was described in Chapter 11. For a really great operating depth, the changes may become not tolerable. To reduce effect of limiting the operating depth by the hydrostatic pressure and to eliminate dependence of hydrophone parameters on the depth, the hydrostatic pressure compensation technique considered in Chapter 12 can be employed in the hydrophone designs.

14.3.3.2 Hydrostatic Pressure Compensation in the Hydrophone Designs

Peculiarities of employing hydrostatic pressure compensation will be illustrated with example of the flexural plate hydrophones that are the most typical in this respect. The function of compensator may be carried out in this case by the internal cavity that communicates with the external environment through a cylindrical hole. The corresponding transducer design is shown schematically in Figure 14.12 (a), and the equivalent circuit of the transducer is presented in Figure 12.6 (b).

At very low frequencies the circuit in Figure 12.6 (b) can be simplified to the circuit in (b), where $C_{\tilde{V}ac}$, $M_{ac\,eff}$, R'_{ac} are the parameters of the cavity with a hole that have to be determined by formulae given in Section 12.2.1 and $C^E_{eqvac} = C^E_{eqv} S^2_{av}$, $n_{ac} = n / S_{av}$ are the equivalent acoustic parameters of the hydrophone. Frequency response of the hydrophone in this range depends on the values of the resonance frequency of the compensator, f_c, and on the quality factor of the resonator formed by the cavity and the cylindrical hole. Qualitatively the frequency response of the hydrophone is shown in Figure 14.12 (c) below frequency f_L. The dashed lines correspond to different values of quality factor of resonator. The resonance frequency f_c is determined by relation

$$f_c = 1/2\pi\sqrt{(C_{\tilde{V}ac} + \tilde{C}^E_{eqvac})M_{ac}} = f_{\tilde{V}} / \sqrt{1 + \tilde{C}^E_{eqvac} / C_{\tilde{V}ac}}, \tag{14.33}$$

where $\tilde{C}^E_{eqvac} = C^E_{eqvac}(1 - k^2_{eff})$ is the acoustic compliance of the open circuited hydrophone, and $f_{\tilde{V}} = 1/2\pi\sqrt{C_{\tilde{V}ac} M_{ac\,eff}}$ is the natural frequency of the resonator under the condition that the mechanical system of the hydrophone is clamped. Generally, compensator is designed in such a way, as to have the frequency f_c below the lower frequency of operating range, f_L. Besides, the settling time for a compensator should be matched with the rate of changing the external pressure. If the frequency f_L is very low, these conditions may become contradictory. Regarding the transient process in a resonator see Section 12.3.2.1.

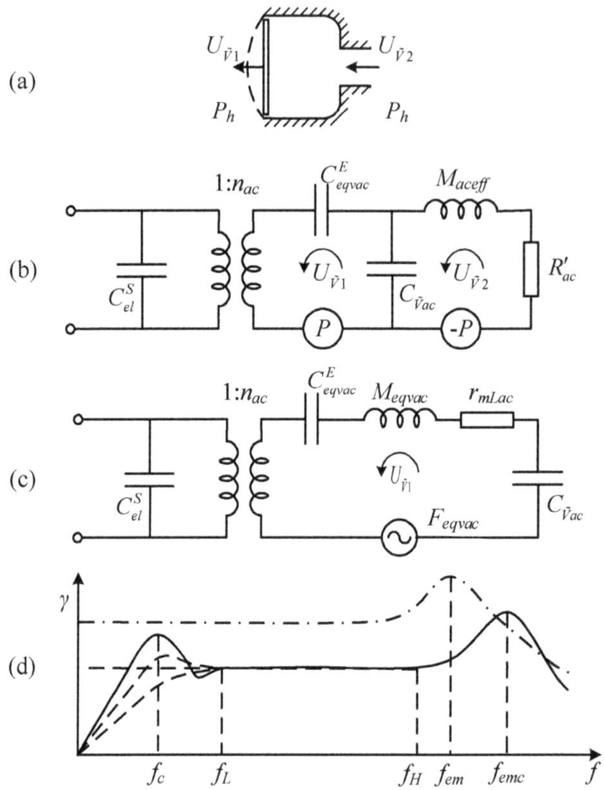

Figure 14.12: Equivalent circuit and the frequency response of a hydrophone with a compensator (solid and dashed lines) and without (dash-dotted line).

In the operating range of the hydrophone at $f_H > f > f_L$ the equivalent circuit can be presented like it is shown in Figure 14.12 (c). If to denote

$$\sqrt{1+\tilde{C}^{E}_{eqvac} / C_{\tilde{V}ac}} = \chi , \tag{14.34}$$

then the equivalent compliance of its mechanical system with the compensator will be

$$\tilde{C}^{E}_{eqvac\,c} = \tilde{C}^{E}_{eqvac} / \chi^2 , \tag{14.35}$$

and frequency of the electromechanical resonance of the sensor with compensator will be

$$f_{em\,c} = f_{em} \cdot \chi . \tag{14.36}$$

Following formula (14.23) for the reduced sensitivity $\gamma_{rd\,c}$, where C^E_{eqv} in the case with the compensator must be replaced by $\tilde{C}^E_{eqv\,ac\,c}$, we obtain

$$\gamma_{rd\,c} = \gamma_{rd} / \chi . \tag{14.37}$$

Thus, the sensitivity of the hydrophone with compensator decreases, while the resonance frequency increases in the factor of χ. This is shown in Figure 14.12 (d). The limitation imposed on the thickness of mechanical systems of the hydrophones by the hydrostatic pressure now does not exist, and we have to return to formula (14.26) for determining the product of sensitivity and bandwidth. As follows from relations (14.36) and (14.37),

$$\gamma_{rd\,c} \cdot f_{rd\,c} = \gamma_{rd} \cdot f_{rd} , \tag{14.38}$$

and formula (14.26) can be rewritten in the form

$$\gamma_{rd\,c} \cdot f_{emc} = Ff \cdot k_{eff} \sqrt{1 - k^2_{eff}} \cdot \frac{1}{\sqrt{\rho t}} . \tag{14.39}$$

The thickness now can be reduced seemingly to its value that is determined by the technological considerations, but its value may be limited since coefficient χ depends on the thickness if the liquid filled compensator is used. In this case the term $\tilde{C}^E_{eqv\,ac} / C_{\tilde{V}ac}$ in formula (14.34) for χ can be represented as

$$\frac{\tilde{C}^E_{eqv\,ac}}{C_{\tilde{V}ac}} = \frac{C^E_{eqv} \cdot S^2_{av}}{(\tilde{V} / B)} \cdot (1 - k^2_{eff}) = \chi' \tag{14.40}$$

(see formulas (12.37) for $C^E_{eqv\,ac}$, (12.5) for $C_{\tilde{V}ac}$, and relation $\rho c^2 = B$, where B is the bulk modulus of liquid used). With this notation introduced the expression (14.34) becomes

$$\chi = \sqrt{1 + \chi'} . \tag{14.41}$$

Values of coefficient χ' for the hydrophones of different type are presented in Table 14.1. and of effective coupling coefficients for the spherical, cylindrical and circular flexural plate hydrophones are presented in Table 14.1.

The calculations in the Table are produced for the hydrophones made of PZT-4 (Y^E_1 = 81 GPa, $\sigma^E_1 = 0.33$, $k_{31} = 0.33$, $k_p = 0.58$) with internal volume filled with Mineral oil ISO 32 (B = 1.8 GPa). (Note that the bulk moduli of the usable liquids have close values, as shown in Table 12.2.) In the variant of the flexural circular double plate hydrophone with separation Δ

Table 14.1 Coefficients χ' for hydrophones of different type.

	Sphere	Cylinder	Circular plate
χ'	$2.2\cdot10^{-2}(a/t)$	$4.4\cdot10^{-2}(a/t)$	$1.4\cdot10^{-2}(a^4/t^3\Delta)$
k_{eff}	k_p	k_{31}	$0.68k_p/\sqrt{1-0.54k_p^2}$

between the plates the total compliance of transducer must be used that equals to doubled compliance per one plate. The maximum practical radius to thickness ratio of the mechanical systems may be estimated as $a/t \approx 5$.

The data show that the low frequency sensitivity and the resonance frequency of the spherical and cylindrical hydrophones practically do not depend on the compensation. Parameters of hydrophones of the flexural type are especially vulnerable to liquid compensation. Even at maximum value of ratio $a/t \approx 5$ the coefficient χ becomes $\chi = \sqrt{1+a/\Delta}$ according to formula (14.41) and without an unreasonable increase of separation between the plates change of hydrophone parameters is hardly acceptable. For achieving a large sensitivity of the flexural type hydrophones intended for operation at great hydrostatic pressures the air compensation is the option though its employment encounter significant technical complications.

14.3.4 Pressure Gradient Hydrophones

14.3.4.1 Requirements for Sensitivity of the First Order Hydrophones

In determining required sensitivities for the hydrophones produced in Section 3.2.2 they were assumed to be omnidirectional or of the zero order. In the case of the first order hydrophones their directional (actually, super directional) property must be considered. The difference in the signal to noise situation, which arises in the super directional system, and the typical approach to its analysis, may be illustrated with example of the dipole hydrophone.

In considering the dipole type pressure gradient receiving systems made of two closely spaced zero order hydrophones connected in opposite phase two options exist that are illustrated in Figure 14.13.

In variant (a) the two hydrophones are connected prior to amplification. In variant (b) the preamplifiers are incorporated with the hydrophones and the receiving channels are connected in opposite phase after amplification.

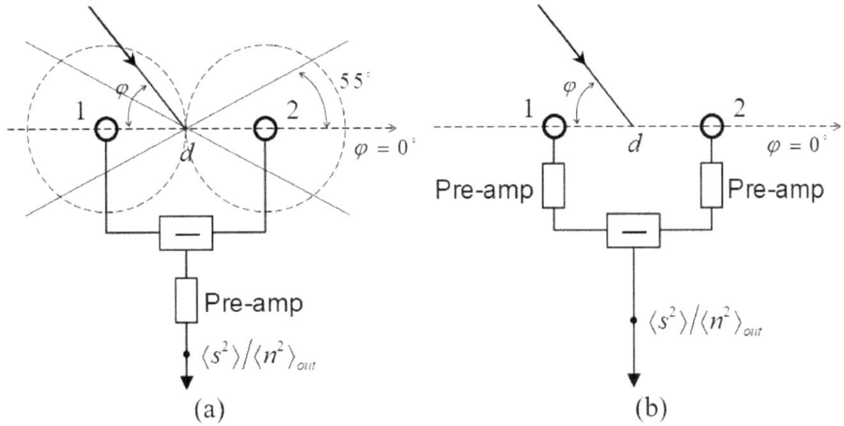

Figure 14.13: Illustration of two hydrophone systems: (a) with a single preamplifier, (b) with separate preamplifiers.

The average signal to noise power ratio at the output of the system in Figure 14.13 (a) can be represented as follows

$$\left.\frac{\langle s^2\rangle}{\langle n^2\rangle}\right|_{out} = \frac{\left\langle (p_{s1} - p_{s2})^2 \right\rangle \cdot \gamma^2}{\left\langle (p_{n1} - p_{n2})^2 \right\rangle \cdot \gamma^2 + 2\langle e_{sn}^2\rangle + \langle e_{amp}^2\rangle}. \tag{14.42}$$

Here p_{s1}, p_{s2} and p_{n1} p_{sn2} are the sound pressures of signal and noise accordingly in the points 1 and 2 of acoustic field, brackets $\langle\ \rangle$ indicate the mean square value per 1 Hz bandwidth. The total amplifier noise, $\langle e_{amp}^2\rangle$, is

$$\langle e_{amp}^2\rangle = \langle e_R^2\rangle + \langle e_{sc}^2\rangle, \tag{14.43}$$

where (see Section 3.2.3) e_{sc} represents the noise voltage of the amplifier with input short-circuited, and e_R is the voltage of thermal noise of equivalent input resistance of the amplifier.

It is considered that sources of internal (self) noise of the hydrophones, e_{sn}, are incoherent and their energies add independently.

The signal to noise power ratio in acoustic field related to the receiving dipole system input is

$$\left.\frac{\langle s^2\rangle}{\langle n^2\rangle}\right|_{in} = \frac{\left\langle (p_{s1} - p_{s2})^2 \right\rangle}{\left\langle (p_{n1} - p_{n2})^2 \right\rangle}. \tag{14.44}$$

In this relation,

$$\left\langle (p_{s1} - p_{s2})^2 \right\rangle = \left\langle p_{s1}^2 \right\rangle - 2\left\langle p_{s1} \cdot p_{s2} \right\rangle + \left\langle p_{s2}^2 \right\rangle, \tag{14.45}$$

$$\left\langle (p_{n1} - p_{n2})^2 \right\rangle = \left\langle p_{n1}^2 \right\rangle - 2\left\langle p_{n1} \cdot p_{n2} \right\rangle + \left\langle p_{n2}^2 \right\rangle. \tag{14.46}$$

When considering the right-hand part of the relation (14.44), the effects of the signal and noise coherence that depend on the statistical structure of the fields of signal and noise, must be taken into account for the particular operating conditions. We will assume that field of the acoustic noise is isotropic, and the signal is a sound pressure in the plane wave propagating in direction of the dipole acoustic axis $\varphi = 0$.

In the isotropic noise field

$$\left\langle p_{n1}^2 \right\rangle = \left\langle p_{n2}^2 \right\rangle = \left\langle p_n^2 \right\rangle, \tag{14.47}$$

$$\left\langle p_{n1} \cdot p_{n2} \right\rangle = \rho_n(d) \cdot \sqrt{\left\langle p_{n1}^2 \right\rangle \cdot \left\langle p_{n2}^2 \right\rangle} = \rho_n(d) \cdot \left\langle p_n^2 \right\rangle, \tag{14.48}$$

where $p_n(d)$ is the spatial correlation coefficient between the noise pressure in the points separated by length *d.* It is known (see, for example, Ref. 4) that in the isotropic noise field

$$\rho_n(d) = \frac{\sin kd}{kd}. \tag{14.49}$$

Thus, according to relations (14.47) - (14.49) the expression (14.46) becomes

$$\left\langle (p_{n1} - p_{n2})^2 \right\rangle = 2\left\langle p_n^2 \right\rangle \cdot (1 - \sin kd / kd). \tag{14.50}$$

In the plane wave of signal, we have

$$\left\langle p_{s1}^2 \right\rangle = \left\langle p_{s2}^2 \right\rangle = \left\langle p_s^2 \right\rangle, \tag{14.51}$$

$$\left\langle p_{s1} \cdot p_{s2} \right\rangle = \rho_s(\bar{k} \cdot \bar{d}) \cdot \sqrt{\left\langle p_{s1}^2 \right\rangle \cdot \left\langle p_{s2}^2 \right\rangle} = \rho_s(\bar{k} \cdot \bar{d}) \cdot \left\langle p_s^2 \right\rangle, \tag{14.52}$$

where $p_s(\bar{k} \cdot \bar{d})$ is the spatial correlation coefficient between the sound pressure in the points separated by segment $\bar{d}$, $\bar{k}$ is the wave vector directed along the direction of wave propagation, and $\bar{k} \cdot \bar{d} = kd\cos\varphi$ (see Figure 14.13). Thus, in case of the plane wave we have

$$\rho_s(\bar{k} \cdot \bar{d}) = \cos(kd\cos\varphi). \tag{14.53}$$

For the wave propagating in direction $\varphi = 0$ it simplifies to

$$\rho_s(\bar{k}\cdot\bar{d}) = \cos kd\,. \tag{14.54}$$

After substituting the corresponding terms from relations (14.51) - (14.53) into relation (14.45), we obtain

$$\left\langle (p_{s1}-p_{s2})^2 \right\rangle = 2\left\langle p_s^2 \right\rangle \cdot [1-\cos(kd\cos\varphi)]\,. \tag{14.55}$$

Now the relation (14.44) for the signal to noise ratio at the dipole system input becomes

$$\left.\frac{\langle s^2\rangle}{\langle n^2\rangle}\right|_{in} = \frac{\langle p_s^2\rangle}{\langle p_n^2\rangle}\frac{1-\cos(kd\cos\varphi)}{1-\sin kd/kd}\,. \tag{14.56}$$

The first factor in the right-hand side of the expression is the signal to noise power ratio measured by the omnidirectional hydrophone. For the dipole pressure gradient system condition $kd \ll 1$ holds. Considering that at this condition

$$\frac{\sin kd}{kd} \approx 1-\frac{(kd)^2}{6},\ \cos(kd\cos\varphi) \approx 1-\frac{(kd)^2\cos^2\varphi}{2},\ \frac{1-\cos(kd\cos\varphi)}{1-\sin kd/kd} \approx 3\cos^2\varphi\,, \tag{14.57}$$

the relation (14.56) becomes

$$\left.\frac{\langle s^2\rangle}{\langle n^2\rangle}\right|_{in} = 3\cos^2\varphi\cdot\frac{\langle p_s^2\rangle}{\langle p_n^2\rangle}\,. \tag{14.58}$$

In fact, this approximation is valid to a sufficient accuracy at $kd < 0.3$.

If to assume that the smallest signal to noise ratio, at which a signal still can be detected, is that at the input of omnidirectional hydrophone, then it follows from relation (14.58) that in the isotropic noise field the dipole hydrophone cannot detect signals beyond the sector of angles, in which $3\cos^2\varphi \ge 1$ that is $|\varphi| \le 55°$ (see Figure 14.13 (a)). Improving the dipole hydrophone or system design cannot change this fact. The gain in the input signal to noise power ratio of dipole in the sector $|\varphi| \le 55°$ by comparison with a single hydrophone is due to directivity of the dipole. In the isotropic noise field, the gain in the direction $\varphi = 0$ is equal to the directivity factor of the dipole in this direction, which is known to be 3. The directivity index is $10\log 3$ or 4.8 dB. The sector in which the minimum signal can be detected increases only in the case that the signal to noise ratio in the sound field increases as measured by an omnidirectional hydrophone.

Taking into account the relations (14.50), (14.54) and (14.55) for the direction $\varphi = 0$, the relation (14.42) may be transformed to

$$\left.\frac{\langle s^2\rangle}{\langle n^2\rangle}\right|_{out} = 3\frac{\langle p_s^2\rangle}{\langle p_n^2\rangle}\frac{1}{1+3\dfrac{2\langle e_{sn}^2\rangle+\langle e_{amp}^2\rangle}{\langle p_n^2\rangle\gamma^2(kd)^2}}. \tag{14.59}$$

In order to realize the gain due to the directivity, the same decrease of the signal to noise ratio as in the case of a single omnidirectional hydrophone must be allowed. And thus, by analogy with Equations (3.196) and (3.197) the condition

$$3\frac{2\langle e_{sn}^2\rangle+\langle e_{amp}^2\rangle}{\langle p_n^2\rangle\gamma^2(kd)^2} \le \beta \tag{14.60}$$

should be fulfilled. Thus, the minimum required sensitivity, $\gamma_{\min}$, will be found as

$$\gamma_{\min}^2 = \frac{6}{(kd)^2}\frac{\langle e_{sn}^2\rangle+\langle e_{amp}^2\rangle/2}{\beta\langle p_n^2\rangle}. \tag{14.61}$$

In the variant that the preamplifiers are integrated with hydrophones, as shown in Figure 14.13 (b), the evaluation procedure for the output signal to noise ratio is almost the same as in the preceding case. The only difference is that in the starting relation (14.42) the amplifier noise power $\langle e_{amp}^2\rangle$ has to be doubled, because the noise of the amplifiers is incoherent. Thus, the signal to noise ratio and minimum sensitivity required will be determined as follows

$$\left.\frac{\langle s^2\rangle}{\langle n^2\rangle}\right|_{out} = 3\frac{\langle p_s^2\rangle}{\langle p_n^2\rangle}\frac{1}{1+6\dfrac{\langle e_{sn}^2\rangle+\langle e_{amp}^2\rangle}{\langle p_n^2\rangle\gamma^2(kd)^2}}, \tag{14.62}$$

$$\gamma_{\min}^2 = \frac{6}{(kd)^2}\frac{\langle e_{sn}^2\rangle+\langle e_{amp}^2\rangle}{\beta\langle p_n^2\rangle}. \tag{14.63}$$

Compare the minimum required sensitivities of a hydrophone integrated with preamplifier in the case that it is intended to be used as a single omnidirectional unit (denote this sensitivity as γ_0) and in the case that it is intended to be used in the dipole system (denote this sensitivity as γ_1). Considering formulas (3.197) and (14.63) we obtain

$$\gamma_1 \geq \frac{\sqrt{6}}{kd}\gamma_0 . \tag{14.64}$$

If f_L and f_H are the lowest and highest frequencies of the operating frequency range, then condition $(kd)_H \leq 0.3$ must be fulfilled at the highest frequency of the range, and at the lowest frequency it will be $(kd)_L = 0.3 f_L / f_H$. Given that required sensitivity of the zero order hydrophone should be determined at the lowest frequency and remains linear in the operating range $(\gamma_0 = const)$, the maximum sensitivity γ_1 of the hydrophone intended for the dipole system should be

$$\gamma_1 > 8\frac{f_H}{f_L}\gamma_0 . \tag{14.65}$$

Strictly speaking, this sensitivity should be determined under the angle $\varphi = 55°$, if the dipole system is used for detecting signals at small signal to noise ratios in the broadest possible sector of forward view.

Comparison between the required sensitivities γ_1 in the variants of a single amplifier and two individual amplifiers shown in Figure 14.13 (a) and (b), respectively, leads to relation

$$\frac{(\gamma_1)_b}{(\gamma_1)_a} = \sqrt{\frac{1+\langle e_{amp}^2\rangle / \langle e_{sn}^2\rangle}{1+\langle e_{amp}^2\rangle / 2\langle e_{sn}^2\rangle}} . \tag{14.66}$$

Thus, using the combined hydrophone-amplifier units requires some increase in hydrophone sensitivity compared with the case of the direct connection of the hydrophones and a single amplifier. But the variant (b) is preferable anyway because it allows optimal matching of the hydrophones with preamplifiers without loss of sensitivity on the connecting cables. This is especially advantages in view of the harsh requirements formulated by relation (14.65).

The above analysis shows that requirements for sensitivity of the zero order hydrophones comprising the dipole pressure gradient system intended to be used in arrays for detecting signals at small signal to noise ratios are extremely demanding especially at low frequencies and at significant depths. The dipoles can be successfully used for operating with strong signals, in particular, for measuring properties of acoustic fields, for short distance navigation and so on.

Remember that the above results are obtained under the assumption that the field of the acoustic noise is isotropic. Considerations regarding using dipoles and vector hydrophones in receiving arrays under assumption of different noise models can be found in Ref. 4.

14.3.4.2 Pressure Gradient Hydrophones of the Fixed Diffraction Type

The pressure gradient hydrophone of the diffraction type can be realized as the bimorph flexural disk with simply supported edge, or as rectangular plate with simply supported opposite edges, as shown in Figure 14.14. The edge-supporting boundary conditions can be provided by foundations in the form of a metal ring, or of two bars in the case of the rectangular plate. It is assumed that the foundations are heavy enough to consider that they do not move under the action of acoustic wave. It may seem that this assumption is not very real. At least it must be estimated, what "heavy enough" is. This will be done in Section 14.3.5.1, where it will be shown that this assumption is not very demanding.

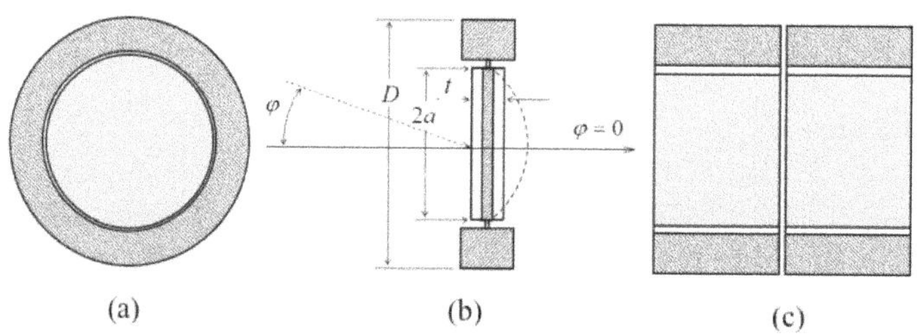

Figure 14.14: Examples of the diffraction type hydrophones: (a) edge-supported disk, (b) side view of the edge-supported disk (rectangular plates) and (c) the edge-supported rectangular plates.

The same considerations on the sensitivity calculation and optimizing as in the case that the plates comprise the sound pressure hydrophone are relevant, except that the corresponding diffraction coefficient must be used in the expression for the equivalent force and form factor in formula (14.25).

Important inherent advantage of the diffraction type hydrophones is that the plates experience action of all around static pressure and the mechanical stresses in piezoelectric material do not exceed the hydrostatic pressure itself, that is the strength of ceramics is no longer a limiting factor. The diffraction type hydrophones can be used to full ocean depth practically without change of their properties. Because of this, expression (14.26) for estimating product $\gamma_{rd} \cdot f_{em}$ of the reduced sensitivity and bandwidth of the hydrophone is valid without restrictions on value of the thickness that can be imposed by mechanical strength or by changing ceramic parameters at operating depth. Thus, the optimized fully active bimorph or trilaminar plates shown in Figure 14.10 may be used. In the last variant the substrate can be made of a dielectric plastic material, which makes it easier to manipulate the electrode segmentation to change parallel and series connections for increasing sensitivity and amplifier matching. In order to increase the specific sensitivity, the resonance frequency of a single transducer unit can be shifted as close to an operating frequency range as is acceptable with respect to specifications of the frequency response. However, operating close to the resonance frequency can result in additional noise contributions and therefore a reduction in the dynamic range. For diminishing this effect, damping the resonance to an acceptable level without significant loss of the sensitivity in the operating frequency range can be used (although an added internal noise due to damping also must be considered).

Another consequence of independence of the hydrostatic pressure is that the rectangular plates can be used as well as circular disks for realizing the diffraction type hydrophone. Whereas it is hard to employ the rectangular plates for pressure hydrophone designs because of the necessity of sealing their free edges. In the case of the diffraction type operation, a single plate is used, and there is no internal volume to be sealed. The same considerations on optimizing the rectangular plates as for the circular disks are valid. The optimum length of the active part of a rectangular plate is $l_{el} = 0.75l$. It is of note that at $l_{el} = 0.5l$ the specific sensitivity remains the same as in the case that the active plate is of the full size. The comparison of the specific sensitivities of hydrophones based on the circular and rectangular plates with transverse polarization under the conditions of equal surfaces and resonance frequencies shows a slight advantage for the rectangular plates. It is of note, that it is possible and advantageous to use the tangential polarization of rectangular plates, in which case the coupling coefficient is proportional to k_{33eff} (see Section 7.2.2.1.14) instead of k_{31}. Thus, the additional increase of the hydrophone sensitivity can be achieved. Approximately the same estimation can be used for the diffraction coefficient of the rectangular plates (especially, when their assembly has the square shape), as for the circular disk with similar dimensions.

The equivalent force acting on the circular disk is

$$F_{eqv} = -j\frac{4}{3\pi}k\frac{D}{2}\cos\varphi \cdot 2S_{pl}P_o, \tag{14.67}$$

where D is diameter of the foundation and $S_{pl} = \pi a^2$. (Remember that the formula is accurate at (kD/2) <0.5, as noted regarding (14.14)) The expression for sensitivity γ of the pressure gradient hydrophone at frequencies significantly below the resonance is

$$\gamma(\varphi) = -j\frac{4}{3\pi}kD\cos\varphi \cdot \frac{nS_{pl}C^E_{eqv}}{C_{Lf}}, \qquad (14.68)$$

instead of the expression by formula (14.16) at $k_{dif} = 1$ for sensitivity of the double circular plate sound pressure hydrophone, γ_p. Thus, the sensitivity of the pressure gradient hydrophone in the direction $\varphi = 0$ may be represented as

$$\gamma_{pg}(0) = -j\frac{4}{3\pi}kD \cdot \frac{nS_{pl}C^E_{eqv}}{C_{Lf}} = -j\frac{4}{3\pi}kD \cdot \gamma_p . \qquad (14.69)$$

Therefore, all the expressions for the specific sensitivity of a double flexural disk hydrophone must be multiplied by the diffraction constant $k_{dif} = -j4kD/3\pi$ to obtain sensitivity of the pressure gradient hydrophone of diffraction type that employs the same disk.

It must be noted that in calculating the low frequency sensitivities effect of radiation impedance of the oscillating disk is neglected for simplicity, as it results in some shift of resonance frequency only.

14.3.4.3 Pressure gradient Hydrophones of the Motion Type

Although motion type hydrophones may employ accelerometers of different kind (some of their configurations are shown in Figure 14.7(d), in this section we will assume for illustration that this is a uniform flexural disk mounted inside of a spherical shell, as shown schematically in Figure 14.15.

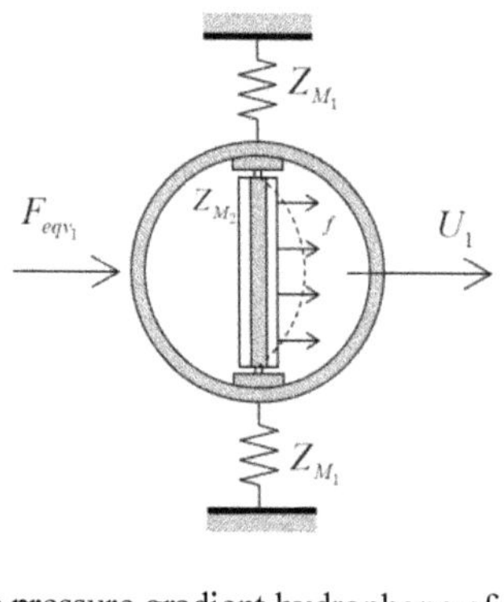

Figure 14.15: Illustration of the pressure gradient hydrophone of the motion type.

The alternative variants of accelerometers that can be used for this purpose will be considered in Section 14.3.5. In Figure 14.15 Z_{M1} is impedance of a mounting structure that supports the shell, Z_{M2} is impedance of the flexural disk mounted to the case. We assume that the resonances of the mechanical system of the hydrophone itself are above the operating frequency range. The motion of the system can be represented as superposition of motion as a rigid body with velocity U_1 under action of the acoustic wave with equivalent force F_{eqv1} and vibration of the mechanical system of the accelerometer under the action of inertial forces. Thus, the velocity U_1 is

$$U_1 = \frac{F_{eqv1}}{Z_{M1} - j\omega M_{\Sigma} + Z_{ac}}, \tag{14.70}$$

where M_{Σ} is the total mass of the system (it includes the mass of piezoelement) and Z_{ac} is the radiation impedance. For the oscillating sphere the acoustic mass M_{as} is

$$M_{ac} = \frac{2}{3}\pi\rho_w a^3, \tag{14.71}$$

where ρ_w is the density of water. This equates to one half the mass of a water filled sphere of the same volume. Note that the impedance of mounting can depend on the direction of motion, i.e., in general $Z_{M1} = Z_{M1}(\varphi)$. In this case the modulus of velocity U_1 will be angular dependent and the directional factor will be distorted from the ideal figure of eight response in spite of the spherical symmetry of the body. For simplicity we will assume that the mounting impedance is the reactance $Z_{M1} = j/\omega C_{M1}$. Thus,

$$U_1 = \frac{F_{eqv1}}{\dfrac{j}{\omega C_{M1}} - j\omega(M_{\Sigma} + M_{ac})}. \tag{14.72}$$

The velocity U_1 as a function of frequency is qualitatively shown in Figure 14.16.

In the frequency range well above the resonance frequency f_M,

$$f_M = 1/2\pi\sqrt{C_{M1}(M_{\Sigma} + M_{ac})}, \tag{14.73}$$

the system is mass controlled and velocity U_1 is frequency independent. At these frequencies impedance $Z_{M1} = j/\omega C_{M1}$ may be neglected in relation (14.72) and we arrive at expression

$$U_1 = \frac{F_{eqv1}}{-j\omega(M_\Sigma + M_{ac})}. \tag{14.74}$$

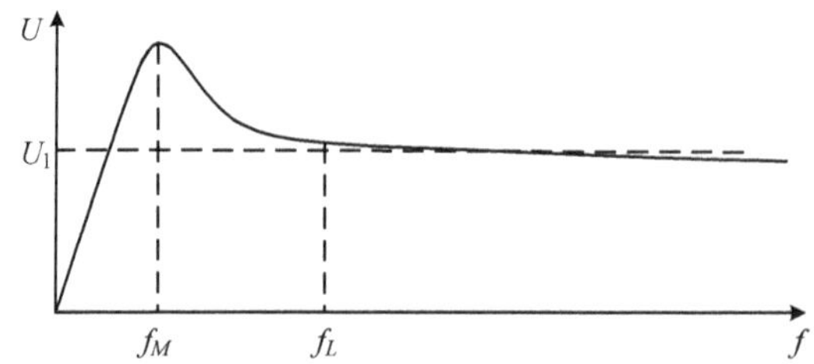

Figure 14.16: Typical velocity frequency response of the compliant mounted hydrophone body. Above the frequency f_L velocity of vibration of the body in direction of wave propagation is frequency independent.

For the spherical shell of small wave size (at $ka < 0.3$) the equivalent force at $\varphi = 0$ is

$$F_{eqv1} = -j\frac{ka}{2}\cdot 4\pi a^2 P_o, \tag{14.75}$$

and the acoustic mass M_{ac} is given by formula (14.71). Upon substituting F_{eqv1} and M_{ac} into expression (14.74) we arrive at

$$U_1 = \frac{2\pi a^3 P_o}{c_w(M_\Sigma + \frac{2}{3}\pi a^3 \rho_w)}. \tag{14.76}$$

Here and further the value of velocity of vibration in direction $\varphi = 0$ will be considered only. If the hydrophone has neutral buoyancy, i.e.,

$$M_\Sigma = \frac{4}{3}\pi a^3 \rho_w, \tag{14.77}$$

the velocity is

$$U_1 = \frac{P_o}{(\rho c)_w}. \tag{14.78}$$

Thus, in this case the shell vibrates as the volume of water in the shape of the shell vibrates in the plane wave.

The distributed inertia forces acting over the mechanical system of an accelerometer mounted inside the shell are

$$f(\boldsymbol{r}_\Sigma) = \rho(\boldsymbol{r}_\Sigma)\cdot \dot{U}_1, \tag{14.79}$$

where $\rho(\boldsymbol{r}_\Sigma)$ is the density distribution in the mechanical system. If the mechanical system has one degree of freedom characterized by the mode of vibration $\theta(\boldsymbol{r}_\Sigma)$, then the equivalent force acting on its mechanical input is

$$F_{eqv} = \dot{U}_1 \int_{\tilde{V}} \rho(\boldsymbol{r}_\Sigma)\theta(\boldsymbol{r}_\Sigma) d\tilde{V}, \tag{14.80}$$

where $\tilde{V}$ is the volume of the mechanical system. In our case of a uniform bimorph plate the density $\rho(\boldsymbol{r}_\Sigma) = \rho$ and $d\tilde{V} = t \cdot d\Sigma$. Thus, the equivalent force is

$$F_{eqv} = S_{av}\rho t \cdot \dot{U}_1. \tag{14.81}$$

Using this expression for F_{eqv}, we obtain the open circuit output voltage of the motion type hydrophone at frequencies much below its resonance as

$$E_{oc} = \frac{nC^E_{eqv}S_{av}}{C_{Lf}} \cdot \rho t \dot{U}_1. \tag{14.82}$$

In the general case expressions (14.72) or (14.74) for U_1 have to be used for calculating the output voltage. For the motion type hydrophone with the spherical shell the sensitivity in terms of sound pressure in harmonic plane wave will be obtained using expression (14.76) for U_1 in the form,

$$\begin{aligned} \gamma_{pg} &= \frac{E_{oc}}{P_o} = -j\omega \cdot \frac{2\pi a^3}{c_w\left(M_\Sigma + \frac{2}{3}\pi a^3 \rho_w\right)} \frac{nC^E_{eqv}S_{av}}{C_{Lf}} \cdot \rho t = \\ &= -j\omega \cdot \frac{2\pi a^3}{c_w\left(M_\Sigma + \frac{2}{3}\pi a^3 \rho_w\right)} \gamma_p \cdot \rho t \end{aligned}, \tag{14.83}$$

where

$$\gamma_p = \frac{nC^E_{eqv}S_{av}}{C_{Lf}} \tag{14.84}$$

is the open circuit sensitivity per one plate of the double plate sound pressure hydrophone made of the same bimorph plates (see (14.16) at $k_{dif} = 1$).

In the variant of neutrally buoyant, the sensitivity of the pressure gradient motion hydrophone may be expressed as

$$\gamma_{pg} = -jk \cdot \frac{\rho}{\rho_w} \cdot \frac{nC^E_{eqv} S_{av}}{C_{Lf}} \cdot t. \tag{14.85}$$

Alternatively, by using expression (14.82) we obtain that the sensitivity of the hydrophone to acceleration is

$$\gamma_{\dot{U}} = \frac{E_{oc}}{\dot{U}_1} = \frac{nC^E_{eqv} S_{av}}{C_{Lf}} \cdot \rho t = \gamma_p \cdot \rho t, \tag{14.86}$$

Obviously, the same relation exists between the specific sensitivities by $\gamma_{sp\dot{U}} = \gamma_{sp\,p} \cdot \rho t$. If to use formula (14.17) for $\gamma_{sp\,p}$ remembering that , we obtain expression for the specific sensitivity of a motion hydrophone to acceleration as

$$\gamma_{sp\dot{U}} = k_{eff} \frac{S_{av}}{S_{eff}} \cdot \frac{\sqrt{M_{eqv}}}{\omega_r}, \tag{14.87}$$

where ω_r is the resonance frequency of the flexural disk. Formula (14.87) shows that the specific sensitivity of accelerometer can be increased by increasing the equivalent mass of its mechanical system and lowering the resona $M_{eqv} = \rho t S_{eff}$ nce frequency. Thus, one of the possible effective ways to increase sensitivity of accelerometer is to attach an additional mass M_{add} to its mechanical system as is shown in Figure 14.17. If M_{add} is attached to the reference point on the

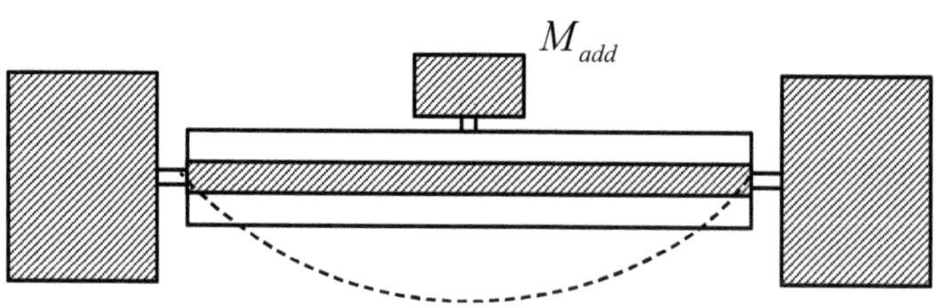

Figure 14.17: Illustration of a mass loaded accelerometer.

plate surface, the equivalent mass of the plate becomes $M'_{eqv} = M_{eqv} + M_{add}$, and the specific sensitivity $\gamma'_{sp\dot{U}}$ and resonance frequency ω'_r become

$$\gamma'_{sp\dot{U}} = \gamma_{sp\dot{U}} \cdot \sqrt{1 + M_{add} / M_{eqv}}, \tag{14.88}$$

$$\omega_r' = \omega_r / \sqrt{1 + M_{add} / M_{eqv}}\,. \tag{14.89}$$

(Related explanations see in the next section). The downside of this way of increasing sensitivity of the accelerometer is that adding mass to the mechanical system of piezoelement makes it more vulnerable to the action of shocks.

The depth limitation of a hydrophone of the motion type can be imposed due to a limited strength of the enclosing accelerometer case under the action of hydrostatic pressure. Increasing the strength usually tends to increase of the mass of the case and thus to decreasing the sensitivity γ_{pg}.

Correlation between sensitivities of the motion type pressure gradient hydrophone in terms of the sound pressure in plane wave, γ_{pg}, and as an accelerometer, $\gamma_{\dot{U}}$, is based on the relation (14.1), or $\dot{U} = -j\omega P_o / (\rho c)_w$. Therefore, the following correlation between the sensitivities may be used

$$\gamma_{pg}[\mu\text{V/Pa}] = 4.2 \cdot 10^{-5} f_{\text{kHz}} \cdot \gamma_{\dot{U}}[\text{mV/g}]\,, \tag{14.90}$$

where f_{kHz} is the frequency measured in kHz. Thus, for example, if the sensitivity of motion type hydrophone as an accelerometer is $\gamma_{\dot{U}}$, = 400 mV/g, then the corresponding sensitivity as the pressure gradient hydrophone in terms of the sound pressure is $\gamma_{pg} = 170 \cdot f_{\text{kHz}}\ \mu\text{V/Pa}$.

14.3.4.4 Comparison of Sensitivities of the Motion and Fixed Type Hydrophones

It is informative to compare sensitivities of the pressure gradient hydrophones of the motion (Figure 14.15) and fixed diffraction (Figure 14.1(b)) types that employ the same simply supported circular bimorph disk under the assumption that the spherical shell and foundation of the fixed type hydrophone have the same diameter.

In the most favorable case that the motion type hydrophone is neutrally buoyant its sensitivity, $\gamma_{pg\,m}$, is given by formula (14.85) and the sensitivity of the fixed type hydrophone, $\gamma_{pg\,f}$, is given by formula (14.69). Thus, the ratio of the sensitivities is

$$\frac{\gamma_{pg\,f}}{\gamma_{pg\,m}} = \frac{4}{3\pi} \cdot \frac{\rho_w}{\rho} \cdot \frac{S_{pl}}{S_{av}} \psi \frac{2a}{t} \approx 0.4 \frac{a}{t}\,. \tag{14.91}$$

For the quantitative estimation in relation (14.91) coefficient $\psi = D / 2a$ is introduced that characterizes size of the plate supporting elements. It is assumed that a reasonable estimate for its value is $\psi \approx 1.6$. Besides, the ratio of the densities of piezoelectric ceramics and water, $\rho / \rho_w \approx 7.5$, and $S_{av} = 0.45 S_{pl}$ are used. Practical low frequency transducer designs usually have the radius to thickness ratio $a/t > 5$. For example, a simply supported circular plate with thickness $t = 3$ mm and diameter $2a = 30$ mm made of PZT-4 has resonance frequency $f_r \approx 10$ kHz. The lower the operating frequency range, the larger the ratio a/t must be. Thus, the pressure gradient hydrophones of the diffraction type may have some advantage in terms of sensitivity over the hydrophones of the motion type having the same cross-section area. Their frequency responses may be similar, if avoiding complications due to mounting conditions of the motion type hydrophone. Though, the resonance frequency of diffraction type hydrophone may be lower because the flexural disk is exposed to acoustic medium and experiences reactive loading. Also, being potted in polyurethane the hydrophone will have lower Q due to a higher damping.

The conclusion regarding comparison of the hydrophone types is not that straightforward for practical transducer designs. For example, the assumption of the neutral buoyancy of the motion type hydrophone may contradict requirements for its operating depth, which may result in reducing the hydrophone sensitivity. On the other hand, the value of coefficient ψ may be overestimated. The mass of supporting element itself can be smaller for the motion type hydrophone, because the mass of the shell adds up to it.

14.3.5 Accelerometers as Parts of the Motion Type Pressure Gradient Hydrophones

Although in the context of this Chapter accelerometers are considered as a part of the motion type pressure gradient hydrophones, the analysis of their properties is applicable to the accelerometers as separate devices for general applications. Several variants of accelerometer designs are presented in Figure 14.8 (c).

As acceleration is a vector, it may be necessary to measure all its three components. In this case the accelerometer that measures any of them must be insensitive to the others, for which purpose special measures must be taken in the accelerometer design. Some of the measures will be discussed in Section 14.4. Here we will assume that accelerometer is intended for measuring the component of acceleration that is normal to the accelerometers' base plane, and it should not react to the components that are acting in the plane. Under this condition the accelerometer has figure of eight directivity relative to the acceleration vector with sensitivity $\gamma_{\dot{U}}$ in the direction of its maximum.

In order not to distort a mode of vibration of the object (source of vibration) accelerometer is installed on, the condition should be fulfilled $|Z_m| \ll |Z_{m.in}|$, where Z_m is the input impedance of the accelerometer at its area of contact, and $Z_{m.in}$ is the internal mechanical impedance of the object. Under this condition the source of vibration can be considered as the "current generator" with respect to the accelerometer, and result of measuring acceleration does not depend on the accelerometer properties.

Low frequency accelerometer designs mainly employ the flexural circular plate or rectangular beam (plates composed out of the beams) piezoelements (Figure 14.8 (c.1), (c.2)). For frequencies that are below about 10 kHz they allow achieving an acceptable sensitivity at sufficiently small mass. For higher frequencies the length expander piezoelements with additional masses are almost exclusively used for accelerometer designs (Figure 14.8 (c.3), (c.4)). Consider the typical versions of these two design types.

14.3.5.1 Accelerometers of the Flexural Type

Example of calculating sensitivity of the accelerometer employing the rectangular bimorph plate or beam is considered in Section 2.6.2. In the more general case, a concentrated mass can be applied to the plate for increasing sensitivity of the accelerometer, as it is shown in Figure 14.8 (c.1) and 14.17. Besides, the input mechanical impedance of the accelerometer must be calculated. We will perform a general analysis of this variant of design. Moreover, that the procedure and results of the analysis will be valid for all the designs involving flexural beams and plates after straightforward substituting appropriate equivalent parameters of piezoelements of different configuration in the finite formulas.

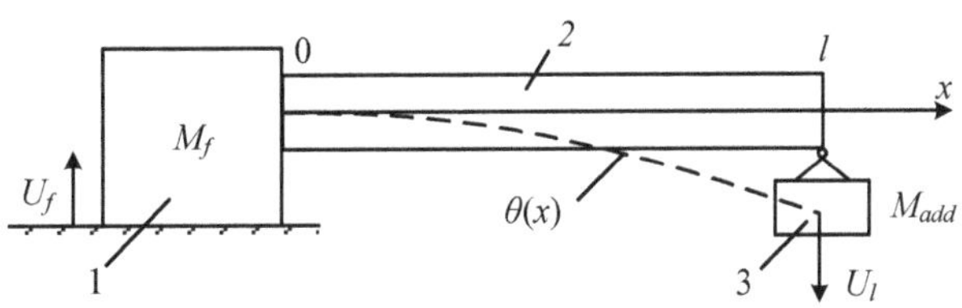

Figure 14.18: Cantilever accelerometer of a general type: 1- foundation having mass M_f, 2- piezoceramic beam, 3- additional mass M_{ad}.

Consider accelerometer made as the cantilever flexural bimorph beam that is clamped on one end to foundation having mass M_f and loaded on the other end by an additional lamped mass M_{ad} (Figure 14.18).

The foundation makes the mechanical input of the accelerometer. In the frequency range up to the first resonance, the motion of the mechanical system of the accelerometer may be represented as superposition of its vibration as a whole with velocity U_f and vibration of the beam relative to the foundation. Namely,

$$U(x) = U_f + U_l\theta(x) . \tag{14.92}$$

Here $\theta(x)$ is the mode of the static deformation of the beam. For the case that one end of the beam is clamped, while the other end is under the action of a concentrated force $\theta(x) = (x^2 / 2l^2)(3 - x/l)$ (see (4.494)).

The equivalent parameters of the cantilever beam in its vibration relative to the foundation are (see (2.131)):

$$M_{eqv} = 0.22M, \quad K_{eqv}^E = 0.27\frac{wt^3}{l^3 s_{11}^E}, \quad S_{av} = 0.38S_{pl},$$
$$n^* = 0.33\frac{wd_{31}t}{ls_{11}^E}, \quad C_e^{S_1*} = 2\varepsilon_{33}^T(1-k_{31}^2)wl/t. \tag{14.93}$$

Here n^* and $C_e^{S_1*}$ are given for a half-beam, $S_{pl} = wl$. Note that the equivalent parameters, which are determined using the mode of the static deformation under action of force on the end and under the action of a distributed load (2.129), differ insignificantly. The latter case corresponds to a greater extent to the design version with $M_{ad} = 0$.

To obtain equations of the accelerometer motion in the generalized velocities U_f and U_1 (in the generalized displacements ξ_f and ξ_l), expressions for the kinetic and potential energies of the system must be considered.

$$W_{kin} = M_f U_f^2/2 + M_{ad}(U_f + U_l)^2/2 + \frac{\rho wt}{2}\int_0^l [U_f + U_l\,\theta(x)]^2 dx =$$
$$= (M_f + M_{ad} + M_{pl})U_f^2/2 + (\rho t S_{av} + M_{ad})U_f U_l + (M_{eqv} + M_{ad})U_l^2/2 = \tag{14.94}$$
$$= M_1 U_f^2/2 + M_{12} U_f U_l + M_2 U_l^2/2.$$

In this expression $M_{pl} = \rho t S_{pl}$ is the mass of the beam (in general, mass of plate made of beams), $\rho t S_{av} = (S_{av} / S_{pl}) M_{pl}$, and the following notations are introduced for brevity:

$$M_1 = (M_f + M_{pl}) + M_{ad}, \; M_2 = M_{eqv} + M_{ad}, \; M_{12} = (S_{av} / S_{pl}) M_{pl} + M_{ad}. \qquad (14.95)$$

The potential energy W^E_{pot} possesses only the deformed beam, and

$$W^E_{pot} = K^E_{eqv} \xi_l^2 / 2. \qquad (14.96)$$

The force that is applied to the foundation at the area of contact with an object of measurement and generates velocity of vibration, U_f, will be denoted as F_f.

Finally, Euler's equations of general type (5. 222) for the case under consideration will be obtained in the following form that is typical for a coupled system,

$$j\omega M_1 U_f + j\omega M_{12} U_l = F_f, \qquad (14.97)$$

$$j\omega M_{12} U_f + j\omega M_2 (1 - \omega_{em}^2 / \omega^2) U_l = 0, \qquad (14.98)$$

where

$$\omega_{em} = \sqrt{\frac{K^E_{eqv} + n^2 / C_e^{S_1}}{M_2}} = \omega_{em0} \frac{1}{\sqrt{1 + M_{ad} / M_{eqv}}}. \qquad (14.99)$$

Here ω_{em0} is the frequency of electromechanical resonance of the beam at $M_{ad} = 0$.

The equivalent circuit that corresponds to equations (14.97) and (14.98) along with the equation of type (5.229) for the electrical coordinate V_{oc} is presented in Figure 14.19 (a).

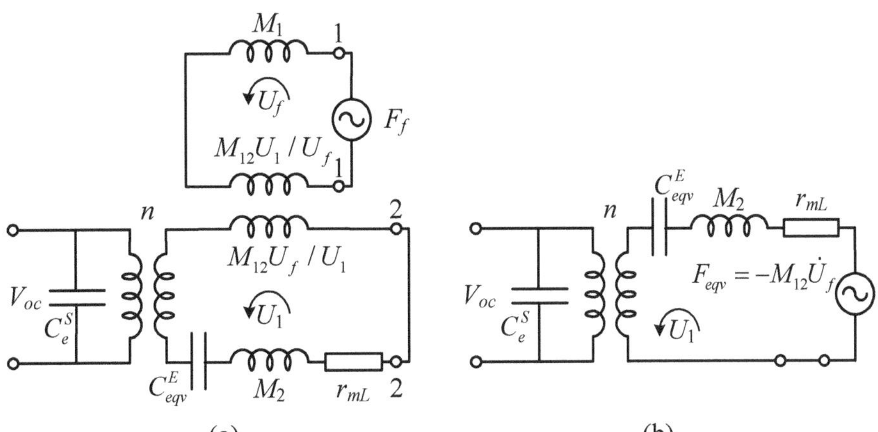

Figure 14.19: (a) Equivalent circuit of the accelerometer as a coupled system, (b) single contour representation of the equivalent circuit.

The equation for the electrical coordinate (output of the transducer) in this case has the form

$$V_{oc} = nU_l / j\omega C_e^{S_1}. \tag{14.100}$$

Values of the electromechanical transformation coefficient and capacitance in the expressions (14.99) and (14.100) depend on the electrical connection of the halves of the beam. For their parallel connection $n = 2n^*$, $C_e^{S_1} = 2C_e^{S_1 *}$; for the series connection $n = n^*$, $C_e^{S_1} = C_e^{S_1 *}/2$.

The input mechanical impedance of the accelerometer, Z_m, is the impedance between the points 1, 1 in the equivalent circuit. Upon determining from equation (14.98) velocity U_l as

$$U_l = -M_{12}U_f / M_2(1 - \omega_{em}^2/\omega^2), \tag{14.101}$$

and after substituting this value into equation (14.97), we arrive at

$$Z_m = F_f / U_f = j\omega M_1[1 - M_{12}^2 / M_1 M_2 (1 - \omega_{em}^2/\omega^2)]. \tag{14.102}$$

The mechanical system of the accelerometer can be considered as a coupled system with two degrees of freedom (see Section 4.6), for which the partial systems are the foundation with the beam vibrating as a whole and the beam with mass M_{add} at the end that vibrates relative to the foundation. In this case in accordance with formula (4.528), $M_{12}/\sqrt{M_1 M_2}$ is the coefficient γ_1 of the inertial coupling between the partial systems. Thus,

$$Z_m = j\omega M_1[1 - \gamma_1^2 / (1 - \omega_{em}^2/\omega^2)]. \tag{14.103}$$

The frequency dependence of the input impedance is qualitatively the same as for the input impedance on the contour of a circular disc (see Figure 4.55). It is shown in Figure 14.20.

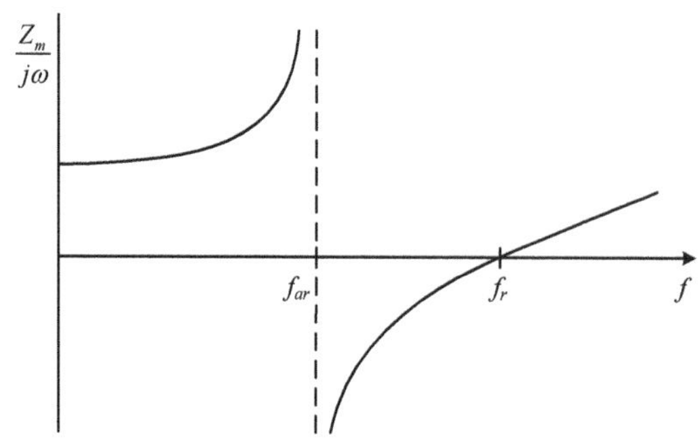

Figure 14.20: Input impedance of the accelerometer.

At frequency $f_{ar} = f_{em}$ the input impedance $Z_{in} \to \infty$. At frequency $f_r = f_{em} / \sqrt{1-\gamma_1^2}$ the input impedance is $Z_{in} = 0$.

The equivalent circuit of the accelerometer for calculating its sensitivity can be represented as shown in Figure 14.19 (b). The equivalent force in this circuit is due to inertia coupling with the first partial system, $F_{eqv} = -M_{12}\dot{U}_f$. The sensitivity of the accelerometer at frequencies below the frequency of electromechanical resonance will be found as

$$\gamma_{\dot{U}} = \frac{V_{oc}}{\dot{U}_f} = \frac{nU_l}{j\omega C_e^{S_1}\dot{U}_f} = -\frac{nM_{12}}{4\pi^2 f_{em}^2 M_2 C_e^{S_1}} = -\gamma_{\dot{U}0}(1+M_{ad} / \rho t S_{av}). \tag{14.104}$$

The sign minus indicates that the measured acceleration and voltage at accelerometer output are in antiphase. As can be seen, in this frequency range the sensitivity in terms of acceleration does not depend on frequency. Namely by this reason the sensor of this kind is called an accelerometer, although, by virtue of the relations $\dot{U}_f = U_f / j\omega = -\xi_f / \omega^2$, the result of the measurements could be attributed to velocity of vibration and to the displacement. The inconvenience due to the frequency dependence of the values $\gamma_U = j\omega\gamma_{\dot{U}}$, $\gamma_\xi = -\omega^2\gamma_{\dot{U}}$ can be avoided by performing operations of integration in a receiving channel. In terms of measuring the displacements, the major drawback of this design is that its sensitivity sharply decreases with frequency. Therefore, for measuring the slowly changing displacements a different sensor design is usually used that will be discussed below.

The equivalent circuits in Figure 14.19 and formula (14.104) for the sensitivity are valid for all the accelerometer designs involving beams and plates that experience flexural deformation, if to use appropriate equivalent parameters of the piezoelements in the above expressions. So far as the additional mass M_{ad} does not change noticeably the mode of vibration being applied at the reference point on the surface of a piezoelement (which is usually the case), the values of the equivalent parameters can be used that are determined without the mass. In this case M_{ad} can be just added to the equivalent mass of the piezoelement. Thus, for the simply supported on a foundation circular plate and rectangular beam the equivalent parameters are:

for circular plates

$$M_{eqv} = 0.3M_{pl}, \quad K_{eqv}^E = 2t^3 / a^2 s_{11}^E(1 - s_{12}^{E2} / s_{11}^{E2}), \quad S_{av} = 0.45\pi a^2,$$
$$n^* = 0.75\pi t d_{31} / (s_{11}^E + s_{12}^E), \quad C_e^{S_{1,2}*} = 2\pi a^2 \varepsilon_{33}^T (1-k_p^2) / t; \tag{14.105}$$

for rectangular beams (plates made of the beams) supported on the opposite ends

$$M_{eqv} = 0.5M_{pl}, \quad K^E_{eqv} = 4wt^3 / l^2 s^E_{11}, \quad S_{av} = 0.64S_{pl},$$
$$n^* = wtd_{31} / s^E_{11}l, \quad C^{S_1 *}_e = 2wl\varepsilon^T_{33}(1 - k^2_{31}) / t. \qquad (14.106)$$

It is noteworthy that in the variant of rectangular beam the stripe-electroded (tangentially polarized) piezoelements can be employed. In this case the corresponding equivalent electro-mechanical parameters (see Section 7.2.2.1.2) d_{33eff}, $s^E_{33\,eff}$, $\varepsilon^T_{33\,eff}$ and $k_{33\,eff}$ must be used in formulas (14.106). The ability of employing the close to longitudinal piezoeffect is to advantage of designs with rectangular vs. circular plates.

14.3.5.2 Effects of Foundation Vibration on the Fixed Diffraction Type Hydrophones

The pressure gradient hydrophones of the fixed diffraction type were considered under the assumption that their foundation does not move, e.g., it has infinite mass. In the real designs the mass of foundation has a finite value. It is interesting to estimate how this may influence parameters of the hydrophone, and how big the mass must be to minimize this influence.

The logic of considering these issues among the sections related to accelerometers is due to the fact that analysis performed in the preceding Section can be almost completely applied to calculating characteristics of the hydrophone of the fixed diffraction type without assumption that its foundation does not move. The calculations will be produced for a circular plate piezoelement, given that analogous results for the rectangular beam (plate) can be obtained in a straightforward way.

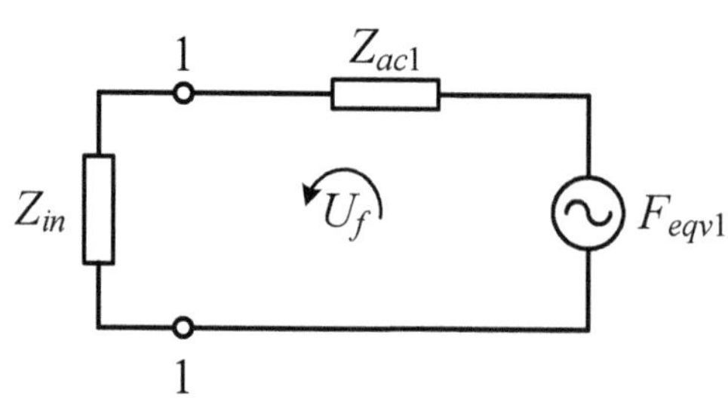

Figure 14.21: Equivalent acoustomechanical generator that produces oscillation of a rigid disk. Z_{in} is the input impedance of the mechanical system that is subjected to action of the acoustic field.

The peculiarity of this case is that the source of the force that is acting between terminals 1, 1 of the equivalent circuit in Figure 14.19 (a) and produces oscillating motion of the first partial system, i.e., of the rigid disk having diameter of the foundation, *D*, is acoustic field. This source can be represented by the equivalent acoustomechanical generator shown in Figure 14.21 (previously was introduced by Eq. (1.63) and by Figure 1.8). The force acting between terminals 1,1 is

$$F_{1,1} = F_{eqv1} - Z_{ac1} U_f , \tag{14.107}$$

where

$$F_{eqv1} = -j(kD^3/3) \cdot P_o , \tag{14.108}$$

following formula (14.67) at $\varphi = 0$, and Z_{ac1} is the radiation impedance of the oscillating disk. According to expressions (6.311), for the oscillating disk of small wave size (practically up to $kD < 0.6$) $Z_{ac1} \approx j\omega m_{ac1}$, where

$$m_{ac1} = \frac{1}{6} D^3 \rho_w . \tag{14.109}$$

In this case the acoustic field is acting also on the second partial system (bimorph plate), and the acoustomechanical generator that has the corresponding parameters F_{eqv2} and Z_{ac2} must be included between points 2, 2 of the equivalent circuit in Figure 14.19.(a). The equivalent force is given by general formula (14.67). For the simply supported circular plate at $\varphi = 0$

$$F_{eqv2} = -j1.3 \cdot ka^2 D \cdot P_o . \tag{14.110}$$

The radiation impedance (inertia reactance due to small wave size), $Z_{ac2} \approx j\omega m_{ac2}$, was neglected when considering the diffraction type sensor in Section 14.3.4.2. The equivalent acoustic mass in this case can be estimated by formula analogous to (14.109). Considering the nonuniform distribution of velocity over the surface it is

$$m_{ac2} = 0.6 a^3 \rho_w . \tag{14.111}$$

Equations (14.97) and (14.98) for the diffraction type hydrophone become with these additions

$$j\omega(M_1 + m_{ac1})U_f + j\omega M_{12} U_o = F_{eqv1} , \tag{14.112}$$

$$j\omega M_{12}U_f + j\omega M_2[1+(m_{ac2}/M_2)](1-\omega_{em\,ac}^2/\omega^2)U_o = F_{eqv2}\,. \tag{14.113}$$

Here velocity U_l is replaced by U_o - velocity of the reference point at the center of the circular plate, and

$$\omega_{em\,ac}^2 = (K_{eqv}^E + n^2/C_e^{S_1})/M_2(1+m_{ac2}/M_2)\,. \tag{14.114}$$

Eq. (14.100) for output voltage V_{oc} remains the same.

We will further assume that no additional mass is attached to the plate, and hence $M_{ad} = 0$ in expressions (14.95) for M_1, M_2, and M_{12}. Our goal is in estimating validity of approximations made in Section 14.4.2.2 in course of calculating sensitivity of the hydrophone. Namely, in determining the value of the relative mass of foundation, M_f/M_{pl}, that is sufficient for neglecting effect of vibrating foundation of hydrophone. Another goal is in estimating the contribution of the acoustic mass that is associated with oscillating the piezoelement.

When considering effect of finite value of the mass of foundation on vibration of a circular plate, it must be remembered that expression of type (14.92) for velocity distribution over surface of the circular plate will be used, in which U_l is replaced by U_o and the expression for the mode shape is approximated by formula for static deflection

$$\theta(r/a) \approx \left(1-\frac{r^2}{a^2}\right)\left(1-\frac{r^2}{4a^2}\right). \tag{14.115}$$

It is interesting to estimate the accuracy of the results obtained by employing this approximation by comparing with results of exact solutions in case that they are known. Thus, it is known (see Section 4.4.3) that the resonance frequency of a passive circular plate with free boundary, f_{free}, is related to the resonance frequency of the same plate simply supported, f_{ss}, as $f_{free} = 1.82 f_{ss}$, and the radius of the nodal circle of the free plate is $r_{nl} = 0.68a$. In the case under consideration the free plate conditions are fulfilled at $M_f = 0$. Likewise, when the mass of foundation is increased up to $M_f \to \infty$, the conditions for simply supported plate are achieved.

The resonance frequency of the plate with a finite mass of foundation, f_{rM_f}, can be derived from the condition that determinant of the system of Equations (14.112) and (14.113) for the case of free vibration of the plate in air (at $F_{eqv1} = F_{eqv2} = 0$ and $m_{ac1} = m_{ac2} = 0$) must be zero. And the values for the relative radius of the nodal line r_{nl}/a can be found using expression

(14.92) from the equation

$$U_f + U_0\,\theta(r_{nl}/a) = 0\,, \tag{14.116}$$

where the ratio U_f/U_0 may be determined from Eq. (14.112) as $U_f/U_o = -M_{12}/M_1$. After some manipulations the following equations for calculating the resonance frequency and radius of the nodal circle will be obtained as

$$f_{rM_f} = f_{ss}\sqrt{\frac{1+(M_f/M_{pl})}{0.3+(M_f/M_{pl})}}\,, \tag{14.117}$$

$$\theta(r_{nl}/a) = \frac{0.46}{1+(M_f/M_{pl})}\,. \quad (14.118))$$

For the plate with free boundary (at $M_f = 0$) from these equations will be found that $f_{fm} = 1.826 f_{ss}$ and $r_{nl} = 0.69a$, i.e., nearly the exact values predicted for these quantities. The plot for the relative resonance frequency vs. relative mass of foundation is shown in Figure 14.22. Thus, the minimal value of ratio M_f/M_{pl}, at which a circular plate can be considered as simply supported, may be estimated as about 6~7.

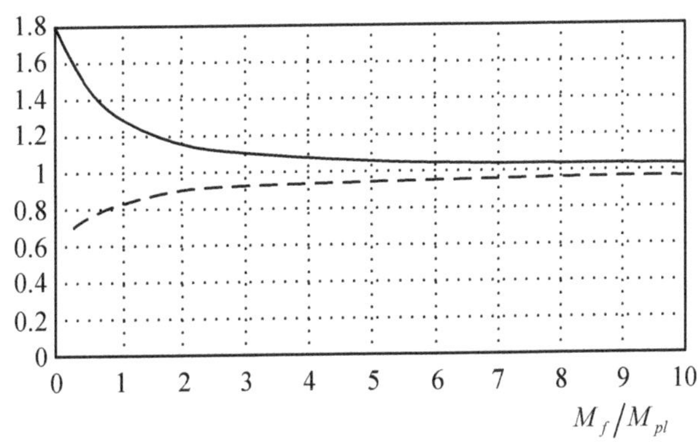

Figure 14.22: Plots of the functions: f_{free}/f_{ss}, (solid line), r_{nl}/a (dash line) vs. value of ratio mass of the foundation to mass of the plate, M_f/M_{pl}.

Shift of the resonance frequency of the hydrophone due to contribution of the acoustic mass can be significant. Besides, the movement of the foundation affects the sensitivity of the hydrophone. For approximate numerical estimations of effect of the finite mass of foundation on sensitivity of the hydrophones we will assume that coefficient $\psi = D/2a$ is 1.6, as it was previously suggested, and the radius to thickness ratio for the plate is $a/t \approx 5$. These assumptions seem to be reasonable for practical design. Moreover, we pursue approximate estimations, and the effect of different dimensions can be considered in the same way.

After representing

$$M_1 + m_{ac1} = M_f \left[1 + \frac{M_{pl}}{M_f} \left(1 + \frac{m_{ac1}}{M_{pl}} \right) \right], \tag{14.119}$$

and estimating m_{ac1} / M_{pl} with help of relation (14.109) as

$$\frac{m_{ac1}}{M_{pl}} \approx 0.2 \cdot \frac{a}{t} = 1.0, \tag{14.120}$$

we obtain

$$M_1 + m_{ac1} \approx M_f \left(1 + 2\frac{M_{pl}}{M_f} \right). \tag{14.121}$$

The term $(m_{ac2} / M_2) = (m_{ac2} / M_{eqv})$ in Equations (14.113) and (14.114) can be represented as

$$\frac{m_{ac2}}{M_{eqv}} \approx 0.1\frac{a}{t}. \tag{14.122}$$

The expressions (14.121) and 14/122) are components of Equations (14.112) and (14.113) that include acoustic masses. Eq. (14.112) depends on ratio / M_f / M_{pl}. Variation of the resonance frequency at a certain value of the relative mass of foundation can be found from these equations. According to expression (14.114) the ratio of value of the resonance frequency determined with acoustic mass taken into account, $\omega_{em\,ac}$, to the resonance frequency of the plate itself (at $M_f / M_{pl} \to \infty$) is

$$\frac{\omega_{em\,ac}}{\omega_{em}} = \frac{1}{\sqrt{1 + 0.1(a/t)}}. \tag{14.123}$$

At $a/t = 5$ the shift of resonance frequency is about 20%. This effect reduces operating range, in which response of the hydrophone is described by formula (14.68) that was obtained under the condition that $M_f \to \infty$, though this shift does not change the formula itself.

For estimating the relative mass of foundation required for validity of this relation determine the low frequency sensitivity considering that mass of the foundation is finite. After substituting U_f from Eq. (14.112) into Eq. (14.113) and taking into account expression (14.121) at $a/t = 5$ will be obtained that

$$j\omega \cdot 0.45 M_{pl} \left[\frac{0.45\alpha_f}{1+2\alpha_f} - \left(1 - \frac{\omega_{em\,ac}^2}{\omega^2}\right)\right] \cdot U_o = \frac{0.45\alpha_f}{1+2\alpha_f} F_{eqv1} - F_{eqv2}, \tag{14.124}$$

where it is denoted for brevity $\alpha_f = M_{pl} / M_f$.

Output of the sensor, V_{oc}, at low frequencies (at $\omega \ll \omega_{em\,ac}$) is proportional to velocity U_o by formula (14.100). Compare values of the velocity from Eq. 14.124 for the cases that M_f is finite (will be denoted U_{oM_f}) and $M_f \to \infty$ (denoted $U_{o\infty}$). At $\omega \ll \omega_{em\,ac}$ Eq. (14.124) becomes

$$j\omega \cdot 0.45 M_{pl} \frac{\omega_{em\,ac}^2}{\omega^2} \cdot U_o = \frac{0.45\alpha_f}{1+2\alpha_f} F_{eqv1} - F_{eqv2}, \tag{14.125}$$

where from

$$\frac{U_{oM_f}}{U_{o\infty}} = 1 - 0.45 \frac{\alpha_f}{1+2\alpha_f} \cdot \frac{F_{eqv1}}{F_{eqv2}}. \tag{14.126}$$

Following expressions (14.108) and (14.110)

$$\frac{F_{eqv1}}{F_{eqv2}} = 1.1 \left(\frac{D}{2a}\right)^2. \tag{14.127}$$

Under the assumption that $D/2a = 1.6$ this ratio is $F_{eqv1} / F_{eqv2} = 2.8$, and

$$\frac{U_{oM_f}}{U_{o\infty}} \approx 1 - 1.2 \frac{\alpha_f}{1+2\alpha_f} = \frac{0.8 + M_f / M_{pl}}{2 + M_f / M_{pl}}. \tag{14.128}$$

Thus, at $M_f / M_{pl} = 6$ the loss of sensitivity in comparison with the situation that the hydrophone does not move ($M_f \to \infty$) is about 15%.

Equating expression in the brackets in Eq. (14.124) to zero results in some increase of the resonance frequency of the plate supported by the finite mass up to ω_{rM_f}, namely,

$$\omega_{rM_f} = \omega_{em\,ac} \sqrt{\frac{1+2\alpha_f}{1+1.55\alpha_f}} = \omega_{em\,ac} \sqrt{\frac{M_f / M_{pl} + 2}{M_f / M_{pl} + 1.55}}. \tag{14.129}$$

At $M_f / M_{pl} = 6$ the frequency increases less than by 3%.

The following example may illustrate design configuration of a real fixed diffraction type pressure gradient hydrophone. Bimorph piezoelement: circular plate made of PZT-4, $2a = 30$ mm, $t = 3$ mm, $a/t = 5$. Foundation: ring made of brass ($\rho = 8500$ kg/m³), $D/2a = 1.6$, $t_f = 10$ mm. Thus, $M_f / M_{pl} \approx 6$ and all the above estimations of parameters are valid.

14.3.5.3 Accelerometers of the Length Expander Type

Accelerometers of this type are shown in Figure 14.8 (c.3) and (c.4). They can be considered based on the general equivalent circuit for the length expander transducer presented in Figure 10.7. In the case of accelerometer of the type (c.3) this circuit can be modified as shown in Figure 14.23.

In the circuit F_0 and Z_0 between terminals 1, 1′ belong to a source of vibration (structure under investigation) that generates vibration of the end with velocity U_f. (In the case of calibration F_0 and Z_0 are the mechanomotive force and internal impedance of a shaker-table).

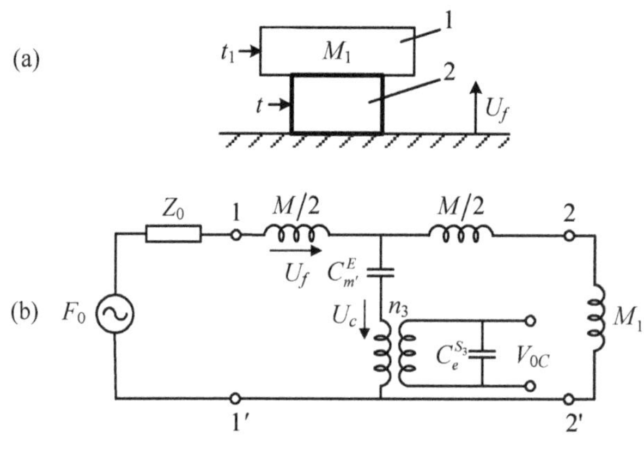

Figure 14.23: (a) Schematic representation of the length expander accelerometer: 1-passive block (M_1), 2-piezoelement. (b) The equivalent electromechanical circuit of the accelerometer.

Another end of the piezoelement is loaded by a passive block. We assume that in the operating range of an accelerometer, which is significantly lower than frequency of its electromechanical resonance, the size of accelerometer is small compared with the wavelength, i.e. $kt \ll 1$ and $kt_1 \ll 1$, where t and t_1 are the thicknesses of the piezoelement and of the passive block. Therefore, the piezoelement can be represented by its lamped parameters and the passive block by its mass M_1, as this is shown in Figure 14.23, and expressions for the parameters in the circuit are (see Section 10.2.3.2.3):

$$C_m^E = ts_{33}^E / S_{cs},\ N_3 = d_{33}S_{el} / s_{33}^E t,\ C_e^{S_3} = C_e^T(1-k_{33}^2) = C_{Lf}(1-k_{33}^2). \tag{14.130}$$

For determining the input mechanical impedance of the accelerometer, it must be suggested that force F_f is applied between terminals 1, $1'$ of the equivalent circuit. Then, $Z_{in} = F_f / U_f$ may be calculated. This will result in

$$Z_{in} = \frac{F_f}{U_f} = j\omega M_1\left(\frac{M}{2M_1} + \frac{1+M/2M_1}{1-\omega^2/\omega_{em}^2}\right), \tag{14.131}$$

where

$$\omega_{em}^2 = 1/C_m^E(1-k_{33}^2)(M_1 + M/2), \tag{14.132}$$

The output voltage of the accelerometer is

$$V_{oc} = \frac{N_3 U_c}{j\omega C_e^{S_3}} = \frac{N_3 U_f}{j\omega C_e^{S_3}} \cdot \frac{1}{1-\omega_{em}^2/\omega^2}. \tag{14.133}$$

At frequencies, for which $\omega^2/\omega_{em}^2 \ll 1$, after some manipulations that involve expression (14.132) for ω_{em} will be obtained

$$V_{oc} = \frac{N_3 C_{eqv}^E}{C_{Lf}} \cdot M_2(1+M/2M_1) \cdot \dot{U}_f, \tag{14.134}$$

$$Z_{in} \approx j\omega(M_1 + M). \tag{14.135}$$

Thus, the sensitivity

$$\gamma_{\dot{U}_f} = V_{oc}/\dot{U}_f = \frac{N_3 C_{eqv}^E}{C_{Lf}} \cdot M_1(1+M/2M_1) \tag{14.136}$$

does not depend on frequency. We can see that with increase of mass M_1 the sensitivity of accelerometer increases, but the operating bandwidth reduces. This is the same tendency as in the previous case of the flexural type accelerometer.

The design shown in Figure 14.8 (c.4) is symmetrical relative to element that mounts the accelerometer to an enclosing case. Due to the symmetry the halves of the design can be calculated independently with the above results. Electrically they must be connected in antiphase in respect to the signs of polarization. This will result in summarizing effects of acceleration, $\dot{U}_f$

, and increasing the specific sensitivity of the device in factor of $\sqrt{2}$. But the main advantage of the symmetrical design is that it must be not sensitive to possible deformations of the inclosing case under unwanted action of the sound pressure under the condition that the halves are electromechanically identical.

14.4 Response of Sensors to Unwanted Actions

14.4.1 Notations for Sensitivities to Unwanted Actions

The general formulation of problem of hydrophones sensitivity to unwanted actions was made in Section 3.2.4. The concept of the noise immunity (*NI*) was introduced therein and illustrated with an example of a single plate vs. symmetrical double plate bimorph transducer. Here some harmful effects that may be produced by unwanted actions on characteristics of the hydrophones and recommendations on reducing these effects will be considered. In particular, the sensitivity of hydrophones of the zero and first order to unwanted actions may be the main cause for distortion of their expected ideal directional factors. Effects of the unwanted actions on characteristics of accelerometers will be addressed as well.

At first the notations for sensitivities of the sensors to the most likely unwanted actions must be introduced. The real hydrophones and accelerometers both move and deform in the acoustic field, which can lead to unwanted electrical output. For example, in case of the zero order (sound pressure) hydrophone the output voltage caused by its motion or deformation due to the pressure gradient in the sound field is unwanted. We will denote this voltage by subscript that indicates the useful effect and by the superscript that indicates the unwanted action, as follows

$$V_p^{\nabla P} = \gamma_p^{\nabla P} \cdot \nabla P, \tag{14.137}$$

where $\gamma_p^{\nabla P}$ is the sensitivity of the sound pressure hydrophone to the unwanted action of the pressure gradient.

For the first order (pressure gradient) hydrophone the output voltage caused by deformations due to the action of the sound pressure is unwanted. Similarly, we will denote this voltage as

$$V_{\nabla P}^{P} = \gamma_{\nabla P}^{P} \cdot P \,, \tag{14.138}$$

where $\gamma_{\nabla P}^{P}$ is the sensitivity of the pressure gradient hydrophone to the unwanted action of the sound pressure.

For the motion type pressure gradient hydrophone, the output voltage caused by movement in the perpendicular to the acoustic axis of the hydrophone direction is also unwanted. This unwanted voltage can be denoted as

$$V_{\dot{U}}^{\dot{U}_{\uparrow}} = \gamma_{\dot{U}}^{\dot{U}_{\uparrow}} \cdot \dot{U}_{\uparrow} \tag{14.139}$$

where $\gamma_{\dot{U}}^{\dot{U}_{\uparrow}}$ is the sensitivity of an accelerometer in the direction that is perpendicular to the acoustic axis (the acceleration and sensitivity in direction of acoustic axis will be denoted as $\dot{U}_{\rightarrow}$ and $\gamma_{\dot{U}_{\rightarrow}}$).

Acceleration of a hydrophone due to structural vibration propagating directly through its mounting elements also produces an unwanted action. We will denote the voltage output due to actions of structural vibration born acceleration as

$$V_{p}^{\dot{U}} = \gamma_{p}^{\dot{U}} \cdot \dot{U} \tag{14.140}$$

and

$$V_{\nabla P}^{\dot{U}} = \gamma_{\nabla P}^{\dot{U}} \cdot \dot{U} \tag{14.141}$$

for the sound pressure and pressure gradient hydrophones accordingly, where $\gamma_{p}^{\dot{U}}$ and $\gamma_{\nabla P}^{\dot{U}}$ are the sensitivities to acceleration due to structural vibration. The effect of structural vibration on a hydrophone is of the same physical nature as the effect due to its movement in the acoustic field, therefore hydrophones of the motion type are especially vulnerable to an unwanted action of this kind.

In order to quantitatively characterize vulnerability of a sensor to unwanted actions the notion of the Noise Immunity factors, (*NI*), can be introduced that are analogous to those presented by relation (3.205) in Section 3.2.4. Namely, the factors

$$NI_{P}^{\nabla P} = \gamma_{P} \,/\, \gamma_{P}^{\nabla P} \text{ and } NI_{P}^{\dot{U}} = \gamma_{P} \,/\, \gamma_{P}^{\dot{U}} \tag{14.142}$$

characterize the noise immunity of the pressure hydrophone to action of the pressure gradient and acceleration, respectively. Analogous notations will be used for other combinations of the measured and unwanted actions. Quantity that is opposite to the Noise Immunity can be called the Noise Susceptibility, *NS*. Thus, for example,

$$\frac{1}{NI_{\nabla P}^{P}} = \frac{\gamma_{\nabla P}^{P}}{\gamma_{P}} = NS_{\nabla P}^{P}. \tag{14.143}$$

is the susceptibility of the pressure gradient hydrophone to the sound pressure.

The same effect as caused by unwanted actions in the acoustic field can be produced by the electrical crosstalk between receiving channels *in the case of combined several first order hydrophones or combined zero and first order hydrophones incorporated in one small unit. Even small spurious electrical coupling between channels can lead to large distortions of the* channel outputs or directional characteristics in case that signal in one of the channels is much larger than in another.

14.4.2 Immunity of Sensors to Unwanted Actions

Immunity of the hydrophones to unwanted actions can be increased by a proper hydrophone designing and precise manufacturing. Some practical considerations on this issue can be illustrated with examples of typical hydrophone designs.

14.4.2.1 Sound Pressure hydrophone

The sound pressure hydrophone of the double plate flexural type that is shown in Figure 14.24 was considered in Section 3.2.4 as an example of the hydrophone that may be insensitive to action of acceleration under the ideal condition that the plates are identical and have equal sensitivities to sound pressure, i.e., $\gamma_{1p} = \gamma_{2p} = \gamma_p$.

In reality this condition can be fulfilled only to some degree of accuracy, and the hydrophone is characterized by a finite value of the noise immunity factor, $NI_{p}^{\dot{U}}$. Here we consider situation regarding a real design, in which case the sensitivities may be not equal, $\gamma_1 \neq \gamma_2$. We will assume that the dimensions of the hydrophone are small in comparison with wavelength, and the mounting elements are located within the plane of symmetry of the hydrophone design. The sound pressure on the surfaces of plates can be represented as

$$P_1 = P + \Delta P, \quad P_2 = P - \Delta P, \tag{14.144}$$

Where

$$P = (P_1 + P_2)/2\,,\ \Delta P = (P_1 - P_2)/2\,. \tag{14.145}$$

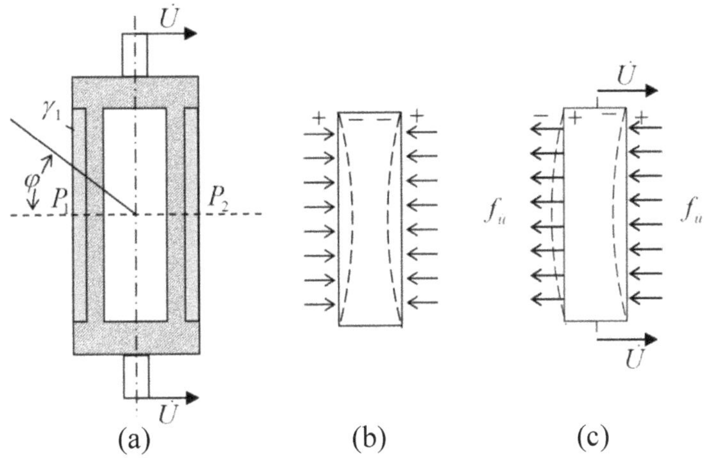

Figure 14.24: Illustrations of: (a) sound pressure hydrophone of the double plate flexural type, (b) the symmetric mode of the plates vibration under action of sound pressure, (c) the anti-symmetric mode of vibration under action of unwanted acceleration.

For the incident plane wave, we have

$$\Delta P = k_{dif} \cdot P = -j\frac{4\pi}{3}ka\cos\varphi \cdot P\,. \tag{14.146}$$

Let the piezoceramic plates be connected electrically in series and in such a way that their outputs due to the plate deformation in the symmetrical mode are in phase, as shown in Figure 14.24 (b). Then the total output voltage is

$$\begin{aligned} V_\Sigma &= V_1 + V_2 = (P + \Delta P)\gamma_1 + (P - \Delta P)\gamma_2 = \\ &= (\gamma_1 + \gamma_2)P + (\gamma_1 - \gamma_2)\Delta P = V_P + V_P^{\nabla P} = V_P[1 + (1/NI_P^{\nabla P})]. \end{aligned} \tag{14.147}$$

Considering relation (14.146), expression for the Noise Immunity factor, $NI_P^{\nabla P}$, is

$$NI_P^{\nabla P} = -j\frac{4\pi}{3}ka\cos\varphi \cdot \frac{\gamma_1 + \gamma_2}{\gamma_1 - \gamma_2}\,. \tag{14.148}$$

The last term in brackets of expression (14.147) determines quantitatively the distortions of the omnidirectional sound pressure output by unwanted contribution due to the pressure gradient. At $\gamma_1 = \gamma_2$ the noise immunity is absolute ($NI_P^{\nabla P} \to \infty$).

In general, the hydrophone may experience acceleration $\dot{U}$ due to its movement in the sound field and/or due to a structural vibration transmitted through a mounting of the hydrophone. If the mounting elements are located within the plane of symmetry of the hydrophone design, the acceleration of the plates is the same. The inertial forces due to the acceleration, $F_{\dot{U}} = m_s \cdot \dot{U}$, are uniformly distributed assuming that the specific mass of the plate, m_s, is uniform. They generate deformation of plates in the asymmetrical mode (Figure 14.24 (c)). Electrical outputs of the plates are in phase opposition, the total output voltage due to the acceleration in case that $\gamma_1 \neq \gamma_2$ is

$$V_P^{\dot{U}} = V_{1P}^{\dot{U}} - V_{2P}^{\dot{U}} = f_{\dot{U}} \cdot (\gamma_1 - \gamma_2) = m_s(\gamma_1 - \gamma_2) \cdot \dot{U}, \tag{14.149}$$

and the unwanted contribution of acceleration to the hydrophone output is characterized by the term

$$\frac{V_P^{\dot{U}}}{V_P} = m_s \frac{\gamma_1 - \gamma_2}{\gamma_1 + \gamma_2} \cdot \frac{\dot{U}}{P}. \tag{14.150}$$

By the definition (14.142)

$$\frac{\gamma_1 + \gamma_2}{m_s(\gamma_1 - \gamma_2)} = NI_P^{\dot{U}}. \tag{14.151}$$

Thus, under the conditions that the mounting elements are located within the plane of symmetry and the plates are identical ($\gamma_1 = \gamma_2$) the hydrophone is insensitive to acceleration.

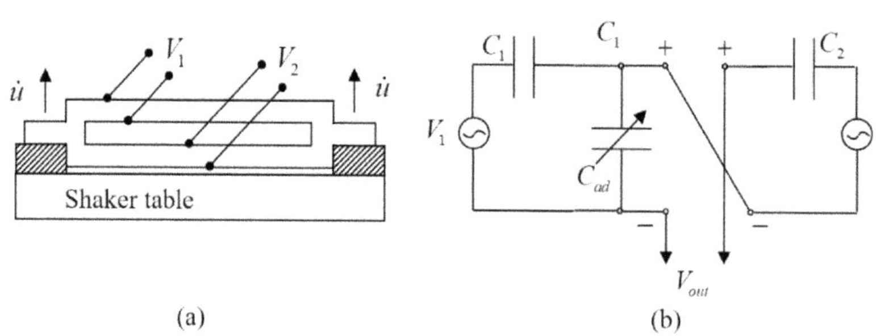

Figure 14.25: Illustration of method of equalizing the sensitivities of individual plates by external capacitive tuning.

The immunity of a pressure hydrophone to acceleration as well as to pressure gradient can be increased by equalizing sensitivities of the plates. A possible procedure of equalizing sensitivities of the plates to a uniformly distributed force may be accomplished by external tuning, as illustrated in Figure 14.25.

By changing the capacitance C_{ad} connected in parallel to the plate having larger sensitivity, the output voltage can be reduced theoretically to zero (and practically to the noise level of instrumentation used). This will be reached at

$$C_{ad} = C_1(\gamma_1 - \gamma_2)/\gamma_1 . \tag{14.152}$$

Note that contribution of acceleration by formula (14.150) depends on the relation between acceleration and total sound pressure generated by different sources including structural vibration. If the ratio $\dot{U}/P$ is small enough, the acceleration canceling feature may not be required. For example, if we assume that the hydrophone is mounted on a plane surface vibrating uniformly, then the sound field generated by the vibration is $P = \dot{U}(\rho c)_w/\omega$, and $\dot{U}/P = \omega/(\rho c)_w = 4\cdot10^{-3} f_{\text{kHz}}$. Under this condition acceleration canceling obviously does not make sense.

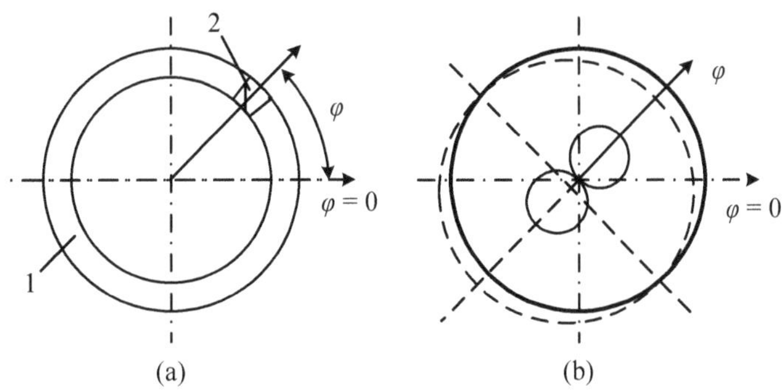

Figure 14.26: Illustration of distortion of omnidirectionality of a hydrophone (a) due to nonuniformity of density (b) of the comprising piezoelement. Projected omnidirectional pattern-dashed line, real pattern distorted by contribution of the pressure gradient response -solid line.

It is noteworthy that sensitivity to acceleration can be a reason for distortion of directionality of small compared to wavelength sound pressure hydrophones that otherwise should be omnidirectional. Thus, if the spherical or cylindrical piezoelement has a nonuniformity (increased or reduced density) in some part of the shell, as is shown schematically in Figure 14.256 (a), the hydrophone becomes sensitive to vibration under action of the pressure gradient with maximum sensitivity in direction of location of the nonuniformity. This will result in distorting the presumable omnidirectionality of the hydrophone by contribution of figure of eight response, as is illustrated in the Figure 14.26 (b).

This contribution may change by magnitude over a frequency range. This circumstance must be taken into consideration when calibrating the measurement hydrophones, which are theoretically assumed to be omnidirectional due to their small wave size. In order to increase accuracy of measurement with a real hydrophone, its directional pattern must be measured, direction of maximum sensitivity has to be marked. Calibration of the hydrophone must be performed in this direction that further must be used for reference.

14.4.2.2 Pressure gradient Hydrophones of the Fixed Diffraction Type

If the halves of a bimorph flexural disk hydrophone are electromechanically identical, then an output under the action of the acoustic field is proportional to the pressure gradient only, because effects of in-phase deformations of the comprising half plates due to the action of sound pressure, V_{1P} and V_{2P}, cancel each other. In the case that the plates are not identical, the output due to sound pressure is $V_{\nabla P}^{P} = V_{1P} - V_{2P} \neq 0$. This output can be reduced by using the above considered procedure of equalizing sensitivities of the half plates.

However, the major unwanted action for hydrophones of this kind is an acceleration $\dot{U}_s$ due to a structural vibration in direction of acoustic axis. (Note that uniform circular plates must be insensitive to vibration in plane due to symmetry. Though, precautions must be taken for a uniform mounting the boundary of the plate.) Contribution of the structural acceleration induced output to the total output of the hydrophone in this case is

$$\frac{V_{\nabla P}^{\dot{U}}}{V_{\nabla P}} = \frac{f_{\dot{U}} S_{av}}{F_{eqv}} = j\frac{3\pi}{4k}\frac{\rho_{av} t}{D}\cdot\frac{\dot{U}_s}{P}, \tag{14.153}$$

where ρ_{av} and t are the average specific density of the flexural plate and its thickness, and D is the diameter of foundation in the case that the plate is circular, or the equivalent dimension in the case of a rectangular plate. (Expressions for F_{eqv} and $f_{\dot{U}} = m_s \dot{U} = \rho_{av} t \cdot \dot{U}$ see in regard to formulas (14.67) и (14.149).)

Except for manipulating the parameter $\rho_{av} t / D$ to some extent the sensitivity to structural vibration can be reduced only by using vibration isolation.

14.4.2.3 Pressure gradient Hydrophones of the Differential Dipole Type

Provided that the individual sound pressure hydrophones are identical, the electrical output of the pressure gradient differential hydrophone is proportional to the pressure gradient only, and for the plane wave we have,

$$V_{\nabla P} = \gamma_P d \cdot \nabla P = jkd\gamma_P \cdot P \tag{14.154}$$

If the sensitivities are not identical, $\gamma_{1P} \neq \gamma_{2P}$ (let $\gamma_{1P} > \gamma_{2P}$), then for the output voltage of the dipole we will obtain

$$V_{out} = \frac{\gamma_{1P} + \gamma_{2P}}{2} d \cdot \nabla P + (\gamma_{1P} - \gamma_{2P})P. \tag{14.155}$$

The second term in this relation is the unwanted contribution, $V_{\nabla P}^P$, of the sound pressure to the first order hydrophone output. Thus, the measure of the first order dipole type hydrophone sensitivity to sound pressure in the plane wave is

$$\frac{V_{\nabla P}^P}{V_{\nabla P}} = -j\frac{2}{kd}\frac{\gamma_{1P} - \gamma_{2P}}{\gamma_{1P} + \gamma_{2P}} \tag{14.156}$$

Usually $d \leq \lambda_H / 4$ where λ_H is the wavelength at the higher frequency of the operating range, and formula (14.156) can be transformed to

$$\frac{V_{\nabla P}^P}{V_{\nabla P}} \approx -j\frac{4}{\pi}\frac{f_H}{f}\frac{\gamma_{1P} - \gamma_{2P}}{\gamma_{1P} + \gamma_{2P}}. \tag{14.157}$$

From this expression follows that contribution of the sound pressure to the first order hydrophone output can be significant at low frequencies even in the case that the difference between sensitivities of comprising dipole hydrophones is not that large. For example, if the difference is about 5% and f_H = 10 kHz, then

$$\frac{V_{\nabla P}^P}{V_{\nabla P}} \approx -j \cdot \frac{1}{\pi f_{kHz}}. \tag{14.158}$$

This means that the omnidirectional output of relative magnitude $1/\pi f_{kHz}$ should be added in quadrature to the directional figure of eight output. To increase the immunity of the first order hydrophone of the dipole type to sound pressure, the same procedure of sensitivity equalization as in the previous case can be used. In particular, this can be done by minimizing output of the dipole in direction $\varphi = \pi / 2$ in the course of the sensitivities equalizing. $\varphi = \pi / 2$

Sensitivity of the dipole type hydrophone to acceleration depends on the sensitivity to acceleration of the individual hydrophones that is considered in Section 14.4.2.1.. However, the requirements for immunity to acceleration of the hydrophones intended to be used in a dipole are higher than for zero order hydrophone. Indeed, the output of a dipole due to acceleration is

$$V_{\nabla P}^{\dot{U}} \leq V_{P_1}^{\dot{U}} + V_{P_2}^{\dot{U}} \approx 2V_P^{\dot{U}}, \tag{14.159}$$

because the phase between $V_{P_1}^{\dot{U}}$ and $V_{P_2}^{\dot{U}}$ is unknown (it cannot be predicted beforehand due to a spurious nature of sensitivity with respect to acceleration). The output due to pressure gradient is $V_{\nabla P} = jkd \cdot V_P$. Therefore, measure of contribution of acceleration to the total output,

$$\left| \frac{V_{\nabla P}^{\dot{U}}}{V_{\nabla P}} \right| \leq \frac{2}{kd} \cdot \frac{V_P^{\dot{U}}}{V_P}, \tag{14.160}$$

is $2/kd$ times larger than in the case of the zero-order hydrophone.

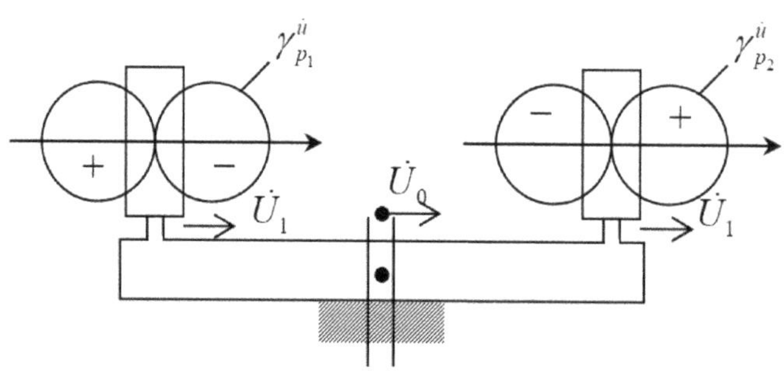

Figure 14.27: Illustration of the procedure for increasing the immunity of the dipole hydrophone to acceleration.

The immunity to acceleration of the dipole design can be increased by using the following procedure based on the idea of compensation of unwanted effects. This is illustrated with Figure 14.27. At first the sensitivities of the comprising hydrophones to acceleration must be measured, and the polarity of a hydrophone regarding the acceleration has to be indicated on the hydrophone surface.

The hydrophones must be symmetrically mounted on a rigid platform in such a way that the hydrophone outputs that are due to the common acceleration are in phase opposition. In this case the negative sign must appear in the formula (14.159) and the contributions due to the acceleration will be reduced. Theoretically very deep acceleration canceling can be achieved in this case by equalizing sensitivities of the comprising hydrophones to acceleration. The sensitivity of a hydrophone to acceleration can be changed independently of its sensitivity to sound pressure by applying an additional mass to its surface. Thus, sensitivities of the hydrophones to acceleration can be equalized, if to apply a proper mass to the surface of the hydrophone that has the smaller sensitivity to acceleration.

14.4.2.4 Pressure gradient Hydrophones of the Motion Type

The sensitivity of the pressure gradient hydrophone to sound pressure can be caused by deformations of the inclosing case that may be transformed into motion of accelerometer. The sensitivity to sound pressure can be reduced if to use the symmetry considerations, when mounting the accelerometer inside the case, as illustrated in Figure 14.8 (c.4).

However, the main possible source of unwanted action for a hydrophone of the motion type is the structural vibration born motion of its body. Obviously, in this case

$$\frac{V_{\nabla P}^{\dot{U}}}{V_{\nabla P}} = \frac{\dot{U}_s}{\dot{U}_{\nabla P}}, \tag{14.161}$$

where $\dot{U}_s$ is unwanted acceleration due to structural vibration and $\dot{U}_{\nabla P}$ is the acceleration in the sound field proportional to the pressure gradient. Under the condition of neutral buoyancy $\dot{U}_{\nabla P} = -j\omega P/(\rho c)_w$, and it follows that

$$\frac{V_{\nabla P}^{\dot{U}}}{V_{\nabla P}} = j\frac{(\rho c)_w}{\omega}\cdot\frac{\dot{U}_s}{P}. \tag{14.162}$$

This quantity does not depend on the hydrophone design unless the buoyancy becomes negative. The only way to reduce the unwanted contribution of acceleration $\dot{U}_s$ is to use vibration isolation. Comparison between expressions (14.153) and (14.162) for the fixed and motion type hydrophones at the same environmental conditions ($\dot{U}_s / P$ = const) results in the relation

$$\frac{(V_{\nabla P}^{\dot{U}}/V_{\nabla P})_{fixed}}{(V_{\nabla P}^{\dot{U}}/V_{\nabla P})_{motion}} = 2.5\frac{\rho_{av}}{\rho_w}\cdot\frac{t}{D} \approx 20\frac{t}{D}, \tag{14.163}$$

if to assume that the plate is fully made of ceramics ant $\rho_{av} = 7.6\rho_w$. In the case that $D/t > 20$, which is very likely for low frequency hydrophone designs, and the density ρ_{av} is decreased due to optimization of the flexural plate, then the fixed type hydrophone can be slightly more favorable than the motion hydrophone in terms of immunity to the sound pressure.

14.4.2.5 Regarding Deepness of Nulls in Directional factor of the First Order Hydrophones

Insufficient immunity of the first order hydrophones to sound pressure is the most probable reason for distortions of their directional factor, which ideally would be $H(\varphi) = \cos\varphi$. It was shown in Section 14.3.4.1 that the first order hydrophone cannot be used for detecting extremely small signals (on the threshold of noise) beyond the sector of angles $|\varphi| < 54°$, because its output signal/noise ratio becomes smaller than it is for the omnidirectional sound pressure hydrophone. In this sense it does not matter how the directional factor behaves in the vicinity of "nulls", as the signal to noise is greatly reduced. Moreover, the actual values of directional factors of ideal hydrophone at the angles close to $\varphi = \pi/2$ may be measured only by using the strong signals compared with the electrical and environmental noise levels. The greater the signal/noise ratio the lower levels of directional factor are attainable. And theoretically the infinitely large signal/noise ratio should exist for approaching value $H(\pi/2) \approx 0$. But this situation is not realistic due to a finite unwanted sensitivity of the first order hydrophone to the sound pressure, and the actual minimum level of the directional factor (so called "deepness of nulls") depends on this sensitivity. The "deepness of nulls", which may be measured by using correspondingly strong signals, characterizes the immunity of the first order hydrophone to sound pressure.

The lack of immunity to sound pressure also deteriorates the signal/noise ratio at the hydrophone output in the operating sector (at $\varphi = 0$ in particular). Thus, the "deepness of nulls" although not practically important for actual operation may be important in terms of characterization of the immunity of the first order hydrophone to sound pressure, which in turn affects the signal/noise ratio at the output of the hydrophone in the operating sector of directional factor. To evaluate the deterioration of the signal/noise ratio due to insufficient immunity to sound pressure (insufficient "deepness of nulls" measured in a strong sound field) we will consider the dipole hydrophone as a typical example of the first order hydrophone.

Let the dipole hydrophone is made of two ideal sound pressure hydrophones having different sensitivities, namely, γ and $(1+\alpha)\gamma$. We will call such a dipole hydrophone "real" to distinguish it from an "ideal" dipole with identical pressure hydrophones. We assume that $\alpha << 1$ and therefore the directional factor outside vicinity of $\varphi \approx 0$ does not change in a frequency range, so far as the wave size of the dipole remains sufficiently small. (Note that for validity of relation (14.58) it must be approximately $kd < 0.3$.)

Sensitivity of the dipole to the sound pressure is due to the difference in sensitivities of the comprising pressure hydrophones. Output of the hydrophone can be represented as

$$U_\Sigma(\varphi) \sim j2\sin[(kd/2)\cos\varphi] + \alpha e^{j(kd/2)\cos\varphi} \quad (14.164)$$

At small *kd*,

$$U_\Sigma(\varphi) \sim jkd\cos\varphi + \alpha = jkd \cdot U'_\Sigma(\varphi), \quad (14.165)$$

where

$$U'_\Sigma(\varphi) \sim \cos\varphi - j\frac{\alpha}{kd} \quad (14.166)$$

is the directional function of the real dipole, which at $\alpha = 0$ becomes directional function of the ideal dipole, $U'_\Sigma(\varphi) \sim \cos\varphi$. Moduli of both functions that are proportional to the directional factors qualitatively are presented in Figure 14.28.

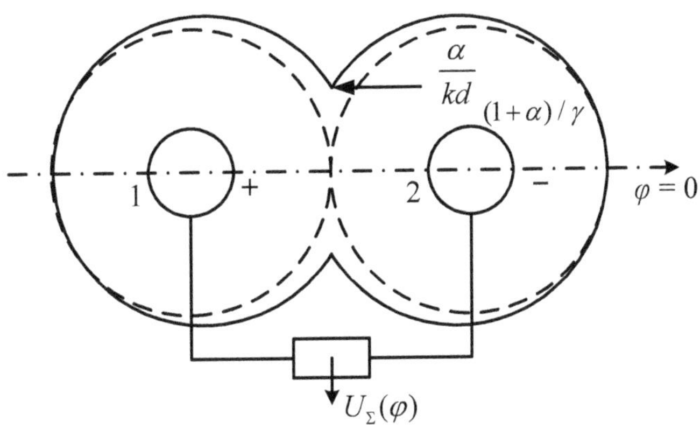

Figure 14.28: Qualitative distortions of the directional factor of ideal dipole hydrophone, $|H(\varphi)| = |\cos\varphi|$ (dashed line), in case that a "real" dipole is composed of the pressure hydrophones having different sensitivities (solid line). The sensitivities of the pressure hydrophones are γ_1 and $\gamma_2 = (1+\alpha)\gamma_1$.

The value of $1/|U'_{\Sigma}(\pi/2)|$ that follows from relation (14.166) can be called the "Deepness of Null" (*DN*). Thus,

$$DN = kd/\alpha\,. \tag{14.167}$$

It is important to estimate an acceptable deepness of null from the point of view of reducing the signal/noise by the real pressure gradient hydrophone vs. ideal one, when formulating requirements for the hydrophones intended for populating arrays that operate in a real signal to noise situation. With this goal let us evaluate the time averaged signal to noise power ratio at the output of the real hydrophone in the same way as it was done for the ideal hydrophone in Section 14.3.4.1.

For a real hydrophone we have

$$\left.\frac{\langle s^2\rangle}{\langle n^2\rangle}\right|_{real} = \frac{\langle [p_{1s} - (1+\alpha)p_{2s}]^2\rangle}{\langle [p_{1n} - (1+\alpha)p_{2n}]^2\rangle}\,. \tag{14.168}$$

We may assume that the noise field is isotropic, and the signal field is that of the plane wave propagating in the direction of dipole axis $\varphi = 0$. In this case Eq. (14.168) may be represented after manipulations analogous to those made in Section 14.3.4.1 as

$$\left.\frac{\langle s^2\rangle}{\langle n^2\rangle}\right|_{real} = \left.\frac{\langle s^2\rangle}{\langle n^2\rangle}\right|_0 \cdot \frac{1+[(1+\alpha)/\alpha^2]\cdot(kd)^2}{1+[(1+\alpha)/\alpha^2]\cdot(kd)^2/3}\,. \tag{14.169}$$

As it was also shown in Section 14.3.4.1, for the ideal hydrophone (at $a = 0$) we have

$$\left.\frac{\langle s^2\rangle}{\langle n^2\rangle}\right|_{ideal} = 3\left.\frac{\langle s^2\rangle}{\langle n^2\rangle}\right|_0\,. \tag{14.170}$$

Thus, we arrive at the relation

$$\frac{\left.\langle s^2\rangle/\langle n^2\rangle\right|_{real}}{\left.\langle s^2\rangle/\langle n^2\rangle\right|_{ideal}} = \frac{1+[(1+\alpha)/\alpha^2]\cdot(kd)^2}{3+[(1+\alpha)/\alpha^2]\cdot(kd)^2}\,. \tag{14.171}$$

The degradation of signal/noise by the real hydrophone in comparison with the ideal hydrophone will be denoted as

$$\frac{\langle s^2\rangle/\langle n^2\rangle\big|_{ideal} - \langle s^2\rangle/\langle n^2\rangle\big|_{real}}{\langle s^2\rangle/\langle n^2\rangle\big|_{ideal}} = \beta . \tag{14.172}$$

From Eq. (14.171) follows that

$$\beta = \frac{2}{3+[(1+\alpha)/\alpha^2]\cdot(kd)^2} . \tag{14.173}$$

Thus, for each value of permissible degradation β of signal/noise the tolerance for sensitivity of the hydrophones comprising the dipole, α, and the corresponding "deepness of nulls", $DN = kd/\alpha$, may be found. Conversely, if DN is determined, then for each kd the resulting degradation of the signal/noise may be found. If the permissible signal/noise degradation is prescribed and it is achieved at the highest frequency of an operating range, f_H, then the lowest frequency, f_L, depends on the tolerance for the pressure hydrophones sensitivity. The closer the tolerance the broader the attainable frequency range. Ratio f_H/f_L can be obtained from Eq. (14.173) as

$$(f_H/f_L)^2 = [\beta/(2-3\beta)]\cdot[DN_H^2 + (kd)_H DN_H] . \tag{14.174}$$

To achieve a broader frequency range, value of the spacing related term at the highest frequency, $(kd)_H$, must be chosen as large as possible. According to our assumption, it is $(kd)_H \approx 0.3$.

For example, in the case that $\beta = 0.1$ and the deepness of nulls at the high frequency is DN_H = 10 (20 dB), the frequency ratio becomes $f_H/f_L = 2.6 DN_L = DN_H/10$. To get a broader operating range, the deepness of nulls at the high frequency must be increased. Thus, at $f_H/f_L = 10$ it should be DN_H = 40 (32 dB). Correspondingly, at the lowest frequency of the operating range the acceptable in terms of the signal to noise ratio "deepness of nulls" in this case will be according to formula (14.167) = 4 (12 dB). But at the low frequencies of operating range the directional factor of the hydrophone becomes significantly corrupted due to contribution of the omnidirectional term in expression (14. 166).

Considering this example, it must be remembered that for achieving value DN_H=40 difference between sensitivities of the pressure hydrophones must be according to formula (14.167) less than 1% under the condition that $(kd)_H \approx 0.3$. Thus, the requirement for the pressure hydrophones identity is very demanding.

As it follows from comparison of formulas (14.167) and (14.156), the coefficients *DN* and noise immunity, $NI^{P}_{\nabla P}$, are proportional, i. e., $DN \sim NI^{P}_{\nabla P}$. Therefore, the results of this section are applicable to the first order hydrophones of different type, if to use these coefficients interchangeably. The conclusion can be made that a tradeoff exists between value of the noise immunity factor, $NI^{P}_{\nabla P}$, and operating frequency range, f_H / f_L. Achieving a broad frequency range requires very careful pressure gradient designing and manufacturing in terms of noise immunity.

14.5 References

1. B. Aronov and D. Brown, *Acoustic Pressure Gradient Sensors of the Motion and Fixed Type*, Technical Report for Office of Naval Research Grant Number N00014-97-1-079, BTech Acoustics LLC, December 2003.
2. *Acoustic particle velocity sensors: Design, Performance, and Applications*, AIP Conference Proceedings 368, Editors: M. J. Berliner, J. F. Lindberg, AIP Press, Woodbury, New York, 1996.
3. G. K. Screbnev, *Combined hydroacoustic sensors*, Sankt- Petersburg, Elmor, 200 p., 1997 (in Russian).
4. B. F. Cron and C. H. Sherman, "Spatial correlation functions for various noise models," J. Acoust. Soc. Am., **34**, 1732-1736, (1962)

CHAPTER 15

COMBINING FINITE ELEMENT ANALYSIS WITH ANALYTICAL METHOD

15.1 Introduction

Application of the energy method for solving electroacoustic transducer problems demonstrated in the preceding content assumed that boundary conditions for the transducer mechanical systems are ideal and their modes of vibration can be found analytically, or the modes of static deformation of the mechanical systems are known, or the direct methods can be used for determining the modes of vibration by choosing appropriate systems of the trial functions. The analytical solutions obtained for the considered transducer types and realized in practical designs showed good enough agreement with experimental verifications. But in some cases, the boundary conditions for the mechanical systems may be inherently not ideal and vibrations of the system hardly can be satisfactory approximated by convenient for analytical treatment functions. Such, for example, are vibrations of mechanical systems of the flextensional transducers, which represent passive shells of various configuration, generated by the piezoceramic drivers. In this and in more general cases the problems of vibration of complicated mechanical systems can be solved using Finite Element Analysis (FEA). Powerful computer-based technologies allow simulating virtually any well-defined linear system (not only mechanical, but electromechanical and acoustic as well) and obtaining its natural frequencies, modes of vibration and frequency responses without using analytical treatment. Though the results thus obtained are purely numerical, and a physical insight into the nature of the solutions anyway is required through theoretical considerations, possibilities of using FEA for calculating the overall performance of the electroacoustic transducers may seem attractive. However, the following reasoning regarding this issue deserves to be considered. While calculating parameters of a transducer of a given configuration (solving the direct problem for the transducer) using FEA is more or less straightforward, solving the reverse problem of designing the transducer that meets certain requirements under real environmental conditions is problematic. Moreover, as it was articulated in Chapter 11, properties of the piezoelectric ceramics are not known to a great accuracy and subject to changing under operating environmental conditions. Therefore, it would be advantageous to obtain results of analyzing a transducer in a form with parameters of the ceramics explicitly presented. This can be achieved by combining the FEA technique and the analytical energy method for calculating transducers. The goal of this Chapter is to illustrate how this can be done.

The most crucial and necessary for application of the energy method step in the vibration analysis of the mechanical systems of the transducers with complex boundary conditions is in determining vibrational mode shape of a passive mechanical system having the same configuration This part of the problem can be especially successfully resolved by the FEA. After this is done, the energy method can be used for calculating transducer parameters in a straightforward way. Using FEA in such capacity regarding electroacoustic transducers was noted in Ref. 3: "Computer programs are very useful for looking in detail at components of the transducer. Thus, one might use them advantageously to find the natural vibrational modes of the transducer and its housing, or to investigate an unusual hinge structure" (read "boundary conditions", the Author's remark).

We will consider the proposed approach with example, in which the flextensional transducer of Class IV is involved. The flextensional transducers are one of the types of low frequency transducers. Theoretical treatments, practical designing and applications of the flextensional underwater transducers have a long history (see Refs. 1 and 2) and are widely discussed in literature. Typical of the flextensional mechanism is generating flexural vibrations of shells made of passive materials by extensionally vibrating piezoelements. Analyzing the flextensional transducer types per se is out of scope of this chapter. The flextensional transducer that realizes flexural vibration of a passive oval shell (Class IV transducer) will be only considered. The goal is in comparing its operating characteristics with those of the low frequency bender transducer previously described in Chapter 9 and of the transducer that employs direct electromechanical conversion in the oval shells by building them from piezoceramic elements (so called oval flexural transducer). The comparison will be made under the condition that all the transducers have approximately the same geometry and amounts of piezoceramic materials used. This comparison makes sense because, supposedly, the hydrostatic and dynamical strength of the oval shells must be greater than for analogous benders utilizing vibration of plates due to transformation of hydrostatic pressure into compression stress in circumferential direction of the shells.

Calculations of operating characteristics of the flextensional transducers were made in literature, predominantly, by using FEA. Therefore, it is logical to apply the same approach for achieving the intended goal of comparing the traditional flextensional transducer with transducer that employs electromechanically active shell having the same geometry as the passive shell of the flextensional transducer has. In distinction from the previous works the FEA modeling is used for determining the natural frequencies and modes of vibration of the oval shells only. After this is done, the equivalent parameters of the transducers are calculated from considering energies associated with electromechanical conversion in the corresponding transducers, and equivalent circuits are presented that allow further analysis of operating characteristics and their comparison. Brief description of this approach was previously reported in Ref. 4.

In order to make the approach to calculations more physically clear, and in order to be able to verify results of calculations at least for one particular case, prototypes of the transducers were built (courtesy of BTech Acoustics), and FEA was done for these particular cases. Note that according to expert recommendation on application of FEA (Ref. 5): "Modeling requires that the physical action of the problem be understood well enough to choose suitable kinds of analysis. Once the results have been calculated, we must check them to see if they are reasonable. Checking is very important, because it is very easy to make mistakes when we rely upon FEA software to solve complicated problems."

Formulating problem for calculating free vibration of a prototype oval mechanical system is done in Section 15.2. The general case of the combined FEA and single degree of freedom equivalent circuit involving treatment of Class IV flextensional transducer and analogous oval shell bender transducer is performed in Sections 15.3 and 15.4. This could be possible under the condition that results of FEA can be scaled for a range of values that characterize geometry of the shell. Verification of this possibility is done in Section 15.3.3.1, and further the equivalent parameters of the equivalent circuits are determined. Comparison of the operating characteristics of the flextensional, oval bender transducers, and rectangular plate bender transduces is made in Section 15.5

15.2 Determining Mode Shapes of the Prototypes Vibration

15.2.1 Formulating the Problem

Flextensional transducers of Class IV employ mechanical systems in the shape of oval shells. Usually, the mechanical system of an entire column like transducer is composed of several single oval (elliptical) shell units that have height, *h,* relatively small compared with their meridional circumference. Therefore, the shell unit can be considered as a ring. This makes the problem of a shell unit vibration one dimensional. Another assumption will be made that the elliptical configuration will be approximated by segments of a circular ring like it is shown in Figure 15.1.

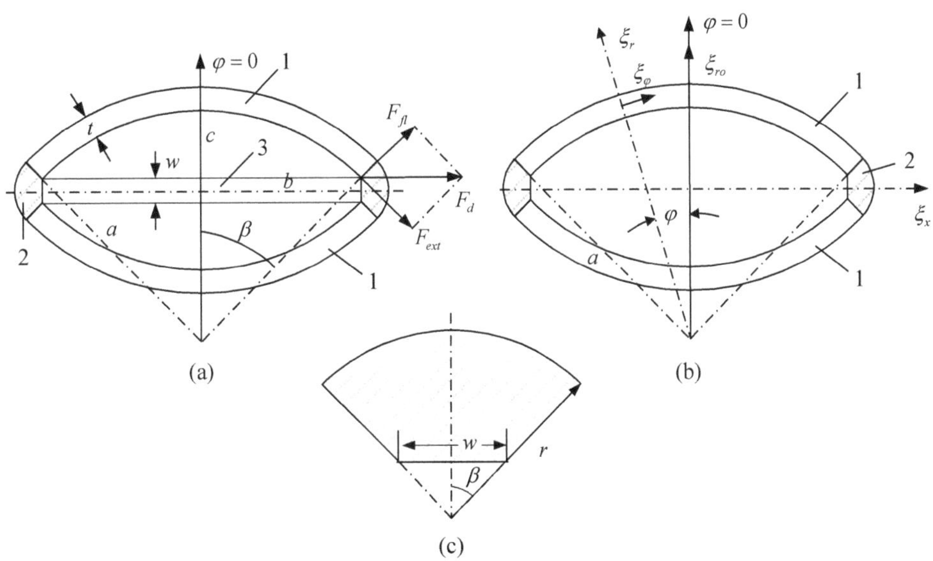

Figure 15.1: (a) Geometry of the mechanical system of a flextensional transducer with oval ring replaced by segments of a circular ring: 1 – segments of a circular ring, 2 – additional parts, 3 – extensionally vibrating piezoceramic driver. (b) Mechanical system of the oval bender. (c) Geometry of the additional part 2.

The practical justification for replacing the elliptical ring by combination of segments of the circular ring is in the fact that it is easier to build out of piezoelements in case of manufacturing the oval bender transducer, in which case the shell is active. Besides, this simplifies treatment of the problem without losing its essence. The radius, *a*, and opening angle, β, of the replacement segments are related to the major, $2b$, and minor, $2c$, axes and their ratio m of the original elliptical shell. From the geometry considerations following from Figure 15.1 (a) these relations are:

$$m = \frac{c}{b} = \frac{1-\cos\beta}{\sin\beta} = \tan\frac{\beta}{2}, \quad b = a\sin\beta, \tag{15.1}$$

Outer radius of the additional part in Figure 15.1 (c) is

$$r = (w/2\sin\beta) + t, \tag{15.2}$$

where w is the width of the driver, and t is the thickness of the shell. To keep amount of the ceramic material in the flextensional and analogous active oval shell transducer the same, we will assume that $w \approx 1.5t$. Thus, the outer radius of the additional part will be

$$r = t\left(\frac{0.75}{\sin\beta} + 1\right). \tag{15.3}$$

The symmetrical forces, F_d, that are generated by the driver on the ends can be decomposed into two components: F_{ext}, which is tangent to its middle line, and F_{fl}, which is normal to the middle line of the ring. The force $F_{ext} = F_d \cos\beta$ produces predominantly extensional deformation of the ring segments, whereas the force $F_{fl} = F_d \sin\beta$ produces predominantly their flexure. Conditions, under which vibrations of the rings are predominantly flexural or extensional, are considered in Ref. 6 and in Section 4.4.4. The resonance frequencies of the extensional (membrane) and flexural vibrations of an incomplete ring may differ practically by an order of magnitude (this will be illustrated below). Therefore, they can be treated separately as independent vibrations, and their effects on performance of the flextensional transducers can be superposed in result. It must be noted that except for the first symmetrical flexural mode of vibration, which is the intended operating mode, the next symmetrical flexural mode of the segment of a circular ring is also generated. The resonance frequency of this mode is about three times higher than that of the first mode (by analogy with the straight beams at different ideal boundary conditions), besides this mode of vibration is not suited for effective radiation, because it has multiple nodes.

Therefore, its contribution to operating characteristics in vicinity of the first resonance frequency is negligible. But the input mechanical impedance related to this mode below its resonance frequency may have some effect on vibration of the driver rod. Although the main attention in calculations will be given to the range of operation around the resonance frequency of the first flexural mode, the contributions of the next flexural and extensional modes of the shell vibration will be also considered.

15.2.2 Results of Calculations and Experimental Verification.

Prior to performing general analysis of the problem, calculating was performed by FEA parameters of vibration of the prototypes that imitated structures of real transducer designs. This was done for experimental verification of results of calculations. Two transducer prototypes were built that corresponded to configurations of transducers shown in Figure 15.1 (a) and (b). The curved parts 1 were built as bilaminar structures out of segments of rings made of aluminum (outer layer) and radially polarized piezoceramics (inner layer). This allowed generating the flexural and extensional vibrations of the segments using direct "internal" electromechanical conversion. Extensionally vibrating transversely polarized bar was used as the driver 3 of the flextensional prototype. The designated opening angle (which is in fact half of full opening angle) of the segments $\beta = \pi / 4$; mean radius of the segments a = 54.5 mm; thicknesses of the aluminum and piezoceramic layers $t_{al} = t_c$ = 3.5 mm.

A finite element model was performed for analyzing the structural vibration of the flextensional transducer and oval bender. The "Piezoelectric Device" under "Structural Mechanics Module" in COMSOL was used for the modeling. A frequency domain solver was used for calculating resonance frequencies, mode shapes, potential and kinetic energies for the transducer from 100 Hz up to 25 kHz. A static solver was used for calculating the displacements and stress inside the mechanical systems under the hydrostatic pressure.

Results of calculations made for the prototypes were compared with results of measuring the following characteristics of both prototypes: resonance frequencies, mode shapes at the resonance frequencies (radial components of the displacements), ratios of radial displacement of the reference point and of the ends, ξ_{r0} / ξ_x. Results of measuring and calculating the resonance frequencies of the oval segments of the flextensional and oval driver prototypes are summarized in Table 15.1. Results of measuring and calculating the mode shapes of the first and second flexural, and the first extensional modes of vibration are presented in Figure 15.2 - Figure 15.4

for the oval driver prototype.

Table 15.1. Measured and calculate values of the resonance frequencies and ξ_{r0} / ξ_x ratios for different modes of oval segments vibration

	1st Flexural	ξ_{r0} / ξ_x	2nd Flexural	ξ_{r0} / ξ_x	1st Extensional	ξ_{r0} / ξ_x
Measured	3 kHz	2.4	14.3 kHz	4.2	20 kHz	0.78
Calculated	2.9 kHz	2.1	14.2 kHz	4.3	19.4 kHz	0.75

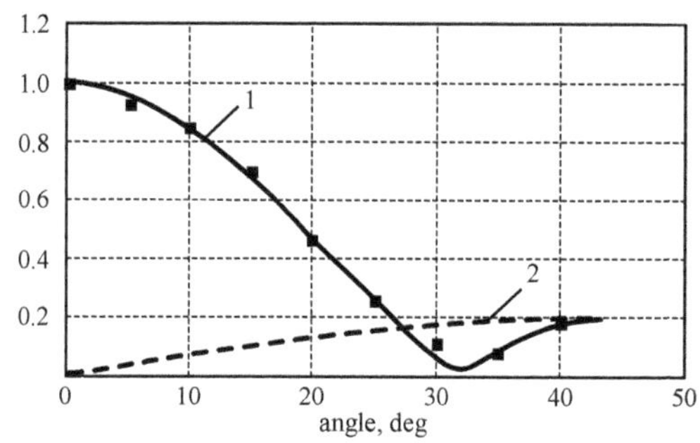

Figure 15.2: Oval driver prototype. Calculated mode shapes of the first flexural mode (1 – radial, 2 – circumferential) and results of measuring (squares).

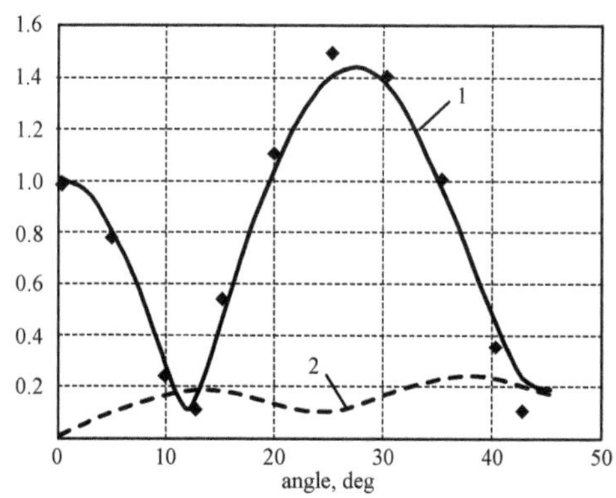

Figure 15.3: Oval driver prototype. Calculated mode shapes of the second flexural mode (1 – radial, 2 – circumferential) and results of measuring (diamonds).

The first flexural and extensional modes of vibration for the flextensional prototype with driver active and the oval segments short circuited are presented in Figure 15.5 and Figure 15.6. The second flexural mode practically can't be generated by the extensional driver, and as result is not visible both in calculating and measuring. Note that the modes of vibration of the oval shell with and without the extensional driver are close.

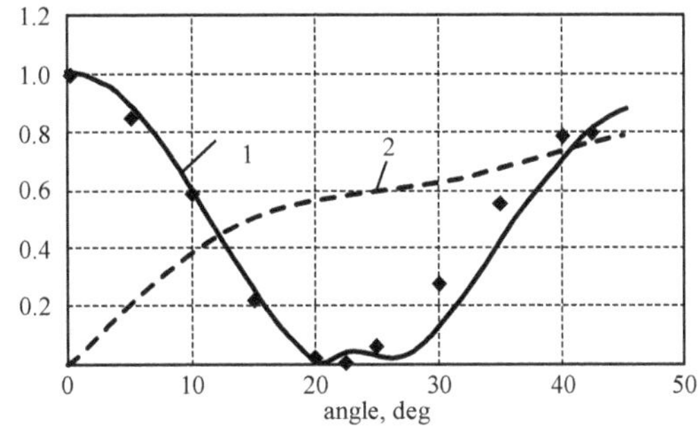

Figure 15.4: Oval driver prototype. Calculated mode shapes of the extensional mode (1 – radial, 2 – circumferential) and results of measuring (diamonds).

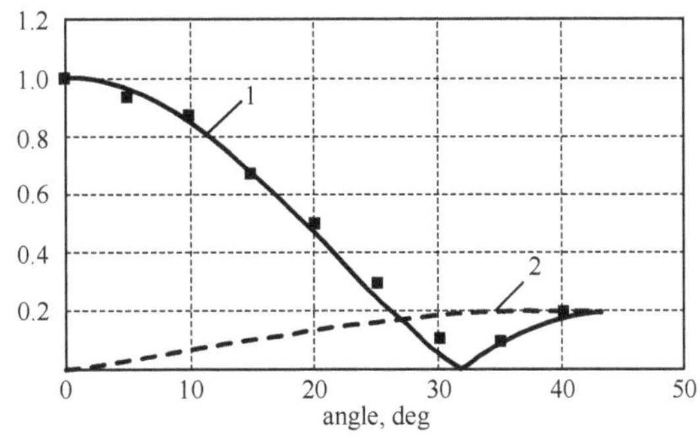

Figure 15.5: Flextensional prototype, the driver is active. Calculated mode shapes of the first flexural mode (1 – radial, 2 – circumferential) and results of measuring (squares).

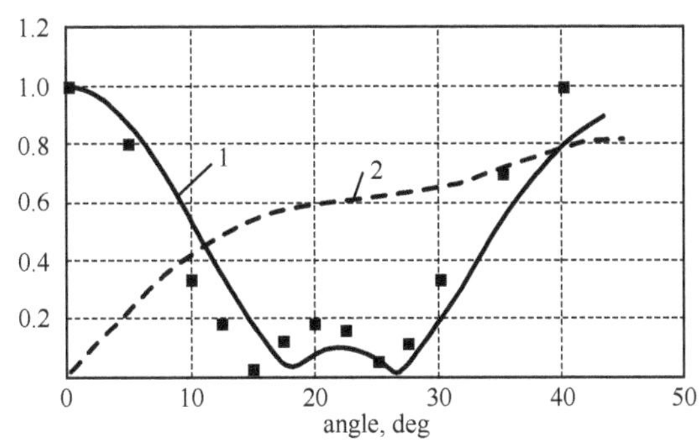

Figure 15.6: Flextensional prototype, the driver is active. Calculated mode shapes of the extensional mode (1 – radial, 2 - circumferential) and results of measuring (squares).

The essential quantities, values of which is necessary to know for determining the input impedances of the oval segments, are the ratios ξ_{r0} / ξ_x. Calculated and measured values of these ratios are presented in Table 15.1. The calculations showed that deformation in the circumferential direction for the 1st flexural mode of vibration,

$$S_\varphi(\varphi) = -\frac{z}{a^2}\left(\frac{d^2\xi_r}{d\varphi^2} + \xi_r\right), \tag{15.4}$$

has zero points at $\varphi \approx \pm 30°$. At these points the electromechanically generated moment changes sign, and hence the full-size electrodes don't produce maximum possible effect. Experiment confirmed this prediction. After the electrodes were split at $\varphi \approx \pm 30°$, and their end parts were connected in antiphase with the central part of the electrodes (by using their series connection), the magnitude of the radial displacement ξ_{r0} increased in factor of 6.5 compared with effect of the full size electrodes (with these parts connected in parallel) at the same electric field in the ceramics (see results of calculating the electromechanical transformation coefficients for such connections in Table 15.5, where from follows that this ratio calculated is 7).

Thus, the conclusion can be made from the results of comparison presented in this Section that calculated by FEA and measured characteristics of the prototypes are in a good agreement.

15.3 Equivalent Circuit of the Oval Transducer

15.3.1 About Scaling Results Obtained by FEA.

It is essential to make sure whether results of FEA obtained for a certain configuration of the mechanical system of a transducer can be scaled, i.e., can be used for the mechanical systems of the same configuration that have different size. In fact, it is advantageous to have configurations of the vibrating shells scalable even from consideration of manufacturing transducers of a particular type. In the case under consideration the ratio of minor to major axes, *m,* and opening angle β of the oval shell are related by Eq. (15.1). Under the assumption that width of the driver, *w*, is in a certain relation with thickness of the oval segment, *t*, (for example, $w = 1.5t$) all the geometry of the oval ring is determined by values of the radius and thickness. Thus, it is possible to express equivalent parameters of the flextensional and oval driver transducers for each value of angle β in a general form as functions of radius a and thickness *t*.

Results of calculating by FEA can be considered scalable, if to make sure that the mode shapes of vibration do not change (to a certain accuracy), when radius and/or thickness change within margins required for practical transducer designing, and the ratio t/a remains within range of applicability the thin shell theory. This assertion is because all the equivalent parameters of transducers can be calculated using expressions for the mode shapes of vibration. With this goal FEA calculations were undertaken of the mode shapes for $\beta = \pi / 4$ and the values of dimensions that are presented in Table 15.2. Based on the results of calculations it may be concluded that the mode shapes of vibration remain virtually unchanged within the range of dimensions presented in Table 15.2. Example of deviation of the 1st flexural mode shapes calculated between the marginal values of dimensions from the Table are shown in Figure 15.7.

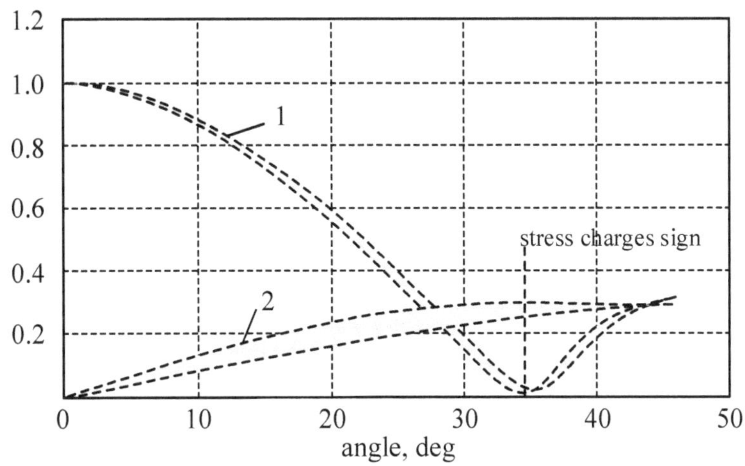

Figure 15.7: Deviations of the 1st flexural mode shapes calculated for the range of changing values of the radius and thickness of the oval segments at $\beta = \pi / 4$ (1 – radial, 2 – circumferential) presented in Table 15.2.

Table 15.2. Dimensions of the oval segments at $\beta = \pi / 4$, for which the mode shapes of vibration were calculated.

a, mm	50	50	50	100	100	100	150	150	150
t, mm	2.5	5.0	7.5	5.0	10	15	7.5	15	22.5
t/a	0.05	0.10	0.15	0.05	0.10	0.15	0.05	0.10	0.15

The analogous procedure can be used regarding the different values of angle β. After the mode shapes are determined, all the equivalent parameters that correspond to calculated modes of vibration can be obtained in the general forms that depends on the thickness and radius of the ring only and on the properties of materials, of which they are built. This will be done within this Chapter for $\beta = \pi / 4$ and predominantly for $t/a = 0.1$.

15.3.2 Equivalent Mechanical Parameters of the Oval Shells for $\beta = \pi / 4$

The equivalent mechanical parameters of an oval segment (per one side of the double-sided structure) may be obtained from the following expressions for the potential and kinetic energies

$$W_{pot} = \frac{1}{2} ha \int_{-t/2}^{t/2} \int_{-\beta}^{\beta} S_\varphi T_\varphi d\varphi dz \,, \tag{15.5}$$

$$W_{kin} = \frac{1}{2} hta\rho \int_{-\beta}^{\beta} (\dot{\xi}_r^2 + \dot{\xi}_\varphi^2) d\varphi + \frac{1}{2} M_2 \dot{\xi}_{2x}^2 \,. \tag{15.6}$$

Here M_2 represents the mass of the end part 2 of the shell in Figure 15.1 (c). From geometry of the part and formula (15.3) follows that $\tilde{V}_2 \approx (\pi/4)t^2 h$ and $M_2 = \rho_2 \tilde{V}_2$ (it is assumed that $w = 1.5t$ and $\beta = \pi / 4$).

After introducing the mode shapes of vibration by formulas

$$\theta_r(\varphi) = \frac{\xi_r(\varphi)}{\xi_r(0)} = \frac{\xi_r(\varphi)}{\xi_{r0}} \,, \; \theta_\varphi(\varphi) = \frac{\xi_\varphi(\varphi)}{\xi_{r0}} \,, \tag{15.7}$$

the expressions for strain will be represented as (see Section 4.4.4)

$$S_{\varphi e} = \frac{1}{a}\left(\frac{\partial \xi_\varphi}{\partial \varphi} + \xi_r\right) = \frac{\xi_{r0}}{a}\left(\frac{\partial \theta_\varphi}{\partial \varphi} + \theta_r\right) \tag{15.8}$$

for the extensional mode, and

$$S_{\varphi f} = -\frac{z}{a^2}\left(\frac{\partial^2 \xi_r}{\partial \varphi^2} + \xi_r\right) = -\frac{z\xi_{r0}}{a^2}\left(\frac{\partial^2 \theta_r}{\partial \varphi^2} + \theta_r\right) \tag{15.9}$$

for the flexural mode. It must be remembered that expressions for the displacement distributions and mode shapes differ for different modes of vibration, though they are not marked by corresponding subscripts for brevity.

The expressions (15.5) for the potential energy now can be represented as follows.

For the extensional mode

$$W_{pot} = \frac{1}{2} htaY \int_{-\beta}^{\beta} S_{\varphi e}^2 d\varphi = \frac{1}{2} \xi_{r0}^2 \frac{2htY}{a} \int_0^{\beta} \left(\frac{\partial \theta_\varphi}{\partial \varphi} + \theta_r \right)^2 d\varphi = \frac{1}{2} \xi_{r0}^2 K_e . \tag{15.10}$$

Here K_e is the equivalent rigidity for the extensional mode that will be represented as

$$K_e = K_e' \cdot \Omega_{Ke}(\beta) , \tag{15.11}$$

where

$$K_e' = \frac{2htY}{a} , \tag{15.12}$$

$$\Omega_{Ke}(\beta) = \int_0^{\beta} \left(\frac{\partial \theta_\varphi}{\partial \varphi} + \theta_r \right)^2 d\varphi . \tag{15.13}$$

For the flexural mode

$$W_{pot} = \frac{1}{2} htaY \int_{-\beta}^{\beta} S_{\varphi f}^2 d\varphi = \frac{1}{2} \xi_{r0}^2 \frac{ht^3 Y}{6a} \int_0^{\beta} \left(\frac{\partial^2 \theta_r}{\partial \varphi^2} + \theta_r \right)^2 d\varphi = \frac{1}{2} \xi_{r0}^2 K_f . \tag{15.14}$$

The equivalent rigidity K_f for the flexural mode will be represented as

$$K_f = K_f' \cdot \Omega_{Kf}(\beta) , \tag{15.15}$$

where

$$K_f' = \frac{ht^3 Y}{6a^3} , \tag{15.16}$$

and

$$\Omega_{Kf}(\beta) = \int_0^{\beta} \left(\frac{\partial^2 \theta_r}{\partial \varphi^2} + \theta_r \right)^2 d\varphi . \tag{15.17}$$

Expressions for the kinetic energy and equivalent masses are

$$\begin{aligned} W_{kin} &= \frac{1}{2} hta\rho \int_{-\beta}^{\beta} (\dot{\xi}_r^2 + \dot{\xi}_\varphi^2) d\varphi + \frac{1}{2} \dot{\xi}_x^2 M_2 = \\ &= \frac{1}{2} \dot{\xi}_{r0}^2 \, 2hta\rho \int_0^{\beta} \left(\theta_r^2 + \theta_\varphi^2 \right)^2 d\varphi + \frac{1}{2} \dot{\xi}_x^2 M_2 = \frac{1}{2} \dot{\xi}_{r0}^2 \left(M_{eqv} + M_2 \frac{\dot{\xi}_x^2}{\dot{\xi}_{r0}^2} \right) . \end{aligned} \tag{15.18}$$

Here the equivalent mass M_{eqv} is

$$M_{eqv} = M_{eqv}' \Omega_M(\beta) , \tag{15.19}$$

where

$$M'_{eqv} = 2hta\rho \tag{15.20}$$

and

$$\Omega_M(\beta) = \int_0^\beta \left(\theta_r^2 + \theta_\varphi^2\right)^2 d\varphi . \tag{15.21}$$

Factors $\Omega_M(\beta)$ have different values for the flexural and extensional modes, due to different dependencies $\theta_r(\varphi)$ and $\theta_\varphi(\varphi)$. They will be distinguished for these modes of vibration as Ω_{Mf} and Ω_{Me}, respectively. Values of all the factors Ω_M and Ω_K are presented in Table 15.3 and Table 15.4.

It has to be noted that expressions (15.16) for K'_f and (15.20) for M'_{eqv} are accurate so far as the thin shell theory is applicable. With increase of the ratio *t*/*a* some correction factors have to be introduced in these expressions that are responsible for shear deformations and rotary inertia in course of the shell flexural vibration. Such factors named $\gamma(t/a)$ are included in Table 15.3 for the flexural modes.

Expressions for the equivalent parameters depend on how position of the reference point is chosen. These points can be $\varphi = 0$ or $\varphi = \beta$ depending on whether the radiation problem or determining the input impedances of the shell are the issues. The reference point is chosen at $\varphi = 0$ for calculating factors in Table 15.3 and Table 15.4, and will remain the same until it is noted otherwise. Important quantity for recalculating equivalent parameters between the reference points at $\varphi = 0$ and at the end of the oval segment is ratio ξ_{r0}/ξ_x that will be denoted for brevity as

$$(\xi_{r0}/\xi_x) = \Omega_\xi . \tag{15.22}$$

We will denote Ω_ξ for the first flexural mode as $\Omega_{\xi f1}$, for the second flexural mode as $\Omega_{\xi f2}$, and for the extensional mode as $\Omega_{\xi e}$. One more quantity, which is important for determining radiation resistance of a projector having small wave dimensions (as in the case under consideration), is the volume velocity,

$$U_{\tilde{V}} = 2ha\int_0^\beta \dot{\xi}_r(\varphi)d\varphi = \dot{\xi}_{r0}\cdot 2ha\int_0^\beta \theta_r(\varphi)d\varphi = \dot{\xi}_{r0}S_{av} . \tag{15.23}$$

The average surface area S_{av} will be denoted as

$$S_{av} = 2ha\Omega_S , \tag{15.24}$$

where

$$\Omega_S = \int_0^{\beta} \theta_r(\varphi) d\varphi . \tag{15.25}$$

Factors Ω_ξ and Ω_S are presented in Table 15.3 and Table 15.4 for corresponding modes of vibration at $\beta = \pi / 4$.

Table 15.3. Factors for calculating the equivalent parameters for the flexural mode at $\beta = \pi / 4$.

t/a	Ω_S	$\Omega_{\xi f1}$	Ω_{Mf1}	$\gamma(t/a)$	$\Omega_{\xi f2}$	Ω_{Mf2}	$\gamma(t/a)$	Ω_{Kf1}	$\gamma(t/a)$	Ω_{Kf2}	$\gamma(t/a)$
0.05		2.28		1.00	3.06		1.00	30.20	1.00	1162.40	1.00
0.1	0.32	2.13	0.33	1.06	3.12	0.46	1.12		0.87		0.85
0.15		2.0		1.11	8.62		1.89		0.76		0.74

Table 15.4. Factors for calculating the equivalent parameters for the extensional mode.

t/a	S_{av}	$\Omega_{\xi e}$	Ω_{Me}	Ω_{Ke}
0.05	0.23	4.91	0.26	0.80
0.1	0.35	0.78	0.65	2.57
0.15	0.14	0.87	0.46	1.88

15.3.3 Electromechanical Conversion in the Oval Shells Made of Piezoceramics

If to assume that the oval segments are made from piezoceramic elements designed to produce a bending moment (bimorph or bilaminar piezoelements that employ the transverse or longitudinal effect) then the electromechanical transformation coefficients that characterize the electromechanical energy conversion must be determined vs. opening angle β. So far as the oval segments represent parts of a circular ring, all the treatment in terms of electromechanical conversion may be performed in the way analogous to those used for considering the circular ring flexural transducer in Section 7.3.1. The main difference is that in the current case the modes of vibration are different. For convenience of reading some parts of this section will be repeated here.

Status of an oval segment that is made of piezoelectric ceramics is governed by the piezoelectric equations

$$S_i = s_{ii}^E T_i + d_{3i} E_3, \tag{15.26}$$

$$D_3 = d_{3i} T_i + \varepsilon_{33}^T E_3, \tag{15.27}$$

where $i = 1, 3$ for the case that ceramics is polarized in the radial or circumferential directions. It follows from Equations (15.26) and (15.27) that

$$T_i^E = (1/s_{ii}^E) S_i = Y_i^E S_i,\; D_3^E = (d_{3i}/s_{ii}^E) S_i,\; \varepsilon_{33}^{S_i} = \varepsilon_{33}^T (1 - k_{3i}^2). \tag{15.28}$$

Thus, in all the above expressions for mechanical equivalent parameters Young's modulus Y must be replaced by $Y_i^E = (1/s_{ii}^E)$. The general formula (9.7) for the electromechanical energy associated with the flexural vibration will be specified in this case as

$$W_{em} = \frac{1}{2} ha \int_{-\beta}^{\beta} \int_{-t/2}^{t/2} D_3^E E_3 dz d\varphi = \frac{1}{2} \frac{ahd_{3i}}{s_{ii}^E} \int_{-\beta}^{\beta} \int_{-t/2}^{t/2} S_{\varphi f} E_3 dz d\varphi = \frac{1}{2} V \xi_{ro} n. \tag{15.29}$$

The variants of configurations of piezoelements that may produce the flexural deformation, out of which segments of fully active oval rings can be build, are shown in Figure 15.8.

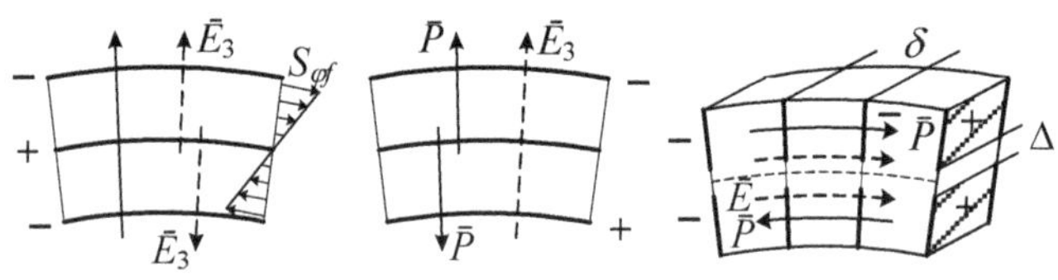

Figure 15.8: Configuration of piezoelements that produce the flexural deformations, and variants of electrodes connections: (a) transverse piezoeffect, parallel connection of electrodes; (b) transverse piezoeffect, series connection of electrodes; (c) longitudinal piezoeffect.

Also widely used are the half passive (bilaminar) designs of the rings, in which case one of the layers is made of passive material. Peculiarities of calculations for all these possible variants of piezoelements are the same as were considered in Chapter 7 regarding cylindrical transducers of the flexural type. The peculiarity of the oval type transducers is in the special mode shape configuration that is determined with help of FEA. Here we will consider the variant of employing the transverse piezoeffect with parallel connection of the layers (Figure 15.8

(a)).

The electric field in the oval segment that is comprised of two layers of ceramics may be represented in general form as

$$E_3(z,\varphi) = (2V/t)\Omega(z,\varphi), \tag{15.30}$$

where V is the voltage applied and $\Omega(z,\varphi) = \Omega_1(z)\Omega_2(\varphi_{el})$ is the function that depends on configuration and on the way of connection of electrodes located in the volume of the ring: $\Omega_1(z)$ - through the thickness ($\Omega_1(z) = 1$ at $z > 0$, $\Omega_1(z) = -1$ at $z < 0$), $\Omega_2(\varphi_{el})$ - along the circumference. After substituting function (15.30), expression for the electromechanical energy becomes

$$W_{em} = \frac{1}{2}V\xi_{r0}\frac{4d_{31}h}{s_{11}^E at}\int_{-t/2}^{t/2}\Omega_1(z)zdz\int_0^\beta \Omega_2(\varphi_{el})\left(\frac{\partial^2\theta_r}{\partial\varphi^2}+\theta_r\right)d\varphi = \frac{1}{2}V\xi_{r0}n. \tag{15.31}$$

Here will be denoted:

$$n = n'\cdot\Omega_n(\beta), \tag{15.32}$$

$$n' = \frac{4d_{31}h}{s_{11}^E at}\int_{-t/2}^{t/2}\Omega_1(z)zdz = \frac{d_{31}th}{s_{11}^E a} \tag{15.33}$$

in the variant of parallel connection of ceramic layers (variant shown in Figure 15.5 (a)), and it is twice smaller for the half passive design; and

$$\Omega_n(\beta) = \int_0^\beta \Omega_2(\varphi_{el})\left(\frac{\partial^2\theta_r}{\partial\varphi^2}+\theta_r\right)d\varphi. \tag{15.34}$$

Coefficient n' depends on the dimensions of the mechanical system, parameters of materials and on the design of the piezoelement. Coefficient $\Omega_n(\beta)$ depends on the angle β (on the mode shape of vibration) and on the status of the electrodes. Factor $\Omega_2(\varphi_{el}) = 1$ at the angles, at which the electrodes have a basic polarity; $\Omega_2(\varphi_{el}) = -1$ at the angles, at which polarity of the electrodes have opposite sign; and $\Omega_2(\varphi_{el}) = 0$ at the angles, at which electrodes don't exist. Decision on how to divide and electrically connect the electrodes depends on how the function looks

$$\left(\frac{\partial^2\theta_r}{\partial\varphi^2}+\theta_r\right) \sim T_\varphi(\varphi). \tag{15.35}$$

Table 15.5. Electromechanical transformation coefficients for different electrode configurations.

	Full Electrode	Central Electrode	End Parts are Connected in Opposite to Central
Ω_n	0.5	2.0	3.5

At the angles, at which the stress in circumferential direction changes sign, the electrodes must be split, and the obtained parts of electrodes may be connected in antiphase in order to realize the maximum possible value of the electromechanical transformation coefficient. For the first flexural mode $T_\varphi = 0$ at $\varphi \approx \pm 30°$. Values of factor Ω_n determined for the variants that electrodes have full size, when they are split at angles $\varphi \approx \pm 30°$ and the central part of the electrodes is used while the remaining parts are short circuited, and when the end parts are connected to the central part in series in opposite polarity are presented in Table 15.5.

15.3.4 Equivalent Circuit of the Oval Bender Transducer

In the context of this Chapter that is devoted to combining FEA and energy method technique the results related to the oval transducers are obtained as a "byproduct" of analysis intended for the flextensional transducers of Class IV, for which vibrating (and radiating) mechanical system has the same configuration. For the oval bender transducers per se it is not necessary in principle to have the central rod, as shown in Figure 15.1 (b), although the variant of design with passive central rod may also have sense, if increase of static strength of the transducer is needed. In this section the basic design of the oval bender transducer without the central rod will be considered. All the equivalent electromechanical parameters for the first flexural mode will be determined. This mode can be considered as isolated, because the resonance frequency of the next flexural mode is higher approximately in factor of 4 (see Table 15.1 and expressions for $M_{eqv\,f}$ and $K_{eqv\,f}$). Therefore, the common one Z_{ac} dimensional equivalent circuit shown in Figure 15.9 is representative, where expressions for the parameters are given by above formulas, and the numerical values of the corresponding coefficients Ω are determined for the opening angle $\beta = \pi / 4$ per one half of the oval mechanical system.

The note must be made regarding value of the radiation impedance introduced in the equivalent circuit. The wave size of the oval transducers is comparable with the wave size of the rectangular benders having approximately the same operating resonance frequencies. The related results that are presented in Section 9.4 are applicable for the oval transducers. Predominantly the transducers behave like three or two dimensional simple sources, in which case their radiation characteristics depend on the volume velocity $U_{\tilde{V}} = S_{av}\dot{\xi}_{ro}$, where S_{av} is expressed by formula (15.24). For the case that the opening angle is $\beta = \pi / 4$ $S_{av} = 0.64ha$

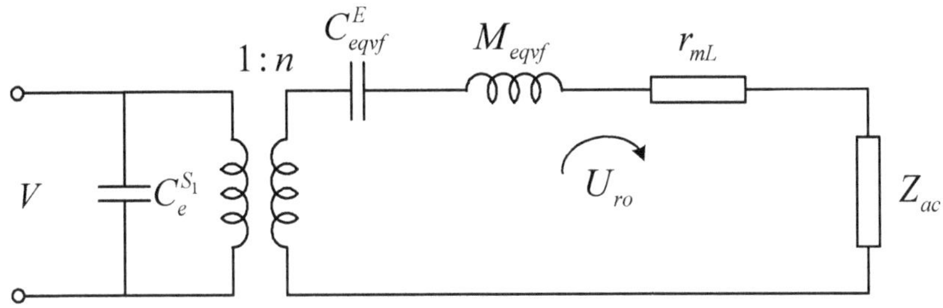

Figure 15.9: The equivalent electromechanical circuit of the oval driver transducer per one half of the mechanical system, $C^E_{eqvf} = 1 / K^E_{eqvf}$.

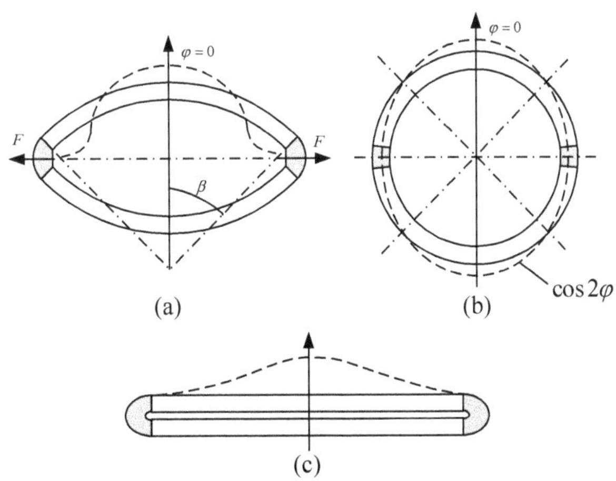

Figure 15.10: Illustration of geometries of the oval benders at limiting values of the opening angle: (a) arbitrary angle $\beta < \pi / 2$; (b) $\beta \to \pi / 2$, circular ring, $\xi_r = \xi_{ro} \cos 2\varphi$; (c) $\beta \to 0$ rectangular bender with ends clamped.

For different values of the opening angles β new values of the parameters must be calculated. Although the value $\beta = \pi / 4$ seems to be close to optimum by the following considerations. Because of symmetry the boundary conditions for the half oval segment at angles $\pm\beta$ are

$$\xi_z(\pm\beta) \approx 0\,, \quad \left.\frac{d\xi_z}{d\varphi}\right|_{\varphi=\pm\beta} \approx 0\,, \tag{15.36}$$

as is illustrated in Figure 15.10 (a).

At $\beta \to \pi/2$ these conditions correspond to the circular ring transducer undergoing flexural vibrations (Fig. 15.10 (b)) that is considered in detail in Section 7.3.1. Thus, with increase of angle β the mode shape tends to $\xi_r = \xi_{ro}\cos 2\varphi$, $S_{av} \to 0$, and the transducer becomes not efficient projector. In the limiting case at $\beta \to 0$ (in formula (15.1) radius $a \to \infty$ at b = const) configuration of the oval transducer tends to become rectangular bender made of beams with clamped ends (Fig. 15.10 (c)). The properties of such transducers are considered in Chapter 9, where it is shown that they have no advantages compared with the benders with simply supported ends.

It is noteworthy that transducer analogous by configuration to the oval transducer was considered in Ref. 7 under the title "A Conformal Driving Class IV Flextensional Transducer." In our notation this transducer presents a particular variant of the oval flexural transducer with half passive piezoelements employing transverse piezoeffect. Treating the transducer is produced in Ref. 7 by the FEA in full in contrast to the current approach. As the result, the information presented is of a qualitative nature and cannot be used without repeating procedures of FEA for designing transducers that employ different variants of the piezoelements or/and meet certain requirements. This example clearly illustrates the advantages of the adopted combined FEA and the energy method technique.

15.4 Equivalent Circuit of the Flextensional Transducer

The equivalent circuit of the flextensional Class IV transducer can be represented as the equivalent circuit of piezoceramic extensionally vibrating rod loaded on the ends by the input impedances of the oval shell. Therefore, at first the input impedances of the oval shell at the area of contact between the rod and the shell must be determined.

15.4.1 Input Impedance of the Oval Shell

For determining the input impedances of the oval shell, we assume that force F is applied to the shell in x direction, as shown in Figure 15.10 (a). As the first step, the reference point that was previously used for determining equivalent parameters of the oval shell must be placed at the point, to which the force F is applied, and displacement of this point in the x direction must be expressed through the displacement of the previous reference point, i.e., through ξ_{r0}. The ratio of the displacements is known as relation (15.22), i.e., $(\xi_{r0i} / \xi_{xi}) = \Omega_{\xi i}$ for each mode of vibration. Here and further, we will use numerical notations for the modes of vibration for brevity as follows: the first flexural mode – 1, the second flexural mode – 2, the extensional mode – 3. The values of Ω_i are given in Table 15.3 for the flexural and Table 15.4 for the extensional modes.

Thus, the total displacement in x direction can be represented as

$$\xi_x = \sum_{i=1}^{3} \xi_{xi} = \sum_{i=1}^{3} (\xi_{roi} / \Omega_i) . \tag{15.37}$$

The energy that is delivered by an external source of force F to vibrating oval segment (per one end) is

$$W_{ext} = F\xi_x = F\sum_{i=1}^{3} (\xi_{r0i} / \Omega_{\xi i}) . \tag{15.38}$$

The equivalent mechanical parameters that were determined regarding the reference point at $\varphi = 0$ and the corresponding generalized coordinates ξ_{r0i} must be recalculated to the new reference point and to the generalized coordinates ξ_{xi}. As the total values of the potential and kinetic energies should not change at this transformation, the equivalent rigidities and masses K_i and M_i per one end of the segment may be found from the equalities $2K_i(x)\xi_{xi}^2 = K_{eqvi}\xi_{r0i}^2$ and $2M_i\dot{\xi}_{xi}^2 = M_{eqvi}\dot{\xi}_{r0i}^2$ as

$$K_i = 0.5K_{eqvi}\Omega_{\xi i}^2 \text{ and } M_i = 0.5M_{eqvi}\Omega_{\xi i}^2 . \tag{15.39}$$

From the Lagrange's equations regarding the independent generalized coordinates ξ_{xi},

$$\frac{d}{dt}\left(\frac{\partial W_{kin}}{\partial \dot{\xi}_{xi}}\right) + \frac{\partial W_{pot}}{\partial \xi_{xi}} = \frac{\partial W_{ext}}{\partial \xi_{xi}}, \ i = 1, 2, 3 . \tag{15.40}$$

after substituting expressions for the energies involved, we arrive at the set of equations in the complex form

$$\left(j\omega M_i + \frac{K_i}{j\omega}\right)\dot{\xi}_{xi} = F\text{ , } i = 1, 2, 3. \tag{15.41}$$

With the mechanical losses and losses on the acoustic radiation included by observation it will be obtained that

$$\left(j\omega M_i + \frac{K_i}{j\omega} + r_{mLi} + Z_{aci}\right)\dot{\xi}_{xi} = F\text{ .} \tag{15.42}$$

Here the values of the radiation impedances, Z_{aci}, must be recalculated from the radiation impedances $Z_{ac\,oi}$ related to the reference point at $\varphi = 0$ in the same way as the mechanical equivalent parameters, namely, $Z_{aci} = 0.5 \cdot Z_{acoi}\Omega_{\xi i}^2$.

The input impedance per one end of the oval segment for each mode of vibration will be found from Eq. (15.42) as

$$Z_{in\,i} = \frac{F}{\dot{\xi}_{xi}} = jM_i + \frac{K_i}{j\omega} + r_{mL} + Z_{aci}\text{ .} \tag{15.43}$$

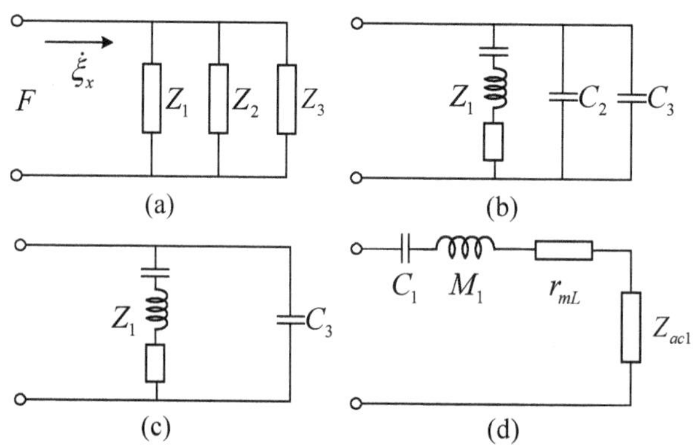

Figure 15.11: The equivalent circuits of the input impedance per one end of an oval segment: (a) for a broad frequency range, (b) for the frequency range around the resonance frequency of the 1st flexural mode, (c) with 2nd flexural mode neglected, (d) to the first approximation.

Considering the expression (15.37) for the total vibration velocity of the end and relation $\dot{\xi}_{xi} = F / Z_{in\,i}$, the total input impedance will be found as the impedance of the parallel connection of the partial impedances Z_i, and its equivalent circuit for a broad frequency range can be represented as shown in Figure 15.11 (a). The general representation of the input impedance may be simplified depending on the operating frequency range. Thus, in the frequency range around the resonance frequency of the 1st flexural mode the impedances of the 2nd flexural and extensional modes are governed by their rigidities, and the equivalent circuit of the input impedance can be represented, as shown in Figure 15.11 (b). Comparing rigidities K_2 and K_3 that can be made by using data presented in Table 15.3 and Table 15.4 results in

$$\frac{K_2}{K_3} = \frac{1}{12}\left(\frac{t}{a}\right)^2 \frac{\Omega_{kf2}\Omega_{\xi 2}^2}{\Omega_{ke}\Omega_{\xi 3}^2} = 12\,. \tag{15.44}$$

Thus, the contribution of impedance of the 2nd flexural mode can be neglected, and the equivalent circuit of the input impedance can be presented as shown in Figure 15.11 (c). Comparing the rigidities K_1 and K_3 results in their ratio

$$\frac{K_1}{K_3} = \frac{1}{12}\left(\frac{t}{a}\right)^2 \frac{\Omega_{kf1}\Omega_{\xi 1}^2}{\Omega_{ke}\Omega_{\xi 3}^2} = 0.07\,. \tag{15.45}$$

Thus, contribution of the extensional mode to the input impedance can be also neglected to the first approximation, though it has tendency of increasing with raising the thickness to radius ratio. The equivalent circuit of the input impedance to the first approximation is presented in Figure 15.11 (d).

This analysis shows that there is no need in calculating parameters of the second flexural and extensional modes of the oval segment to a great accuracy, as far as the frequency range around and below the resonance frequency of the first flexural mode is of interest, which is usually the case. For practical transducers designing their influence even can be neglected, and the equivalent circuit of the input impedance per one end of the oval segment shown in Figure 15.11 (d) can be used to the first approximation.

15.4.2 Equivalent Circuits of the Extensionally Vibrating Piezoceramic Bar

The well-known and widely used equivalent T- network circuit representation of longitudinally vibrating piezoceramic bar having two mechanical ports and electrical port that is described in Sections 5.8.3 and 10.2.3 is the most appropriate for use in the case under consideration. The circuit, as shown in Figure 15.12, is valid for both variants of the piezoceramic bar design: side electroded and segmented that realize the transverse and longitudinal piezoeffect under the assumption that number of segments comprising the bar is more than 6 or its wavelength is small.

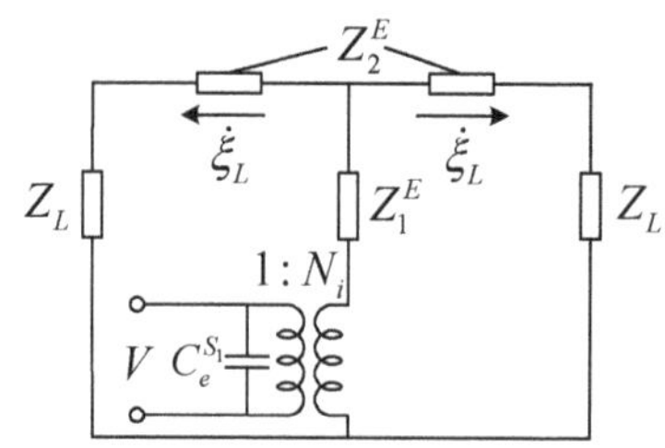

Figure 15.12: The equivalent circuit representation of longitudinally vibrating piezoceramic bar loaded at the ends by the impedances Z_L.

All the impedances for both cases have the same expressions for the values, namely,

$$Z_1^E = -j\rho c_i^E S_{cs} / \sin k_i^E l,\ Z_2^E = j\rho c_i^E S_{cs} \tan(k_i^E l/2). \tag{15.46}$$

Here S_{cs} is cross section area of the bar; l is the length of the bar; $c_i^E = \sqrt{Y_i^E / \rho}$, $i = 1, 3$ for the transverse and longitudinal piezoeffect, respectively. Parameters on the electrical side and electromechanical transformation coefficients differ and have the following values. For the transverse piezoeffect (assuming that electrodes are applied to the surfaces having width w)

$$N_i = N_1 = wd_{31} / s_{11}^E,\ C_e^{S_i} = C_e^{S_1} = \varepsilon_{33}^T (1 - k_{31}^2) wl / h, \tag{15.47}$$

and for the longitudinal piezoeffect of a segmented bar at parallel connection of the segments (under condition that number of segments is more than six)

$$N_i = N_3 = whd_{33} / s_{33}^E \delta,\ C_e^{S_i} = C_e^{S_3} = \varepsilon_{33}^T (1 - k_{33}^2) wlh / \delta^2, \tag{15.48}$$

where $\delta = l / N_\Sigma$, and N_Σ is number of the segments.

15.4.3 Equivalent Electromechanical Circuit of the Flextensional Transducer

In the case that piezoceramic bar is used as a driver of the flextensional transducer its ends must be loaded by the input impedances of the oval segments that were determined in Section 15.4.1 (impedances of two symmetrical segments for each end of the bar). Thus, in the circuit shown in Figure 15.12 the impedance Z_L must be replaced by $2Z_{in}$ and velocity $\dot{\xi}_L$ by $\dot{\xi}_x$, where Z_{in} is the input impedance on the end of the shell that must be determined from one of the circuits in Figure 15.11 depending on the approximation accepted. After this is done, the circuit in Figure 15.12 can be represented as shown in Figure 15.13

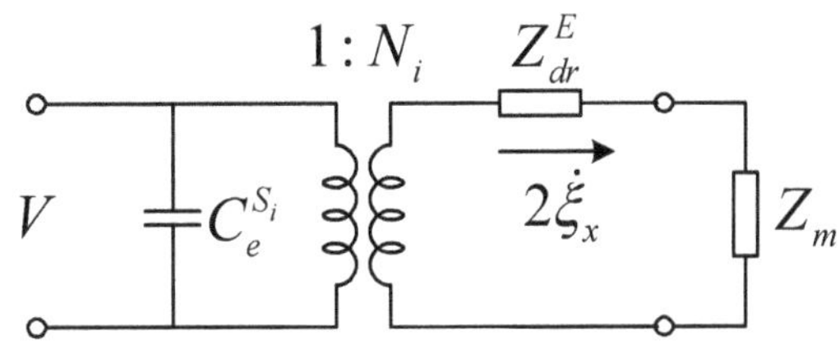

Figure 15.13: The one-dimensional representation of equivalent circuit of flextensional transducer for determining velocity $\dot{\xi}_x$.

In this variant of the equivalent circuit

$$Z_{dr}^E = -j\frac{\rho c_i^E S_{cs}}{\tan(k_i^E l)}. \tag{15.49}$$

At $k_i^E l \ll 1$, (which is usually the case for the flextensional transducers),

$$Z_{dr}^E = K_{dr}^E / j\omega + j\omega M_{dr} / 4, \tag{15.50}$$

where $K_{dr}^E = Y_i^E S_{cs} / l$ is the static rigidity of the bar and $M_{dr} = \rho whl$. At $\omega \to 0$

$$Z_{dr}^E = K_{dr}^E / j\omega. \tag{15.51}$$

After velocity $\dot{\xi}_x$ is found from the equivalent circuit of Figure 15.13, all the operating parameters of the flextensional transducer can be calculated so far as the mechanical equivalent and radiation related parameters of the oval segment reduced to the displacement ξ_x are already known. Distribution of radial velocity of the oval shell vibration is known as $\dot{\xi}_r(\varphi) = \dot{\xi}_{r0}\theta_r(\varphi)$, where $\xi_{r0} = \xi_x \Omega_{\xi 1}$, thus all the data needed for calculating the acoustic field radiated by the vibrating shell are available.

15.5 Example of Comparing the Operational Characteristics

With the equivalent electromechanical circuits known a comparison can be made between operating characteristics of the flextensional and oval bender transducers that use the vibrating systems of the same geometry. The goal of this Section is to illustrate how physically clear information regarding important properties of the transducers can be obtained, as result of calculating mode shapes of vibration by FEA in combination with energy method approach. The comparison will be made by the effective coupling coefficients, resonance frequencies and by the mechanical strength limited maximum operating depth of the transducers.

The calculations in this Section will be made using prototype examples of the transducers with only purpose to illustrate how the combined technique can be practically applied. The numerical results obtained must be considered as illustration of tendencies rather than characterization of real transducer designs. According to assumptions made regarding the prototypes geometry the following set of their relative dimensions will be used: $\beta = \pi/4$, $w/t = 1.5$, $t/a = 0.1$, $l/a = \sqrt{2}$.

15.5.1 The Effective Coupling Coefficients

The general definition for the effective coupling coefficient (2.88),

$$k_{eff}^2 = \frac{\text{energy stored in mechanical form at the working mode of vibration}}{\text{total input energy}}, \qquad (15.52)$$

can be directly used for estimating the effective coupling coefficients of the flextensional and analogous oval bender transducers at $\omega \to 0$.

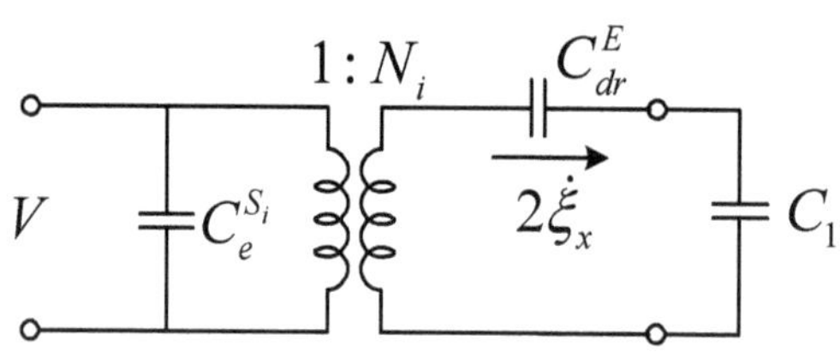

Figure 15.14: Equivalent electromechanical circuit of the flextensional transducer at $\omega \to 0$. $C_{dr}^E = 1/K_{dr}^E$, $C_1 = 1/K_1$.

15.5.1.1 The Flextensional Transducer

The equivalent circuit of a flextensional transducer in Figure 15.13 simplifiers at $\omega \to 0$ to the circuit in Figure 15.14. According to expression (15.43) $Z_{in} \to K_1/j\omega$, where K_1 is determined as (15.39), and $Z_{dr}^E \to K_{dr}^E/j\omega$ by expression (15.51).

The total input energy supplied to the transducer can be presented using this circuit as

$$W_{total} = W_{el}^{S_i} + W_{em}, \qquad (15.53)$$

where

$$W_{el}^{S_i} = V^2 C_e^{S_i}/2 \text{ and } W_{em} = (2\xi_x V N_i)/2. \qquad (15.54)$$

On the other hand

$$W_{em} = W_{mech} = (2\xi_x)^2 (K_{dr}^E + K_1)/2 = (2\xi_x V N_i)/2 \,.. \tag{15.55}$$

where K_{dr}^E is the static rigidity of the driving bar (15.51), and K_1 is the rigidity of the oval segment per one end from Eq. (15.39) (after the input impedance of the segment is reduced to impedance of the first flexural mode only, there is no need in subscript 1 in ξ_{x1}). From this equation follows that

$$\frac{V}{2\xi_x} = \frac{K_{dr}^E + K}{N_i} \,. \tag{15.56}$$

The total energy supplied to the transducer is

$$W_{total} = \frac{1}{2}[V^2 C_e^{S_i} + (2\xi_x)^2 (K_{dr}^E + K_1)] \,, \tag{15.57}$$

where the energy stored in the mode of vibration of the oval segment (the working mode of vibration) is

$$W_{fl1} = \frac{1}{2}(2\xi_x)^2 K_1 \,. \tag{15.58}$$

After substituting expressions (15.56) - (15.58) into expression(15.52) for k_{eff}^2, it becomes

$$k_{eff}^2 = \frac{1}{(K_{dr}^E C_e^{S_i} / N_i^2)\cdot(1 + K_1 / K_{dr}^E) + 1} \cdot \frac{1}{1 + K_{dr}^E / K_1} \,. \tag{15.59}$$

Summary of parameters involved in formula (15.59) for k_{eff}^2 is: $K_{dr}^E = Y_1^E wh/l$ from (15.51); $N_1 = wd_{31}/s_{11}^E$, $C_e^{S_i} = C_e^{S_1} = \varepsilon_{33}^T(1-k_{31}^2)wl/h$ from (15.47); $K_1 = 0.5(ht^3 Y / 6a^3)\Omega_{K1}\cdot\Omega_{\xi 1}^2$ from (15.15) and (15.39); $\Omega_{K1} = 30.2$, $\Omega_{\xi 1} = 2.0$ from Table 15.3. Note that $(K_{dr}^E C_e^{S_i} / N_1^2) = (1-k_{31}^2)/k_{31}^2$. For PZT-4 ceramics $k_{31} = 0.33$.

After substituting the above expressions for the parameters into formula (15.59). we arrive at expression for the effective coupling coefficient

$$k_{eff1}^2 = \frac{k_{31}^2}{7.8 - 1.2k_{31}^2} \,, \tag{15.60}$$

where from $k_{eff1} \approx 0.36k_{31} = 0.12$.

15.5.1.2 The Oval Transducer

Expressions for the energies involved in the definition (15.52) for the oval transducer may be obtained from the equivalent circuit in Figure 15.9. (Though the circuit is presented for one oval segment, the result will be the same as for the entire transducer, as all the energies in this case double due to symmetry). Namely,

$$W_{total} = \frac{1}{2}(V^2 C_e^{S_i} + \xi_{ro}^2 K_{eqvi}^E), \tag{15.61}$$

$$W_{em} = W_{mech} = \frac{1}{2}\xi_{ro}^2 K_{eqvi}^E = \frac{1}{2} V \xi_{ro} n. \tag{15.62}$$

From the last equality follows that

$$\frac{V}{\xi_{ro}} = \frac{K_{eqvi}^E}{n}. \tag{15.63}$$

After substituting expressions for the energies into (15.52), the effective coupling coefficient will be obtained as

$$k_{eff}^2 = \left(\frac{K_{eqvi}^E C_{el}^{S_i}}{n^2} + 1\right)^{-1}. \tag{15.64}$$

Note that $n^2 / K_{eqvi}^E C_e^{S_i} = \alpha_c$, and (15.64) is the common expression (2.93) for k_{eff}^2 of one degree of freedom piezoceramic transducer. Parameters involved in formula (15.64) are: $K_{eqv1}^E = 30.2ht^3 / 6s_{11}^E a^3$ from (15.15) and Table 15.3; $n = (d_{31}th / s_{11}^E a)\Omega_n$ from (15.32). If the electrodes are divided at angles $\varphi = \pm 30°$ and the end parts are connected in antiphase to the central part, from Table 15.5 follows that $\Omega_n = 3.5$. In this case $C_{el}^{S_1} = \varepsilon_{33}^T(1 - k_{31}^2)\pi ah / t$.

After substituting expressions for the parameters into formula (15.64) we obtain

$$k_{eff1}^2 = \frac{k_{31}^2}{1.3 - 0.3k_{31}^2}, \tag{15.65}$$

where from $k_{eff1} \approx 0.89k_{31} = 0.29$.

It is noteworthy that with replacing the transverse poled piezoelements by longitudinally poled segmented piezoelements (common for the bar, and such as shown in Figure 15.8 (c) for the oval segments) the effective coupling coefficients can be calculated from formulas (15.60) and (15.65) by changing k_{31}^2 to k_{33}^2 with negligible error.

The difference in the effective coupling coefficients of the flextensional and oval transducers is partially because in case of the flextensional transducer the active piezoelement is shunted by the passive shell. The difference may change to some extent, if in the flextensional transducer the ratio w/t is increased. Besides the amount of active material of the driver in the flextensional transducer can be increased in comparison with the oval segments in practical designs. But the main inaccuracy of the comparison made is that the transducers have different strength under hydrostatic pressure. The advantage of the flextensional transducer is that its shell can withstand greater hydrostatic pressure due to reinforcing function of the driver bar (putting aside that the shell can be made of a material stronger than ceramics). Introducing the reinforcing central passive bar into the oval transducer for achieving comparable static strength conditions will result in a significant drop of the effective coupling coefficient of the oval transducer. As it will be shown in Section 15.5.3, instead of $k_{eff1} = 0.29$ for the oval transducer without reinforcing bar by formula (15.65) it will be $k_{eff1} = 0.17$ with the reinforcing bar having the same rigidity as the driver of flextensional transducer has.

15.5.2 The Resonance Frequencies

The resonance frequencies of the oval and flextensional transducers can be determined using the equivalent circuits in Figure 15.9 and Figure 15.13. Thus, for the oval transducer

$$f_{ro} = \frac{1}{2\pi}\sqrt{\frac{K^E_{eqvi}}{M_{eqvi}}} . \tag{15.66}$$

When determining the resonance frequency of the flextensional transducer, the impedance Z^E_{dr} in Figure 15.13 must be used in the form of expression (15.50). This will result in

$$f_{rf} = \frac{1}{2\pi}\sqrt{\frac{K_1 + K^E_{dr}}{M_1 + M_{dr}/4}} = \frac{1}{2\pi}\sqrt{\frac{K_1}{M_1}}\sqrt{\frac{1+K^E_{dr}/K_1}{1+M_{dr}/4M_1}} . \tag{15.67}$$

Note that the equivalent rigidity and mass in formula (15.66) are for the piezoceramics and in formula (15.67) K_{eqv} and M_{eqv} are for the passive material, of which the oval segments are made. At the same dimensions of the oval shell

$$\frac{f_{rf}}{f_{ro}} = \sqrt{\frac{Y_p\rho_c}{Y^E_i\rho_p}}\sqrt{\frac{1+K^E_{dr}/K_1}{1+M_{dr}/4M_1}} , \tag{15.68}$$

where subscript p stays for "passive". Given that for PZT-4 ceramics, Y^E_1 = 81 GPa and ρ_c =

7500 kg/m³, and for aluminum, $Y_p = 70$ GPa and $\rho_p = 2700$ kg/m³, in the example considered this ratio is

$$f_{rf} = 3.2 f_{ro}. \tag{15.69}$$

The flextensional transducer has higher resonance frequency than oval transducer having the same dimensions partially because its shell is made of the lighter material, but mainly due to increase of the potential energy of its mechanical system due to additional rigidity of the driver bar.

The above estimations do not mean that the practical designs of the oval transducers may have advantages over the flextensional transducers in underwater applications, because no real environmental and operating conditions were taken into consideration. The most challenging among these conditions is the hydrostatic pressure and related issues of the static mechanical strength of the transducer designs. The low frequency transducers of flextensional type were introduced as alternative to the flexural plate drivers, which are extremely vulnerable to the hydrostatic pressure, as it was articulated in Chapter 14. The oval driver transducers may be considered as possible improvement of the rectangular benders (see the considerations related to Figure 15.10), if they can withstand greater hydrostatic pressures, and remain sufficiently electromechanical effective. But a fair comparison between the flextensional and oval transducers must be produced with accounting for their strength under the hydrostatic pressure.

15.5.3 Mechanical Strength of the Transducers Under the Hydrostatic Pressure

The static strain in the oval shell can be determined by formula (4.221) presented for the circular ring in Section 4.4.4,

$$S_\varphi = \frac{1}{a}\left(\frac{\partial \xi_\varphi}{\partial \varphi} + \xi_r\right) - \frac{z}{a^2}\left(\frac{\partial^2 \xi_r}{\partial \varphi^2} + \xi_r\right), \tag{15.70}$$

or

$$S_\varphi = \xi_{r0}\left[\frac{1}{a}\left(\frac{\partial \theta_{\varphi h}}{\partial \varphi} + \theta_{rh}\right) - \frac{z}{a^2}\left(\frac{\partial^2 \theta_{rh}}{\partial \varphi^2} + \theta_{rh}\right)\right] = S_{\varphi e} + S_{\varphi f}. \tag{15.71}$$

Here θ_{rh} and $\theta_{\varphi h}$ are the distributions of displacements under the hydrostatic pressure in the radial and tangential directions normalized to displacement ξ_{r0} of the reference point; $S_{\varphi e}$ and $S_{\varphi f}$ are the strains that correspond to the extensional and flexural deformations, respectively. φ The corresponding stresses are,

$$T_{\varphi e}(\beta,\varphi)=\xi_o\frac{Y}{a}\left(\frac{\partial\theta_{\varphi h}}{\partial\varphi}+\theta_{rh}\right)=\xi_o\frac{Y}{a}\Omega_{he}(\beta,\varphi) \tag{15.72}$$

and

$$T_{\varphi f}(\beta,\varphi)=\xi_o\frac{zY}{a^2}\left(\frac{\partial^2\theta_{rh}}{\partial\varphi^2}+\theta_{rh}\right)=\xi_o\frac{zY}{a^2}\Omega_{hf}(\beta,\varphi)\,. \tag{15.73}$$

Here the functions $\Omega_{eh}(\beta,\varphi)$ and $\Omega_{fh}(\beta,\varphi)$ are introduced for brevity.

The distributions of static displacement were determined by FEA for $\beta=\pi/4$ and for the range of ratios t/a = 0.05 to 0.15. Their plots are presented in Figure 15.15.

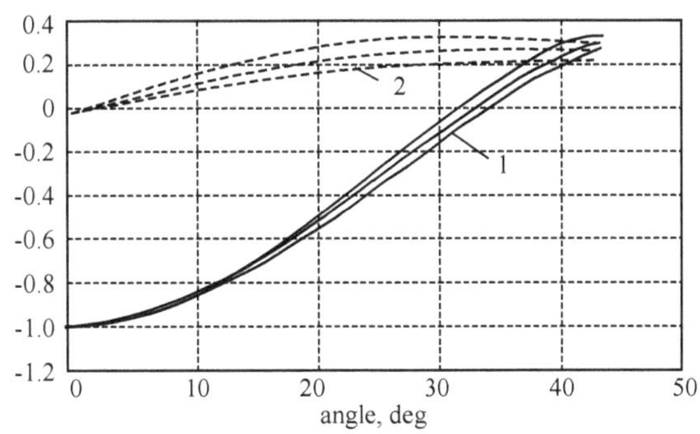

Figure 15.15: Deviations of the normalized displacements in radial and tangential direction for oval shells with t/a ranging between 0.05 and 0.15. Normalizing coefficient for t/a = 0.1 is ξ_{r0} = 0.11 mm per 1 Pa of hydrostatic pressure.

As within this range of thickness to radius ratio the distributions do not deviate noticeably from those at t/a = 0.1, further analysis will be restricted to this ratio. Thus, the functions in (15.71) and (15.72) will depend on the angle φ only.

It is noteworthy that the mode of static deflection practically coincides with the first flexural dynamic mode of vibration that is shown in Figure 15.7 (just in different scale). This is in line with the rule that the mode shapes of vibration at different boundary conditions can be determined as static deflections under uniform load (hydrostatic pressure in particular).

Distribution of displacements in the same oval shell are shown in Figure 15.16 for the case that a central bar is installed into the shell analogous to the driver of flextensional transducer with goal of reducing stress in the shell. In this case it is assumed that the shell is made of aluminum and the bar is made of steel. The distributions of displacement changed to some extent, but what is more important the normalizing coefficient in the case with the central bar is reduced in factor of six, which means that the stresses in the shell are reduced accordingly.

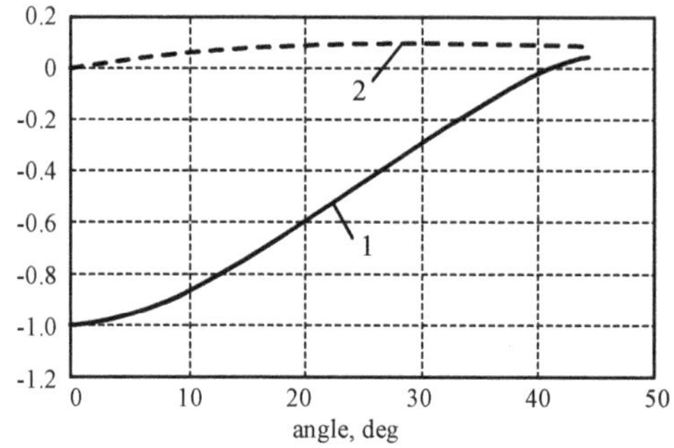

Figure 15.16: Normalized displacement in radial (1) and tangential (2) direction for the oval shell made of aluminum at $t/a = 0.1$ with the steel central bar. The normalizing coefficients is ξ_{r0} = 0.016 mm per 1 Pa of the hydrostatic pressure.

The corresponding distributions of stress in the circumferential direction are shown in Figure 15.17 and Figure 15.18 for the oval driver transducer without and with the central bar, respectively, at thickness to radius ratio $t/a = 0.1$. Remember that according to the adopted rule of signs the tension is positive, and compression is negative.

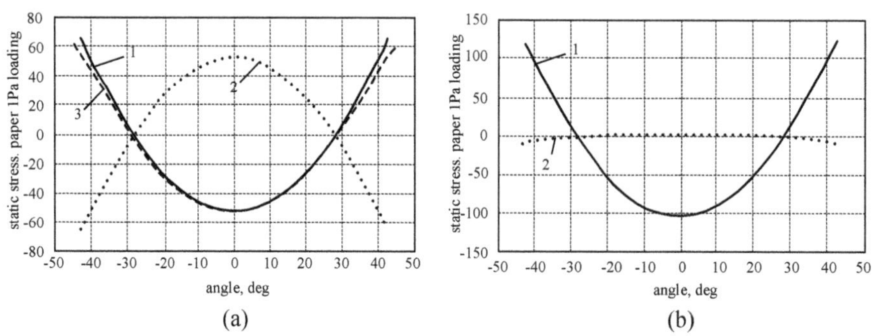

Figure 15.17: Stress developing in the oval transducers without central bar per 1 Pa of hydrostatic pressure: (a) flexural stress (top half (1) and bottom half (2)) and extensional stress (3); (b) sum of the flexural and extensional stress (top half (1) and bottom half (2)).

Thus, the central bar has a significant influence on the magnitude and distribution of the static stress in the shell. Quantitatively this effect depends on the rigidity of the bar, which is proportional to Young's modulus of material used. The plots in Figure 15.18 are obtained for the extreme case that bar is made of steel with $Y = 210$ GPa to clearly illustrate the tendency. In reality choosing the material for the strengthening bar is a tradeoff because its effect does not come for free. Introducing the central bar results in an increase of the resonance frequency and reducing the effective coupling of the oval transducer.

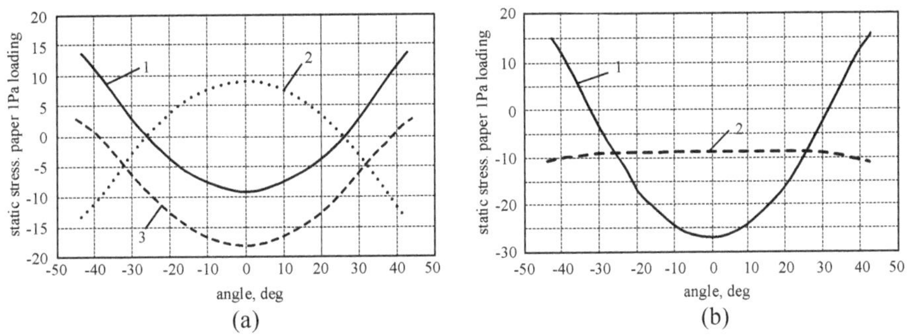

Figure 15.18: Stress developing in the oval shell with a central bar per 1 Pa of hydrostatic pressure: flexural stress (top half (1) and bottom half (2)) and extensional stress(3); (b) sum of the flexural and extensional stress (top half (1) and bottom half (2)).

Both effects are due to the appearance of an additional potential energy of deformation of the electromechanically passive bar. This can be illustrated with the example considered in Section 0 with help of the equivalent circuit of the oval transducer shown in Figure 15.9. We will assume that the bar has the same rigidity as the piezoceramic bar of the flextensional transducer ($Y = 81$ GPa). The additional rigidity, K_{ad}, must be introduced into the equivalent circuit to account for the potential energy of deformation of the bar, which is

$$W_{pot} = \frac{1}{2} K_{bar} \xi_x^2 = \frac{1}{2} \frac{K_{bar}}{\Omega_\xi^2} \xi_{ro}^2 = \frac{1}{2} K_{add} \xi_{ro}^2 . \tag{15.74}$$

Here $K_{add} = K_{bar} / \Omega_\xi^2$, $K_{bar} = YS_{cs} / l$ is the static rigidity of the bar and $\Omega_\xi = 2.13$ from Table 15.3.

The effective coupling coefficient of the oval transducer with the additional rigidity of the central bar, k'_{effi}, can be calculated by formula that may be obtained by including the additional passive rigidity into expression (15.64) in the following way,

$$k'^2_{effi} = \frac{1}{(K^E_{eqvi} C^{S_i}_e / n^2_i)(1 + K_{add} / K^E_{eqvi}) + 1} = \frac{k^2_{effi}}{1 + (K_{add} / K^E_{eqvi})(1 - k^2_{effi})}. \tag{15.75}$$

In the example that we consider $K_{add} / K^E_{eqv} = 2.3$ and $k'^2_{effi} = k^2_{effi} / \left(3.3 - 2.3k^2_{effi}\right)$.

Thus, instead of $k_{eff1} = 0.29$ for the oval transducer without reinforcing bar (formula (15.65)) it will be $k_{eff1} = 0.17$ with the bar, i.e., the drop of effective coupling is dramatic.

So much attention to distribution and magnitude of stress in the volume of the oval transducers is justified, because they can be considered as an alternative to the rectangular benders, in which ceramics is especially vulnerable to tension that is developing under the hydrostatic pressure in half of their volume. As it was previously mentioned and illustrated with Figure 15.15, configuration of the rectangular bender can be imagined as extreme case of the oval shell at $\beta \to 0$. Another extreme case of the oval shell at $\beta \to \pi / 2$ being circular ring vibrating in flexure along the circumference. The case at $\beta = \pi / 4$ can be regarded as intermediate and close to optimal in terms of its electroacoustic parameters. The worst in terms of the static strength is the rectangular bender (at $\beta \to 0$), in which case half of the volume experiences tension. The best is the flexural ring transducer, in which case all the volume is under compression. And the oval transducer at $\beta = \pi / 4$ is somewhere in the middle. Given that the flexural cylindrical transducer without baffling parts of the surface vibrating in opposite phase is not effective low frequency projector (with $S_{av} = 0$), the oval transducer at $\beta = \pi / 4$ may be considered as a real alternative to the rectangular bender for underwater applications that require operating at a moderate depth. It is remarkable that all the volume of the oval shell except for the segments of upper half beyond the angles $\varphi = \pm 30^o$ does not experience tension under the hydrostatic pressure, as it follows from plots in Figures 15.17 (b) and 15.18 (b). Moreover, in the variant with the central bar all this volume is under compression.

In addition to the distrbutions of stress formulas for the maximum values of the stress at the center of oval segment (at $\varphi = 0$) must be provided. For the designs without central bar theycan be obtained from expressions (15.72) and (15.73). Dependence of the maximum compression stress in the extensional mode, $T_{\varphi e}(0)$, is

$$T_{\varphi e}(\beta, 0) = p_h \frac{a}{t} \Omega_{he}(\beta, 0), \tag{15.76}$$

where p_h is the hydrostatic pressure, and coefficient $\Omega_{he}(\beta, 0)$ at $\beta = \pi / 4$ is $\Omega_{he} \approx 5$.

Express $T_{\varphi e}(\beta,0) = p_h \frac{a}{t}\Omega_{he}(\beta,0)$ ion for the stress $T_{\varphi f}(0)$ can be obtained in the following way. As it was noted, the mode of displacement under the hydrostatic pressure may be approximated by the first resonance mode of the flexural vibration. Therefore, the displacement ξ_{r0} under the hydrostatic pressure can be found as

$$\xi_{r0} = p_h S_{av1} / K_{eqv1}, \tag{15.77}$$

where K_{eqv1} and S_{av1} are determined by formulas (15.15) and (15.24). After substituting this expression for ξ_{r0} into (15.73), we arrive at the formula for the maximum stress at $z = t/2$,

$$T_{\varphi f}(\beta,0) = p_h \frac{6a^2}{t^2}\frac{\Omega_{S1}(\beta)}{\Omega_{Kf}(\beta)}\Omega_{hf}(\beta,0). \tag{15.78}$$

By calculation from the flexural mode shape at $\beta = \pi/4$, $\Omega_{hf}(\pi/4,0) = 8.3$. From Table 15.3 follows that $\Omega_s(\pi/4) = 0.32$ and $\Omega_{Kf} = 30.2$. Therefore,

$$T_{\varphi f\,\max}(0) = -0.5\frac{a^2}{t^2}p_h. \tag{15.79}$$

Now all the information needed for estimating effects of hydrostatic pressure on parameters of an oval transducer, and for taking decision regarding rational configuration of active parts of its mechanical system is available. Together with the equivalent electromechanical circuits and expressions for the equivalent parameters of the oval and flextensional transducers derived above this makes possible analytical designing the transducers that meet certain operating requirements. Detailed analyzing properties of transducers of these types can be performed in the same way, as it was illustrated with numerous examples in the preceding chapters.

Summarizing the results presented in this chapter it may be concluded that combining the FEA technique for calculating mode shapes of vibration of mechanical systems and energy method of analysis the transducer properties based on using the obtained mode shapes may provide a physical clear analytical approach to designing the transducers.

15.6 References

1. K. D. Rolt, "The history of the flextensional electroacoustic transducer", J. Acoust. Soc. Am, 87,1340-1349, (1990).
2. C. H. Sherman and J. L. Butler, *Transducers and Arrays for Underwater Sound*, (Spinger, New York, 2007), pp. 126-142.
3. Ralf S. Woollett, *The Flexural Bar Transducer*, p. 250, Published by Naval Underwater System Center, New London, Connecticut,
4. David A. Brown, Xiang Yan and Boris Aronov, "The energy method for solving electroacoustic problems using combining finite element analysis and analytical methods", J. Acoust. Soc. Am. 141, 3745, (A), (2017).
5. Randy H. Shin, *Introduction to Finite Element Analysis using Solid Works Simulation*, Oregon Institute of Technology,
6. B. S. Aronov, "Piezoelectric circular ring flexural transducers," J. Acoust. Soc. Am., 134 (2),1021-1030, (2013)
7. Tianfang Zhou, Yu Lan, Quicheng Zhang, Jingwen Yuan, Shichang Li, and Wei Lu, "A Conformal Driving Class IV Flextensional Transducer", Sensors, 2018, 18, 2102

AFTERWORD

The energy methods used throughout this treatment are inherently approximate. Real distributions of displacements in the mechanical systems of the transducers (the mode shapes) are represented as modes of displacements under static loads, or as expansions in a finite series of supporting functions. This allows simplifying the problems and achieving physically clear results by relatively elementary means. But this simplicity does not come for free. The question arises regarding accuracy of the finite results of calculations. To answer this question and to estimate an acceptable accuracy of the approximations it must be known first, what values of parameters of transducers under consideration may be considered as accurate. In fact, the accurate (true) values of parameters are not known a priory and it must be established what values may be regarded as such. The most convincing and close to the true values may be considered those obtained as result of physical experimental verification. But results of an experiment are not accurate themselves. Thus, by estimations made in the book of R Bobber, *Underwater Electroacoustic Measurement,* accuracy of measurements of acoustic parameters may be on the order of (5-10) %. Probably, the most accurate may be results of measuring the resonance frequencies of the mechanical systems in air. The results of calculating this very important characteristic are the most vulnerable to not appropriate approximations made. Other widely used "true values" of parameters are those obtained analytically under ideal assumptions regarding dimensional aspect ratios for the mechanical systems: $t/l \ll 1$ for longitudinally vibrating rods; $t/a \ll 1$ ($t/l \ll 1$) for flexural vibration of plates (beams) according the elementary theory of bending. Numerous data thus obtained for the mechanical systems under different boundary conditions can be found in literature, for example, in the book of Leissa, *Vibration of Plates*, though no estimations are provided on how small this aspect ratios must be for considering the results to be sufficiently accurate. Real mechanical systems have finite aspect ratios, and one must be very cautious using this data, moreover, that in the most cases they are presented for several high modes of vibration. Thus, for example, in Table 2.3 of referred book values of the resonance frequency related parameter $\lambda^2 = \omega a^2 \sqrt{\rho / D}$ are given for the first three consecutive modes of vibration of the simply supported circular plate as 4.977, 13.94, 25.65 without presenting information on how small the aspect ratio for the plate must be to insure four-digit accuracy of the results. At the same time, if to consider corrections due to rotary inertia and shear deformations presented in Section 9.4.3 for a plate with finite aspect ratio to results of the elementary theory, f_r, the valuer of resonance frequency for the first mode of vibration, f'_r, must be calculated by formula

$$\frac{f'_r}{f_r} = \sqrt{\frac{1-0.5[(1+\sigma)/(1-\sigma^2)](t/a)^2}{1+0.5(t/a)^2}}\ .$$

(See formula (9.268), where for passive elastic materials $Y^E_{\sigma a} = Y/(1-\sigma^2)$ and s^D_{44} must be replaced by $1/\mu = 2(1+\sigma)/Y$). For fairly thin plate at $t/a = 0.2$ the ratio is 0.97. To ensure the same level of accuracy of approximation to results of the elementary theory for the higher modes of vibration, the thickness to radius aspect ratio must be reduced roughly in number of the mode square times (see S. P. Timoshenko, *Vibration Problems in Engineering*.) Thus, for the second and third modes of a circular plate t/a must be less than 0.05 and 0.02, respectively. Such thin plates are hardly practical. This example shows that in case of the simply supported plates presenting four-digit accurate values for the resonance frequencies related parameter is misleading, moreover that no restrictions on the aspect ratio of the plates are imposed. Analogous reasonings are applicable for results of calculations that are based on elementary theory of longitudinal vibration. Extent of the aspect ratios, to which the results can be considered accurate enough, can be estimated using data regarding the coupled vibrations in the corresponding systems that are presented in Section 4.6.

Besides of the systematic errors due to not appropriate aspect ratios of the vibrating bodies it must be remembered that parameters of materials used in the transducer designs are determined in experimental way and their values as a rule are presented in specifications with not more than three-digit accuracy.

In summary, it can be concluded that the upper level of accuracy in determining transducer related quantities may not be more than three-digits. Presenting results of calculating with bigger precision does not make sense and may even be misleading.

LIST OF SYMBOLS

Symbol	Description
A	radius
B	bulk modulus
c, c_c, c_w	sound speed, speed of sound in ceramic composition and in water
c_{mi}^E	elastic stiffness of a piezoceramics at constant electric field
C, C_e^S	capacitance, capacitance of blocked transducer
C, C_{eqv}^E	compliance, equivalent compliance of a mechanical system at constant electric field
d, d_{mi}	separation, distance; piezoelectric constant
D	diameter, flexural rigidity $D = Yh^3/12(1-\sigma^2)$
D_i, D_i^E	charge density, charge density at constant electric field
e_{mi}^E	piezoelectric constant, $e_{mi} = d_{mj}c_{ji}^E$, j =1…6
E, E_{op}, E_p	electric field, operating field, permissible field
Ef	effectiveness
f, f_r, f_{ar}, Δf	frequency, resonance frequency, antiresonance frequency, deviation of frequency
f_{ip}	partial resonance frequencies of a coupled system
F, F_{eqv}	force, equivalent force
G	torsional rigidity
h	height
$H(\theta,\varphi)$	directional factor
I	current
I_L, I_C, I_m	current through inductance, current through capacitance, motional current
J, J_p	moment of inertia, polar moment of inertia
k; k_c, k_{eff}; k_{dif}	wave number $k = \omega / c$; electromechanical coupling coefficient, effective coupling coefficient; diffraction coefficient
k_E, k_T	reserves of the electrical and mechanical strength coefficients
K, K_{eqv}^E, K_{il}	rigidity, equivalent rigidity of a mechanical system, mutual rigidity of coupled systems
ΔK	additional rigidity term that characterizes electrical interaction between elements in nonuniformly deformed piezoelectric body

Symbol	Description
$l,\ t,\ w$	length, thickness, width
$L\ ;\ L_p,\ L_s$	Lagrangian, inductance; parallel and series inductances
ms_w	Mismatch coefficient, $ms_w = r_{ac} / r_{opt}$
ms_i	mode shape coefficient
$M\ ;\ M_{eqv},\ M_{il}$	Moment, total mass; equivalent mass, mutual mass of coupled systems
n	turns ratio, electromechanical transformation coefficient,
$N,\ N_i$	Number of segments in segmented mechanical system, electromechanical transformation coefficients, $i = 1,3$.
o	subscript that denotes a reference point
$P\ ,\ P_o\ ;\ P_h$	sound pressure, sound pressure of simple source; hydrostatic pressure
$Q,\ Q_e,\ Q_m$	quality factor, $Q = W_{kin} / W_{Loss}$; electrical and mechanical quality factors
$r,\ \boldsymbol{r}$	distance, radius vector
$r,\ r_{mL};\ r_{ac},\ r_{opt}$	resistance, resistance of mechanical loss; radiation resistance, optimal value of the radiation resistance
$R,\ R_{eL}$	resistance, resistance of electrical loss
s_{mi}^{E}	elastic compliance of piezoceramics at constant electric field
$S\ ,S_{ik},\ S_i$	deformation, tensor of deformation $(i,k = 1,2,3)$, tensor of deformation $(i = 1,..,6)$
$S_{\Sigma},\ S_{av},\ S_{eff}$	surface area, average surface area, effective surface area
$T\ ,\ T_{ik},\ T_i$	stress, stress tensor $(i,k = 1,2,3)$, stress tensor $(i = 1,..,6)$
$T_{op},\ T_p$	operating stress, permissible stress
$u,\ U\ ;\ U_o,\ U_i$	Velocity; velocity of reference point, velocity of reference point in i^{th} mode of vibration
$U_{\tilde{V}}$	volume velocity
$v,\ V$	voltage
$\tilde{V}$	volume
$w;\ w_{\text{int}},\ w_e,\ w_{mch},\ w_{em}$	width, energy density; densities of the internal, electrical, mechanical, and electromechanical energies
$W\ ,\ \dot{W}\ ,\bar{\dot{W}}$	energy, energy flux (power), complex power
$W_{el},\ W_e^S$	total electrical energy, electrical energy stored in a blocked piezoelement
$W_{\text{int}},\ W_m,\ W_{em},$	internal, mechanical, electromechanical, and acoustic energies

Symbol	Description
W_{ac}	
W_{kin}, W^E_{pot}	kinetic energy, potential energy at constant electrical field
W_{eL}, W_{mL}	energies of electrical and mechanical W_{kin} l loss
$\dot{W}_{mE}$, $\dot{W}_{mT}$	maximum power electric field limited and mechanical stress limited
ΔW	additional energy term that characterizes electrical interaction between elements in nonuniformly deformed piezoelectric body
$x;\ x_{ac}$	coordinate; reactance of acoustic radiation
$y;\ y=\delta/t$	coordinate; ratio of thickness of active layer to total thickness of mechanical system
$Y,\ Y_i^E=1/s_{ii}^E$	Young's modulus, Young's modulus of piezoceramics (i =1, 3)
Y_a^E, Y_p	Young's moduli of active and passive materials
Y_σ	$Y_\sigma = Y/(1-\sigma^2)$
$z;\ z_{il}$	Coordinate; mutual impedance between modes of vibration
$Z,\ Z_{il}=z_{il}U_i/U_l$	impedance, introduced impedance
Z_m, Z_m^E, Z_{in}	mechanical impedance, impedance at constant electric field, input impedance
Z_{ac}	radiation impedance
α_{ac}	nondimensional coefficients of the radiation resistance
$\alpha_c = n^2C_m^E/C_e^S$	coefficient related to effective coupling coefficient, $k_{eff}^2=\alpha_c/(1+\alpha_c)$
β_{ac}	nondimensional coefficient of the radiation reactance
$\beta = f_{1p}/f_{2p}$	detuning factor between partial frequencies of a coupled system
γ, γ_m, γ_k,	coefficient of coupling between partial systems, coefficients of inertial and elastic coupling
γ_Y	$\gamma_Y = Y_p/Y_a^E$
γ_ρ	$\gamma_\rho = \rho_p/\rho_a$
η ; η_{em}, η_{ma}, η_{ea}	efficiency; electromechanical, mechanoacoustic, electroacoustic efficiencies
δ	separation between electrodes,
δ_e, δ_m	angles of dielectric and mechanical losses, $\tan\delta_e = 1/Q_e$, $\tan\delta_m = 1/Q_m$
ε ; ε_{ik}^T, ε_{ik}^S	dielectric constant; tensors of dielectric constants of piezoceramics at free and clamped conditions
θ ; $\theta(\boldsymbol{r})$	angle, mode shape

Symbol	Description
λ	wavelength, Lame constant
μ	Lame constant (share modulus)
ξ ,	displacement, displacement of reference point
ρ , ρ_a , ρ_p	density, density of the active and passive materials
σ , σ_i^E	Poisson's ratio; Poisson's ratio of piezoceramics, $\sigma_1^E = -s_{12}^E / s_{11}^E$, $\sigma_3^E = -s_{13}^E / s_{33}^E$
Σ	surface in general
φ	angle
χ	diffraction function
ω , ω_r , ω_{ar}	angular frequency, resonance and antiresonance frequencies
$\Omega = f^2 / f_{1p}^2$	nondimensional frequency factor
$\Omega = 2\Delta f / f_r$	normalized bandwidth

1. Vectors are displayed in bold letters.
2. Low case letters denoting the time dependent quantities indicate instantaneous values; the capital letters are values in rms.
3. An overbar on a capital letter denotes a complex quantity.

APPENDIX A. Properties of Passive Materials

Table A.1: Elastic properties of the passive materials *).

Material	Y (GPa)	σ	$\rho\cdot10^{-3}$ (kg/m^3)	c (m/s)	$\rho c\cdot10^{-6}$ (kg/m^{2}s)
Aluminum	71	0.33	2.7	5130	13.5
Alumina	300	0.21	3.7	9000	33.3
Beryllia, BeO	345	0.26	3.0	10,700	32.1
Beryllium Cu	125	0.30	8.2	3900	32,0
Brass	97	0.31	8.5	3400	29.0
Corprene	0.23	0.43	1.1	460	0.51
Glass	62	0.24	2.3	5200	12
G-10	24	0.14	1.8	3600	6,6
Invar	148	0.3	8.0	4300	34
Lead	16.5	0.44	11,3	1200	13.6
Macor	67	0.29	2.5	5180	13
Pyrex	64	0.24	2.3	5300	12
Stainless steel	193	0.28	7.9	4940	39
Tin	50	0.36	7.3	2600	19
Titanium	104	0.36	4.5	4810	21.6

*) Bulk modulus $B = Y/3(1-2\sigma)$. Shear modulus $\mu = Y/2(1+\sigma)$

Table A.2: Properties of the fluids at room conditions

Liquid	Air	Water	Seawater	Castor oil	Motor oil SAE-30	Hydraulic fluid ISO 32	Silicon oil
B, GPa	$142\cdot10^{-6}$	2.15	2.34	2.1	1.5	1.8	2.1
$\rho\cdot10^{3}$ kg/m3	$1.2\cdot10^{-3}$	1.0	1.02	0.96	0.88	0.86	0.97
c, m/s	340	1500	1500	1470	1300	1450	1500

Table A.3: Properties of the polyurethanes

Property		ρ, kg/m3	c, m/s	B, GPa	G, MPa
PR1547	4°C	1.05	1650	2.9	6
	34°C		1500	2.3	4
GS960PU, 20°C		1.08	1700	3.3	1.2

APPENDIX B. Properties of Piezoelectric Ceramics

Table B.1: Piezoelectric constants

Property	PZT-4 Type I	PZT-5A Type II	PZT-8 Type III	PZT-5H Type VI
s_{11}^E, 10^{-12} m²/N	12.3	16.4	11.5	17.7
s_{33}^E	15.5	18.8	13.5	21.9
s_{13}^E	−5.31	−7.22	-4.8	–
s_{12}^E	−4.05	−5.74	-3.7	-5.7
s_{44}^E	39.0	47.5	31.9	–
s_{11}^D	10.9	14.4	10.1	15.5
s_{33}^D	7.9	9.46	8.5	10.5
s_{13}^D	−2.1	−2.98	-2.5	–
s_{12}^D	−5.42	−7.71	-4.5	-7.6
s_{44}^D	19.3	25.2	22.6	–
s_{66}	32.7	44.3	30.4	48.5
c_{11}^E, 10^{10} N/m²	13.9	12.1	14.9	–
c_{33}^E	11.5	11.1	13.2	–
c_{13}^E	7.43	7.52	8.11	–
c_{12}^E	7.78	7.54	8.11	–
c_{44}^E	2.56	2.11	3.13	–
c_{11}^D	14.5	12.6	15.2	–
c_{33}^D	15.9	14.7	16.9	–
c_{13}^D	6.09	6.52	7.03	–
c_{12}^D	8.39	8.09	8.41	–
c_{44}^D	5.18	3.97	4.46	–
c_{66}	3.06	2.26	3.40	–
d_{33}	289	374	225	620
d_{15}	496	584	330	–
e_{31}, C/m²	−5.2	−5.4	-4.1	–
e_{33}	15.1	15.8	14.0	–
e_{15}	12.7	12.3	10.3	
K_{11}^S	730	916	900	–
$\varepsilon_{33}^S / \varepsilon_0$ [4)]	635	830	600	–

Property	PZT-4 Type I	PZT-5A Type II	PZT-8 Type III	PZT-5H Type VI
$\varepsilon_{11}^T / \varepsilon_0$	1475	1730	1290	–
$\varepsilon_{33}^T / \varepsilon_0$	1300	1700	1000	–
k_{31}	0.334	0.344	0.30	0.35
k_{33}	0.7	0.705	0.64	0.72
k_{15}	0.71	0.685	0.55	–
k_p	0.58	0.60	0.51	0.60
k_t	0.513	0.486	0.48	0.50
ρ, 10^3 kg/m^3	7.5	7.75	7.6	7.4
$\tan\delta_{e0}$ [1]	0.005	0.015	–	0.018
$\tan\delta_{eE}$	0.01	0.08	–	–
Q_{m0} [2]	500	75	1000	65
Q_{mT}	140	20	–	–
ΔK_{33}^T [3]	0.1	0.3	–	–
$\Delta d_{31} / d_{31}$	0.12	0.06	–	–

[1] $\tan\delta_{e0}$, $\tan\delta_{eE}$ are the $\tan\delta_e$ in weak field and at $E = 1$ kV/cm.

[2] Q_{m0}, Q_{mT} are the Q_m values at low stress and at $T = 10$ MPa.

[3] The parameter changes at temperatures 0-60°C. [4] $\varepsilon_0 = 8.85 \cdot 10^{-12}\ F/m$.

APPENDIX C. Special Functions

In the Appendix some data regarding the properties of special functions that are required for treating the radiation and vibration problems related to the cylindrical and spherical transducers are summarized. More details regarding properties of the functions and their numerical values can be found [1, 2], which are the primary sources of the information and where these functions are tabulated. Some of the integral relations that include the special functions are presented from a source [3] where much more particular useful relations can be found.

C.1 Cylindrical Bessel Functions

Definition

Cylindrical functions $Z_n(x)$ are the solutions to Bessel equation

$$\frac{d^2 Z_n}{dx^2} + \frac{1}{x}\frac{dZ_n}{dx} + \left(1 - \frac{n^2}{x^2}\right) = 0 . \tag{C.1}$$

Partial solutions to this equation are the Bessel functions (cylindrical functions of the first kind) $J_n(x)$, Neumann functions (cylindrical functions of the second kind) $N_n(x)$, and Hankel functions (cylindrical functions of the third kind) $H_n^{(1)}(x)$ and $H_n^{(2)}(x)$, where $H_n^{(1)}(x) = J_n(x) + jN_n(x)$ and $H_n^{(2)}(x) = J_n(x) - jN_n(x)$. The functions $H_n^{(1)}(x)$ or $H_n^{(2)}(x)$ are used alternatively according to the time dependence $e^{-j\omega t}$ or $e^{j\omega t}$ (the later is accepted in our treatment). In course of this treatment, it will be assumed that n is the natural integer number and for the cylindrical coordinates $x = kr$. Thus, it will be used form of

$$H_n^{(2)}(kr) = J_n(kr) - jN_n(kr) . \tag{C.2}$$

Properties

$$H_{-n}^{(2)}(x) = J_{-n}(x) - jN_{-n}(x) , \tag{C.3}$$

where

$$J_{-n}(x) = (-1)^n J_n(x), \quad N_{-n}(x) = (-1)^n N_n(x) . \tag{C.4}$$

Series representation

$$J_n(x) = \frac{1}{0!n!}\left(\frac{x}{2}\right)^2 - \frac{1}{1!(n+1)!}\left(\frac{x}{2}\right)^{n+2} + \frac{1}{2!(n+2)!}\left(\frac{x}{2}\right)^{n+4} - \dots \quad \text{(C.5)}$$

$$J_0(x) = 1 - \frac{x^2}{2^2} + \frac{x^4}{2^2 \cdot 4^2} - + \frac{x^6}{2^2 \cdot 4^2 \cdot 6^2} - \dots \quad \text{(C.6)}$$

$$J_1(x) = \frac{x}{2} - \frac{2x^3}{2 \cdot 4^2} + \frac{3x^5}{2 \cdot 4^2 \cdot 6^2} - \dots \quad \text{(C.7)}$$

Approximations at small argument $x < 1$ *(low frequency approximations at* $x = kr \ll 1$*)*

$$J_0(x) \approx 1 - \frac{x^2}{4}, \quad J_1(x) \simeq \frac{x}{2} - \frac{x^3}{16}, \quad \text{(C.8)}$$

$$N_0 \approx \frac{2}{\pi}(\ln x - 011), \quad N_1 \approx -\frac{2}{\pi} \cdot \frac{1}{x}, \quad \text{(C.9)}$$

$$H_1^{(2)}(x) \approx \frac{x}{2} + j\frac{2}{\pi x}, \quad H_1^{(2)\prime}(x) \approx -j\frac{2}{\pi (x)^2}. \quad \text{(C.10)}$$

At large arguments $x \gg 1$ (high frequency approximation, large distances from a cylinder at $x = kr \to \infty$)

$$H_n^{(2)}(x) \to \sqrt{\frac{2}{\pi x}} e^{-j\left(x - \frac{\pi n}{2} - \frac{\pi}{4}\right)}, \quad \text{(C.11)}$$

$$J_n(x) \to \sqrt{\frac{2}{\pi x}} \cos\left(x - \frac{n\pi}{2} - \frac{\pi}{4}\right), \quad \text{(C.12)}$$

$$N_n(x) \to \sqrt{\frac{2}{\pi x}} \sin\left(x - \frac{n\pi}{2} - \frac{\pi}{4}\right). \quad \text{(C.13)}$$

Functional equations

$$Z_{n-1}(x) + Z_{n+1}(x) = \frac{2n}{x} Z_n(x) \quad \text{(C.14)}$$

$$N_{n-1} J_n - N_n J_{n-1} = \frac{2}{\pi x} \quad \text{(C.15)}$$

Differential formulas

$$\frac{dZ_n}{dx} = -\frac{n}{x} Z_n + Z_{n-1} = \frac{n}{x} Z_n - Z_{n+1} = \frac{1}{2}(Z_{n-1} - Z_{n+1}) \quad \text{(C.16)}$$

$$Z_0' = -Z_1, \quad Z_1' = Z_0 - \frac{1}{x} Z \quad \text{(C.17)}$$

Integral formulas

$$\int x^{-n+1} Z_n(x)dx = -x^{-n+1} Z_{n-1}(x), \quad \int x^{n+1} Z_n(x)dx = x^{n+1} Z_{n+1}(x) \tag{C.18}$$

$$\int Z_1(x)dx = -Z_0(x), \quad \int xZ_0(x)dx = xZ_1(x) \tag{C.19}$$

$$\int J_n^2(x)xdx = \frac{x^2}{2}[J_n^2(x) - J_{n-1}(x)J_{n+1}(x)] \tag{C.20}$$

Integral representation

$$J_n(x) = \frac{1}{2\pi j^n}\int_0^{2\pi} e^{jx\cos\varphi}\cdot e^{jn\varphi}d\varphi \tag{C.21}$$

$$J_0(x) = \frac{1}{\pi}\int_0^{\pi} e^{jx\cos\varphi}d\varphi = \frac{1}{\pi}\int_0^{\pi}\cos(x\sin\varphi)d\varphi = \frac{2}{\pi}\int_0^{\pi/2}\cos(x\sin\varphi)d\varphi \tag{C.22}$$

Also tabulated are functions Struve that are solutions to one of variations of the Bessel equation [1, 2]:

$$S_0(x) = \frac{2}{\pi}\int_0^{\pi/2}\sin(x\cos\varphi)d\varphi\,, \tag{C.23}$$

$$S_1(x) = \frac{4}{\pi}\int_0^{\pi/2}\sin(x\cos\varphi)\sin^2\varphi d\varphi\,. \tag{C.24}$$

There series representations are

$$S_0(x) = \frac{2}{\pi}\left[x - \frac{x^3}{1^2\cdot 3^2} + \frac{x^5}{1^2\cdot 3^2\cdot 5^2} - \ldots\right], \tag{C.25}$$

$$S_1(x) = \frac{2}{\pi}\left[\frac{x^2}{1^2\cdot 3} - \frac{x^4}{1^2\cdot 3^2\cdot 5} + \frac{x^6}{1^2\cdot 3^2\cdot 5^2\cdot 7} - \ldots\right], \tag{C.26}$$

$$\int xS_0(x)dx = xS_1(x)\,. \tag{C.27}$$

Modified Bessel functions (Bessel functions of imaginary values of argument), $I_n(x)$ *and* $K_n(x)$

The modified functions are the partial solutions to the equation (Compare with Eq. (C.1))

$$\frac{d^2Z_n}{dx^2} + \frac{1}{x}\frac{dZ_n}{dx} - \left(1 + \frac{n^2}{x^2}\right) = 0\,. \tag{C.28}$$

The modified functions are defined by equations:

$$I_n(x) = j^n J_n(-jx) \tag{C.29}$$

for the first kind,

$$K_n(x) = -\frac{j\pi}{2} e^{-j\frac{\pi n}{2}} H_n^{(2)}(-jx) \tag{C.30}$$

for the second kind, with

$$K_0(x) = -\frac{j\pi}{2} H_0^{(2)}(-jx), \quad K_1(x) = -\frac{\pi}{2} H_1^{(2)}(-jx)\,. \tag{C.31}$$

The properties of these functions can be obtained from formulations of the corresponding properties of functions $J_n(x)$ and $H_n^2(x)$ by replacing $x \to -jx$ and introducing the factors from Eqs. (C.24) and (C.25). In particular

$$I_{-n}(x) = I_n(x), \quad K_{-n}(x) = K_n(x)\,, \tag{C.32}$$

$$K_0'(x) = -K_1(x). \tag{C.33}$$

C.2 Spherical Bessel Functions

The partial solutions to equation

$$\frac{d^2R}{dz^2} + \frac{2}{z}\frac{dR}{dz} + \left[1 - \frac{m(m+1)}{z^2}\right] R = 0\,, \tag{C.34}$$

where $z = kr$, are the spherical Bessel functions (or *Bessel* functions for the spherical coordinates). Spherical Bessel functions of order m of the first kind are defined as

$$j_m(z) = \sqrt{\pi / 2z} J_{m+1/2}(z)\,; \tag{C.35}$$

of the second kind (spherical Neumann functions) as

$$y_m(z) = \sqrt{\pi / 2z} N_{m+1/2}(z)\,; \tag{C.36}$$

and of the third kind (spherical Hankel functions) as $h_m(z)$. For outgoing wave

$$h_m^{(2)}(z) = j_m(z) - j y_m(z) = \sqrt{\pi / 2z} H_{m+1/2}^{(2)}(z)\,. \tag{C.37}$$

In particular,

$$\left.\begin{aligned}
&j_0(z)=\frac{\sin z}{z},\quad y_0(z)=-\frac{\cos z}{z};\\
&j_1(z)=\frac{\sin z}{z^2}-\frac{\cos z}{z},\quad y_1(z)=-\frac{\sin z}{z}-\frac{\cos z}{z^2};\\
&j_2(z)=\left(\frac{3}{z^3}-\frac{1}{z}\right)\sin z-\frac{3}{z^2}\cos z,\quad y_2(z)=-\frac{3}{z^2}\sin z-\left(\frac{3}{z^3}-\frac{1}{z}\right)\cos z.
\end{aligned}\right\}\tag{C.38}$$

Functions j_m and y_m are tabulated [1] at $z<0.3$ as

$$\left.\begin{aligned}
&j_m(z)\approx\frac{(z)^m}{1\cdot3\cdot5\cdots(2m+1)},\quad y_m(z)\approx-\frac{1\cdot3\cdot5\cdots(2m-1)}{(z)^{m+1}},\\
&h_1^{(2)}\approx\frac{1}{3}\left[z+j\frac{3}{(z)^2}\right],\quad h_1^{(2)\prime}(z)\approx\frac{1}{3}\left[1-j\frac{6}{(z)^3}\right];
\end{aligned}\right\}\tag{C.39}$$

and at $z\to\infty$

$$\left.\begin{aligned}
&j_m(z)\to\frac{1}{z}\cos\left(z-\frac{m+1}{2}\pi\right),\quad y_m(z)\to\frac{1}{z}\sin\left(z-\frac{m+1}{2}\pi\right),\\
&h_m^{(2)}\to\frac{1}{z}e^{-j\left(z-\frac{m+1}{2}\pi\right)},
\end{aligned}\right\}\tag{C.40}$$

$$y_{m-1}(z)j_m(z)-y_m(z)j_{m-1}(z)=z^{-2}.\tag{C.41}$$

The following properties are the same for the functions j_m, y_m and h_m that will be collectively denoted as f_m.

Recurrent relations

$$f_{m-1}(z)+f_{m+1}(z)=(2m+1)z^{-1}f_m(z)\tag{C.42}$$

$$mf_{m-1}(z)-(m+1)f_{m+1}(z)=(2m+1)\frac{d}{dz}f_m(z)\tag{C.43}$$

$$h_1^{(2)\prime}(z)=\frac{1}{3}[h_0^{(2)}(z)-2h_2^{(2)}(z)]\tag{C.44}$$

$$\frac{d}{dz}[z^{m+1}f_m(z)]=z^{m+1}f_{m-1}(z),\quad\frac{d}{dz}[z^{-m}f_m(z)]=-z^{-m}f_{m-1}(z)\tag{C.45}$$

Integral formulas

$$\int f_1(z)dz=-f_0(z),\quad\int f_0(z)z^2dz=z^2f_1(z)\tag{C.46}$$

$$\int f_m^2(z)z^2dz=\frac{z^2}{2}[f_m^2(z)-f_{m-1}(z)f_{m+1}(z)]\tag{C.47}$$

C.3 Legendre Polynomials

The partial solutions to Legendre equation

$$\frac{d}{dx}\left[(1-x^2)\frac{dP}{dx}\right]+m(m+1)P=0 \text{ or } (x^2-1)\frac{d^2P}{dx^2}+2x\frac{dP}{dx}-m(m+1)P=0\,. \tag{C.48}$$

at m integer and $x=\cos\theta$ are the Legendre polynomials of the order m

$$P_m(x)=\frac{1}{2^m m!}\frac{d^m}{dx^m}(x^2-1)^m\,. \tag{C.49}$$

In particular,

$$\left.\begin{aligned}&P_0(x)=1,\\&P_1(x)=x=\cos\theta,\\&P_2(x)=\frac{1}{2}(3x^2-1)=\frac{1}{4}(3\cos 2\theta+1),\\&P_3(x)=\frac{1}{2}(5x^2-3z)=\frac{1}{8}(5\cos 3\theta+3\cos\theta).\end{aligned}\right\} \tag{C.50}$$

$$P_m(-x)=(-1)^m P_m(x),\quad P_m(x)=P_{-(m+1)}(x) \tag{C.51}$$

Recurrent relation

$$P_{m-1}(x)=\frac{2m+1}{m}xP_m(x)-\frac{m+1}{m}P_{m+1}(x) \tag{C.52}$$

Differential formulas

$$mP_m(x)=xP_m'(x)-P_{m-1}'(x) \tag{C.53}$$

$$(2m+1)P_m(x)=\frac{d}{dx}[P_{m+1}(x)-P_{m-1}(x)] \tag{C.54}$$

Orthogonality

$$\int_{-1}^{1}P_n(x)P_m(x)dx=\begin{cases}0 & n\neq m\\ 2/(2m+1) & n=m\end{cases} \tag{C.55}$$

(Any function of x in the range from $x=1$ to $x=-1$ can be expanded in terms of series of these functions.)

Useful integrals with Legendre polynomials

$$\int_0^1 P_m(x)P_n(x)dx = \begin{cases} 1/(2m+1) & m = n \\ \dfrac{(-1)^{0.5(m+n+1)}}{2^{m+n+1}} \times \ldots & m \neq n, \\ \qquad \dfrac{m!n!}{(m+n)(m+n+1)\{[(m/2)![(n-1)/2]!\}^2} & m > n, \\ & (m-n) \text{ even} \\ 0 & m \text{ even}, \\ & n \text{ odd} \end{cases} \tag{C.56}$$

$$\int_0^1 P_{2m}(x)dx = 0, \quad \int_0^1 P_{2m+1}(x)dx = \frac{(-1)(-3)\cdots(-2m+1)}{(2m+2)\cdot 2m\cdot(2m-1)} \tag{C.57}$$

$$\int_{-1}^1 x^b P_m(x)dx = 0 \ \text{ at } \ b < m \tag{C.58}$$

$$\int_{-1}^1 [P_m'(x)]^2 dx = m(m+1) \tag{C.59}$$

$$\int_{-1}^1 (1-x^2)[P_m'(x)]^2 dx = \frac{2m(m+1)}{2m+1} \tag{C.60}$$

$$\int_{-1}^1 (1-x)^{-1/2} P_m(x)dx = \frac{2^{3/2}}{2m+1} \tag{C.61}$$

C.5 References

1. M. Abramowitz, I. A. Stegun (ed.), *Handbook of Mathematical Functions* (National Bureau of Standards Applied Mathematical Series, Washington, DC, 1964)
2. G. N. Watson, *A Treatise on the Theory of Bessel Functions* (Cambridge Univ. Press, Cambridge, 1995)
3. I. S. Gradshtein, I. M. Ryzhik, *Tables of integrals, Series, and Products* (Elsevier/Academic Press, Amsterdam, 2007)

www.ingramcontent.com/pod-product-compliance
Ingram Content Group UK Ltd.
Pitfield, Milton Keynes, MK11 3LW, UK
UKHW050057260226
468387UK00009B/128

9 781644 698235